# INTRODUCTION TO BIOPHYSICS

**For Graduate and Postgraduate Students of Biophysics, Microbiology, Biotechnology and Biological Sciences**

**Dr. PRANAB KUMAR BANERJEE**

M.Sc. (C.U.), Ph.D. (C.U.), FZS, FZSEI

*Associate Professor and Chairperson*
*Department of P.G. Studies in Zoology*
*Serampore College, Serampore, Hooghly*

*Guest Professor*
*Department of Microbiology*
*R. K. Mission Vidyamandir*
*Belur Math, Howrah*

**S Chand And Company Limited**

**(ISO 9001 Certified Company)**

**S Chand And Company Limited**

**(ISO 9001 Certified Company)**

***Head Office:*** Block B-1, House No. D-1, Ground Floor, Mohan Co-operative Industrial Estate, New Delhi – 110 044 | Phone: 011-66672000

***Registered Office:*** A-27, 2nd Floor, Mohan Co-operative Industrial Estate, New Delhi – 110 044 Phone: 011-49731800

www.**schandpublishing.com**; e-mail: **info@schandpublishing.com**

**Branches**

| | | |
|---|---|---|
| Chennai | : | Ph: 23632120; chennai@schandpublishing.com |
| Guwahati | : | Ph: 2738811, 2735640; guwahati@schandpublishing.com |
| Hyderabad | : | Ph: 40186018; hyderabad@schandpublishing.com |
| Jalandhar | : | Ph: 4645630; jalandhar@schandpublishing.com |
| Kolkata | : | Ph: 23357458, 23353914; kolkata@schandpublishing.com |
| Lucknow | : | Ph: 4003633; lucknow@schandpublishing.com |
| Mumbai | : | Ph: 25000297; mumbai@schandpublishing.com |
| Patna | : | Ph: 2260011; patna@schandpublishing.com |





*First Edition 2008*
*Reprints 2010, 2012*
*Second Revised and Enlarged Edition 2014*
*Reprints 2015, 2017, 2018 (Twice), 2019*
*Reprint 2021*

**ISBN:** 978-81-219-3016-1 **Product Code:** H6BPH68BIOL10ENAB14O

PRINTED IN INDIA

By Vikas Publishing House Private Limited, Plot 20/4, Site-IV, Industrial Area Sahibabad, Ghaziabad – 201 010 and Published by S Chand And Company Limited, A-27, 2nd Floor, Mohan Co-operative Industrial Estate, New Delhi – 110 044

Dedicated to the
Departed Souls of My
Father Shri Panchanan Banerjee
and
Uncle Shri Ajit Kumar Banerjee

**RAMAKRISHNA MISSION VIVEKANANDA UNIVERSITY**

(Declared by Government of India Under Section 3 of UGC Act)

**PO: Belur Math, Dist, Howrah, West Bengal – 711 202, India**

# FOREWORD

It gives me immense pleasure to write a foreword for Dr. Pranab Kumar Banerjee's book titled ***Introduction to Biophysics.*** I have known Dr. Banerjee for more than two decades and have had the opportunity of writing a Foreword to one of his earlier book containing problems on genetics.

Biophysics is no longer a young discipline it used to be. Over the decades it has grown into a full-fledged independent discipline although it began as an inter-disciplinary one. Physics is proverbially known as the science of matter, while biology is the science of living beings. Biophysics, therefore, is an interesting discipline which applies the principles and methods of physics to living beings and organisms.. The present century is being spoken of as the century of life sciences even as the last century was said to be century of physics. Hence the immense possibility of the coming together of these two sciences in the 'new' discipline of biophysics needs to be appreciated and understood. Books on biophysics by persons belonging to the discipline of physics will naturally have a slant towards physics while those written by biologists will lean more towards biology. It is therefore good that both physicists as well as biologists write books on biophysics in order that the subject be understood from either point of view. Given the rich experience of Dr. Banerjee in teaching zoology to undergraduate students at the Honours level under Calcutta University, and the experience that he has been gaining recently in writing lucid textbooks on biostatistics, genetics, etc. I am sure that this book on biophysics too will be found to be lucid in presentation and clear in concepts, and that the students and teachers alike would find it useful to gain an entry into the fascinating world of biophysics.

Belur Math
26 September 2007

**Swami Atmapriyananda**
Vice Chancellor
Ramakrishna Mission Vivekananda University
Howrah, West Bengal

# PREFACE TO THE SECOND REVISED AND ENLARGED EDITION

I feel great pleasure to place before students and academics this thoroughly revised enlarged, elegant and appropriately updated edition of the book **"Introduction to Biophysics"**. To enhance the utility of this book, a thorough revision and recast along with numerous numerical solved problems at the end of each chapter have been incorporated. The second major innovation of this edition is the introduction of a new chapter on Nanotechnology. Theoretical discussion along with a good number of assorted worked-out numerical problems has been represented in a simple and lucid manner so as to cater the needs of the students of all streams of Biological Sciences (Zoology, Botany, Physiology, Microbiology, Biotechnology, Pharmacy, and Medical Science, etc.).

I claim no originality for the matter presented in this book but the method of representation is my own. All available standard books allied to this subject have been freely consulted during the preparation of the revised edition of this book. I acknowledge gratefully my indebtedness to the authors and publishers of those books.

I am continuing to welcome constructive criticism and healthy suggestions from all students who use this book as a part of their studies and also the academics who adopt the book to complement their teaching.

I convey my respect and pronams to Swami Atmapriyanandaji Maharaj (Vice Chanceller, R.K. Mission Vivekananda University, Belur Math, Howrah). I extend my respect and gratitude to my teachers (Prof. R.N. Chatterjee, Genetics Research Unit Dept. of Zoology, University of Calcutta), Prof. S.K. Ghosal (Ex-teacher, University of Burdwan) and well-wishers Prof. Dhruobojyoti Chatterjee (Pro Vice Chancellor Academic, University of Calcutta), Prof. Chandra Shekar Chakraborty (Vice Chancellor, West Bengal University of Animal and Fishery Sciences) and Dr. Tarit Kumar Banerjee (Ex-Associate Professor), Dept. of Zoology, Raja Pearymohan College, Uttarpara.

I also extend my thanks to my departmental colleagues (both teaching and non-teaching) for their encouragement and constant inspiration. My sincere thanks are forwarded to the Management and the Editorial Team of S.Chand & Company Pvt. Ltd. for their encouragement and neat execution to revised and enlarged edition of this book in suitable form.

Last but not the least, I extend my sincere thanks to Mrs. Mandira Banerjee (Headmistress, Baidyabati Charusila Bose Balika Vidyalaya) and Miss Debdatta Banerjee (Daughter) for their endurance and active assistance during the preparation of this book.

**Dr. PRANAB KUMAR BANERJEE**

# PREFACE TO THE FIRST EDITION

**"Biophysics"** is an independent as well as an interdisciplinary and emerging subject in the field of Biological Sciences in the recent years. It is a hybrid science which deals with Physics, Chemistry and Biology. The concept of Biophysics is the application of ideas and methods of both Physics and Chemistry in living beings. It is an introductory source to the various physical activities occur inside the living beings and also to the various techniques employed to bringing out the essential structural and functional aspects of biological activities.

The book entitled **"Introduction to Biophysics"** is comprehensively written to meet the curriculum requirements for the under-graduate and post-graduate students of Biophysics, Microbiology, Zoology, Botany, Physiology, B. pharm, Biotechnology, Medical Sciences and Agricultural Sciences. The book contains 14 chapters covering many aspects of Physics, Chemistry and Biology. I claim no originality for the matter presented in the text but the method of presentation is my own. This book has been written in a clear and lucid manner. Attempts have been made to discuss the relevant topics (both theoretical and practical) with simple diagrams. The subject matter is discussed in a very simple and precise language. At the end of each chapter, some numerical problems are included to get a sense of satisfaction and involvement both in Physics and Chemistry learning.

I convey my respect and pronams to Swami Atmaprinandaji Maharaj (Vice Chancellor, Ramakrishna Vivekananda University, Belur Math, Howrah) and extend my respect and gratitude to Prof Rabindra Nath Chatterjee (Genetics Research Unit, Dept. of Zoology, C.U.) and Dr. Tarit Kumar Banerjee (R.P.M. College, Uttarpara) for their enthusiastic inspiration.

I also extend my thanks to my colleagues Dr. Subhash Mitra (retired), Dr. Kripa Nath Patari (retired), Dr. Kanchan Kumar Mukherjee and my best friend Mrs. Urmi Ganguly for their encouragement and inspiration.

I am grateful to my colleagues Dr. Suchandra Chatterjee, Dr. Kunal Kumar Kamar, Mr. Debasish Dhak, Mr. Prosenjit Sarkar, Mr. Sankha Das and Mr. Sovan Kumar Bera (Headmaster, Vivekananda Institution, Howrah) for their active co-operation for the preparation of this book. I am also grateful to Mrs. Banhisikha Ghatak, Librarian and other staff of Serempore College Library for their active co-operation by providing relevant books and journals.

Attempts have been made to make the book free from all kinds of errors and printing mistakes, but some unwanted errors and printing mistakes may crop up. Comments and positive criticism and healthy suggestions from students as well as some erudite teachers for corrections and improvement of this book will be highly appreciated and incorporated in the future edition of this book.

It is indeed a pleasure to thank my well-wishers Dr. Saroj Kumar Paul, Mr. Tarak Nath Mukherjee, Mr. Debal Kishore Chakraborty, Ritabrata Banerjee, Mr. Siddhartha Sankar Banerjee, Mrs. Nivedita Pande, Mrs. Jharna Bhadra and Mrs. Dipanwita Goswami (Sheoraphuli S.N.V. for Girls) who have directly or indirectly helped me in the preparation of this book.

Last but not the least, I extend my sincere thanks to Mrs. Mandira Banerjee (Head mistress, Baidyabati Charusila Bose Balika Vidyalaya) and Miss Debdatta Banerjee (Daughter) for their endurance and active assistance during the preparation of this book.

**Dr. PRANAB KUMAR BANERJEE**

# Introduction to Biophysics

● **Physics :** It is perhaps the most fundamental scientific discipline such as light, heat, sound, etc. It concerns matter and energy and its theories can be applied in every other scientific discipline.

● **Chemistry :** It deals with the composition, characteristic properties, methods of preparation of substances and their reaction when combined or in contact with one another.

● **Biophysics :** It is an intradisciplinary as well as an emerging hybrid science which is developed in collaboration with physics, chemistry and biology.

## Objectives of Biophysics :

(*i*) The use of the ideas and methods of physics, chemistry and to study and explain the structures of living organisms and mechanism of life process.

(*ii*) It is predominantly devoid of general physico-biological problems and physio-mathematical modelling of biological process.

(*iii*) The molecular biophysics discloses the physical mechanism responsible for the biological function of molecule.

(*iv*) Both physical and biological knowledge are essential for the study of biophysics.

## Branches of Biophysics :

### 1. Molecular Biophysics

(*i*) It is the most developed branch of biophysics.

(*ii*) It rests on biochemical discipline as well as on physics of small and large systems.

(*iii*) It is concerned with the study of large molecules and particles of comparable size.

### 2. Mathematical Biophysics

(*i*) It is known as *theoretical biophysics.*

(*ii*) Here the behaviour of living organism is explained on the basis of mathematics and physical theory.

(*iii*) Here biological processes are being examined in terms of the thermodynamics, hydrodynamics and statistical mechanics.

(*iv*) It encompasses the application of computer technology and quantitative theories in biological system.

### 3. Physiological Biophysics

(*i*) It is known as *classical biophysics.*

(*ii*) It deals with physical mechanism for explaining the behaviour of living organism.

(*iii*) Bioelectricity, membrane potential, electrical potential related to the activity of living tissues are included in the physiological biophysics.

### 4. Radiation Biophysics

(*i*) The study of ionizing radiation such as $\alpha$, $\beta$ or X-ray and UV ray on living organism and its effects are concerned with radiation biophysics.

(*ii*) The biological responses are manifested either in the form of cellular death of tissues or somatic or genetic mutation in the living organism.

**Applications of Biophysics :**

(*i*) Classical Biophysics has been greatly used to the behaviour and functional activity of living organism.

(*ii*) Diffussion, osmosis like biological process within the living body can be explained with the help of biophysical principle.

(*iii*) Biological instruments (electron microscope, phase contrast microscope, ultra centrifuge, etc.) greatly facilitate biophysical research. These tools are peculiarly adapted to the study of problems related to viral disease heart disease and cancer.

(*iv*) Separation techniques and physiochemical techniques are used to study biomolecular structure.

(*v*) X -ray crystallography and NMR (Nuclear Magnetic Resonance) spectroscopy have been used in determining biomolecular structure in prominently.

(*vi*) The radiation biophysics has been greatly used in cancer research.

(*vii*) Space biology is an emerging concept of biophysics which is concerned with the study of living things in the space environment.

(*viii*) Mathematical biophysics covers all applications of mathematics, computer technology to biological system.

(*ix*) The main objectives of Nanobiotechnology are using biological components to achieve nanoscale task.

# CONTENTS


| | | |
|---|---|---|
| **1.** | Properties of Matter | 1 — 26 |
| **2.** | Properties of Solution | 27 — 45 |
| **3.** | Colloids | 46 — 57 |
| **4.** | pH and Buffer Solution | 58 — 88 |
| **5.** | Intra- and Intermolecular Interactions in Biological System | 89 — 111 |
| **6.** | Microscopy | 112 — 168 |
| **7.** | X-ray Crystallography | 169 — 201 |
| **8.** | Radioactivity | 202 — 246 |
| **9.** | Colorimetry, Photometry, Polarimetry and Spectrophotometry | 247 — 277 |
| **10.** | Separation Techniques | 278 — 330 |
| **11.** | Thermodynamics | 331 — 352 |
| **12.** | Nuclear Magnetic Resonance (NMR) | 353 — 367 |
| **13.** | Biomechanics | 368 — 377 |
| **14.** | Neurobiophysics | 378 — 396 |
| **15.** | Nanotechnology | 397 — 402 |
| | Appendix 1: Some Useful Tables | 403 — 405 |
| | Appendix 2: Greek Letters | 406 |
| | Bibliography | 407 — 408 |

CHAPTER 1

# Properties of Matter

- **Phases :**

These are forms of matter and are distinguishable by definite boundaries.

- **System :**

(*i*) In the light of Biophysics, system encompasses the group of materials and their activities within the living system.

(*ii*) A system consists of one or more than one phases.

- **Homogeneous System :**

(*i*) It consists of single phase.

(*ii*) It has no distinct boundaries between its particles.

**Example :** A solution of glucose and sucrose in water.

- **Heterogeneous System :**

(*i*) It consists of more than one phases.

(*ii*) It has distinct boundaries between its components.

**Example :** The cellular membrane consists of lipid phase surrounded by aqueous phases of extracellular and intracellular fluids.

## • A. KINETIC THEORY OF GASES :

**Bernoulli, Clausisus, Maxwell, Boltzmann** and others developed kinetic theory of gases or kinetic molecular theory of gases. According to this theory gases are made up of molecules which are in continuous random motion (kinetics).

In a sample of a liquid or gas, the motion of the molecules is random and in all directions. The movement of gaseous molecules was first observed by **Robert Brown** (1827). The movement of the molecules is called **Brownian movement.** The movement of the molecule is a zig-zag type motion. Brownian movement has been accepted as a strong evidence to support the validity of the kinetic molecular theory of liquid and gases.

**Main postulates of kinetic theory of gases**

(*i*) A gas is composed of a large number of minute, discrete, spherical particles of identical mass and size and are not visible even with the help of most powerful microscope. These particles are called **molecules.**

(*ii*) There being no intermolecular forces of attraction or repulsion between the molecules, they are in a continuous, rapid, random movement in all possible directions. Their **motion** is along **straight line** until they collide with another molecule of gas or wall of the containing vessel.

(*iii*) The molecules of a gas behave as **perfect elastic bodies.** While moving, the molecules collide against each other and with the walls of the container. During these collisions, they change their direction of motion without any loss of their kinetic energy. This also means that the energy of the molecules before and after collision is the same *i.e.* the collisions are perfectly elastic.

(*iv*) The pressure exerted by the gas is due to the collision of gas molecules against the walls of the container. The larger the number of collisions, the greater is the pressure of the gases.

(*v*) The volume occupied by the gaseous molecules is negligibly small as compared to the total volume occupied by the gas.

(*vi*) The average kinetic energy of a gas molecule is directly proportional to the absolute temperature of the gas.

(*vii*) The pressure exerted by the gas does not depend on the gravitational force.

## • Velocity of Gas Molecules

**(*a*) Average velocity :**

(*i*) All the molecules in a gas do not have the same velocity.

(*ii*) The velocity of a particular molecule is always changing direction and magnitude due to collision.

(*iii*) If at any instant, '*n*' individual molecules have velocities $C_1, C_2, C_3, C_4, C_5$ ...................... $C_n$, the arithmetic mean of the individual velocities of the molecules is called **average velocity ($\overline{C}$) or ($C_A$).**

$$\overline{C} \text{ or } C_A = \frac{C_1 + C_2 + C_3 + C_4 + C_5 \ldots\ldots\ldots\ldots C_n}{n}$$

**(*b*) The root mean square velocity (r.m.s.):**

(*C*) is the square root of the mean of the squares of individual velocities of the gas molecules.

$$C = \sqrt{\frac{C_1^2 + C_2^2 + C_3^2 + C_4^2 + C_5^2 \ldots\ldots\ldots + C_n^2}{n}}$$

It has been found that $C \simeq 1.085\, C_A$.

## • Deduction from Kinetic Theory of Gases

**1. Kinetic theory and pressure of the gas :**

The pressure exerted by a gas is equal to the force per unit area. The pressure exerted by the gas is due to the collisions of the molecules of the gas with the walls of the container. The pressure exerted by the gas is expressed by the following equation :

$$P = mNC^2/3V$$

where $P$ = Pressure exerted by the gas

$V$ = Volume of the gas

$N$ = Number of molecules present in volume (V).

$m$ = Mass of a molecule.

$C$ = Root mean square velocity.

The equation $P = m\,NC^2/3V$ is known as **kinetic gas equation.**

**2. The kinetic energy and temperature of a gas :**

The kinetic gas equation is $P = \frac{1}{3}\frac{mNC^2}{V}$

or $$PV = \frac{1}{3} mNC^2 \text{ [Here } mN : \text{total mass of the gas } (M)]$$

$\therefore$ $$PV = \frac{1}{3} MC^2$$

or $$PV = \frac{2}{3}.\frac{1}{2} MC^2$$

but $\frac{1}{2} MC^2$, the average kinetic energy of the gas. For one mole of an ideal gas, $PV = RT$.
Putting the value of $PV$, we get

$$\frac{2}{3}.\frac{1}{2} MC^2 = RT$$

Now M = molecular weight of the gas in this equation. $\frac{1}{2} MC^2$ is the kinetic energy, $E_k$ for one mole of the gas or $\frac{2}{3}$ $E_k$ = RT or $E_k = \frac{3RT}{2}$

Since R is constant, $E_k \propto T$.

Now kinetic energy $E_k$ for one mole $= \frac{3RT}{2}$

$\therefore$ Kinetic energy for $n$ moles $= \frac{3nRT}{2}$

As one mole of a gas contains Avagadro's number $N_A = 6.023 \times 10^{23}$

$\therefore$ Kinetic energy $E_k$ of 1 molecule $= \frac{3RT}{2} \times 6.023 \times 10^{23}$.

**3. Kinetic equation and Boyle's law :**

**(*a*) Boyle's Law :**

**At a constant temperature, the volume of a given mass of a gas is inversely proportional to the pressure of the gas.**

$$V \propto \frac{1}{P}; \text{if } T \text{ is constant} \quad \left[\begin{array}{l} \text{P = Pressure} \\ \text{V = Volume} \\ \text{T = Temperature} \end{array}\right.$$

or $PV = k$ (constant)

(*i*) Thus the product of pressure ($P$) and volume ($V$) of a gas remains constant if its temperature ($T$) is kept constant ($k$).
If the volume of a gas under pressure $P_1$ is $V_1$ and under pressure $P_2$ it is $V_2$ then at constant temperature, $P_1V_1$ = constant, $P_2V_2$ = constant. So $P_1V_1 = P_2V_2$.

(*ii*) From kinetic equation, we get $PV = \frac{1}{3} m NC^2$ or $PV = \frac{2}{3} N. \frac{1}{2} m C^2$

Now for a fixed amount of gas, $N$ is constant. Hence, for a fixed amount of gas and at a constant temperature 2/3 $N. \frac{1}{2} m C^2$ is constant.

$\therefore$ $PV$ = is constant and this is Boyle's law.

**4. Charles' Law :**

**At constant pressure, the volume of a definite mass of a gas increases or decreases by 1/273 part of its volume at 0 °C, for each degree Celsius rise or fall in temperature.**

(*i*) Let the volume of given mass of a gas at constant pressure and at 0°C be $V_0$ ml.

Now if the temperature of the gas is increased keeping the pressure constant, then according to Charles' law–

For 1°C rise of temperature, the increase in volume will be $\frac{V_0}{273}$ ml.

(*ii*) $\therefore$ At 1°C the volume of gas will be $\left(V_0 + \frac{V_0}{273}\right)$ ml $= V_0\left(1 + \frac{1}{273}\right)$ ml.

Again for the increase of $t$°C, the increase in volume will be $\frac{V_0 t}{273}$ ml.

(*iii*) At t°C the volume of the gas will be $\left(V_0 + \frac{V_0 t}{273}\right)$ ml $= V_0\left(1 + \frac{t}{273}\right)$ ml. Again for the decrease of $t$°C, the decrease in volume of the gas will be $\left(V_0 - \frac{V_0 t}{273}\right)$ ml $= V_0\left(1 - \frac{t}{273}\right)$ ml.

(*iv*) Let $V_1$ and $V_2$ be the volume of a given mass of gas at $t_1$°C and $t_2$°C respectively under constant pressure and $T$ is the absolute temperature in degree Kelvin ($T = t°C + 273$)

$$V_1 = V_0\left(1 + \frac{t_1}{273}\right) \quad V_2 = V_0\left(1 + \frac{t_2}{273}\right) \text{ ml.}$$

$$\therefore \quad \frac{V_1}{V_2} = \frac{V_0\left(1 + \frac{t_1}{273}\right)}{V_0\left(1 + \frac{t_2}{273}\right)} = \frac{\left(\frac{t_1 + 273}{273}\right)}{\left(\frac{t_2 + 273}{273}\right)} = \frac{T_1}{T_2} \quad [\because T = t°C + 273]$$

or $$\frac{V_1}{V_2} = \frac{T_1}{T_2}$$

Thus it may be stated that the volume ($V$) of a given mass of gas is directly proportional to the absolute temperature ($T$) if the pressure remains constant.

$$V \propto T$$

or $$\frac{V}{T} = K \text{ (constant)}$$

The value of constant '$K$' depends on the mass and pressure of the gas.

**Temperarture-Volume Data for 0.2 mole of a gas at 1 atm pressure**

| *T* in Kelvin | *V* in Litre | *V/T* | |
|---|---|---|---|
| 200 | 3.2 | 0.016 | Constant |
| 300 | 4.8 | 0.016 | |
| 400 | 6.4 | 0.016 | |
| 500 | 8.0 | 0.016 | |

If the volumes of a given mass of gas are plotted against absolute temperatures at constant pressure, the plots will give a straight line which confirms that the gas obeys Charles' law. Such a straight line is called **isobar** as the pressure remains constant or the **Gay-Lussac's Law isobar** for that pressure.

Similarly, the pressure of a given mass of an ideal gas at constant volume bears a linear relationship with the absolute temperature. This straight line is called the **Gay-Lussac's Law isochore** and its slope rises with rise in volume.

## 5. Avogadro's Law :

**Avogadro's law states that under the same conditions of temperature and pressure, equal volumes of all gases contain equal number of molecules.**

Let the two gases A and B of equal volumes be taken at the same temperature and pressure. Let the number of molecules in A and B be $N_A$ and $N_B$ respectively. From kinetic gas equation for equal volumes of two gases at a fixed temperature,

$$PV = \frac{1}{3} M_A . N_A C_1^2 = \frac{1}{3} M_B N_B C_2^2$$

or

$$\frac{2}{3} N_A \frac{1}{3} M_A C_1^2 = \frac{2}{3} N_B \frac{1}{3} M_B C_2^2 \qquad ....(i)$$

(*i*) where $M_A$ and $C_1$ represent the mass number and root mean square velocity of $N_A$ molecules of gas A.

(*ii*) On the other hand, $M_B$ and $C_2$ represent the mass number and root mean square velocity of $N_B$ molecules of gas B.

At a given temperature the average kinetic energy of a molecule is constant.

$$\therefore \qquad \frac{1}{2} M_A C_1^2 = \frac{1}{2} M_B C_2^2$$

From equation (1) we get $\frac{2}{3} N_A = \frac{2}{3} N_B$.

or $N_A = N_B$ *i.e.* the number of molecules in gas A is equal to the number of molecules in gas B. This is **Avogadro's Law**.

## 6. Gas equation :

**Boyle's Law, Charles' Law** and **Avogadro's Law** express the changes in volume of a gas as each of the three quantities *viz*., *n*, *p*, and *T* is varied separately. These law combine to obtain an equation which connects all the four variables *P*, *V*, *n*, and *T* with each other. This equation is called **Ideal Gas Equation** and is expressed as

$$PV = n\,RT$$

where R is a constant named as **Molar gas constant** or **Gas constant.**

1. From the kinetic gas equation, $PV = \frac{1}{3}\, m\, NC^2$

Now $m$N = total mass of the gas = $M$. For 1 mole, $M$= Mol. wt.

$$\therefore \quad PV = \frac{1}{3} MC^2 \quad \text{when } M = \text{molecular weight of the gas.}$$

$$\text{or} \quad PV = \frac{2}{3}\frac{1}{2} MC^2 \quad \frac{1}{2} MC^2 \text{ is the kinetic energy, } E_k \text{ of one mole of the gas.}$$

Now $E_k$ is directly proportional to the absolute temperature.

$$\text{or} \quad E_k \propto T \text{ or } E_k = K_1 T_1 \quad K_1 = \text{constant.}$$

$$PV = \frac{2}{3} E_k = \frac{2}{3} K_1 T$$

$$\text{or} \quad PV \propto T \quad \therefore \frac{PV}{T} = K \text{ (constant)}$$

The value of $K$ depends on the mass of the gas. For one mole of an ideal gas, $K$ is expressed as $R$.

$$\text{Hence} \quad \frac{PV}{T} = R \text{ or } PV = RT$$

For $n$ moles $\boldsymbol{PV = nRT}$. This is **ideal gas equation.**

Values of $R$ in other different units

| Units of $PV$ | Values of $R = \frac{PV}{T}$ |
|---|---|
| Torr (76 cm Hg.). Litre | 62.36 litre-Torr/K/mole |
| Erg | $8.314 \times 10^7$ erg/K/mole |
| Joule | 8.31 Joule/K/mole |

## • Law of Partial Pressure

**John Dalton** (1901) enunciated the law of partial pressure which deals with the pressure exerted by a mixture of non-reacting gases. The law states that

**At constant temperature when two or more gases or vapours, which do not react chemically with one another, are mixed together in a vessel of fixed volume, the total pressure of mixture of gases or vapours is the sum of the partial pressure of each of the constituent gases or vapours.**

$$p_t = p_1 + p_2 + p_3 + \ldots\ldots\ldots\ldots$$

[$P_t$ = total pressure, $p_1, p_2, p_3$ etc. represent the partial pressure of each gas].

**Explanation :**

Let us take several gases of pressures $p_1, p_2, p_3 \ldots\ldots p_n$ at same temperature and volume and if the total pressure of them at the same temperature and volume be $P$, then according to **Dalton's law of partial pressure**

$$P = p_1 + p_2 + p_3 \ldots\ldots\ldots\ldots\ldots\ldots p_n$$

**Derivation of Dalton's Law from Ideal Gas equation**

(*i*) Dalton's law can also be derived from an ideal gas equation. The ideal gas equation can be applied to calculate the partial pressure exerted by any gas.

(*ii*) If $n_1, n_2, n_3$ ......... represent the number of moles of the component gases of gaseous mixture, then by the application of ideal gas equation, the values $P_1, P_2, P_3$................. etc.

$$P_1 = n_1 \frac{RT}{V}$$

$$P_2 = n_2 \frac{RT}{V}$$

$$P_3 = n_3 \frac{RT}{V}$$

Sum of the partial pressures = $P_1 + P_2 + P_3 + ..............$

$$= n_1 \frac{RT}{V} + n_2 \frac{RT}{V} + n_3 \frac{RT}{V} + ..........$$

$$= \frac{RT}{V}(n_1 + n_2 + n_3 ..........) \quad .....(i)$$

Now the total pressure ($P_t$) exerted by the gaseous mixture contained in the same volume $V$ and at the same temperature $T$ will be given by

$$P_t = n_m \frac{RT}{V} \quad ........(ii)$$

[$n_m$ = total number of moles of the gaseous mixture]

We know $n_m = n_1 + n_2 + n_3$ ..... (*iii*)

Substituting the values of $n_m$ from equation (*iii*) in equatiion (*ii*).

$$P_t = (n_1 + n_2 + n_3 ............) \frac{RT}{V} \quad ...... (iv)$$

Comparing the equation (*i*) and (*iv*) we find that

Sum of the partial pressures = Total pressures exerted by the gaseous mixture.

or $P_1 + P_2 + P_3 + .................. = P_t$

**Ideal gas :**

(I) A gas which obeys Boyle's law, Charles law etc. under all conditions of temperature and pressure is known as ideal gas.

(II) A gas which obeys the gas equation ($PV = nRT$) under all conditions of temperature and pressure is called ideal gas.

It is observed that there is no gas which obeys the gas laws or gas equation under all conditions of temperature and pressure. The gases are found to obey the gas laws fairly well if the pressure is low or the temperature is high.

**Real gas:**

The real gas is one which obeys the gas laws fairly well under low pressure or high temperature.

All gases are real gases. They show more and more deviations from the gas laws as the pressure is increased or temperature is decreased.

## B. DIFFUSION :

**Definition: Diffusion is a physical process in which molecules or ions of a matter by virtue of their inherent random translational kinetic motion migrate from a region of higher concentration to a region of lower concentration.**

### Characteristic Features :

(*i*) In diffusion, the net movement of particles would occur from higher to lower concentration.

(*ii*) Diffusion continues until the two substances are uniformly mixed to produce a homogeneous mixture. However, the molecular motion in the mixture does not stop even after the cessation of diffusion.

(*iii*) Diffusion may occur between gas and gas, liquid and liquid or solid and liquid, *i.e.*, diffusion occurs in **liquid** or **gaseous medium.**

(*iv*) **Diffusion** is a **passive process** because it is not driven by any external force or pressure and it is dependent on the inherent movement or motion of the particle.

(*v*) According to the second law of thermodynamics, the molecular movement of a substance takes place from an area of higher free energy of molecules to an area of lower free energy. Thus, diffusion is due to the **difference in free energy of the molecules of a substance.**

**Causes of Diffusion :**

Molecules of all substances always remain in a state of random linear motion due to the inherent kinetic energy in them.

(*i*) Movement of solute molecules from a concentrated solution into the solvent.

(*ii*) Movement of solvent molecules into the concentrated solution.

**• Diffusion Pressure (DP) :**

**The potential ability of a gas, liquid or solid to diffuse from an area of its greater concentration to an area of its lesser concentration is termed as Diffusion pressure (DP).**

It may be stated that the pressure created by the motion of diffusible particles is called diffusion pressure.

(*i*) **Diffusion pressure** (D.P.) is directly proportional to the concentration of diffusible material.

(*ii*) Diffusion is driven by the difference of **diffusion pressure** (*i.e.* diffusion gradient) between two regions.

(*iii*) Diffusion pressure is the cause rather than the effect of diffusion.

**• Diffusion Pressure Deficit (DPD).**

The amount by which the diffusion pressure of water (or a solvent) in a solution is lower than that of pure water (or pure solvent) is called diffusion pressure deficit.

**Factors Influencing Diffusion :**

The rate or velocity of diffusion is influenced by the following factors.

**1. Temperature :**

Diffusion is directly proportional to the temperature because rise of temperature increases the kinetic energy of the molecules and hence the molecular motion.

**2. Medium :**

(*i*) The rate of diffusion is inversely proportional to the concentration of molecules or particles present in the medium.

(*ii*) The more is the number of medium molecules, the lower is the rate of diffusion and vice versa.

(*iii*) Gas molecules diffuse more rapidly in vacuum than in air.

**3. Solubility :**

(*i*) The rate of diffusion of molecules is directly proportional to the degree of solubility of the molecules of the medium.

(*ii*) The more soluble is the substance in medium, the faster it will diffuse.

**4. Viscosity of the solvent :**

(*i*) A higher viscosity of the solvent increases the frictional resistance against free movement of solute particles and consequently lower the diffusion rate.

(*ii*) The diffusion coefficient ($D$) of the solute is inversely proportional to the viscosity ($\eta$) of the solvent.

$$D = \frac{RT}{6\pi r\eta N}$$ [$R$ = Molar gas constant, $T$ = absolute temperature]

[$N$ = Avogadro number, $r$ = radius of the solute]

**5. Density of Diffusing molecules :**

The rate of diffusion of molecules is inversely proportional to the square root of their densities.

$$r \propto \frac{1}{\sqrt{d}} \quad \left[\begin{array}{l} r = \text{rate of diffusion} \\ d = \text{density of diffusing material} \end{array}\right]$$

[This is known as Graham's law of diffusion]

**Graham's Law of Diffusion :**

At a given temperature and pressure, the rates of diffusion of gases are inversely proportional to the square root of their densities.

$$r \propto \frac{1}{\sqrt{d}}$$

If $r_1$ and $r_2$ represent the rates of diffusion of two gases whose densities are $d_1$ and $d_2$ respectively, the law takes the form

$$r_1 \propto \frac{1}{\sqrt{d_1}} \quad \text{or} \quad r_1 = \frac{k}{\sqrt{d_1}}$$

$$r_2 \propto \frac{1}{\sqrt{d_2}} \quad \text{or} \quad r_2 = \frac{k}{\sqrt{d_2}}$$

$$\frac{r_1}{r_2} = \sqrt{\frac{d_2}{d_1}}$$ [Provided temperature and pressure are the same].

**Relation between the rate of diffusion and molecular weight**

Molecular weight of a gas = 2 × D

Let $M_1$ and $M_2$ be the molecular weights of two gases A and B respectively.

Then $$d_1 = \frac{M_1}{2} \text{ and } d_2 = \frac{M_2}{2}$$

Since $$\frac{r_1}{r_2} = \sqrt{\frac{d_2}{d_1}} = \sqrt{\frac{\frac{M_2}{2}}{\frac{M_1}{2}}} = \sqrt{\frac{M_2}{M_1}}$$

$$\frac{\text{Rate of diffusion of A}}{\text{Rate of diffusion of B}} = \sqrt{\frac{\text{Molecular weight of A}}{\text{Molecular weight of B}}}$$

or $$r \propto \frac{1}{\text{M}} \text{ for any gas.}$$

**6. Molecular size :**

(*i*) The rate of diffusion is inversely proportional to the size and weight of the diffusible molecules.

(*ii*) Larger the size, slower the rate of diffusion.

**7. Distance :**

Diffusion is inversely proportional to the square of the distance ($l$) to be traversed.

**8. Mean molecular velocity and mean free path :**

(*i*) In cases of solutes and also for gases, diffusion varies directly with the mean molecular velocity. All the molecules do not move at the same velocity, they may passess an average velocity called **mean molecular velocity** ($\bar{c}$) and mean free path [average linear distance traversed by each molecule between two successive collisions is called mean free path ($\lambda$)].

(*ii*) The diffusion rate ($r$) across a unit area per second and the diffusion coefficient ($D$) are consequently related to the mean molecular velocity ($\bar{c}$) and the mean free path ($\lambda$) of diffusing material.

$$r = -\frac{1}{3}\lambda\,\bar{c}\,\frac{dc}{dx} \qquad\qquad r = -D\frac{dc}{dx}$$

Puting the value of $r$, $-D\dfrac{dc}{dx} = -\dfrac{1}{3}\lambda\,\bar{c}\,\dfrac{dc}{dx}$

or $$D = -\frac{1}{3}\lambda\bar{c}$$

**9. Concentration gradients and Fick's Law :**

**Concentration gradient :**

**The rate of change in the concentration of molecules or ions on two sides of the plasma membrane *i.e.,* between E.C.F. (Extra Cellular Fluid) and I.C.F. (Intra Cellular Fluid) per unit distance is called concentration gradient.**

The diffusion rate for a solute under concentration gradient is given by Fick's equation

$$\frac{ds}{dt} = DA\frac{C_1 - C_2}{x}$$

$\dfrac{ds}{dt}$ = the number of moles of solute $S$.

$dt$ = time interval of diffusion

$D$ = Diffusion coefficient of solute

$A$ = Cross-sectional area through which diffusion occurs

$C_1 - C_2$ = Difference in solute concentration of two sides

$x$ = Distance

**• Fick's Law :**

**Diffusion of a solute is directly proportional in rate to the magnitude of its concentration gradient and occurs down the concentration gradient.**

**or**

**It may be stated that the amount of solute (*S*) diffusing across the area (*A*) in a period of time (*dt*) is directly proportional to the concentration gradient $\left(\dfrac{dc}{dx}\right)$ at that point.**

$\frac{ds}{dt} = -DA\frac{dc}{dx}$ (D = Diffusion coefficient).

- **Diffusion Coefficient :**

**It is the quantity of solute diffusing across 1cm² area in 1 sec down the concentration gradient of unity.**

Negative sign indicates the diffusion from a higher to a lower concentration.

**Factors :**

(*i*) Size and shape of the molecule

(*ii*) Viscosity of the solvent.

**10. Pressure gradient :**

**The rate of change in the partial pressure of the diffusible molecules on the two sides of the plasma membrane per unit distance is called pressure gradient.**

The rate of diffusion is proportional to the magnitude of difference $(P_1 - P_2)$ between its own partial pressure in two regions.

The diffusion rate for a solute or gas under pressure gradient is given by Fick's equation

$$V_G = \frac{A}{T} \times D.(P_1 - P_2)$$

$V_G$ = Volume of gas
$A$ = Total surface area of diffusion
$T$ = Thickness of the membrane
$D$ = Diffusion coefficient of gas
$P_1 - P_2$ = Difference of pressure of gas *G*.

**11. Electrical gradients :**

The term electrical gradient means the number of different charged particles on the two sides of the plasma membrane.

(*i*) It helps diffusion of permeable ions from the side of higher potential to the side of lower potential across the plasma membrane.

(*ii*) It also generates ATP.

**Biological applications or Significance of diffusion :**

1. **Absorption of food :**
Transport of glucose, amino acids from intestine into blood.
2. **Respiration :**
Exchange of $O_2$ and $CO_2$ between air and blood in the lung capillaries.
3. **Transport :**
Exchange of foodstuffs, $O_2$ etc. between the tissue and red blood cells.
4. **Ion exchange :**
Exchange of ions ($HCO_3^-$, $Cl^-$) between plasma and red cells during the transfer of $Cl^-$ ions.
5. **Drainage of materials :**
Drainage of waste materials and $CO_2$ from the tissue to blood.
6. **Reabsorption :**
Reabsorption area in the renal tubules.

7. **Placental diffusion :**

   Placenta serve as a foetal lung. Respiratory gases are exchanged between foetus and maternal component by diffusion.

8. **Return of CSF and Aqueous humor**

   Cerebrospinal fluid (CSF) and aqueous humor is returned to the circulation (blood) by diffusion.

## C. OSMOSIS :

**The net movement of solvent molecules from a solution of lower concentration to a solution of higher concentration through a semipermeable membrane is called osmosis.**

### Types of membrane :

1. **Impermeable membrane :**

   A membrane which does not permit the free movement of both solute and solvent molecules is called impermeable membrane.

   **Example :** a sheet of rubber

2. **Permeable membrane :**

   A membrane which allows free movement or diffusion of both solvent and solute molecules is called permeable membrane.

   **Example :** filter paper, cell wall of plant cell.

3. **Semipermeable membrane :**

   A membrane which allows free movement or diffusion of only solvent molecules through it is called semipermeable membrane .

   **Example :** Artificial : parchment paper, cellophane paper.
   Natural (Biological) : cell membrane, egg membrane.

4. **Selectively permeable :**

   A membrane which allows free movement or diffusion of solvent molecules and also some selected solute molecules or ions through it is called selectively permeable membrane.

   **Example :** Plasmalemma.

### Types of Osmotic Solution :

(*a*) **Isotonic solution :**

A solution which exerts no osmotic effect on the cells ( *i.e.* the cell neither takes in nor gives out water when suspended in this solution) is termed as **isotonic solution.**

**Example :** 0.9 gm % NaCl solution in water is isotonic for mammalian cells.

(*b*) **Hypertonic solution :**

A solution which causes exosmosis in a cell (*i.e.,* draws out water from it) when the cell is suspended in the solution is termed as **hypertonic solution.**

**Example :** NaCl solution above 0.9 gm% is hypertonic to mammalian cell.

(*c*) **Hypotonic solution:**

A solution which causes endosmosis ( or inflow of water) in the cell when the cell is suspended in the solution is known as **hypotonic solution.**

**Example :** NaCl solution below 0.9 gm% is hypotonic solution to mammalian cell.

- **Osmotic Pressure (O.P.)**

**The excess hydrostatic pressure which must be applied on the solution to prevent the inflow of solvent (water) to the solution when they are separated by semipermeable membrane and establish an equilibrium is called osmotic pressure of the solution.**

- **Osmotic Potential :**

**It is the amount by which free energy or a chemical potential has been reduced as compared with pure water under ideal osmotic condition or it is the amount by which water potential is reduced due to the presence of solutes in a solution.**

It is directly related to the concentration of solute molecules in solution.

**Factors Affecting Osmotic Pressure :**

1. **Temperature :**

   Osmotic pressure is directly proportional to the absolute temperature.

2. **Concentration :**

   (*i*) Osmotic pressure is directly proportional to the molar concentration of the solute in it.

   (*ii*) Equimolar solutions of non-ionised solutes exert the same osmotic pressure.

3. **Ionisation of solute :**

   (*i*) An ionised solute exerts more osmotic pressure than an equimolar solution of non-ionisable solution.

   (*ii*) This is because each ion exerts osmotic pressure as a separate solute particle.

4. **Molecular weight of the solute:**

   (*i*) Osmotic pressure is inversely proportional to the molecular weight of the solute.

   (*ii*) Low molecular weight of a solute exerts more osmotic pressure than high molecular weight of solute.

**UNITS :**

Osmotic pressure is expressed either in mm Hg or in atm or dynes $cm^{-2}$.

**Osmole (osm) :**

The amount of osmotic pressure exerted by a solute is proportional to the concentration of the solute in numbers of molecules or ions and the concentration is expressed in terms of number of particles. The unit is called **osmole.**

The number of osmotically active particles per mole of a solute is termed as **osmole.**

**Osmolarity :**

The solute concentration of a solution, expressed in osmole per litre of solution, is termed as **osmolarity.**

**Osmolality :**

The solute concentration in osmole (OSM) per kg of solvent is known as **osmolality.**

- **Van't Hoff Laws of Osmotic Pressure**

Like gases, laws of osmotic pressure are applicable only for dilute solution but not for concentrated solution.

**Law 1 :**

**The osmotic pressure (π) of a solution is directly proportional to the molar concentration (C) of the solute as long as the temperature remains constant.**

$$\pi \propto C$$

or $\pi = K_1 C \qquad K_1 = \text{constant}$

If $V$ litres of the solution contain one mole of the solute, then

$$C = \frac{1}{V} \;\therefore\; \pi = K_1 C = K_1 \frac{1}{V}$$

or $\pi V = K_1$

**Law 2 :**

**The osmotic pressure (π) of a solution is directly proportional to the absolute temperature (T) as long as its concentration remains constant.**

$$\pi \propto T$$

or $\pi = K_2 T \qquad K_2 = \text{constant.}$

or $\frac{\pi}{T} = K_2$

**Law 3 :**

**Equimolecular quantities of different solutes dissolved in the same volume of a solvent exert equal osmotic pressure under identical conditions of temperature.**

**Van't Hoff equation of osmotic pressure:**

It emerges from a combination of two laws.

$$\pi V = nRT \qquad \text{or } \pi = \frac{n}{V} RT \quad \text{or } \pi = CRT$$

[$R$ = Molar gas constant, π = osmotic pressure, $T$ = Absolute temperature, $C$ = Molar concentration of solute, $n$ = number of moles of solute, $V$ = volume containing $n$ moles of solutes.]

- **Relation between O.P., T.P. and D.P.D.**

**Turgor Pressure :** It can be defined as the hydrostatic pressure which presses the protoplasm against the cell wall. It is due to entry of water into cell sap *i.e.* decrease in concentration of cell sap.

**Wall Pressure :** It can be defined as the pressure exerted by the cell wall against the turgor pressure.

**Diffusion Pressure Deficit (D.P.D.) or Suction Pressure (S.P.)**

(*i*) Diffusion Pressure Deficit (D.P.D.) is the pressure difference of diffusible molecules of two regions.

(*ii*) Diffusion pressure (D.P.) of the solvent in a solution is inversely proportional to the concentration of the solution. It means the higher is the concentration of a solution, the lower will be the D.P. of its solvent and as a greater will be the D.P.D. of the solution.

Thus, D.P.D. of a solution is directly proportional to its concentration *i.e.* D.P.D. of a solution increases with increase in concentration of the solute particles in the solution.

(*iii*) The term D.P.D. refers to the amount by which the D.P. of the solvent in the solution is lower than the D.P. of its pure solvent at the same temperature.

(*iv*) Thus D.P.D. of a solution = D.P. of pure solvent – D.P. of solvent in the solution.

or D.P.D. = Osmotic potential of a cell – Wall pressure or inward pressure exerted by cell wall.

or D.P.D. = O.P. – W. P./T.P.

- **Osmoregulation :**
  The process by means of which an organism maintains the osmotic concentration of its body fluid against change of its environment is called *osmoregulation.*
- **Osmoregulator :**
  Osmoregulators are those organisms which are osmotically stable (independent) and are able to maintain their internal osmotic concentration at constant level or nearly so, despite changes in their external environment.
- **Osmoconformer :**
  Osmoconformers are those organisms which are osmotically labile (dependent) and whose body fluid concentration changes with the medium.
- **Stenohaline :**
  Stenohaline organisms are those organisms which have limited tolerance to changes in the osmotic concentrations of the external environment.
- **Eurohaline :**
  Eurohaline organisms are those organisms which can tolerate a wider range of osmotic concentrations, although the degree of tolerance depends on the length of exposure age and temperature etc.
- **Exosmosis :**
  When water moves out of the cell sap in response to an external hypertonic solution, the process is termed as exosmosis.
  Due to exosmosis, the turgor pressure of the cell decreases and the cell becomes flacid.
- **Endosmosis :**
  When water enters in a cell in response to an external hypotonic solution, the process is termed as *endosmosis*.
  Due to endosmosis the turgor pressure of the cell increases and the cell becomes turgid.
- **Plasmolysis :**
  It is a process of protoplasm shrinkage due to water loss in response to external hypertonic solution.
- **Deplasmolysis :**
  Attainment of turgidity of a plasmolysed cell immersed in hypotonic solution is termed as *deplasmolysis* or *reverse plasmolysis.*
- **Incipient Plasmolysis:**
  The initial stage of plasmolysis when the plasma membrane of plant cell just seperates from the cell wall is termed as *incipient plasmolysis.*
- **Crenation :**
  The process of shrinkage of the whole animal cell due to exosmosis in response to an external hypertonic solution is termed as *crenation.*
- **Lysis :**
  Breakdown of an animal cell in response to the built up turgor pressure development due to endosmosis is termed as *lysis.*

**Significance of Osmosis :**

**(*A*) In plants :**

(*i*) Absorbption of water from soil by the root hairs of plants occurs due to osmosis.

(*ii*) Osmosis controls the cell to cell movement of water from root hair through the cortical cells to the xylem vessels and also from xylem vessels to the different parts of the body.

(*iii*) Endosmosis in plant cells develop turgor pressure which helps different functions.

(*a*) Controls the movement of guard cell for stomatal opening and closing.

(*b*) Growth of meristematic tissue.

(*c*) Various types of movement. Example : seismonastis in mimosa.

(*d*) Rigidity and mechanical support to soft plants.

**(*B*) In animals :**

(*i*) Water content of animal cells is maintained by osmosis between ICF (Intra Cellular Fluid) and ECF (Extra Cellular Fluid).

(*ii*) **Transfusion :** Isotonic solutions are used for transfusion in medicine and surgery for dehydration, burn and shock.

(*iii*) **Purgative :** Hypertonic saline purgative draws more water in the large intestine causing purgation.

- **Facilitated Diffusion :**

**Diffusion of particles through a membrane with the help of carrier molecules present in the membrane or carrier mediated transportation of solutes or ions in favour of concentration gradient is known as facilitated diffusion.**

**Example :** Movement of glucose across the membrane of most cells.

**Differences between Osmosis and Diffusion**

| Character | Osmosis | Diffusion |
|---|---|---|
| 1. Membrane | Semi-permeable membrane is required. | No membrane is required. |
| 2. Medium | It occurs only in a liquid medium. | It occurs between gas and gas; liquid and liquid; liquid and gas, solid and gas; solid and liquid. |
| 3. Nature of solvent | Osmosis takes place between similar types of solvent. | Diffusion takes place between different types of solvent. |
| 4. Independence | Only diffusion of solvent molecules takes place but not solute molecules. | Here diffusion of both solvent and solute molecules takes place. |
| 5. Flow of direction | Flow of particles occurs only in one direction. | Flow of particles occurs in all directions. |

**Differences between Simple Diffusion and Facilitated Diffusion**

| Character | Simple Diffusion | Facilitated Diffusion |
|---|---|---|
| 1. Solute | It is not solute specific | It is solute specific |
| 2. Carrier | Not required | Required |
| 3. Rate | Slow | Rapid |
| 4. Inhibitor | Cannot inhibit | Can inhibit |

## • Donnan Membrane Equilibrium

**When two ionized diffusible or non-diffusible ions are separated by a semipermeable membrane, the ratio of equilibrium concentrations of diffusible cation on its two sides equals the reciprocal of the similar ratio of a diffusible anion, provided no active transport is involved.**

The explanation of such phenomenon was theoretically shown by Gibbs and was confirmed experimentally by Donnan (1911). Hence it is called *Gibbs Donnan effect* or *Donnan membrane equilibrium*.

**Explanation :**

**(*A*) Symmetrical Distribution** (Two ionized and Diffusible ions)

If two ionized solutions (M) and (N) are placed in a container of constant volume and separated by a semipermeable membrane, an equilibrium is reached.

I. Solution of both sides (M & N) are electrically neutral because total charges on cations are equal to anions.

Cation (M) = Anions (M)

Cations (N) = Anions (N)

II. The product of diffusible ions of one side of the membrane is equal to the product of the diffusible ions on the other side.

Diffusible cations (M) × Diffusible anions (M) = Diffusible cations (N) × Diffusible anions (N)

III. At equilibrium (for diffusible ions-NaCl) both sides of the membrane have the same concentration of both the ions ($Na^+$ & $Cl^-$) and is called symmetrical distribution.

**(*B*) Asymmetrical Distribution** (Non diffusible & diffusible ions)

When non diffusible ions are present at equilibrium, the distribution of diffusible ions is a symmetrical.

I. A nondiffusible anion ($Pr^-$) of sodium salt i.e. sodium protinate ($Na^+ Pr^-$) is placed in a compartment (M) and diffusible sodium chloride (NaCl) solution in the other compartment (N) separated by a semi-permeable membrane.

II. Sodium ions ($Na^+$) and chloride ions ($Cl^-$) will tend to pass in pair through the membrane from (M) side to the (N) side.

III. The compartment M contain a solution of nondiffusible salt i.e. sodium protinate ($Na^+ Pr^-$) of concentration 'a' where Pr is a large anion (polyelectrolyte in nature). The compartment N contain a solution of NaCl of concentration 'b'.

IV. NaCl now diffuse from N to M and some of it may reversely diffuse back into N. This continues till the system attains equilibrium.

V. Suppose $x$ concentration of NaCl diffuses from N to M then at equilibrium their concentration are as follows:

| M (Compartment) | N (Compartment) |
|---|---|
| $Na^+$ (a mols)<br>$Pr^-$ (a mols) | $Na^+$ (b mols)<br>$Cl^-$ (b mols) |

**Fig.** (*a*) Distribution of ions across semipermeable membrane at zero time.

| M (Compartment) | N (Compartment) |
|---|---|
| $Na^+$ ($a + x$ mols)<br>$Pr^-$ ($a$ mols)<br>$Cl^-$ ($x$ mols) | $Na^+$ ($b - x$ mols)<br>$Cl^-$ ($b - x$ mols) |

**Fig:** (*b*) Distribution of ions across semipermeable membrane at equilibrium.

VI. The rate of diffusion of sodium chloride from (N) to (M) is proportional to the product of concentration of $Na^+$ and $Cl^-$ in (N) and the rate of diffusion of NaCl from (M) to (N) is proportional to the product of the concentration of Na+ and Cl in (M)

| M | | | N |
|---|---|---|---|
| $[Na^+]_M$ | $[Cl^-]_M$ | = | $[Na^+]_N\ [Cl^-]_N$ |
| $(a+x)$ | $(x)$ | = | $(b-x)\ (b-x)$ |

The above mentioned equation reveal that the concentration of Na and $Cl^-$ in the compartment (N) is equal but in the compartment (M) the concentration of $Cl^-$ is less than the concentration of $Na^+$.

VII. The concentration of diffusible anion is less in the compartment (M) containing nondiffusible anion than the compartment (N).

The restriction of the nondiffusible anion at the membrane while the cation is tending to diffuse through it and sets up a potential difference between the two solutions.

VIII. At equilibrium, the solution in the compartment (M) acquires a positive potential with respect to other (N) because of the slight excess of Na. Thus it is seen that the presence of nondiffusible ion in one side of the semipermeable membrane causes unequal distribution (asymmetrical) of every diffusible ion on the two sides of the membrane.

IX. There is a slight excess of cation in the compartment (M) containing non diffusible anion (Pr) and a slight excess of diffusible anions in the opposite compartment (N).

X. It should be emphasized that the difference between the number of anions and the number of cation on either side of the membrane is extremely small relative to the total number of anions and cations present.

XI. Therefore, relative concentration of nondiffusible and diffusible ion is important in the system of Donnan equilibrium. If the concentration of non diffusible ion is increased, a greater inequality in the ion distribution is observed.

**Biological application:**

1. Ionic concentration between plasma, interstial fluid & lymph. The interstial fluid and lymph have lower concentration of inorganic cations ($Na^+$, $K^+$) but higher concentration of anions ($Cl^-$) than plasma. It is maintained by Donnan equilibrium.
2. **Difference in pH between cells & plasma**
   (*i*) Intracellular pH is usually 6-7.4 in different cells, pH of plasma is 7.4.
   (*ii*) Donnan effect causes such difference in pH.
3. **Ionic concentration between Intra Cellular Fluid (ICF) & Extra Cellular Fluid (ECF)**
   (*i*) ICF contains in low concentration of $Na^+$ and high concentration of $K^+$ and low concentration Cl & $HCO_3^-$ and predominantly organic phosphates (ATP) than extra cellular fluid (ECF).
   (*ii*) The ionic difference between I.C.F. & E.C.F. are maintained by Donnan effect.
4. **Chloride-Bicarbonate shift on RBC & plasma**
   It is maintained by Donnan effect.

**Reverse osmosis :**

(*i*) When a solution is separated from pure water by a semipermeable membrane, osmosis of water occurs from water to solution.

(*ii*) The osmosis can be stopped by applying pressure equal to or more than osmotic pressure, on the solution.

(*iii*) If pressure is greater than osmotic pressure is applied; osmosis is made proceed in the reverse direction to ordinary osmosis i.e. from solution to water.

(*iv*) This kind of osmosis take place from solution to pure water by application of pressure greater than osmotic pressure, on the solution, is termed as reverse osmosis.

## PROBLEMS

**On kinetic theory of gases :**

**1.** *Calculate the kinetic energy of 10 moles of an ideal gas at 0°C.*

**Solution :**

From kinetic theory, we know that

$$E_k = \frac{3}{2} RT \text{ for 1 mole of an ideal gas.}$$

For 10 moles $E_k = 10 \times \frac{3}{2} RT$

Now $R = 8.314 \text{ Nm K}^{-1} \text{ mol}^{-1}; T = 273 + 0 = 273 \text{ K}$

$$\therefore \quad E_k = 10 \times \frac{3}{2} \times 8.314 \times 273$$

$$= 10 \times 3 \times 4.157 \times 273 = 34045.8 \text{ joules}$$

**2.** *Determine the kinetic energy of 5 g of ammonia at 27° C.*

**Solution :**

$$5 \text{ g of } NH_3 = \frac{5}{17} = 0.294 \text{ moles}$$

$$T = 273 + 27 = 300 \text{ K}$$

$$\therefore \quad E_k = \frac{3}{2} nRT = \frac{3}{2} \times 0.294 \times 8.314 \text{ J K}^{-1} \text{ mol}^{-1} \times 300 \text{ K}$$

$$= 3 \times 0.294 \times 4.157 \times 300 = 1099.94 \text{ joules.}$$

**3.** *Calculate the average kinetic energy in joules of molecules in 8.0 g of methane at 300 K.* ***[I.I.T.-82]***

**Solution :**

From kinetic theory we know $E_k = \frac{3}{2} nRT.$

$$n = \text{number of moles of the gas} = 8/16 = 0.5 \text{ mole}$$

$$R = 8.314 \text{ joules K}^{-1} \text{ mole}^{-1}, T = 300 \text{ K}$$

$$E_k = \frac{3}{2} \times 0.5 \times 8.314 \times 300 = 0.5 \times 3 \times 4.157 \times 300$$

$$= 1870.65 \text{ joules.}$$

$$\text{Average } E_k = \frac{1870.65}{6.023 \times 10^{23} \times 0.5} = 6.21 \times 10^{-23} \text{ joules}$$

**4.** *Calculate kinetic energy of a mole of $H_2$ at 0°C.*

**Solution :**

The total kinetic energy of a mole of any gas is $\frac{3}{2} RT$. So the kinetic energy of a mole of $H_2$ at 0°C will be

$E_k = \frac{3}{2}\, nRT$ $\quad$ $n$ = number of moles of the g as $= \frac{1}{1} = 1$

$R$ = 8.314 joules $K^{-1}$ $mole^{-1}$

$T$ = 273

$$E_k = \frac{3}{2} \times 1 \times 8.314 \times 273$$

$$= 3 \times 4.157 \times 273 = 3404.583 \text{ joules.}$$

**5.** *Calculate translational kinetic energy of 2 moles of a gas at 27°C.*

**Solution :**

For 1 mole of gas $k_E = \frac{3}{2}\, RT$

$\therefore$ For 2 moles of the gas $k_E = \left(\frac{2}{3}\, RT\right) \times 2$

$$k_E = \left(\frac{3}{2} \times 8.314 \times 300\right) \times 2$$

$R$ = 8.314 joules $K^{-1}$ $mole^{-1}$

$T$ = 273 + 27 = 300 K

$$= (3 \times 4.157 \times 300) \times 2$$

$$= 3741.3 \times 2 = 7482.6 \text{ joules.}$$

**6.** *Calculate the kinetic energy of 16 gm of oxygen and one molecule of $O_2$ at 27°C.* **[*B.U. - 2000*]**

**Solution :**

$k_E = \frac{3}{2}\, nRT$ $\quad$ $n = \frac{16}{32} = \frac{1}{2}$

$R$ = 8.314 $joules^{-1}$ $K^{-1}$ $mole^{-1}$

$T$ = 273 + 27 = 300 K

$$k_E = \frac{3}{2} \times \frac{1}{2} \times 8.314 \times 300$$

$$= \frac{3 \times 8.314 \times 300}{4}$$

$$= \frac{900 \times 8.314}{4}$$

$$= \frac{7320.6}{4} = 1830.15 \text{ joules.}$$

For 1 molecule of $O_2$ the average energy would be

$$\frac{1830.15}{6.023 \times 10^{23}} = 303.86 \times 10^{-23} \text{ Joules.}$$

**7.** *The volume of a certain mass of a gas at a constant pressure and at 15°C is 360 ml. Find the temperature of the gas when its volume will be 480 ml.*

**Solution :**

From Charles' law,

$$\frac{V_1}{T_1} = \frac{V_2}{T_2} \text{ (at constant temperature)}$$

Here $V_1 = 360$ ml $\quad T_1 = 273 + 15 = 288$ K, $\quad V_2 = 480$ ml $\quad T_2 = ?$

Putting the values of $V_1$, $T_1$ and $V_2$

or $$\frac{360}{288} = \frac{480}{T_2}$$

or $$T_2 = \frac{480 \times 288}{360} = 384\text{K}$$

or $$T_2 = 384 - 273 = 111°\text{C}.$$

**8.** *The pressure of 40 ml of a gas is changed from 760 mm to 800 mm. Find the new volume at the same temperature.*

**Solution :**

From Boyle's law,

$$P_1V_1 = P_2V_2 \quad \text{(at constant temperature)}$$

Here $V_1 = 40$ ml. $\quad V_2 = ?$

$P_1 = 760$ mm. $\quad P_2 = 800$ mm.

Putting the values of $V_1$, $P_1$ and $P_2$,

$$760 \times 40 = 800 \times V_2$$

or $$V_2 = \frac{760 \times 40}{800} = 38.0 \text{ mm.}$$

**9.** *What will be the pressure required to reduce 600 ml of a dry gas at 750 mm pressure to 500 ml at the same temperature?*

**Solution :**

From Boyle's law,

$$P_1V_1 = P_2V_2 \quad \text{(at constant temperature)}$$

Here $V_1 = 600$ ml, $\quad P_1 = 750$ mm.

$V_2 = 500$ ml, $\quad P_2 = ?$

Putting the values of $V_1$, $V_2$ and $P_2$,

$$750 \times 600 = P_2 \times 500$$

or $$P_2 = \frac{750 \times 600}{500} = 900 \text{ mm.}$$

**10.** *100 ml (V) of a gas at pressure 740 mm (P) at certain temperature is taken. What will be the volume of the same mass of gas at 780 mm pressure at the same temperature?*

**Solution :**

From Boyle's law,

$$P_1V_1 = P_2V_2 \quad \text{(at constant temperature)}$$

Here $V_1 = 100$ ml, $\quad P_1 = 740$ mm.

$V_2 = ?$ $\quad P_2 = 780$ mm.

Putting the values of $V_1$, $P_1$ and $P_2$.

$$740 \times 100 = 780 \times V_2$$

or $$V_2 = \frac{740\times100}{780} = 94.87 \text{ ml.}$$

- **Problems on Partial Pressure :**

**11.** *A mixture of gases at normal pressure contains nitrogen, chlorine and methane. The percentage of each volume is 65, 15, and 20 respectively. Find the partial pressure of each gas.*

**Solution :**

Normal pressure = 760 mm.

$$\text{Partial pressure of nitrogen} = \frac{65}{100}\times760 = 494 \text{ mm}$$

$$\text{Partial pressure of chlorine} = \frac{15}{100}\times760 = 114 \text{ mm}$$

$$\text{Partial pressure of methane} = \frac{20}{100}\times760 = 152 \text{ mm}$$

**12.** *The partial pressure of oxygen is 56 cm, nitrogen 1.1 atm and that of hydrogen is 360 mm. What is the total pressure of these gases in the flask?*

**Solution :**

Partial pressure of oxygen $P_{O_2}$ = 56 cm = 560 mm.

Partial pressure of nitrogen $P_{N_2}$ = 1.1 atm = 1.1 × 760 = 836 mm

Partial pressure of hydrogen $P_{H_2}$ = 360 mm.

Partial pressure of all gases = $P_{O_2} + P_{N_2} + P_{H_2}$ = 560 + 836 + 360 = 1756 mm.

**13.** *A mixture of gases at 760 mm pressure contains 80% nitrogen and 20% oxygen. What is the partial pressure of the gas?*

**Solution:**

Let $P_N$ = Partial pressure of nitrogen.

$P_O$ = Partial pressure of oxygen.

Total pressure ($P_t$) = $P_N + P_O$ = 760 mm.

$$\therefore \quad P_N = \frac{760\times80}{100} = 608 \text{ mm.}$$

$$P_O = \frac{760\times20}{100} = 152 \text{ mm.}$$

- **Problems on Diffusion**

**14.** *Find out the relative rates of diffusion of hydrogen and oxygen. Molecular weight of hydrogen is 2 and oxygen is 32.*

**Solution :**

$$\frac{r_{H_2}}{r_{O_2}} = \sqrt{\frac{M_{O_2}}{M_{H_2}}} \qquad \text{Here } M_{O_2} = 32 \ \& \ M_{H_2} = 2$$

or $$\frac{r_{H_2}}{r_{O_2}} = \sqrt{\frac{32}{2}} = \sqrt{16} = 4$$

**15.** *100 ml of a gas diffuses in 50 seconds while 100 ml of oxygen diffuses in 80 seconds. Molecular weight of oxygen is 32. Find out the molecular weight of the other gas.*

**Solution:**

Let $r_1$ be the rate of diffusion of unknown gas and $r_2$ the rate of diffusion of oxygen.

Therefore $$r_1 = \frac{100}{50} = 2 \text{ ml /sec}$$

$$r_2 = \frac{100}{80} = 2.5 \text{ ml /sec}$$

We know $$\frac{r_1}{r_2} = \sqrt{\frac{M_2}{M_1}}$$ Here $M_2 = 32$ $r_1 = 2$

$M_1 = ?$ $r_2 = 2.5$

or $$\frac{2}{2.5} = \sqrt{\frac{32}{M_1}}$$

or $$0.8 = \sqrt{\frac{32}{M_1}} \text{ or } .64 = \frac{32}{M_1} \text{ or } M_1 = \frac{.32}{.64} = \frac{32}{64} \times 100 = 50$$

**16.** *432 ml of a gas M takes 36 minutes to diffuse through a porous hole. 288 ml of another gas N takes 48 minutes to diffuse through the same porous hole under same conditions. If the molecular weight of N is 64, what is the molecular weight of M ?*

**Solution :**

Let $r_1$ the rate of diffusion of $M = \dfrac{432}{36 \times 60} = 0.2$ ml/sec.

$r_2$ the rate of diffusion of $N = \dfrac{288}{48 \times 60} = 0.1$ ml/sec.

$$\frac{r_1}{r_2} = \sqrt{\frac{M_2}{M_1}}$$ Here $r_1 = 0.2$, $r_2 = 0.1$ $M_1 = ?$ $M_2 = 64$

$$\frac{0.2}{0.1} = \sqrt{\frac{64}{M_1}} \text{ or } \frac{.04}{.01} = \frac{64}{M_1}$$

or $$4M_1 = 64$$

$$M_1 = 16$$

$\therefore$ Molecular weight of $M = 16$

**17.** *Molecular weights of two gases are 64 and 100 respectively. The rate of diffusion of the first gas is 15 ml/sec. What is the rate of diffusion of the second one?*

**Solution :**

According to Graham's law of diffusion

$$\frac{r_1}{r_2} = \sqrt{\frac{M_2}{M_1}}$$

$r_1$ = rate of diffusion of one gas

$r_2$ = rate of diffusion of other gas

$M_1$ and $M_2$ = Molecular weight of two gases.

Here $r_1 = 15$ ml/sec. $M_1 = 64$

$r_2 = ?$ $M_2 = 100$

Therefore,

$$\frac{15}{r_2} = \sqrt{\frac{100}{64}}$$

or

$$\frac{15}{r_2} = \frac{10}{8}$$

or

$$r_2 \times 10 = 15 \times 10$$

$$r_2 = \frac{15 \times 8}{10} = \frac{120}{10} = 12$$

- **Problems on Osmosis**

**18.** *When a solution A, having osmotic pressure of 10 atmosphere, is separated from solution B, having osmotic pressure of 7 atmosphere, to which direction will the solvent molecules flow? Give reason.*

**Ans :** According to the principle of osmosis, solvent molecules will flow from solution *B* to solution *A* because the former is less concentrated than the latter *i.e.* the former contains relatively higher number of solvent molecules than that in the latter.

**19.** *The cytoplasm of a cell contains water and sugar in the ratio 50 : 50. If it is placed in two solutions having water and sugar ratio of (i) 80 : 20 and (ii) 20 : 80 respectively, what physical change will be observed in the cell and why ?*

**Ans :** (*i*) The first solution having water and sugar ratio of 80 : 20 contains more water and less sugar *i.e.* less concentrated than the cytoplasm.

(*ii*) On the other hand, the second solution having water and sugar in the ratio of 20 : 80 is more concentrated *i.e.* contains less water and more sugar than the cytoplasm.

So if the cell is placed in the first solution, it will swell up by drawing in water due to endosmosis and become turgid.

Conversely in the second solution, the cell will shrink and become crenated due to exosmosis.

**20.** *A solution of sucrose (molecular mass 342) is prepared by dissolving 68.4 g of it per litre of the solution. What is osmotic pressure at 300 K? R = 0.082 litre atm $K^{-1}$ $mol^{-1}$.*

**Solution:**

We know $\pi v = nRT$

Given, molecular mass of the solute = 342

Weight of the solute = 68.4 g

Volume of the solution = 1 litre.

Temperature ($T$) = 300 K

Osmotic pressure = $\pi$ = ?

$R = 0.082$

$$\pi = \frac{nRT}{V} \quad n = \text{the number of moles of the solute} = \frac{\text{wt. of the solute}}{\text{molecular mass of the solute}}$$

$$n = \frac{68.4}{342} = 0.2$$

$$= \frac{0.2 \times 0.082 \times 300}{1}$$

= 4.92 atm.

∴ Osmotic pressure 4.92 atm.

**21.** *Calculate the osmotic pressure of a 5% solution of a cane sugar ($C_{12}H_{22}O_{11}$) at 288 K. R = 0.082 lit atm $K^{-1}$ $mol^{-1}$.*

**Solution :**

Molecular mass of cane sugar ($C_{12}H_{22}O_{11}$)

= (12 × 12 + 22 + 11 × 16)

= (144 + 22 + 176) = 342

∵ The solution is 5%

∴ wt of sugar per lit. = 50 gram

$$\therefore n\text{, the no. of moles} = \frac{\text{wt. of solute}}{\text{molecular mass of solute}} = \frac{50}{342} = 0.146$$

$V$ = 1 litre, $T$ = 288 K

$$\pi = \frac{n}{V}RT = \frac{0.146 \times 0.082 \times 288}{1} = 3.447 = 3.45 \text{ atm.}$$

∴ Osmotic pressure = 3.45 atm.

**22.** *Calculate osmotic pressure of 5% of solution of glucose ($C_6H_{12}O_6$) at 18°C. R = 0.082 lit atm $K^{-1}$ $mol^{-1}$.*

**Solution :**

Molecular mass of glucose = 12 × 6 + 12 + 16 × 6 = 72 + 12 + 96 = 180

∵ The solution is 5%

∴ wt. of glucose per lit = 50 g

$$\therefore n\text{, the no. of moles} = \frac{\text{wt. of solute}}{\text{mol mass of solute}} = \frac{50}{180} = 0.278$$

$V$ = 1 lit. $T$ = 273 + 18 = 291 K $T$ = 0.082

$$\pi = \frac{n}{V}RT = \frac{0.278 \times 0.082 \times 291}{1}$$

= 6.633 = 6.64 atm.

∴ So osmotic pressure = 6.64 atm.

**23.** *Calculate osmotic pressure of 20% anhydrous calcium chloride solution at 273 K, assuming that the solution is completely dissociated (R = 0.082 lit atm $K^{-1}$ $mol^{-1}$).*

**Solution :**

Molecular mass of $CaCl_2$ = 40 + 35.5 × 2

= 40 + 71 = 111

Since solution is 20%

∴ weight of $CaCl_2$ per lit = 200 gr.

$V$ = 1 litre

$T$ = 273 K

$$n = \frac{\text{wt. of solute}}{\text{molecular mass of the solute}} = \frac{200}{111} = 1.8018$$

$$\pi = \frac{n}{V} \times RT = \frac{1.8018 \times 0.082 \times 273}{1} = 40.335 \text{ atm.}$$

Since $CaCl_2$ dissolves to give 3 particles for each molecule on complete dissociation ($CaCl_2 \rightarrow Ca^{++} + 2Cl^-$)

$\therefore$ observed osmotic pressure = Normal osmotic pressure × 3

= 40.335 × 3 = 121.005 atm.

$\therefore$ observed osmotic pressure = 121.005 atm.

**24.** *What is the driving force of osmosis?*

**Ans :**

(*i*) The driving force of osmosis is the chemical potential.

(*ii*) The increase of chemical potential in a solvent solution system decreases the tendency of the solvent to pass to solution. So the driving force is the chemical potential.

**25.** *Calculate the osmotic pressure at 27°C of a solution formed by dissolving 1g glucose and 1 g sucrose in 1 litre of water.* ***[C.U. 1997]***

**Solution :**

Glucose = $C_6H_{12}O_6$, molecular mas = 12 × 6 + 12 × 1 + 16 × 6

= 72 + 12 + 96 = 180

Sucrose = $C_{12}H_{22}O_{11}$ molecular mass = 12 × 12 + 22 × 1 + 16 × 11

= 144 + 22 + 176 = 342

Number of moles of solute $= \frac{1}{180} + \frac{1}{342} = \frac{19 \quad 10}{3420} = \frac{29}{3420}$

$= 0.008479 = 8.48 \times 10^{-3}$

Assuming that volume of the solution = Volume of the solvent

$\pi = (8.48 \times 10^{-3} \text{ mol lit}^{-1}) \times (0.082 \text{ lit atm K}^{-1} \text{ mol}^{-1}) (300 \text{ K})$

$= 8.48 \times 0.082 \times 10^{-3} \times 300$ [$\because$ 273 + 27 = 300]

$= 2.086 \times 10^{-3} \times 10^2$

$= 208.608 \times 10^{-3} = 0.208608 = 0.209$ atm.

CHAPTER 2

# Properties of Solution

**True Solution: The true solution is defined as homogeneous mixture containing solute particles of 1 Å to 10 Å in size distributed uniformly throughout the entire mass of the solvent. Due to the high degree of dispersion, there is no dispersed phase in a true solution and the entire system is one single phase of homogeneous state. Neither the solute particle nor the solvent is observed by human eye or with the aid of microscope. True solution is electrically neutral.**

**Solvent:** The dispersing component which is present in a greater proportion of the solution and determines the phase of the solution is called solvent.

**Solute:** The dispersed component which is present in a smaller proportion of the solution is called solute.

Solution = Solute + Solvent

- **Types of Solution :**

Generally we express solvent as a liquid and the solute as a solid or sometimes as a liquid. In fact, all the three states of matter (gas, liquid or solid) may behave either as solvent or solute.

| Serial No. | Solute | Solvent | Example |
|---|---|---|---|
| 1. | gas | gas | Air-solution of oxygen in nitrogen |
| 2. | gas | liquid | Soda-water ($CO_2 + H_2O$) |
| 3. | gas | solid | Hydrogen in platinum or palladium |
| 4. | liquid | solid | Solution of mercury in gold |
| 5. | solid | liquid | Common salt or sugar in water |
| 6. | solid | solid | Coloured gems, some alloys |
| 7. | liquid | liquid | Solution of alcohol and water |
| 8. | liquid | gas | Fog or mist |
| 9. | solid | gas | Solid aerosol (dust particle in air) |

**Solution of liquids in liquids :**

(*i*) Liquids are completely miscible in one another. **Example:** ethanol and water

(*ii*) Liquids are partially miscible. **Example:** ether and water (*iii*) Liquids are practically immiscible. **Example:** benzene and water

**Solution of Solids in Liquids :**

(*i*) It is of immense importance to the chemists.

(*ii*) Copper sulphate (blue vitriol) is stirred well with water taken in beaker, the solid particles of copper sulphate readily disappear and a clear blue liquid is obtained.

(*iii*) Similarly, sugar dissolves in water.

**Solution of Gases in Liquids :**

(*i*) All gases are found to dissolve to a certain extent in all liquids. The gaseous substances like hydrogen, oxygen, nitrogen, air are only sparingly soluble in water.

(*ii*) Carbon dioxide is fairly soluble and hydrogen chloride, ammonia etc. are highly soluble in the the same solvent under ordinary temperature and pressure.

(*iii*) However, the amount of gas dissolved in a given liquid depends upon the pressure, temperature, the nature of the gas and the nature of the liquid.

**Characteristics of True Solution:**

**(*a*) Homogeneous:**

(*i*) A true solution is homogeneous as each part of it has the same physical properties and invariant composition.

(*ii*) Here the solute or dispersed substance is evenly distributed throughout the solvent or the dissolving or disposing state.

**(*b*) Passage Through Filter:**

(*i*) The constituents (solutes and solvent) of a solution. Each of the constituents will exist as a separate entity and retains its own properties.

(*ii*) The constituents of a solution (both solvent and solutes) can easily be separated from one another either by simple mechanical means like evaporation, filtration, distillation and crystallisation, or through a filter very readily.

**(*c*) Phase:**

It is a single phase system.

**(*d*) Boiling and Freezing Point:**

(*i*) Every pure liquid has fixed boiling and freezing point.

(*ii*) But the boiling and freezing point of a liquid containing a dissolved solute is found to differ from that of the liquid in pure state.

(*iii*) The boiling increases while the freezing point decreases. This increase or decrease depends on the amount of the dissolved solute. Thus an aqueous solution of a salt or sugar boils above 100°C and freezes below 0°C.

**(*e*) Ionic Characters:**

(*i*) An ionic solute dissociating into coloured ions in the solvent forms a solution which assumes the characteristic colour of the ions produced.

(*ii*) Thus aqueous solution of potassium permanganate is violet due to the presence of coloured $MnO_4^-$ in it and for the presence of $Cu^{++}$ ions, a copper sulphate solution is blue in colour.

**(*f*) Boundary Zone:**

There is no boundary zone between solvent and solution.

**(g) Thermal Character:**

(*i*) Usually there is no perceptible thermal change during the preparation of a solution but heat is either evolved or absorbed during the formation of some solutions.

(*ii*) Formation of the solution sulphuric acid or caustic soda in water is usually accompanied by a **considerable production** of **heat.**

Again heat is absorbed during the formation of a solution of ammonium chloride in water.

**(h) Colligative Properties:**

(*i*) These are the properties of a solution which depends on the number of molecules or ions of the solute in the solution but not their nature and shape.

(*ii*) These include the osmotic pressure, the depression of vapour pressure, the elevation of boiling point and the lowering of the freezing point of the solution.

- **Saturated Solution :**

A solution containing the maximum amount of a solute in the dissolved state at a given temperature is said to be saturated with respect to the solute at that temperature.

- **Unsaturated Solution :**

A solution containing minimum amount of a solute in the dissolved state can take more solute into the solution at a given temperature. It is said to be unsaturated solution.

- **Supersaturated Solution :**

A solution is said to be supersaturated when it contains in solution more of the solute than it can hold at that temperature to form a saturated solution.

- **Ideal Solution:**

(*i*) A solution of two components is said to be ideal if each components of solution obeys **Raoult's law** exactly at all concentrations and at all temperatures an ideal solution is one where at all compositions the total vapour pressure of this solution is intermediate between the vapour pressures of the two liquid constituents.

(*ii*) If a solution is formed by mixing the liquids *A* and *B*, $P_Ao$ and $P_Bo$ are their vapour pressures of pure components, $P_A$ and $P_B$ are their partial pressures, $X_A$ and $X_B$ are their mole fraction respectively.

$$P_A = P_Ao \cdot X_A \text{ and } P_B = P_Bo \cdot X_B \text{ thus } P_{\text{total}} = P_A + P_B = P_Ao \cdot X_A + P_Bo \cdot X_B$$

(*iii*) Solution formed by mixing the two liquids is ideal if

- there is no charge on mixing *i.e.* $\Delta V_{\text{mix}} = 0$
- there is no enthalpy change on mixing *i.e.* $\Delta H_{\text{mix}} = 0$

**Non-Ideal Solution :**

A solution obtained on mixing two liquids will be non-ideal if it does not obey **Raoult's Law** is called a non-deal solution.

(*i*) The solute-solvent interactions are weaker or stronger than the solute-solute and solvent-solvent interaction.

(*ii*) The interaction of *A* and *B* molecules in the solution are not similar to those of pure *A* and pure *B* or $\Delta V_{\text{mix}} \neq 0$ and $\Delta H_{\text{mix}} \neq 0$.

- **Azeotropes or Azeotropic Mixtures :**

**Mixture of liquids which boil at constant temperature like a pure liquid such that the distillate has the same composition as that of the constant boiling point are called azeotropic mixture or azeotropes.**

**Conjugate Solution :**

In case of partially miscible liquids, the two solutions (one of the liquid *A* in *B* and the other of liquid *B* in *A*) which are in equilibrium are called *conjugate solutions.*

- **Solution and Equilibrium Concept :**

(*i*) A **saturated solution** is defined as one in which the dissolved solid solute exists in equilibrium with the undissolved pure solute at a specific temperature.

$$\text{Undissolved solid particle} \rightleftharpoons \text{dissolved solute.}$$

(*ii*) A solution in which the dissolved solute is not in equilibrium with the pure solute is said to be **unsaturated.**

(*iii*) In a **supersaturated** solution, the dissolved solute can never be in equilibrium with the pure solute.

**Solution of Gases:**

(*i*) All gases are dissolved to a certain extent by all liquids, the amount dissolved depending upon the pressure, the temperature, the nature of the gas and the nature of the solvent.

(*ii*) When a gas dissolves in a liquid, it may form a physical dispersion or it may react chemically with it.

**Example :**

(*a*) Oxygen ($O_2$) dissolves in water by physical dispersion.

(*b*) Ammonia ($NH_3$) dissolves in water by chemical reaction:

$$NH_3 + H_2O \longrightarrow NH_4OHQ.$$

(*iii*) Carbon dioxide dissolves in water by both physical dispersion and to some extent by chemical reaction with water forming carbonic acid.

$$H_2O + CO_2 \longrightarrow H_2CO_3$$

- **Factors Affecting Solubility of a Gas in Liquid :**

(*a*) Nature of the gas and solvent

(*b*) Temperature

(*c*) Pressure

**(*a*) Nature of the Gas and Solvent :**

(*i*) Solubility of a gas in a given solvent varies considerably with the nature of the gas.

(*ii*) For example, gases like oxygen, nitrogen, and hydrogen are sparingly soluble in water. On the other hand, carbon dioxide, ammonia, hydrochloric acid gas are more soluble.

(*iii*) In general, gases which can be liquefied more easily are more soluble in common solvents. For example, carbon dioxide which can be liquefied more easily is more soluble than hydrogen or oxygen in water or in any other common solvent.

**(*b*) Temperature :**

(*i*) When a gas molecule dissolves in a liquid, the gas molecules come much closer to the solvent molecules and they attract each other, lowering the potential energy.

(*ii*) Hence when a gas is dissolved in a liquid, generally there is an evolution of heat.

(*iii*) In general, solubility of gases decreases as the temperature increases.

**Table: Solubility of gases at different temperatures**

| Gas | 0°C | 20°C | 60°C |
|---|---|---|---|
| Nitrogen | 0.0003 g | 0.0002 g | 0.001 g |
| Carbon dioxide | 0.387 g | 0.200 g | 0.070 g |
| Oxygen | 0.007 g | 0.004 g | 0.002 g |

**(*c*) Pressure :**

(*i*) The relation between pressure and solution was first noticed by Henry (1803).

***Henry Law:* At constant temperature, the solubility of a gas in a liquid is directly proportional to the pressure of the gas above the liquid.**

(*ii*) Let $m$ be the mass of the gas dissolved, $p$ the pressure and $k$ be **Henry's constant**. According to law :

$$m \propto p \text{ or } \frac{m}{p} = k$$

(*iii*) When several gases are being dissolved simultaneously in a solvent, then solubility of each gas forms a mixture of gases which is directly proportional to the partial pressure of the gas in the mixture.

- **Solubility :**

**The solubility of a substance (solute) in a solvent at a particular temperature is the number of grams of the substance (solute) necessary to saturate 100 gms of the solvent at that temperature.**

In other words, the solubility of a substance represents the maximum amount of it (solute) which can be dissolved by 100 grams of a given solvent at a given temperature.

$$\therefore \quad \text{Solubility at a particular temperature} = \frac{wt. \text{ of solute in grams}}{wt. \text{ of solvent in grams}} \times 100$$

- **Solubility Product :**

**The solubility product of a sparingly soluble electrolyte is the maximum product of the concentrations of the ions in its saturated solution at a given temperature and is usually denoted by $K_s$.**

**Explanation**

(*i*) All salts are strong electrolytes. Some of them, such as NaCl, $K_2SO_4$, $Pb(NO_3)_2$ etc. are highly soluble in water white salts like AgCl, $BaSO_4$, $PbSO_4$ are sparingly soluble or apparently insoluble in the same solvent.

(*ii*) When a sparingly soluble electrolyte is agitated with a limited amount of water until the solution gets saturated, a particular type of heterogeneous equilibrium is established between the solid state and its ions produced from the complete ionization of the electrolyte in its saturated solution.

(*iii*) At equilibrium condition, the rate at which the ions pass from the solid electrolyte into the solution is equal to the rate at which ions return to the solid.

Let, MA, the formula of a sparingly soluble binary electrolyte. Each molecule of which dissociates in aqueous solution producing univalent ions, $M^+$ & $A^-$. In saturated solution of MA in contact with the solid, we have the equilibrium

$$\text{MA (Solid)} \rightleftharpoons \text{MA (dissolved)} \rightleftharpoons M^+ + A^-$$

∴ According to law of mass action, the equilibrium constant

$$k = \frac{[M^+]\times[A^-]}{[\text{MA (solid)}]}$$

(Dissolved MA is taken to be completely ionised and the concentrations of $M^+$ and $A^-$ are $[M^+]$ and $[A^-]$ respectively.)

or $$K\,[\text{MA(solid)}] = [M^+][A^-]$$

The concentration of undissolved solid MA is reasonably taken to be constant at a fixed temperature, so the above equation reduces to $K_s = [M^+] \times [A^-]$

or $$[M^+] \times [A^-] = \text{constant.}$$

This constant representing the product of ionic concentrations of MA is denoted by $K_s$ and is known as **solubility product constant,** or simply *solubility product of MA.*

**Example:**

(*i*) Sparingly soluble salt silver chloride (AgCl) on agitation with water forms saturated aqueous solution.

$$\text{AgCl(solid)} \rightleftharpoons \text{AgCl(dissolved)} \rightleftharpoons Ag^+ + Cl^-$$

$$K_s = [Ag^+]\,[Cl^-] = \text{solubility product of AgCl.}$$

(*ii*) In a saturated solution of AgCl, the product of the concentrations of $Ag^+$ and $Cl^-$ ions is a constant (equal to $K_S$) so long as the temperature remains unaltered.

- **Limitation of Henry's Law:**

(*i*) The pressure should be low and the temperature should be high i.e. the gas should be have like an ideal gas.

(*ii*) The gas should not undergo compound formation with the solvent or association or dissociation in the solvent.

(*iii*) Solubility of the gas in the solvent.

- **Solubility coefficient :**

It is defined as the volume of a gas dissolved in a unit volume of the solvent at a given temperature and pressure.

- **Nernst's Distribution Law :**

A mathematically constant ratio exists between the concentrations of a given molecular species in any two phases in contact with each other at a constant temperature.

- **Solubility Curves :**

**The curve which is obtained by plotting solubilities of a given substance in a given solvent against the corresponding temperatures is called solubility curve of that substance.**

**Types :**

**(*a*) Continuous Solubility Curve:**

(*i*) It shows no break or changes in direction but remains in continuous form at the maximum point.

(*ii*) Solubility of calcium salts of fatty acids, potassium chloride etc.

**(*b*) Discontinuous Solubility Curve:**

(*i*) The curve which exhibits sudden changes of direction is called discontinuous solubility curve.

(*ii*) Solubility curve of sodium sulphate and ferric chloride.

**Utility of Solubility Curve:**

1. The solubility of a substance in a given solvent at a particular temperature can be ascertained directly from its solubility curve.
2. A comparison of the solubility of two or more substances in a given solvent at the same temperature may easily be made with the help of their solubility curve.
3. A clear idea of the rate of change of solubility of a substance with temperature can be derived from the shape of the solubility curve.

   A steep curve indicates rapid change while a flat curve shows slow change.

- **Ideal fluid :**

An ideal fluid is fluid which is non-viscous and incompressible.

- **Streamline flow :**

**It may be defined as the path straight or curved, the tangent to which at any point gives the direction of the flow of liquid at that point.**

**Characteristics :**

(*i*) Here every particle follows the path of its preceding particle.

(*ii*) Path followed by the particle is known as *streamline.*

(*iii*) The tangent drawn at any point of the streamline represents the direction of velocity at that point

(*iv*) Two streams do not intersect each other.

(*v*) Pressure over any cross section is constant.

- **Turbulent flow :**

**It is a bundle of streamlines having the same velocity of fluid elements over any cross section perpendicular to the direction of flow.**

- **Characteristics :**

(*i*) It is zig-zag flow of liquid.

(*ii*) The velocity of a particle crossing a particular point of the liquid is not constant in direction and magnitude. It varies with time.

(*iii*) The energy maintaining the flow of a liquid is mainly used in producing eddy currents through this liquid.

(*iv*) A liquid flowing slowly and steadily behaves as to be consisting of a number of layer or laminae one above the other.

(*v*) In the case of a nonviscous liquid, the velocity of all the particles at any section of a tube is the same and the velocity profile is plane. (Fig. 2.1 a).

(*vi*) The velocity profile of a viscous liquid is parabolic i.e. velocity of the layer is maximum at the axis and decreases to zero at the wall of the tube. Such flow of liquid is laminar flow. (Fig. 2.1 b).

**Fig. 2.1** (a) and (b)

- **Critical velocity :**

It is the maximum velocity of a fluid above which a streamline flow changes to a turbulent flow.

- **Energy of a liquid :**

Since liquid has intertia therefore it possesses three types of energy:

(*a*) Kinetic energy
(*b*) Potential energy
(*c*) Pressure energy

If two liquids are allowed to flow through a glass pipettes under identical conditions, it is observed that one liquid flows faster than the other. This difference of rate of flow of liquids are associated with the phenomenon known as *viscosity*.

- **Shearing force :**

When a force is applied at the upper surface of a metal which is held fixed at its lower surface, it will be deformed. The deformation is proportional to the applied force. Such applied force is called *shearing force*.

- **Velocity gradient :**

(*i*) When a liquid flows i.e. flows over another layer and it experiences resistance.

(*ii*) If a shearing force is applied to overcome the attractive forces between the molecules of the adjacent layers, each layer will move with different velocity i.e. the velocity gradient.

(*iii*) The velocity gradient $\frac{dV}{dx}$ so formed can be related to the shearing force applied per unit area by the equation where layer distance ($dx$).

$$\frac{F}{A} = \eta \frac{dV}{dx}$$

$\frac{F}{A}$ = shearing force per unit area
$\eta$ = viscosity of the liquid
$\frac{dV}{dx}$ = differences in velocity between two layers

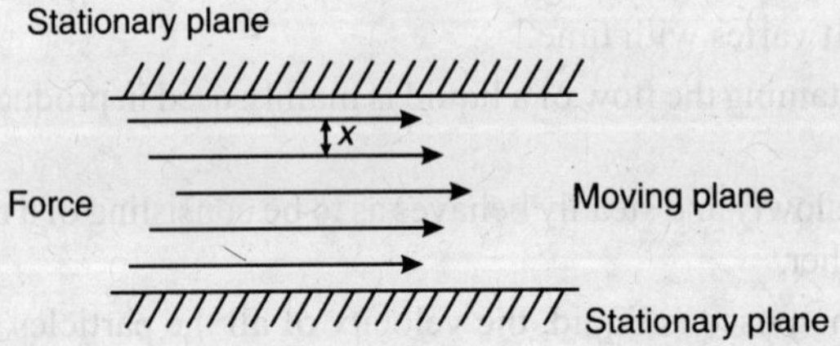

Stationary plane
x
Force
Moving plane
Stationary plane

Fig. 2.2

- **Viscosity**

**The internal resistance force or friction against the free flow of layers of a liquid or gas over each other is termed as viscosity.**

$$f \propto Av/x$$

$A$ = unit area, $x$ = unit distance
$v$ = velocity difference, $f$ = friction
$\eta$ = coefficient of viscosity

$$f = \eta \frac{Av}{n}$$

It is increased by the presence of solute (i.e. the structural properties of solute).

**Coefficient of viscosity (η) :**

**It is the force required to maintain unit difference velocity between two parallel liquid surfaces, unit distance apart and having an unit area.**

(*i*) Unit of viscosity: dynes $sec^{-1}$ $cm^{-2}$

$$\text{Poise (P)} = 10^{-1}\ \text{kg m}^{-1}\ \text{sec}^{-1}$$

After the name of Poiseville who pioneered the study of viscosity.

(*ii*) The reciprocal of viscosity is called fluidity $\phi$ (phi) = $\frac{1}{\eta}$ ($\eta$ = eta)

Fluidity expresses the tendency of a liquid to flow, on the other hand viscosity is a measure of resistance of a liquid offers to this flow.

**Factors affecting viscosity :**

(*a*) **Temperature :** Generally increase of temperature (at a limit) result decrease of viscosity.

(*b*) **pH** : The effect of pH on viscosity of colloidal solutions is very distinct.

(*c*) **Pressure** : Pressure increases viscosity increase.

(*d*) **Chemical composition** : The viscosity of liquid depends upon the size, shape, flexibility of the particle present in solution.

(*e*) **Specific volume** : Specific volume increases, viscosity decreases.

- **Measurement of viscosity :**

Three different instruments are used based on

(*a*) Capillary flow.

(*b*) Rotation of cylinder immersed in the solution.

(*c*) The rate of fall of a ball through solution.

**Viscometer**

A calibrated U-shaped glass shaped glass tube with bulbs and an intervening capillary, used in estimating the viscosity of opaque liquids.

- **Significance of viscosity of Biological system :**

(*i*) The blood, is highly viscous and is affected profoundly by even a small change in the viscosity of the medium. Generally blood flow varies inversely with viscosity of blood.

(*ii*) Besides the hematocrit, the constitutes of blood plasma mainly immunoglobulin and other plasma proteins play significant role in altering the viscosity of blood.

(*iii*) pH of protein solution affects the viscosity of a protein solution by affecting the ionisation of protein.

**Concentration Units**

(*i*) The composition of any solution is expressed in terms of solute dissolved in one unit of the solvent or solution.

(*ii*) The amount of solute dissolved in one unit of solvent or solution is called concentration.

- **Percentage by weight :**

It is defined as the percentage of solute in grams present in 100 gram of the solution.

$$\text{Thus \% by weight} = \frac{\text{weight of the solute}}{\text{weight of the solution}} \times 100$$

- **Strength :**

The strength of solution is defined as the number of grams of the solute dissolved per litre of the solution.

***(a)* Normality (N) :**

It is defined as the number of equivalents of the solute dissolved per litre of the solution.

$$\text{Normality (N)} = \frac{\text{Number of equivalents of the solute.}}{\text{Volume of the solution per litre.}}$$

Or, If the volume of the solution in litres (1000cc) is V and the equivalent mass of solute is E and mass of the solute is denoted by W.

$$\text{Normality (N)} = \frac{\text{Mass of the solute (W)}}{\text{Equivalent mass of solute (E)} \times \text{Volume of the solution in litre (V)}}$$

$$\text{N} = \frac{\text{W}}{\text{E} \times \text{V}}$$

***(b)* Molarity (M) :**

It is defined as the number of moles of the solute dissolved per litre ($dm^3$) of the solution.

$$\text{Molarity of a solution} = \frac{\text{Number of moles of solute}}{\text{Volume of the solution in litres (dm}^3\text{)}}$$

$$\text{M} = \frac{n}{v} \qquad \begin{bmatrix} n = \textit{number of moles of the solute} \\ v = \textit{volume of the solution} \end{bmatrix}$$

Or, If w gram of a substance having molar mass $m$ is dissolved in $V$ $dm^3$ of the solution, then the molarity $M$ of the solution is

$$\text{Molarity (M)} = \frac{\text{w}}{m \times v} \text{ mol or dm}^{-3}$$

***(c)* Molality (m) :**

It is defined as the number of moles of the solute per kg of the solvent,

Or, It means moles or gram molecular weight of solute per 1000 gm of solvent.

$$\text{Molality } (m) : \frac{\text{Number of moles of solute}}{\text{Mass of the solvent in kg}} = \frac{n}{w} \text{ mol kg}^{-1}$$

$$\text{or} \qquad \text{Molality } (m) : \frac{\text{Mass of the solute (w)}}{\text{Molar mass of the solute (M)} \times \text{Mass of the solvent in kg (W)}}$$

$$m = \frac{w}{\text{M} \times w} \text{ mol kg}^{-1}$$

***(d)* Mole Fraction (X) :**

The mole fraction of any component of a solution is defined as the ratio of the number of moles of that component to the total number of moles of all the components of solution.

If a solution contains $n_A$ moles of A and $n_B$ moles of B, then

(*i*) Mole fraction of A ($X_A$) = $\dfrac{\text{Number of moles of A}}{\text{Number of moles of A + Number of moles of B}}$

$$X_A = \frac{n_A}{n_A + n_B}$$

(*ii*) Mole fraction of B($X_B$) = $\dfrac{\text{Number of moles of B}}{\text{Number of moles of A + Number of moles of B}}$

$$X_B = \frac{n_B}{n_A + n_B}$$

Adding the above equation:

$$X_A + X_B = \frac{n_A}{n_A + n_B} + \frac{n_B}{n_A + n_B} = \frac{n_A + n_B}{n_A + n_B} = 1$$

Thus the sum of the fraction of all the components of a solution is equal to one. *i.e.* $X_A + X_B = 1$.

It does not change with temperature.

**Mole :**

**(*i*) The mole is the amount of substance which contains as many elementary entities (ions, atoms, molecules or electrons)**

**(*ii*) Mole is a unit which stands for $6.02 \times 10^{23}$ entities of a substance. This number is called Avogadro's number.**

Hence moles of a substance = $\dfrac{\text{Weight of the substance in grams}}{\text{Molecular weight or atomic weight or ionic weight}}$

- **Colligative Properties :**

The colligative properties of dilute solution are very important as these provide valuable methods of finding the molecular weight of the dissolved substances.

**(*a*) Lowering Vapour Pressure :**

If a non-volatile solute is added to a volatile liquid, the vapour pressure of the solution is lower than the vapour pressure of the pure solvent.

**Explanation :**

(*i*) In a pure liquid, the whole surface of the liquid is occupied by the molecules of the liquid.

(*ii*) In the case of solution, solute molecules bind with some of the solvent molecules to form **solvate** complexes and thereby decrease the number of solvent molecules.

(*iii*) This reduces the solvent activity and consequently decrease the number of solvent molecules at the surface of the solution.

(*iv*) This reduces the escaping tendency of solvent molecules from the liquid phase to the vapour phase at the given temperature and thereby lowers the vapour pressure.

**Raoult's Law :**

The **relative lowering of vapour pressure of any solution containing a non-volatile solute in a volatile solvent is proportional to the mole fraction of the solvent in the solution.**

**Explanation :**

(*i*) Let P be the vapour pressure of the pure solvent and $P_s$ is the vapour pressure of the solution. *n* and *N* the number of moles of solute and solvent respectively.

(*ii*) Lowering the vapour pressure relative to the vapour pressure of the pure solvent is termed the relative lowering of vapour pressure.

$$\text{Relative lowering of vapour pressure} = \frac{P - P_s}{P}$$

(*iii*) Therefore $\dfrac{P - P_s}{P} = \dfrac{n}{n + N}$

**(*b*) Elevation of Boiling Point :**

(*i*) When a liquid is heated, its vapour pressure rises gradually. When the vapour pressure equals the atmospheric pressure, the liquid begins to boil.

(*ii*) The addition of a non-volatile solute lowers the vapour pressure and consequently elevates the boiling point as the solution has to be heated to a higher temperature to make its vapour pressure equal to atmospheric pressure.

(*iii*) If $T_b$ is the boiling point of the solvent and *T* the boiling point of the solution, then the difference of these two boiling points ($T - T_b$ or $\Delta T_b$) is called elevation of boiling point.

$$\Delta T_b = T - T_b$$

**(*c*) Raoult's Law :**

**Law 1 :**

**The elevation of boiling point of a solution is directly proportional to the molar concentration of the solute in the solution.**

$\Delta T_b \propto C_m$ $\quad$ $\Delta T_b$ = elevation of *boiling point*; $C_m$ = molal concentration of solute.

$\Delta T_b = K_b\, C_m$ $\quad$ $K_b$ = Molal boiling point elevation constant.

when $\quad C_m = 1 \quad \Delta T_b = K_b$

$K_b$ is called either molal boiling point elevation constant or **ebulloscopic constant.**

**Molal ebulloscopic constant (*Kb*) may be defined as the elevation of boiling point produced by dissolving one mole of solute in 1000 g of solvent.**

| Solvent | Boiling Point (K) | Kb Value (K kg mol$^{-1}$) |
|---|---|---|
| 1. Water ($H_2O$) | 373° | 0.52 |
| 2. Chloroform ($CHCl_3$) | 334.4° | 3.63 |
| 3. Diethyl ether ($C_4H_{10}O$) | 307.8 | 2.02 |

**Thermodynamic Derivation of Molal Ebulloscopic Constant**

$$K_b = \frac{RT^2}{1000lb} = \frac{0.002}{1000lb} \; : \; K_b = \text{constant},\ T = \text{boiling point of solvent}$$

$R$ = Molar gas constant, $lb$ = latent heat of vaporisation.

**Law 2:**

**Equimolar solutions of different solutes produce an identical elevation of the boiling point of the same solvent.**

**Ebulloscopic determination of Molecular Weight (*MW*):**

If the elevation of boiling point be $\Delta T_b$ ,we have, according to Raoult's law,

$$\Delta T_b = K_b\, C_m$$

[$K_b$ = Ebullioscopic constant
$C_m$ = molality]

If the weight of the solvent is $w_1$ gram and the weight of the solute is $w_2$ gm, m is the molecular weight of the solute : $w_1$ gm is dissolved in $w_2$.

**(*c*) Freezing Point Depression :**

(*i*) The pressure of a solute lowers the freezing point of a solvent. The more concentrated the solution, the greater is the freezing point depression.

(*ii*) Freezing point of a substance is defined as the temperature at which the liquid and the solid states exist at equilibrium.

Or, It is the temperature at which the liquid and the solid states of a substance have the same vapour pressure.

(*iii*) The difference of the freezing point of the pure solvent and the solution is referred to as the depression of freezing point. It is represented by symbol $\Delta T_f$.

If the freezing point of solution is $T'_f$ and the freezing point of the pure solvent is $T'_f$ and the depression of freezing point is $\Delta T_f$.

therefore $\Delta T_f = T'_f - T_f$

**(*d*) Raoult's Law on Freezing Depression :**

**Law 1 :**

**The depression of the freezing point of a solvent by a dissolved solute is proportional to the molal concentration of the dissolved substance.**

$\Delta T_f$ = Freezing Depression, $C_m$ = Molal concentration of solution.

$\Delta T_f \propto C_m$

or $\Delta T_f = K_f \times C_m$ $K_f$ = cryoscopic constant or molal depression constant.

If $C_m = 1$, *i.e.* when 1 mole of the solute is dissolved in 1000 grams of the solvent, then $\Delta T_f = K_f$

**Thus molal depression constant or cryoscopic constant** is defined as **the depression of freezing point which would be produced by dissolving one mole of the solute in 1000 grams of the solvent.**

| Solvent | Freezing point | $K_f$ value |
|---|---|---|
| Water $H_2O$ | 273.0 | 1.86 |
| Diethyl ether $C_4H_{10}O$ | 156.9 | 1.79 |
| Chloroform $CHCl_3$ | 209.6 | 4.70 |

**Law 2:**

**Equimolecular quantities of different solutes dissolved in the same quantity of a particular solvent depress the freezing point to the same extent.**

**Relation between Depression of Freezing Point and Molecular Weight of Solute :**

From Raoult's first law of depression of freezing point,

$\Delta T_f = K_f C_m$ $C_m$ = molal concentration

Let $w_2$ gram solute be dissolved in $w_1$ gram of solvent and the molecular weight of the solute is $M_2$.

Therefore, the number of gram molecule of solute = $\frac{w_2}{M_2}$

$\therefore$ $w_1$ gram solvent contain $\frac{w_2}{M_2}$ (gram molecule of solute) 1 gram solvent contains $\frac{w_2}{M_2 \times w_1}$

or 1000 gram solvent contain $\frac{w_2 \times 1000}{M_2 \times w_1}$

Therefore,
$$C_m = \frac{w_2 \times 1000}{M_2 \times w_1} = \frac{w_2}{w_1} \times \frac{1000}{M_2}$$

$$\Delta T_f = K_f \times \frac{w_2}{w_1} \times \frac{1000}{M_2}$$

$$M_2 = K_f \times \frac{w_2 \times 1000}{w_1 \times \Delta T_f}$$

$M_2$ = Molecular weight of the solute
$K_f$ = Molal depression constant
$w_2$ = Weight of the solute
$w_1$ = Weight of the solvent
$\Delta T_f$ = Depression of freezing point

**Raoult's Law :**

In a solution the vapour pressure of a component at a given temperature is equal to the mole fraction of that component in the solution multiplied by the vapour pressure of that component in the pure state.

(*i*) Vapour pressure of the solvent in the solution
= Mole fraction of the solvent in solution × vapour pressure of the pure solvent.

$$P_s = x_1 \times P^o \quad ...(i)$$

or
$$\frac{P_s}{P^o} = x_1 \quad ...(ii)$$

(*ii*) If the solution contains $n_2$ moles of the solute dissolved in $n_1$ moles of the solvent, we have mole fraction of the solvent in solution ($n_1$)

i.e.
$$x_1 = \frac{n_1}{n_1 + n_2}$$

or
$$\frac{P_s}{P_o} = \frac{n_1}{n_1 + n_1} \qquad \left[\because \frac{P_s}{P^o} = x_1 \quad ...(ii)\right]$$

(*iii*) Subtracting each side from 1, we get

$$1 - \frac{P_s}{P^o} = 1 - \frac{n_1}{n_1 + n_2}$$

$$\frac{P^o - P_s}{P^o} = \frac{n_1 + n_2 - n_1}{n_1 + n_2} = \frac{n_2}{n_1 + n_2}$$

$$\boxed{\frac{P^o - P_s}{P^o} = \frac{n_2}{n_1 + n_2}}$$

(*iv*) $P^o - P_s$ gives the lowering of vapour pressure, $\frac{P^o - P_s}{P^o}$ is called relative lowering of vapour pressure, $\frac{n_2}{n_1 + n_2}$ represent the mole fraction of the solute in the solution.

- **Newtonian fluid and Laminar flow**

If velocity of fluid is always same magnitude and direction at a particular point, the fluid is Newtonian fluid, and the flow is called Laminar flow.

- **How and why does vapour pressure of a liquid depend upon the temperautre?**

**(Guwahati Univ. 2006)**

**Ans.** (i) Vapour pressure of a liquid increases with the increase of temperature ?

(ii) With the increase of temperature, the kinetic energy of the molecules on the surface increases $\left(KE = \frac{3}{2}RT\right)$.

(iii) More and more molecules leave the surface of the liquid and converted into vapours, thereby raising vapour pressure.

## PROBLEMS

**1.** *50 gms of a saturated aqueous solution of $KNO_3$ at 25°C contains 21 gms of the salt. Calculate the solubility of $KNO_3$ at 25°C.*

**Solution :**

Weight of the saturated solution $KNO_3$ = 50 gm.

Weight of solute $KNO_3$ = 21 gm.

∴ Weight of water of the solution = 50 – 21 = 29

∴ Solubility of $KNO_3$ at 25°C = $\frac{\text{Weight of solute}}{\text{Weight of solvent}} \times 100$

$$= \frac{21}{29} \times 100 = 72.41$$

**2.** *At 20°C, 7.6 gms of saturated sugar solution contains 5.1 gm sugar. Determine the solubility of sugar at 20°C.*

**Solution :**

At 20°C, 7.6 gm saturated solution contains 5.1gm sugar

∴ Solute = 5.1 – solvent water = 7.6 – 5.1= 2.5 gm.

∴ Solubility of sugar = $\frac{wt.\text{of solute}}{wt.\text{of solvent}} = \frac{5.1}{2.5} \times 100 = 204$

**3.** *10.5 gms of saturated solution of potassium chloride at 40°C contains 3 gm salt. What is the solubility of KCl at this temperature?*

**Solution :**

Solute = 3 gm; solvent = 10.5 – 3 = 7.5

Solubility (s) of KCl = $\frac{wt.\text{ of solute}}{wt.\text{ of solvent}} \times 100 = \frac{3}{7.5} \times 100 = 40$

**4.** *In a solution 69 gm $KNO_3$ is dissolved in 50 gm water at 70°C. How much $KNO_3$ will crystallise if the solution is cooled to 20°C (solubility of $KNO_3$ is 138 gm at 70°C and 32 gm at 20°C)*

**Solution :**

At 20°C, 100 gm water dissolves 32 gm $KNO_3$ (since its solubility 32)

At 20°C, 50 gm water dissolves 16 gm $KNO_3$.

$\therefore$ The amount of $KNO_3$ crystallised out = 69 – 16 = 53 gm

**5.** *The solubility of a salt in water is 40g at 300K. Calculate the amount of water required to dissolve 120g of the salt at the same temperature.*

**Solution :**

$$\text{Solubility} = \frac{wt.\text{of solute}}{wt.\text{ of solvent}} \times 100$$

$$40 = \frac{40}{wt.\text{of solvent}} \times 100 \text{ or solvent} = \frac{40}{40} \times 100 = 100g.$$

40 g of salt requires 100 g of $H_2O$

1 g of salt requires $\frac{100}{40}$ g of $H_2O$

120 g of salt requires $\frac{100}{40} \times 120 = 300$ g.

**6.** *A sugar syrup of weight 214.2g contains 34.2 g of sugar ($C_{12}H_{22}O_{11}$). Calculate (i) molal concentration and (ii) mole fraction of sugar in syrup.*

**Solution :**

Weight of sugar syrup = 214.2 gram.

Weight of sugar = 34.2

Amount of water = 180.00

$$\text{Molal concentration} = \frac{34.2 \times 1000}{180 \times 342} = 0.555$$

$$\text{Mole fraction of sugar} = \frac{\frac{34.2}{342}}{\frac{34.2}{342} + \frac{180}{18}} = \frac{0.1}{0.1 + 10} = \frac{0.1}{10.1} = 0.0099$$

**7.** *46 grams of ethyl alcohol ($C_2H_5OH$) is dissolved in 36 grams of water. Calculate the mole fraction of each.*

**Solution :**

Molecular mass of ethyl alcohol = 2 × 12 + 1 × 5 + 16 + 1 = 46

Molecular mass of water = 1 × 2 + 16 = 18

(*i*) wt. of $C_2H_5OH$ = 46 $\therefore$ No. of moles of $C_2H_5OH = \frac{wt.\text{of } C_2H_5OH}{\text{g.mol.wt.of } C_2H_5OH}$

$$= \frac{46g}{46g} = 1$$

(*ii*) wt. of water ($H_2O$) = 36 $\therefore$ No. of moles of $H_2O = \frac{\text{wt of } H_2O}{\text{g.mol. wt of } H_2O} = \frac{36g}{18g} = 2$

Total number of moles of $C_2H_5OH + H_2O = 1 + 2 = 3$

$$\therefore \quad \text{Mole fraction of } C_2H_5OH = \frac{1}{1+2} = \frac{1}{3} = 0.33$$

$$\text{Mole fraction of } H_2O = \frac{2}{1+2} = \frac{2}{3} = 0.67$$

**8.** *5.85 g. of NaCl are dissolved in 90 g of water, Calculate mole fraction of NaCl.*

**Solution :**

Molecular mass of NaCl is 23 + 35.5 = 58.5

Mass of NaCl is 5.85

$$\text{Number of moles of NaCl} = \frac{wt.\text{ of NaCl}}{\text{g.mol.}wt.\text{ of NaCl}} = \frac{5.85}{58.5} = 0.1$$

Mass of water 90 g and molecular mass of water ($H_2O$) = $2 \times 1 + 16 = 18$

$$\text{Number of moles of water } \frac{90}{18} = 5$$

$$\therefore \quad \text{Mole fraction of NaCl} = \frac{0.1}{0.1+5} = \frac{.1}{5.1} = 0.196$$

**9.** *The number of moles of a solute in its solution is 20 and total number of moles are 80. Calculate the mole fraction of solute.*

**Solution :**

Here the number of moles of a solute is 20.

$$\text{Mole fraction of solute} = \frac{20}{80} = 0.25$$

**10.** *Generally the addition of a solute increases the boiling point of the solvent. Why?*

**Ans.** (*i*) On adding a solute to a pure solvent, its vapour pressure gets lowered.

(*ii*) Hence in order to raise the vapour pressure of the solution to the level of atmospheric pressure, more heat is to be supplied. Consequently the boiling point will increase.

**11.** *The freezing point of water is lowered by the addition of common salt. Explain.*

**Ans.** (*i*) It is due to the fact that vapour pressure of solution is less than that of pure solvent.

(*ii*) It means that the vapour pressure of the solution will become equal to that of solid solvent only at lower temperature so that it may start freezing.

**12.** *The vapour pressure of an aqueous sugar solution is lower than that of pure water. Why?*

**Ans.** (*i*) In aqueous solution of sugar, mole fraction of solvent is less than that of pure solvent.

(*ii*) Hence lower number of solvent molecules are available for vaporisation and consequently vapour pressure gets lowered.

**13.** *A solution contains 25% of water, 25% ethanol and 50% acetic acid by mass. Calculate mole fraction of each component.*

**Solution :**

Let the total mass of the solution be 100 gram.

Molecular mass of $H_2O = 2 + 16 = 18$

Molecular mass of $C_2H_5OH = 2 \times 12 + 1 \times 5 + 16 + 1 = 46$

Molecular mass of $CH_3COOH = 12 + 3 + 12 + 32 + 1 = 60$

(*i*) Mass of water 25 gram,

$$\text{Number of moles of water} = \frac{25}{18} = 1.39$$

(*ii*) Mass of ethanol 25 gram.

$$\text{Number of moles of ethanol} = \frac{25}{46} = 0.544$$

(*iii*) Mass of acetic acid 50 gram

$$\text{Number of moles of acetic acid} = \frac{50}{60} = 0.833$$

$$\text{Total number of moles of the three substances} = 1.39 + 0.544 + 0.833 = 2.767$$

$$\text{Mole fraction of water} = \frac{1.39}{2.767} = 0.502$$

$$\text{Mole fraction of ethanol} = \frac{0.544}{2.767} = 0.196$$

$$\text{Mole fraction of acetic acid} = \frac{0.833}{2.767} = 0.301$$

**14.** *0.440g of a substance dissolved in 22.2g of benzene lowered the freezing point of benzene of 0.567. Calculate molecular mass of the substance.* ($K_f = 5.12°C\ mol^{-1}$)

**Solution:**

We know

$$M = \frac{1000 \times K_f \times w_2}{\Delta T_f \times w_1}$$

Here, $w_2 = 0.440g$

$\Delta T_f = 0.567$

$K_f = 5.12°C\ mol^{-1}$

$w_1 = 22.2g$

$$M = \frac{1000 \times 5.12 \times 0.440}{0.567 \times 22.2} = 178.9$$

∴ Molecular mass of substance = 178.9

**15.** *Freezing point of naphthalene was 80. 6° C. 0.512 gm of substance was dissolved in 7.03gm of naphthalene and the freezing point of the mixed solution was found to be 75.2°C. Molal freezing depression of naphthalene was 6.8°C. Calculate molecular mass of the solute.*

**Solution:**

We know $M = \dfrac{1000 \times K_f \times w_2}{\Delta T_f \times w_1}$

$w_2 = .512$

$w_1 = 7.03$

$K_f = 6.8°C \qquad \Delta T_f = 80.6 - 75.2$

$= 5.4°C$

$$M = \frac{1000 \times 6.8 \times 0.512}{5.4 \times 7.03} = 91.7$$

$\therefore$ Molecular mass = 91.7.

**16.** *0.30 gm of a substance dissolved in 30gm of benzene and the freezing point of the mixed solution was found to be 5.49°C, Freezing point of pure benzene was 5.7°C. Calculate the molecular mass of the substance ($K_f$ = 5.12).*

**Solution:**

We know :- $M = \dfrac{1000 \times K_f \times w_2}{\Delta T_f \times w_1}$

Here $T_f$ (Depression of freezing point) is

$= 5.7 - 5.49 = 0.21$

$w_1 = 30, \; w_2 = 0.30$

$$M = \frac{1000 \times 5.12 \times 0.30}{0.21 \times 30} = 243.81$$

$\therefore$ Molecular weight of the substance = 243.81.

**17.** *Molal depression of water (Kf) is 1.86°C. 3.33 gm of urea dissolved in 250 gm of water lowered the depression of freezing point 0.413. Calculate the molecular weight of the substance.*

**Solution :**

We know $M = \dfrac{1000 \times K_f \times w_2}{\Delta T_f \times w_1}$

$K_f = 1.86°C$

$\Delta T_f = 0.413$

$w_2 = 3.33$

$w_1 = 250$

$$M = \frac{1000 \times 1.86 \times 3.33}{0.413 \times 250} = 59.99.$$

$\therefore$ Molecular weight of the substance = 59.99.

CHAPTER 3

# Colloids

## Introduction

The foundation of colloids was laid by **Thomas Graham** (1861). He divided soluble substances into two classes, **crystalloids** and **colloids,** on the basis of their power of diffusion across the animal and plant membranes.

Colloid is not a particular class of substance but a substance in a certain state of subdivision. It is more precise to use the term colloid in reference to a state of matter than a kind of matter. In fact, the same substance may act as '**colloid**' in one medium and "**crystalloid**" in other medium. For example, sodium chloride (NaCl) acts as "crystalloid" in its aqueous solution but it forms "colloidal suspension" in benzene. Similarly, soap behaves as "colloid" in water and as "crystalloid" in alcohol. Many naturally occurring substances such as milk, rubber, clouds, fog etc. are colloidal in nature. The term colloid is now applied to a particular state of matter–hence the term colloidal state is now used instead of colloidal substance.

(*i*) **Crystalloid:** Some substances in solution readily diffuse through the parchment paper or through plant and animal membrane. They are called crystalloid.

(*ii*) **Colloid:** Some amorphous substances which exhibit little or no tendency to diffuse through the parchment paper because of their gluey nature are called colloid.

**Example:** Salts, sugar, urea, starch, gelatin, gum etc.

**Differences between Crystalloids and Colloids**

| | Crystalloids | Colloids |
|---|---|---|
| (*i*) | These are readily crystallized from their aqueous solution. | These are not readily crystallized. |
| (*ii*) | These substances diffuse readily through plant and animal membrane. | These substances do not pass through plant and animal membrane. |
| (*iii*) | Usually non sticky. | Sticky. |
| (*iv*) | Low molecular weight. | High molecular weight. |
| | Example: Salt, sugar, urea. | Example: Gum, gelatin, starch. |

- **Dispersion Systems :**

A system which is made up of a substance distributed or scattered as minute particles of a solid, droplet of liquids or tiny bubbles of a gas through another substance is known as **dispersion** system. The distributed substance is known as **'dispersed phase'** (or the internal phase) and the continuous medium around the distributed substance is known as **'dispersion medium'** (or the external phase). Depending on the degree of dispersion, the dispersion system is divided as true solution, colloid and suspension.

(*a*) **True Solution:** The true solution is defined as the solution containing solute particles of 1Å to 10Å in size, distributed uniformly throughout the entire mass of the solvent.

(*i*) Due to the high degree of dispersion there is no dispersed phase in a true solution and the entire system is one single phase.

(*ii*) It forms a **homogeneous** solution. Neither the solute particle nor the solvent is observed by human eye or with the aid of microscope.

(*iii*) A true solution is electrically **neutral.**

(*b*) **Colloids:** A colloid consists of particles of 10Å–1000Å in size and is invisible in ordinary microscope or detained by ordinary filter paper and forms a translucent heterogeneous mixture.

(*i*) It has two distinct phases-disperse phase and dispersion medium.

(*ii*) Colloid particles can be photographed by electron microscope.

(*c*) **Suspension:** A suspension consists of particles of a solid bigger than 1000Å in diameter and is visible under ordinary microscope. Here the relatively heavy particles settle down due to gravitational force but more fine particles remain suspended in water.

(*i*) It has two distinct phases-a **solid disperse phase** and a **liquid dispersion medium.**

(*ii*) They are influenced by the gravitational pull and settle down automatically on standing.

**Comparison of the Characteristic Properties of True Solutions, Colloidal Solutions and Suspensions**

| S. No. | Property | True Solution | Colloidal Solution | Suspension |
|---|---|---|---|---|
| 1. | **Particle size** | Diameter of the particle is in between 1 Å-10 Å | Diameter of the particle 10 Å - 1000 Å | Diameter of the particle is more than 1000 Å |
| 2. | **Nature** | Homogeneous, stable one-phase system. | Heterogeneous, stable, two phase system. | Heterogeneous, metastable, two phase systems. |
| 3. | **Appearance** | Clear and homogeneous. | Clear and homogeneous. | Opaque and heterogeneous. |
| 4. | **Visibility under microscope** | Not visible even after electron microscope. | Visible under electron microscope. | Visible under naked eye or microscope. |
| 5. | **Separation by Filtration** | Not possible. | Not possible. | Possible. |
| 6. | **Sedimentation** | Solute particles do not sediment out from the solvent under gravity alone. | Disperse phase particles not sediment out from the dispersion medium under gravity alone. | Precipitate under the force of gravity. |
| 7. | **Tyndall effect** | Does not exhibit tyndall effect. | Exhibits tyndal effect. | Light cannot pass. |

(*Contd.*)

| S. No. | Property | True Solution | Colloidal Solution | Suspension |
|---|---|---|---|---|
| 8. | **Brownian movement** | Not observable. | Exhibits Brownian movement. | It may occur. |
| 9. | **Dialyzability** | Can pass through dialyser. | Cannot pass through dialyser. | Cannot pass through dialyser. |
| 10. | **Electrolysis** | Undergoes electrolysis. | Undergoes catephoresis. | Undergoes electrolysis. |
| 11. | **Colour** | Depends on the soluble salt or ions. | Depends on the size of the colloidal particle. | Depends on the particle. |

- **Classification of Colloids**

**1. On the basis of states of disperse phase and dispersion medium**

**(*a*) Sols** (lysols) & **gels :**

**Sols.**

(*i*) Sols are relatively low viscose colloid.

(*ii*) It consists of a liquid continuous phase in which particles of **solid disperse** phase remain scattered.

(*iii*) **Example** : Aqueous sols of agar, starch, gum acacia etc.

**Gels :**

(*i*) Gels are semisolid, jelly-like and viscous colloid.

(*ii*) Here liquid disperse phase droplets remain scattered in the meshes of solid continuous phase.

(*iii*) **Example** : Gelatin gel, silica gel, yoghurt

**(*b*) Suspension and Emulsion:**

**Suspensions:**

(*i*) It is a two-phase heterogeneous system consisting of a liquid dispersion medium and a solid disperse phase, having microscopically visible particles of sizes bigger than 1000Å.

(*ii*) They are influenced by the gravitational pull and settle down automatically by standing.

(*iii*) Suspension may be lyophobic and lyophillic.

**Emulsions:**

(*i*) It is a kind of colloidal system with its liquid disperse phase particles scattered in a liquid continuous phase.

(*ii*) Emulsions are in most cases lyophillic.

(*iii*) When a mixture of oil and water is vigorously stirred with a glass rod, in presence of soap emulsion results.

**2. On the basis of nature of disperse phase and dispersion medium.**

**Lyophobic sols:**

(*i*) Lyophobic sols are solvent-hating colloids having a natural tendency to come out of the medium.

(*ii*) Here there is no immobilized layer of the dispersion medium around the disperse phase particles as there is no affinity between the disperse phase and the dispersion medium.

(*iii*) Mixing of substances like metals and their sulphides do not form the colloidal sol.

**Lyophillic sols:**

(*i*) Lyophillic sols are solvent-loving colloids which are not easily coagulated by adding electrolyte.

(*ii*) It exhibits high affinity for the solvent i.e. high degree of solvation. It is characterised by a strong affinity between the disperse phase and dispersion medium.

(*iii*) So, a mixing of substances like gum, gelatin starch, rubber etc. with the suitable liquid dispersion medium forms a lyophillic sol. spontaneously.

(*iv*) In lyophobic sol each disperse phase is enveloped by an immobilized layer of the dispersion medium due to an affinity between the two phases.

(*v*) In lyophillic sols there is an interaction between disperse phase and dispersion medium.

(*vi*) As it forms colloidal sol directly, it is also called **intrinsic colloids.**

**Differences Between Lyophobic and Lyophillic Sols**

| | Property | Lyophobic sols | Lyophillic sols |
|---|---|---|---|
| 1. | **Detection of particles** | The particles are visible or detectable with the aid of ultra microscope. | The particles are not visible or detectable by an ultra microscope. |
| 2. | **Viscosity** | Viscosity hardly differs from that of the dispersion medium. | Viscosity is much higher than that of dispersion medium. |
| 3. | **Surface tension** | It is similar to that of the dispersion medium. | It is often lower than that of the dispersion medium. |
| 4. | **Reversible or irreversible nature** | Lyophobic sols are irreversible. | Lyophillic sols are reversible. |
| 5. | **Stability (precipitation)** | Dispersed particles are precipitated by the addition of a small amount of electrolytes. | Dispersed particles are not precipitated by small amount of electrolytes although large amount of electrolytes cause precipitation. |
| 6. | **Electric charge** | All particles in a lyophobic sols have the same charge resulting from the adsorptions of ions from solutions. | The charge on colloidal particles depends upon the pH of the medium, since the particles readily adsorb $H^+$ or $OH^-$ ions. |
| 7. | **Migration of particles in electric field** | Particles migrate in one direction in an electric field depending upon the charge they bear. | The particles may migrate in either direction or may not migrate in an electric field depending upon the pH of the medium. |
| 8. | **Colligative properties** | Colligative properties are insignificantly small. | Colligative properties are well marked. |

(*Contd.*)

| | Property | Lyophobic sols | Lyophillic sols |
|---|---|---|---|
| 9. | **Tyndall Effect** | Lyophobic sols have stronger tyndall effect. | Lyophillic sols have weaker tyndall effect. |
| 10. | **Law of mixture** | Physical properties follow the law of mixture. | Physical properties do not follow the law of mixture. |
| 11. | **Component system** | It hardly ever exist as two component systems because a protective lyop. | It may have two component system consisting of a single disperse phase component in the dispersion medium. |
| 12. | **Isoelectric precipitation** | It is much easier because of the absence of solvate envelope. | Adjustment of pH to the isoelectric pH of the disperse phase minimizes their net charge and thereby precipitate (isoelectric ppt). |

**Tyndall effect :**

**When a narrowly defined concentrated beam of light is allowed to pass through a true solution. The path of light will not visible. But when the same is passed through a colloidal solution, the path of light becomes illuminated due to the scattering of light by the colloidal particles. This phenomenon is known as *Tyndall effect.***

- **Gel :**

**Sols are colloid which looks like liquid. Many lyophilic sol and few lyophobic sol when coagulated under certain conditions, convert into semi-rigid or gelly like mass. Such product is called gel and the process is known as gelation.**

**Example :** Gels of gelation or agar-agar etc are prepared by cooling their solutions of moderate concentration.

(*i*) It is clear, hyaline, apparently structureless and elastic.

(*ii*) Gels have high viscosity and swells if placed in dispersion medium.

(*iii*) After gel formation, a small amount of liquid diffuses out without any change of the gel volume. This phenomenon is called *syneresis* (aging of gel).

**Example :** During blood clotting, serum squeezes.

(*iv*) Certain gels are easily transformed into sols on shaking when the resulting solutions are allowed to stand undisturbed, they again convert into gel. This type of sol-gel reversible conversion is known as *thixotropy*.

**Example :** Pseudopodia of Protozoa (amoeba).

**Importance of Gels :**

- **Microbiology :** Preparation of culture media of bacteria.
- **Cookery :** Preparation of food jellies, cornstarch, etc.
- **Physiology :** (*i*) Cell structure (*ii*) Clotting of blood.
- **Industry :** Preparation of pastes, glues and various mucilages

### Comparison of Sol and Gel

| | Sol | Gel |
|---|---|---|
| I. | The liquid state of colloidal solution is known as sol. | The solid or semisolid jelly-like colloidal solution is known as gel. |
| II. | It is easily dehydrated. | It cannot be dehydrated. |
| III. | It can be converted to gel by cooling. | It can be converted to sol by heating or warming. |
| IV. | Dispersion medium may be water (Hydrosol) or alcohol (Alcosol) etc. | Dispersion medium is formed by hydrated colloid particles. |
| V. | It has low viscosity. | It has high viscosity. |
| VI. | It has no definite structure. | It is assumed to be honeycomb structure. |

- **Colloidal Dispersions**

It comprises two phases viz the solute phase and the solvent phase.

| Dispersion medium | Dispersed phase | General name of colloid | Examples |
|---|---|---|---|
| Gas | Liquid | Liquid foam | Clouds |
| Gas | Solid | Solid foam | Pumice stone |
| Gas | Gas | Foam | Froth |
| Liquid | Liquid | Emulsion | Milk |
| Liquid | Solid | Solid emulsion | Butter, cheese |
| Solid | Gas | Solid aerosol | Smoke |
| Solid | Liquid | Sol | Starch in water |
| Solid | Solid | Solid sol | Rubby glass |

- **Association colloids :**

(*i*) On increasing the concentration, solute molecules or ions may come closer to form aggregate spontaneously of colloidal size. These are called association colloids, which are thermodynamically stable.

(*ii*) The aggregates are called 'micelles'. **Example:** Fatty acids

**Emulsion :**

**Emulsions are colloidal system in which dispersed phase as well as dispersions medium are liquid. Example: milk (liquid fat dispersed in water).**

(*i*) They have properties (viscosity) similar to those of *lyophilic* colloid.

(*ii*) Emulsions are prepared by mixing and shaking the two liquids, they are less stable.

(*iii*) Small amounts of selected substances (called emulsifying agent or emulsifier) when added to the emulsions can stabilize it greatly. Example: soap, sulphonic acid, albumins, gelatins are emulsifier. In milk casein act as emulsifier.

(*iv*) They show **Tyndall effect** and **Brownian movement** like lyophobic colloid.

(*v*) The liquid droplet dispersed in an Emulsion may exhibit electrophoretic properties.

**Importance of Emulsions :**

- **Cookery :**

(*i*) Preparation of sweet milk and cream for food.

(*ii*) Preparation of gravies and food sauces.

- **Cleaning :**

  For removal of fats and oils from skin, garments and utensils with soap and water by way of an emulsion.

- **Physiological :**

  (*i*) Bile salts stabilize the intestinal emulsion of fat for bringing fat, water and digestive enzyme into contact which enhance digestion.

  (*ii*) Egg albumin stabilizes the emulsion of liquid in egg.

- **Purification of Colloidal Solution**

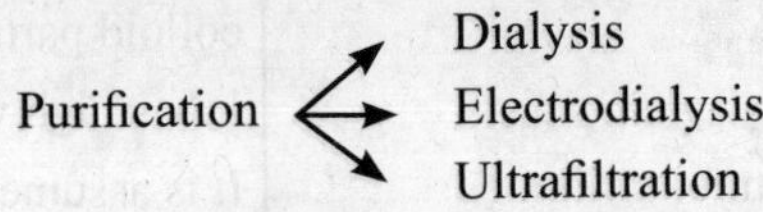


- **Dialysis: It is a process of separation of colloidal substances from small solutes by passing through a semipermeable dialyzer membrane. (Parchment paper or animal membrane).**

(*i*) This method is based on the fact that if a mixture of colloid and crystalloid is kept in a parchment bag (or in other time semipermeable membrane).

(*ii*) The crystolloid (being smaller in size) passes through the membrane out leaving the pure colloidial solution (as the particle bigger in size than crystalloids) inside the bag.

- **Electrodialysis:**

(*i*) The process of dialysis is very slow and takes long time. It can be quickened by passing electric current between two electrodes in between which the membrane containing the mixture of colloid and crystalloid, is kept.

(*ii*) On passing current, the charged crystalloid (anions and cations) migrate quickly to respective electrodes through the pores of the membrane. This process is known as *electrodialysis*.

- **Ultrafiltration: It is the process of separation of large disperse phase particles of a colloid. Sol from small solutes and solvent or electrolytes, water, glucose etc. by filtering the mixed solution under pressure through a membrane of different pore size (graded filter) permeable to the solvent (water) and mineral ions and impermeable to the disperse phase particles of colloid.**

(*i*) Ordinary filter paper is soaked with a solution of gelation or of collodion and hardened by formaldehyde where by pores becomes smaller, so that, colloid particles can not pass through it.

(*ii*) Like ordinary filtration, the mixture is added to the hardened filter paper,when the crystalloids ions impurities pass through it, but the colloidal particles retained. This process is known as **ultrafiltration.**

**Example :** During urine formation in the kidney, the fluid formed in glomerulus is an ultra filtrate.

- **Properties of Colloids:**

**1. Heterogeneity:**

(*i*) Colloids are heterogeneous and remain suspended in the dispersion medium.

(*ii*) A colloidal solution consists of tiny solid particles dispersed in liquid which are visible under an ultra microscope.

**2. Filterability:**

(*i*) Colloidal particles can easily pass through an ordinary filter paper and do not sediment out spontaneously on standing.

(*ii*) They can be filtered out by using ultra filters made of unglazed porcelain or of filter paper impregnated with collodion to reduce the size of the pore.

**3. Kinetic Behaviour:**

(*i*) When colloidal solutions are examined carefully under powerful ultra microscope, it is found that the colloidal particles exhibit continual, random, rapid, zig-zag motion in all directions.

(*ii*) This kinetic behaviour of the colloid particles suspended in the liquid is called **Brownian movements** after the name of the observer **Robert Brown** (1927). The Brownian movement is due to the bombardment of the particles of the dispersion medium on the colloid particles.

(*iii*) As a result, at equilibrium the colloid particles acquire a kinetic energy which is equal to that of the kinetic energy of the particles of dispersing medium. This acquired kinetic energy of the colloid particles is exhibited on their Brownian movements.

(*iv*) The colloid particles move with a speed slower than the particles of the dispersing medium due to their heavier mass.

**4. Optical Properties:**

(*i*) Colloid particles are so small that they are not visible under ordinary microscope, but their presence can be visualized by optical means.

(*ii*) If a strong beam of light is passed through a medium containing particles of large size (10Å-100Å), the medium becomes at once filled with coloured specks of light due to the scattered light from the colloid particles. This phenomenon of scattering of light by particles was studied by **J. Tyndall (1969)** and is called **tyndall effect.**

(*iii*) True solutions do not exhibit this effect–they are optically clear because both solute and solvent particles of true solutions are too small to scatter light.

(*iv*) On the other hand, colloid particles scatter the light rays that fall upon it. It is observed that the intensity of the scattered light depends upon the difference of the refractive indices of the medium and the suspended particles.

**5. Electrokinetic Properties:**

(*i*) Besides Brownian movement, the stability of colloid particles is attributed to the electrical charge contained in them. Unlike electrolyte, colloids are either fully positively charged or negatively charged.

(*ii*) A colloidal solution is electrically neutral because every charge carried on particle is neutralized by a charge opposite value carried by ion in the dispersion medium.

(*iii*) These consist of relative movements of the solid and liquid phases of colloidal systems.

**(*a*) Potential Differences Due to Phase Movements:**

The solid disperse phase and the liquid dispersion medium may move relative to each other and thereby cause the following potential differences between the different areas of the sol.

- **Sedimentation Potential:**

(*i*) It is the potential difference between the superficial layer and the bottom layer of a sol on prolonged standing or ultracentrifugation.

(*ii*) It results from a sedimentation of the solid disperse phase particles bearing fixed charges to the bottom of the liquid dispersion medium, leaving behind the counter ions in the superficial liquid layer.

(*iii*) Thus a movement of the solid phase produces a potential difference.

- **Streaming Potential:**

(*i*) It is a kind of potential difference which results from a movement of the liquid dispersion medium relative to the solid disperse phase.

(*ii*) When a liquid is forced through a capillary tube or through a porous pot, a potential difference i.e. streaming potential is set up.

**(*b*) Phase Movement Due to Potential Differences:**

Imposition of an external potential may cause movement of the phases relative to each other.

- **Electrophoresis:**

(*i*) The particles of a colloidal solution carry an electric charge and therefore move towards one or other electrode when the solution is placed in an electric field. This movement of colloidal particles under the influence of electric field is called **electrophoresis.**

(*ii*) If the movement of colloidal particle is towards cathode, it is referred to as **cataphoresis.** If the movement of colloidal particle is towards the anode, it is referred to as **anaphoresis.**

(*iii*) The rate of movement varies directly with the net amount of fixed charges and inversely with the particle size and the viscosity of the dispersion medium.

- **Electro-osmosis:**

(*i*) The movement of dispersion medium under the influence of an electric field keeping the disperse phase stationary is called **electro-osmosis.**

(*ii*) We know that a colloidal solution is electrically neutral. Consequently it is logical to conclude that the charge on the disperse phase must be balanced by an equal and opposite charge on the dispersion medium.

(*iii*) Here the movement of dispersion medium occurs towards one of the electrodes in an electric field and the solid disperse phase particles remain stationary.

**6. Coagulation or Flocculation:**

(*i*) Coagulation is a phenomenon of precipitation of colloid particles due to aggregation by the addition of an electrolyte. If the precipitated colloids float on the medium instead of settling down, it is called **flocculation.**

(*ii*) Since all colloidal particles in a sol carry the same charge, they repel each other, thus tending to remain scattered or suspended throughout the liquid.

(*iii*) When an electrolyte is added to a colloidal sol, the charged colloidal particles adsorb the oppositely charged ions produced by the electrolyte and the colloidal particles are neutralized. As a result neutral particles coalesce to form a precipitate.

(*iv*) Coagulation means death of a colloid. Thus when a few drops of acid is added to milk, it becomes **casein.** In this case the colloidal milk is being coagulated by the acid to form casein.

(*v*) Coagulation of colloidal particles in a sol is brought about not only by electrolytes but also by other colloids with opposite charges on their particles.

(*vi*) Coagulation of lyophobic sols also occurs when they are boiled. Probably at higher temperatures the colloidal particles are caused to collide together more frequently and vigorously so that they coalesce to form precipitate.

**7. Adsorption:**

(*i*) It is essentially a surface phenomenon where substances in a state of subdivision have a specific ability of attracting and holding other substances. Example: Molecules of gas or molecules and ions of dissolved substances on their surface.

(*ii*) Colloid particles possess high power of adsorption because of their larger surface area for given mass of material.

(*iii*) It may be stated that the phenomenon of concentration of molecules of gas or liquid at a solid surface.

**8. Solvation:**

(*i*) The colloid particles of sols are generally solvated. That is, they are surrounded by an adsorbed layer of the dispersion medium which does not permit them to come together and coagulate.

*Example*: Hydration of gelatin.

(*ii*) There is no solvation of the hydrophobic sol particles for want of interaction with the medium.

**9. Gelation:**

The process of conversion of a sol into a gel so that the solid disperse phase and liquid continuous phase of the former change respectively to the solid continuous phase and the liquid disperse phase of the latter is termed as **gelation.**

(*i*) Gel is a jelly-like colloidal system which has **solid dispersion** medium and **a liquid disperse phase.**

(*ii*) Gelation may be thought of as partial coagulation of a sol. It first unites to form long thread -like chains. These chains are then interlocked to form a solid framework. The liquid dispersion medium gets trapped in the cavities of this framework. This results in semi-solid porous mass known as gel.

**10. Viscosity:**

Viscosity of a liquid is a measure of its frictional resistance.

(*i*) Lyophillic sols of colloid possess higher viscosities than pure solvent because free movements of their dispersion medium are considerably restricted.

(*ii*) Viscosity rises during gelation.

**11. Protection:**

The process by which hydrophobic sol particles are protected from coagulation and precipitation by electrolysis due to previous addition of some hydrophillic sols is termed as **protection** and **hydrophilic** sol which is added to prevent coagulation of hydrophobic sol is known as *protective colloid.*

(*i*) *Example* : Gum arabic is used to prevent the coagulation of inks.

(*ii*) The protective action of **lyophilic sols** is measured is terms of **gold numbers** introduced by **Zsigmondy.**

The less gold number value, the greater is the protective action.

**The gold number is defined as the number of milligrams of a hydrophillic colloid that will just prevent the precipitation of a 10 ml of a gold solution on the addition of 1 ml of 10 per cent (10%) sodium chloride solution.**

***"Gold number of Albumin is 0.15"*** **means that 0.15 mg of Albumin is required to 10 ml of gold solution (containing 1 ml of 10% NaCl) to prevent precipitation or coagulation.**

**Gold number of Some Hydrophilic Colloids**

| Sl. No. | Lyophilic Colloid | Gold Number |
|---|---|---|
| 1. | Haemoglobin | 0.030 – 0.07 |
| 2. | Egg albumin | 0.08 – 0.10 |
| 3. | Gelatin | 0.005 – 0.01 |
| 4. | Dextrin | 0.600 – 12 |
| 5. | Gum arabic | 0.10 – 0.15 |

**Applications of Colloids**

Colloids play an important role in our daily life as well as in industry, agriculture, medicine and biology.

**1. Life Process:**

(*i*) Most components of living organisms such as blood, body fluids, muscles, tissues, skin, hair, etc. involve in colloidal materials.

(*ii*) Harmones, enzymes, and vitamins too form colloidal solutions.

(*iii*) The cells are made up mainly of colloidal particles suspended in water. Any disturbance in the degree of dispersion of colloidal proteins in the organs and tissues of the body will cause certain mental disorders and swelling in the tissues.

**2. Foods:**

(*i*) A number of foods we consume are colloids in nature.

(*ii*) Milk is an emulsion of fat dispersed in water.

(*iii*) Eggs, fruit jellies, tea in water etc. are colloidal origin. Ice cream is a dispersion of ice in cream.

(*iv*) Bread is a dispersion of air in baked dough.

(*v*) The cooking of rice, potato etc. involves the coagulation of starch particles.

(*vi*) Frying and poaching of egg too involves the coagulation of egg albumin by hot oil or hot water as albumin is a lyophillic colloid.

**3. Medicines:**

(*i*) Colloidal medicines finely divided are more effective and are easily absorbed in our system.

(*ii*) Cod-liver oil and halibut-liver oil, emulsion of respective oils in water act as tonic. Colloidal antimony is used in treatment of 'Kalazar':

(*iii*) The gold, silver and calcium are introduced in human system in the form of colloidal solution to raise vitality and for treatment of tuberculosis, gout, etc.

(*iv*) Argyrol and protargol, colloidal sols of silver, are used for the treatment of granulations.

(*v*) Colloidal snipper is used as germ killer.

(*vi*) Many ointments for application to skin consist of physiologically active components dissolved in oil and made into an emulsion with water.

(*vii*) Antibiotics such as penicillin and streptomycin are produced in colloidal form suitable for injections.

**4. Purification of Water:**

(*i*) The municipal water obtained from natural sources often contain colloidal particles. The process of coagulation is used to remove these.

(*ii*) The sols particles carry negative charge. When aluminium sulphate (alum) is added to water, a gelatinous precipitate of hydrated aluminium hydroxide (floc) is formed.

$$Al^{3+} + 3H_2O \rightarrow Al(OH)_3 + 3H^+$$

$$Al(OH)_3 + 4H_2O + H^+ \rightarrow Al(OH)_3(H_2O)_4^{\ +}$$

The positively charged floc attracts to it negative sol particles which are coagulated. The floc along with the suspended matter comes down, leaving the water clear.

**5. Formation of Delta:**

(*i*) Formation of delta at the confluence of sea and river is an application of colloid.

(*ii*) The river water contains colloidal particles of sand and clay which carry negative charge. The sea water on the other hand contains positive charge (ions) such as $Na^+$, $Mg^{2+}$ $Ca^{2+}$

(*iii*) As the sea water meets sea water, these ions discharge the sand or clay particles which are precipitated as **delta.**

**6. Sewage Disposal:**

(*i*) Sewage water contains colloidal impurities which are negatively charged.

(*ii*) These are removed by allowing the dirty water to flow through big sewage disposal tanks fitted with metallic electrodes.

(*iii*) The colloidal dirt particles lose their charge at the oppositely charged electrodes, coagulate, and settle down as **sludge.**

(*iv*) The sludge is used as manure and the clear water is used for irrigation.

**7. Smoke Precipitation:**

(*i*) Smoke is a colloidal dispersion of negatively charged finely divided carbon particles in air.

(*ii*) If this charge is removed, the carbon particles will be aggregated and settle down. This is done by passing smoke through **control precipitator.**

(*iii*) Carbon particles settle down at the bottom while clean gases escape through chimney.

(*iv*) Pollution of air by smoke can be prevented by this way.

**8. Industry:**

(*i*) Rubber is obtained from the sap of certain trees as an emulsion of negatively charged particles.

(*ii*) Soap and detergent provide an excellent example of colloidals at work.

(*iii*) Photographic industry utilizes gelatine and colloidal silver.

(*iv*) In tanning industry, both hides and leather proteins are in the colloidal state.

**9. Colloids in Nature:**

(*i*) **Blue Colour of the Sky:** The colloidal dust particles floating about in the sky scatter blue light, thus making the sky appear blue.

(*ii*) **Rain:** When air saturated with vapour reaches a cold region, the colloidal particles of water are formed due to condensation.

(*iii*) **Blood:** It is a colloidal solution of an albuminoid substance.

(*iv*) **Soil:** Good soils are colloidal in nature which can absorb moisture and nourishing materials.

- **Salting in :**

  Addition of small amount of electrolyte ions particularlly divalent or multivalent ions in the colloidal solution increases the stability of the solution is referred to as salting in.

- **Salting out :**

  The precipitation of emulsoids (lyphophilic colloids) by additing large amounts of soluble salts is termed as salting out.

CHAPTER 4

# pH and Buffer Solution

## Acid and Base

In 1923, **J.N. Bronsted** and **J.M. Lowry** proposed a border concept of acid and base. According to their concept, an *acid* is any molecule or ion which donates a proton ($H^+$) to increase $H^+$ concentration to the solution while a *base* is any molecule or ion that can accept a proton to lower the $H^+$ concentration.

## Conjugate Acid -Base pair :

The acid ($HA$) and the conjugate base ($A^-$) that are related to each other by donating and accepting a single proton are said to constitute a *conjugate acid-base* pair.

In an acid-base reaction the acid gives up its proton ($HA$) and produces new base ($A^-$). The new base that is related to the original acid is called *conjugate* (**meaning related**) *base.* Similarly, the original base ($B^-$) after accepting proton ($H^+$) gives a new acid ($HB$) which is called *conjugate acid.* Thus, an acid is in equilibrium with its conjugate base as below.

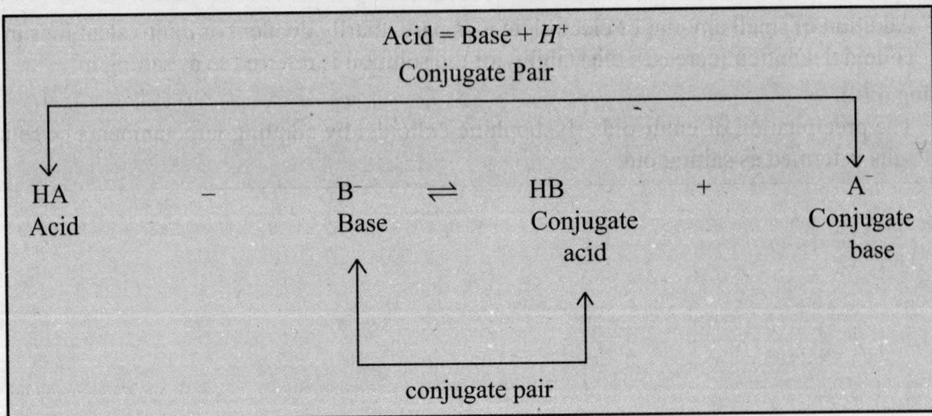


## Strength of acid and base :

The strength of an acid depends on its ability to transfer its proton to a base to form conjugate base or the strength of an acid may be defined as the concentration of $H^+$ ions in its aqueous solution at a given temperature.

When a monoprotic or monobasic acid ($HA$) is dissolved in dilute aqueous solution, it transfers its proton to water to form hydronium ion ($H_3O^+$) and a conjugate base

$$HA + H_2O \rightleftharpoons H_3O^+ + A^-$$

For simplifying we can write $H_3O^+ = H^+$. Thus it can be written as

$$HA + H_2O \rightleftharpoons H^+ + A^-$$

This equation represents the dissociation of the acid $HA$ into $H^+$ ion and $A^-$ ion. Applying law of mass action to the acid dissociation equilibrium,

$$K_a = \frac{[H^+][A^-]}{[HA]}$$

$K_a$ = Acid dissociation constant.

In the dilute solution of acid ($HA$) mass of concentrated water remains essentially constant. Therefore the value of $K_a$ for a particular acid is a measure of its acid strength or acidity. Values of $K_a$ are more conveniently expressed as ionization component of an acid *i.e.*, $pK_a$.

$$pK_a = -\text{Log } K_a = \log (1/K_a)$$

The stronger the acid, the greater is its ionization. Consequently, the higher is its $K_a$, the lower is its $pK_a$.

The strength of a base may be defined as the concentration of $OH^-$ ions in aqueous solution at a given temperature or a base is a substance which produces $OH^-$ ions in aqueous solution.

Let us consider a base $BOH$ whose dissociation can be represented as

$$BOH \rightleftharpoons B^+ + OH^-$$

Applying law of mass action to the basic dissociation equilibrium,

$$K_b = \frac{[B^+][OH^-]}{[B\,OH]}$$

$K_b$ = It is the base dissociation constant and is more conveniently expressed as ionization component of a base *i.e* $pK_b$.

$$pK_b = -\log K_b = \log\left(\frac{1}{K_b}\right)$$

The value of $K_b$ for certain base is a measure of its base strength. In the aqueous solution of a strong base, practically all the original base is dissociated and the value of $K_b$ is large. In the case of weak base, it is dissociated in aqueous solution to a very small extent and the value of $K_b$ is also small.

## pH :

The $pH$ concept is very convenient for expressing hydrogen ion concentration. **In 1909, Sorensen** introduced the term $pH$ as a convenient way of expressing hydrogen ion concentration.

**Definition :**

***pH* is the negative logarithm of the hydrogen ion concentration or negative logarithm to the base 10 of $H^+$ ion concentration or it may be stated as the logarithm to the base 10 of the reciprocal of $H^+$ ion concentration.**

Mathematically, it may be expressed as $pH = -\log [H^+]$, where $[H^+]$ is the concentration of hydrogen ions in moles per litre. Alternative and more useful form of $pH$ is :

$$pH = \log\left(\frac{1}{[H^+]}\right)$$

or

$$[H^+] = \mathbf{10^{-pH}}$$

The $pH$ for pure water or neutral aqueous solution at room temperature is

$$pH = -\log [H^+] = -\log [10^{-7}] = 7$$

*pH* may be defined as the index of the exponential term obtained by writing the molar concentration of $H^+$ as a power of 10, omitting the negative sign of the term.

**Dissociation of water :**

Water dissociates to give $H^+$ and $OH^-$ (ions). Law of mass action can describe the dissociation of water into $H^+$ and $OH^-$ which is an equilibrium process. From the law of mass action, the rate of a chemical reaction is proportional to the active masses of reacting substances.

$$H_2O \rightleftharpoons H^+ + OH^-$$

$$K_a = \frac{[H^+][OH^-]}{[H_2O]}$$

$K_a$ is the acid dissociation constant of water at the given temperature.

Very few molecules of water ($H_2O$) undergo photolysis (dissociation of water) in a mass of water, the mass of undissociated water ($H_2O$) is practically constant. Thus,

$$[H^+]\,[OH^-] = K_a\,[H_2O] = K_w.$$

$K_w$ is the ion product or dissociation constant of water at a given temperature.

**The product of the concentration of $H^+$ and $OH^-$ ions in water at a particular temperature is known as ionic product of water, *Kw.***

Since pure water, on being dissociated, gives same number of $H^+$ and $OH^-$ ions, the concentration of both of these is equal in pure water.

$$[H^+]\,[OH^-] = [H^+]^2 = K_w$$

or

$$[H^+] = \sqrt{K_w}$$

The concentration of each of the two ions of pure water at room temperature (**25°C**) is $10^{-7}$ mole/litre.

$$K_w = 10^{-7} \times 10^{-7} = 10^{-14} = 1.008 \times 10^{-14} = 10^{-14}.$$

The ionic product of water at 25°C is

$$K_w = 10^{-7} \times 10^{-7} = 10^{-14}$$

$$[H^+] = [OH^-] = 10^{-7}$$

$$pH = -\log_{10} H^+ = 7.$$

**Temperature and $K_w$ :**

(*i*) Value of $K_w$ varies widely with temperature.

(*ii*) Even a small change in temperature from 37°C to 40°C causes 8% increase in $H^+$ and $OH^-$ ions which leads to a profound biological change in a living system.

**Ionization constant ($K_w$) of water, and *pH* of neutrality at various temperatures**

| Temperature (°C) | Ionization constant $K_w$ | *pH* of neutrality |
|---|---|---|
| 0 | $0.12 \times 10^{-14} = 10^{-14.74}$ | 7.97 |
| 25 | $1.03 \times 10^{-14} = 10^{-14.00}$ | 7.00 |
| 37 | $2.51 \times 10^{-14} = 10^{-13.60}$ | 6.80 |
| 40 | $2.95 \times 10^{-14} = 10^{-13.53}$ | 6.77 |
| 75 | $16.9 \times 10^{-14} = 10^{-12.77}$ | 6.39 |
| 100 | $48.00 \times 10^{-14} = 10^{-12.32}$ | 6.16 |

***pOH* value of a solution:**

(*i*) *pH* value of a solution expresses the concentration of $H^+$ ions given by it or the degree of its acidic character.

(*ii*) On the other hand, *pOH* value of a solution expresses the concentration of $OH^-$ ions given by it or the degree of its basic character.

$$pOH = -\log_{10}[OH^-]$$

$$= \log_{10}\frac{1}{[OH^-]}$$

$$OH^- = 0^{-pOH}$$

*Example*: Water is neutral solution.

$$[H^+] = [OH^-] = 10^{-7}$$

$$pOH = -\log_{10}[OH^-]$$

$$= -\log_{10}10^{-7}$$

$$= 7$$

Consequently :

(*i*) ***pH* = *pOH* = 7**
(*ii*) ***pH* + *pOH* = 7 + 7 = 14**

**Relationship between *pH* and *pOH* of different solution**

| pH | pOH | pH + pOH = 14 |
|---|---|---|
| 0 | 14 | 0 + 14 = 14 |
| 1 | 13 | 1 + 13 = 14 |
| 3 | 11 | 3 + 11 = 14 |
| 5 | 9 | 5 + 9 = 14 |
| 7 | 7 | 7 + 7 = 14 |
| 9 | 5 | 9 + 5 = 14 |
| 11 | 3 | 11 + 3 = 14 |
| 13 | 1 | 13 + 1 = 14 |
| 14 | 0 | 14 + 0 = 14 |

**pH Scale :**

(*i*) The scale on which *pH* values are computed is called the *pH* scale. It is a logarithmic scale and extends from 0 to 14 for dilute aqueous solutions like biological fluids.

(*ii*) *pH* = 7 is considered as neutral *pH*.

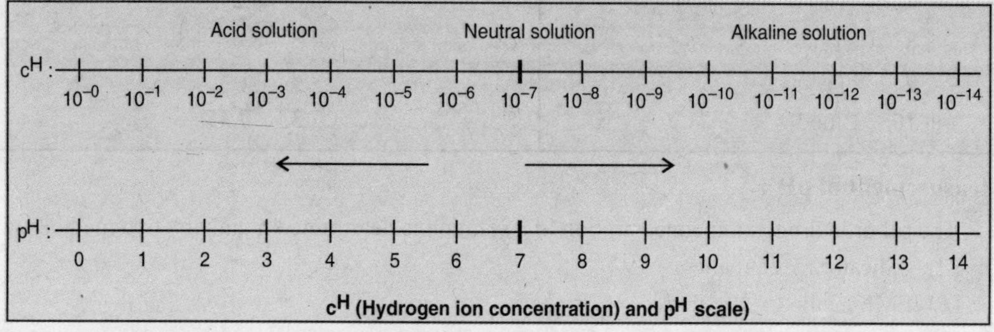

cH (Hydrogen ion concentration) and pH scale)

(*iii*) *pH* lower than 7 indicates $[H^+]$ higher than $10^{-7}$ mol lit$^{-1}$ *i.e.* the solution is acidic. The greater the acidity, the lower the *pH*.

(*iv*) *pH* higher than 7 indicates $[H^+]$ lower than $10^{-7}$ mol lit$^{-1}$ *i.e.* the solution is alkaline. The greater the alkalinity, the higher is the *pH*.

| $[H^+]$ | pH | Nature of solution | Variation in the nature (**acidic/basic**) of solution |
|---|---|---|---|
| $10^0$ | 0 | **A** | ↑ |
| $10^{-1}$ | 1 | **C** | |
| $10^{-2}$ | 2 | **I** | Acidic |
| $10^{-3}$ | 3 | **D** | Character |
| $10^{-4}$ | 4 | **I** | increasing |
| $10^{-5}$ | 5 | **C** | |
| $10^{-6}$ | 6 | ↓ | |
| $10^{-7}$ | **7** | **Neutral** | |
| $10^{-8}$ | 8 | **B** | |
| $10^{-9}$ | 9 | **A** | |
| $10^{-10}$ | 10 | **S** | Basic |
| $10^{-11}$ | 11 | **I** | Character |
| $10^{-12}$ | 12 | **C** | increasing |
| $10^{-13}$ | 13 | ↓ | |
| $10^{-14}$ | 14 | | ↓ |

| | | |
|---|---|---|
| *pH* | = 7 | Neutral |
| *pH* | < 7 | Acidic |
| *pH* | > 7 | Basic |

| Biological fluids | pH |
|---|---|
| 1. Saliva | 5.9 – 7.2 |
| 2. Gastric juice | 0.9 – 3.0 |
| 3. Pancreatic juice | 7.1 – 8.4 |
| 4. Intestinal juice | 6.4 – 9.1 |
| 5. Bile | 6.9 – 8.6 |
| 6. Sweat | 3.8– 7.6 |
| 7. Tear | 5.6 – 8.2 |
| 8. Urine | 4.8 – 8.0 |
| 9. Plasma | 7.4 |
| 10. Blood | 7.35 – 7.45 |

**Measurement of pH :**

The pH or hydrogen ion concentration of a solution is determined mainly by two methods viz

(*A*) Indicator method and

(*B*) EMF method.

## (A) Indicator method :

**It is an organic substance (phenolphthalein, and methyl orange) which change their colour in dilute solution with change of *pH* of the solution are called acid–base indicator or *pH* indicator.**

The intensity of the colour developed by an acid-base indicator is compared to the colour developed by the same amount of indicator in some solution of known *pH*.

### Characteristics of a good indicator:

(*i*) Colour change should be sharp and clear.

(*ii*) Colour should be stable and brilliant and contrasting.

(*iii*) The range of *pH* over which colour change of the indicator takes place should be close to *pH* of solution at the end point of the reaction.

### Chemical nature:

(*i*) They are chiefly salts of either a very weak acid or a weak base or the free acid or base itself.

(*ii*) These may be regarded as a very weak organic acid having capacity of dissociation like all acids into hydrogen and anion expressed as

$$H\text{In} \rightleftharpoons H^+ + In^-$$

(*iii*) The main characteristic is that the colour of the undissociated acid, *HIn*, is different from that of the anion, $In^-$.

### Mechanism of colour change :

(*i*) Each indicator exists as a conjugate pair of either a weak acid and its conjugate base or a weak base and its conjugate acid.

(*ii*) Two members of such a conjugate pair possess different colours.

(*iii*) At acidic *pH*'s, an indicator accepts protons from the medium and exists as protonated or acid form (HIn) with usually a *benzenoid* structure, at alkaline pH's, an indicator (HIn) donates proton to the medium to change into deprotonated or base form ($In^-$) which is usually a *quinonoid* structure.

(*iv*) Due to such structural interconversions, the indicator solution changes colour with pH changes.

$(CH_3)_2N-C_6H_4-N=N-C_6H_4-SO_3^-$ Yellow (Basic/alkaline solution)

$\updownarrow$

$(CH_3)_2N-C_6H_4=N-HN-C_6H_4-SO_3H$ Red (Acid Solution)

Methyl orange-red quinonoid form (acid solution); Yellow – Benzenoid form basic)

Suppose an indicator HIn is a weak organic acid. It is dissociated as follows –

$$HIn \rightleftharpoons H^+ + In^-$$

*HIn* molecules produce a certain colour in an aqueous solutions whereas $In^-$ ions give different colour.

Applying law of mass action –

$$K_{in} = \frac{[H^+][In^-]}{[HIn]}$$

or $$[H^+] = K_{in} \frac{[HIn]}{[In^-]}$$

or $$\log [H^+] = \log K_{in} + \log \frac{[HIn]}{[In^-]}$$

or $$-\log [H^+] = -\log K_{in} - \log \frac{[HIn]}{[In^-]}$$

or $$pH = pK_{in} + \log \frac{[In^-]}{[HIn]}$$

or $$pH = pK_{in} + \log \frac{[\text{Indicator salt}]}{[\text{Indicator acid}]}$$

$pK_{in}$ is a constant called indicator component. The above equation is Henderson-Hasselbalch equation.

(*i*) From the equation it is certain that the colour of the indicator solution changes with *pH*.

(*ii*) When the indicator is present in acid, the solution is **red.**

(*iii*) If a few drops of acid are added to an indicator solution, the acid, due to large $H^+$ ion concentration produced by dissociation, checks the ionisation of the indicator acid. The solution would then contain large amount of *HIn* molecules whose colour will be *acid colour* of the solution.

(*iv*) When only salt is present, the solution is **yellow.**

(*v*) If small amount of *NaOH* is added to the solution, sodium salt of the indicator will be formed. which being a salt will ionise completely giving an excess of $In^-$, whose colour will be *yellow i.e.* alkaline colour of the indicator.

(*vi*) At intermediate *pH* values the colour will be a mixture of yellow and red colour, *i.e.* neutral colour.

**Some common indicators**

| Indicator | Colour change | | *p*H range | $pK_{in}$ |
|---|---|---|---|---|
| | **Acid range** | **Basic range** | | |
| 1. Litmus | red | blue | 4.5 – 8.3 | 7.0 |
| 2. Methyl Orange | red-orange (pink) | yellow | 3.1 – 4.4 | 3.7 |
| 3. Methyl red | red | yellow | 4.2 – 6.3 | 5.1 |
| 4. Phenolphthalein | colourless | pink | 8.2 – 10.0 | 9.4 |
| 5. Thymol blue | red | yellow | 1.2 – 2.8 | 1.5 |
| 6. Bromophenol blue | yellow | blue | 3.0 – 4.6 | 4.0 |

**Biological Application of Indicators :**

(*a*) **Digestive juice :** Methyl red, Thymol blue, Methyl orange

(*b*) **Blood :** (*i*) Total fatty acid : Thymol blue

(*ii*) Blood urea : Methyl red

(*iii*) Serum sodium : Phenolphthalein

(*c*) **Urine :** (*i*) Ammonia : Phenolphthalein
(*ii*) Urea : Methyl red
(*iii*) Urinary NPN = Methyl red.

**(B) EMF method :**

(*i*) The experimental solution is set in equilibrium with an electrode reversible with respect to hydrogen ion. This is then connected to a reference electrode.

(*ii*) The e.m.f.of the assembled cell is then measured by a potentiometer and the pH is calculated.

(*iii*) The electrodes reversible with respect to hydrogen ion may be (*i*) hydrogen electrode, (*ii*) a glass electrode or (*iii*) a quinhydrone electrode. The reference electrode is usually a calomel or a silver-silver chloride electrode.

- **pH meter :**

The *pH* meter is basically an electronic voltameter (or potentiometer) which determines the *pH* of a solution potentiometrically.

(*i*) It consists of a concentration cell (test cell) placed in a potentiometer circuit.

(*ii*) The test cell consists of two half cells, one serving as *reference electrode* and the other as the *indicator electrode.*

- **(A) Reference electrode :**

(*i*) It is an electrode dipped in a standard solution of known *pH* against which the *pH* of a test solution is measured using an indicator electrode.

(*ii*) The reference electrode may be a calomel electrode (*Mercurous chloride is known as a calomel*) or silver-silver chloride electrode.

(*iii*) The basic function of a reference electrode is to maintain a constant electrical potential against which deviation may be measured.

(*iv*) The desirable characters of reference electrode : (*a*) It should be easy to construct ; (*b*) it should develop potential which are reproducible even if small current is passed.

- **(B) Indicator electrode :**

(*i*) It is an electrode dipped in a test solution to measure its *pH* against that of a standard solution of known *pH* in which reference electrode is immersed.

(*ii*) The indicator electrode may be a glass electrode or a hydrogen electrode.

(*iii*) An ion-conducting salt bridge connects the two electrodes.

(*iv*) An EMF is generated in the test cell depending on the difference in $H^+$ ion concentration between two solutions. [EMF (ref) and EMF (glass)].

(*v*) The potentiometer is adjusted to supply an equal and opposite EMF for nullifying the test cell EMF and suspending the current flow in the circuit.

(*vi*) The nullifying EMF gives the test cell EMF. It is used for computing the test cell *pH.*

EMF (Cell) = EMF (Ref.) – EMF (glass)

EMF ( ) = EMF ω ( ) – EMF ( )

**Hydrogen Electrode :**

(*i*) It consists of a piece of platinum foil dipped in a 1M solution of hydrogen ions.

(*ii*) The solution used to provide this hydrogen ion activity is 1.18 M HCl.

(*iii*) To increase its surface area, the platinum foil is coated with platinum black.

(*iv*) Hydrogen gas at one atmosphere pressure is bubbled over the platinum foil. The platinum strip takes up hydrogen gas up to saturation and then functions as a hydrogen electrode just as metal electrode.

(*v*) This arrangement *in toto* is known as the *standard hydrogen electrode.* Its electrode potential is taken arbitrarily as 0.00 volt at all temperature.

**Reaction of hydrogen electrode :**

| Hydrogen adsorbed on the platinum surface | | Hydrogen atom | | Electrons remain on platinum surface | | Pass into liquid |
|---|---|---|---|---|---|---|
| $H_2$ | $\rightleftharpoons$ | $2H$ | $\rightleftharpoons$ | $2e^-$ | $\longrightarrow$ | $2H^+$ ions at surface of electrode. |

(*i*) The rate at which H atoms lose electrons to the platinum and pass into liquid as $H^+$ ions is proportional to the concentration of $H_2$ molecules dissolved in or adsorbed on the platinum black.

(*ii*) The concentration of $H_2$ molecules in platinum is in turn proportional to the pressure of gaseous hydrogen around the electrode.

(*iii*) With the increase of pressure, more hydrogen $H_2$ molecules are taken up by the platinum and more $H^+$ ions pass into solution.

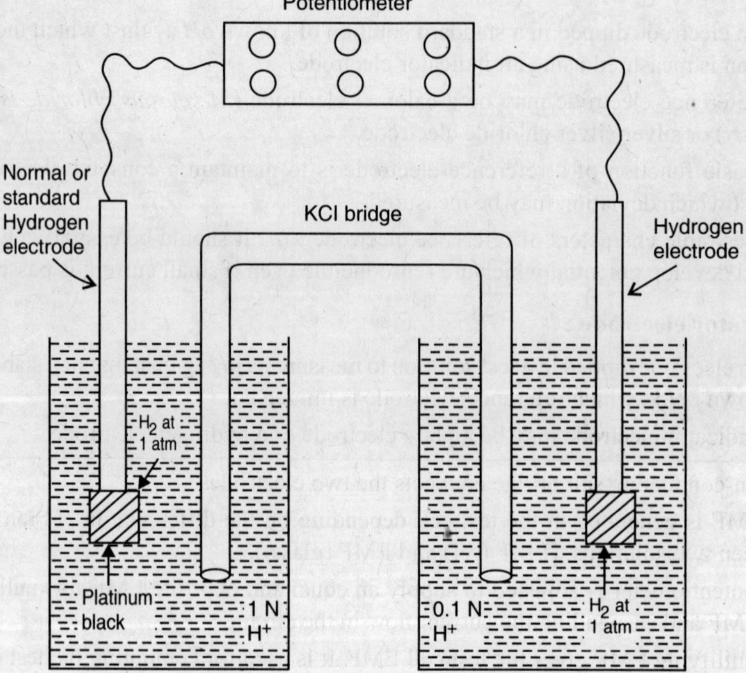


**Fig. 4.1:** Hydrogen concentration cell.

(*iv*) The hydrogen electrode with hydrogen ($H_2$) gas at one atmospheric pressure and dipped in a 1N $H^+$ ion solution *i.e.* one gram equivalent of "active" $H^+$ per litre is called standard hydrogen electrode whose potential is zero and designated as Eho. It may be used as *reference electrode.*

(*v*) A concentration cell (test cell) is constituted by placing the standard hydrogen electrode and an indicator electrode placed in an unknown solution of different hydrogen ion ($H^+$) concentration.

(*vi*) The potential of indicator electrode (hydrogen electrode) of unknown (*test*) solution is designated as Eh.

(*vii*) The EMF ($E$) generated in the concentration cell depending upon the concentration of test solution is measured potentiometrically.

The difference between the electrode potential of normal or reference electrode (Eho.) and indicator or unknown electrode (Eh) generate electomotive force (EMF or E).

$$\text{E (volts)} = \text{Eh} - \text{Eho}$$

If we denote the hydrogen concentration of two solutions viz. solution of standard electrode (reference electrode) as $H_1^+$ (unit hydrogen activity) and test solution electrode (indicator electrode) as $H_2^+$.

According to Nernst equation:-

$$\text{EMF (E)} = Eh - Eh_o = \frac{RT}{nF} \log_e \frac{[H^+]_1}{[H^+]_2} = \frac{\text{(Unit H}^+\text{ activity)}}{\text{(activity of unknown solution)}}$$

$R$ = The gas constant

$T$ = The absolute temperature

$n$ = The valence change

$F$ = Faraday, represent the quantity of electricity per gram ion equivalent (96,500 coulombs)

$\log_e$ = The natural logarithm

$$E = \frac{RT}{nF} \log_e \frac{1}{[H^+]_2} \qquad \because [H^+]_1 = 1$$

Substituting the values of R, $n$ and F in the above equation

$$E = \frac{8.315 \times T}{1 \times 96,500} \log_e \frac{1}{[H^+]_2} \qquad \begin{bmatrix} R = 8.315 \\ n = 1 \\ F = 96,500 \end{bmatrix}$$

Again $\log_e a = \log_{10} a \times 2.30258$

$$\log_b^a = \log_e^a / \log_e^b$$

$$\log_{10}^a \times \log_e^{10} = \log_e^a$$

$$\log_e^{10} = 2.30258$$

$$E = \frac{8.315 \times T \times 2.30258}{1 \times 96,500} \log_{10} \frac{1}{[H^+]_2}$$

$$= 0.00019837\text{T} \log_{10} \frac{1}{[H^+]_2}$$

$\log \frac{1}{[H^+]_2}$ can be written as $\log 1 - \log[H^+]_2$

Since $\log 1 = 0$, we have $- \log [H^+]_2$

As we know $- \log H^+ = pH$

$\therefore$ EMF ($E$) = 0.0019837 T $\times$ $pH$

$$\because T = 273 + 25 = 298K.$$

At 25°C, E = 0.059 $pH$ or $pH = \frac{E}{0.059}$

**Limitations (Disadvantages):**

(*i*) Ultra pure hydrogen is required.

(*ii*) The partial pressure of the gas should be precisely known.

(*iii*) It requires continuous flow of highly inflammable hydrogen gas.

(*iv*) It cannot be used in solutions that contain metal ion.

**Advantages :**

(*i*) It is used exclusively for calibrating other reference electrodes.

(*ii*) It is also used for measuring hydrogen ion ($H^+$) concentration of solution in which no other electrode operates satisfactorily.

- **Calomel Electrode :**

This is commonly used as reference electrode. It is filled with potassium chloride solution which remains in contact with mercurous chloride and mercury. **Mercurous chloride** is known as *calomel* and hence the name of the electrode.

(*i*) A calomel electrode is a standard half cell consisting of metallic mercury overlaid successively by a layer of mercury calomel paste of *KCl* solution (little saturated ) and then a layer of *KCl* solution saturated with calomel ($Hg_2\ Cl_2$)

(*ii*) A strip of platinum is sealed into glass and allowed to dip into mercury. It connects the mercury with the potentiometer circuit.

(*iii*) The standard *KCl* solution of the calomel electrode is in turn connected to the unknown (*test*) solution by a salt bridge made of saturated *KCl* solution.

(*iv*) Three types of standard calomel electrodes carrying 0.1 *N, N* and saturated solutions of *KCl* are widely used as reference electrodes for *pH* measurement.

(*v*) On standardizing them against a standard hydrogen electrode, whose electrode potential is 0 volt, these calomel electrodes show electrode potentials of – 0.336, – 0.280 and – 0.244 volt respectively at 298K.

| Solution (N) | Electrode potential (volt) |
|---|---|
| 0.1N | – 0.336 |
| N | – 0.280 |
| Saturated | – 0.244 |

(*vi*) Now place the indicator electrode (Hydrogen electrode) in the unknown (test ) solution and connecting the latter by a salt bridge to a standard calomel reference electrode in saturated *KCl* solutions.

(*vii*) It generates EMF.

**Explanation and Calculation**

(*i*) The calomel electrode is half cell and can be represented as

$$Pt\ /\ Hg\ /\ Hg_2Cl_2\ /\ KCl\ \text{(saturated)}$$

[Every vertical denotes an interface at which a potential is developed.]

(*ii*) If the *KCl* solution is kept saturated, all the potential developed in the calomel half-cell will be constant.

(*iii*) The potential in a calomel half-cell is derived from the primary reaction

$$Hg_2^{++} + 2e^- \rightleftharpoons 2\ Hg$$

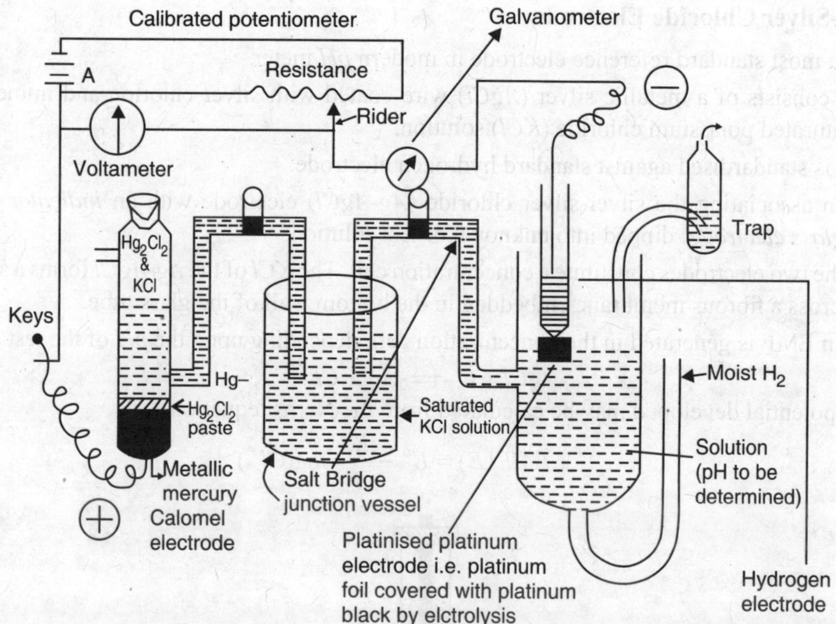


**Fig. 4.2.** Calomel-Hydrogen electrodes for *pH* measurement.

According to Nernst equation,

$$\text{EMF } (E) = E_{Calo} - E_H\left[\frac{\text{RT}}{\text{nF}}\log_e \frac{\text{unit H}^+ \text{ activity}}{\text{activity of unknown solution}}\right]$$

$$= E_H = -\,0.059\ ;\ 25°\text{C}$$

$$= E_{Calo} - \frac{(-0.059)}{2}\log\frac{1}{[Hg^{++}]_2}$$

$$E = E_{Calo} + 0.059\, pH$$

$$0.059\, pH = E - E_{\text{Calo}}$$

$$pH = \frac{E - E_{calo}}{0.059}$$

When we connect calomel and hydrogen electrodes to a voltameter, a potential of 0.246 mV develops (at 25°C). Thus, if we connect a calomel and hydrogen electrode together and hydrogen electrode is dipped into test solution as indicator electrode and the potential will be 0.246 mV.

The equation of *pH* when colomel is used as reference electrode would be –

$$pH = \frac{E - 0.246}{0.059}$$

**Problem :** Using saturated calomel electrode $pH = \frac{E - 0.246}{0.059}$ at 25°C, the electrodes are dipped in a solution. The voltameter reads 0.652. What is the *pH* of the solution?

**Solution** : $pH = \frac{E - 0.246}{0.059}$

$$= \frac{0.652 - 0.246}{0.059} = 6.9$$

- **Silver-Silver Chloride Electrode**

It is a most standard reference electrode in modern *pH* meter.

(*i*) It consists of a metallic silver (*AgCl*) wire coated with silver chloride and immersed in a saturated potassium chloride (*KCl*) solution.

(*ii*) It is standardised against standard hydrogen electrode.

(*iii*) On associating the silver-silver chloride (*Ag-AgCl*) electrode with an *indicator electrode* (*glass electrode*) dipped into unknown or test solution.

(*iv*) The two electrodes constitute a concentration cell. The *KCl* of the *Ag-AgCl* forms a salt bridge across a fibrous membrane embedded in the bottom wall of the glass tube.

(*v*) An EMF is generated in the concentration cell, depending upon the *pH* of the test solution.

$$AgCl + e^- \rightleftharpoons Ag + Cl^-$$

The potential developed may be calculated from the Nernst equation.

$$EMF(E) = E^o - \frac{RT}{F}\log(Cl^-)$$

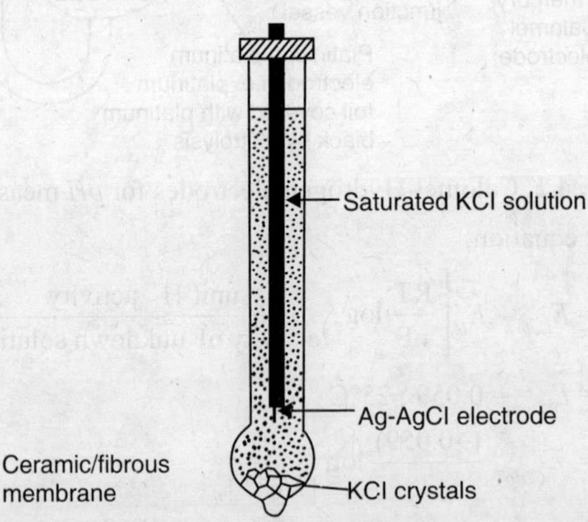


**Fig. 4.3** Silver-silver chloride electrode.

(*vi*) With reference to the standard hydrogen electrode and its electrode potential is taken 0.00 volt, the standard *Ag–AgCl* electrode potential of –0.2223 volt at 298K.

$$E = 0.2223 - 0.059\ (pH)$$

$$\therefore \quad pH = \frac{E - 0.223}{0.059}$$

- **The Glass Electrode:**

The measurement of *pH* by glass electrode involves the use of two reference electrodes separated by a glass membrane. Its function is to establish an electrical potential depending upon the hydrogen ion activity of the solution being tested.

(*i*) A pH meter commonly consists of a glass electrode as the *indicator electrode* and a standard *Ag-AgCl* electrode as the *reference electrode.*

(*ii*) A glass electrode consists of a high resistance glass tube with a thin low resistance glass bulb fused at the bottom.

(*iii*) The bulb is responsible for the *pH* sensitivity, the rest of the electrode is insensitive to $[H^+]$.

(*iv*) The tube is filled up with 0.1M solution of *HCl* of constant $H^+$ activity.

(*v*) An *Ag - AgCl* electrode possessing a constant electrode potential is immersed in that solution.

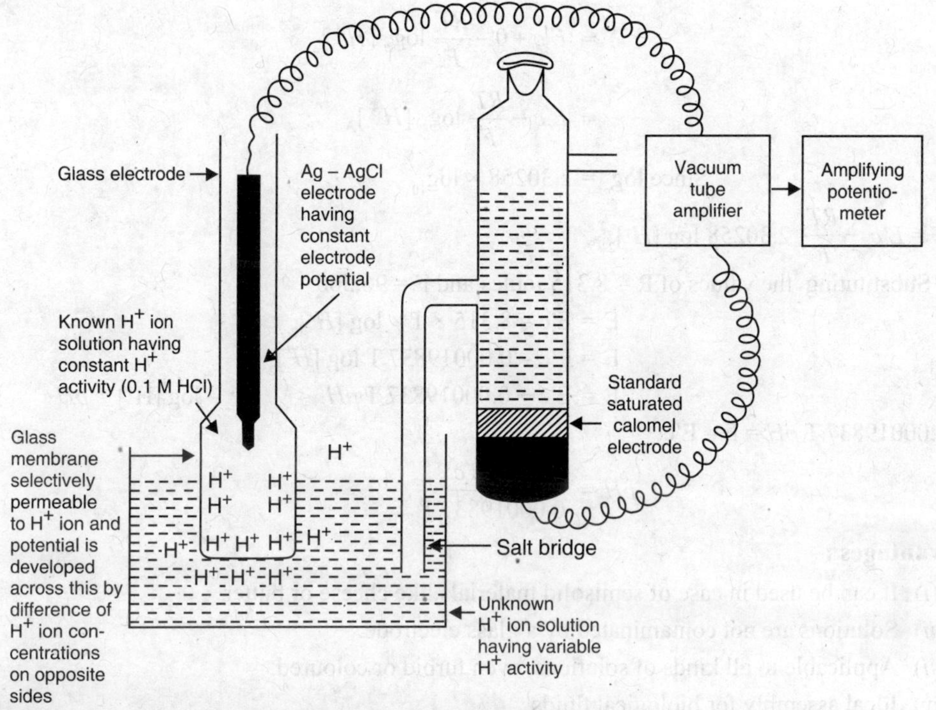


**Fig. 4.4** Glass electrode assembly

(*vi*) When the glass electrode is immersed in the unknown (test) solution differing in $H^+$ concentration from the standard electrolyte solution inside its bulb, only $H^+$ ions tend to flow across glass membrane separating the two solutions whereas other ions cannot pass through the membrane.

(*vii*) This generates a potential difference between the two sides of the glass membrane depending on $H^+$ ion concentration of the test solution.

(*viii*) This electrode potential of the glass electrode is measured with reference to an adjoined reference (calomel) electrode. It is connected with a vacuum tube amplifier to amplify the potential generated because of small current passing through glass membrane and then amplified by potentiometer.

(*ix*) $E = \dfrac{RT}{F} \log_e \dfrac{[H^+]_1}{[H^+]_2}$

The potential due to difference of $H^+$ ion concentration can be expressed as :

| Ag | AgCl, *HCl* solution | Test Solution | *KCl* (saturated), $Hg_2\ Cl_2$ | *Hg* |
|---|---|---|---|---|
| Glass electrode | | | Reference electrode | |

Different glass membranes may have different potential than the equation expressed above, To solve this problem a general constant E′c may be taken.

$$E = E'c + \frac{RT}{F} \log_e \frac{[H^+]_1}{[H^+]_2}$$

If $[H^+]_1$ is = 1.0 = 10°

$$E = E'c + \frac{RT}{F} \log_e 1 - \frac{RT}{F} \log [H^+]_2$$

$$= E'c + 0 - \frac{RT}{F}\log_e [H^+]_2$$

$$= E'c - \frac{RT}{F}\log_e [H^+]_2$$

Since $\log_e = 2.30258 \times \log_{10}$

so $E = E'c - \frac{RT}{F}\ 2.30258 \log [H^+]_2$

Substituting the values of R = 8.315 joules and F = 96,500

$$E = E'c - 8.315 \times T \times \log [H^+]_2$$

$$E = E'c - 0.00019837\ T \log [H^+]_2$$

$$E = E'c + 0.00019837\ T\ pH \qquad \because -\log [H^+] = pH$$

or $0.00019837\ T\ pH = E - E'c$

$$pH = \frac{E - E'c}{0.00019837\ T}.$$

**Advantages :**

(*i*) It can be used in case of semisolid materials like cheese or butter.

(*ii*) Solutions are not contaminated by a glass electrode.

(*iii*) Applicable to all kinds of solutions, even turbid or coloured.

(*iv*) Ideal assembly for biological fluids.

**Disadvantages:**

(*i*) It does not function below $pH = 1$.

(*ii*) It does not function at high *pH*.

(*iii*) The electrical resistance of the glass membrane is high.

**Biological Importance of *pH* :**

**1. Ionization :**

*pH* influences the ionization of side chains and free α-amino and α-carboxyl groups of amino acids.

**2. Three-dimensional forms of DNA and RNA :**

It influences the natural three-dimensional forms of DNA and RNA by destroying ionic and hydrogen bonds.

**3. Enzyme activity :**

(*i*) Each enzyme shows maximum effect at a particular pH known as "*optimum pH.*"

(*ii*) Enzyme activity declines both above and below optimum *pH*.

**4. Charge of some biological organic compounds :**

(*i*) At a specific *pH i.e.* isoelectric *pH* (*pI*) of the molecules, each such molecule exists as dipolar *zwitterion* bearing both anionic and cationic groups and minimum net charge.

(*ii*) Below its isoelectric *pH*, such substance exists as a cation by accepting $H^+$ from the medium above the isoelectric *pH* it exists as anion by donating $H^+$ to the medium.

*Example* : Nucleic acid, phosphoglycerides, sphingolipids.

**5. Separation of proteins and amino acids :**

The *pH* dependence of charges of amino acids and proteins are employed for their isolation and separation from biological materials by paper electrophoresis, ion exchange chromatography and iso-electrophoresis.

**6. Muscle activity :**

(*i*) The *pH* changes affect excitability, membrane depolarisation and action potential of muscles and nerves.

(*ii*) Increase in *pH* causes contraction and fall in *pH* may decrease actin-myosin interaction.

**7. Gibbs-Donnan Effect**

(*i*) *pH* affects the Gibbs-Donnan effect of non-diffusible protein, ion because the ionic state of the protein changes with *pH*.

(*ii*) At isoelectric *pH* of the protein in the Gibbs-Donnan effect will be minimum and the protein remains as an anion.

(*iii*) *pH* influences the Donnan effect by determining the quantity and net charge on the non-diffusible protein ions.

**8. Oxygenation of haemoglobin :**

It has been found that as the *pH* is increased, the percentage of oxygenation of the hemoglobin for a given oxygen ($O_2$) tension increases.

**9. Precipitation of proteins :**

Some high molecular weight proteins can be precipitated by simply adjusting the *pH* of their solution to their isoelectric *pH* (isoelectric precipitation).

$$\log_{10}[OH^-] = \log K_b + \log \frac{[BOH]}{[B^+]}$$

or
$$-\log_{10}[OH^-] = -\log K_b + \log \frac{[B^+]}{[BOH]}$$

or
$$pOH = pK_b + \log \frac{[\text{Salt}]}{[\text{Base}]} \quad \text{.......}(iii)$$

This is called Henderson's equation.

$$pH = pK_w - pOH$$

or
$$pH = 14 - pOH$$

$$pH = 14 - pK_b - \log \frac{[\text{Salt}]}{[\text{Base}]}.$$

• **Buffer capacity :**

**The capacity of the buffer to resist the changes in *pH* is indicated by "Van Slykes buffer value" which is number of moles of a strong monoacidic base required to be added to one litre of the buffer solution to raise its *pH* by 1.'**

**Factors affecting buffer capacity :**

**(*a*) When pH = pK**

(*i*) The ability of a buffer solution (system) to resist changes of *pH* is maximum when *pH* of the solution is identical with that of the *pK* value.

(*ii*) When *pH* = *pK*, then from Henderson-Hasselbach equation

$$pH = pK_a + \log \frac{[A^-]}{[HA]}$$

or
$$\log \frac{[A^-]}{[HA]} = 0$$

Thus [A⁻] = [HA] for log of unity is 0.

**(*b*) Concentration of the weak acid :**

The total concentration of the weak acid determines the buffer capacity.

**(*c*) Quantity of acid or alkali added :**

(*i*) If an acid or alkali is added to a buffer solution in excess amount than the equivalent to the amount of base or acid member, the buffer solution loses buffering capacity.

**(*d*) Concentrated buffer solution:**

More concentrated buffer solution has higher buffering capacity.

**(*e*) Ratio of Salt to Acid concentration :**

The maximum efficiency of a buffer with a given total concentration of salt plus acid is greatest when the ratio of salt to acid is equal to 1.

**(*f*) Addition of strong acid or alkali:**

The smaller *pH* changes caused by the addition of a given amount of acid or alkali, the greater is the buffer capacity.

## • Some Important Buffers

**1. Bicarbonate buffer :**

(*i*) It is the principal buffer in extra cellular fluid such as plasma.

(*ii*) It comprises bicarbonate ($HCO_3^-$) and carbonic acid ($H_2CO_3$) as the base and acid members respectively.

(*iii*) The ionization component ($pK_a$) of bicarbonate is 6.1. It makes the solution more effective in the *pH* range of 5.1 to 7.1.

(*iv*) It functions in close cooperation with the haemoglobin buffer of erythrocytes for buffering $CO_2$.

**2. Phosphate buffer :**

(*i*) It comprises diabasic phosphate ($H_2PO_4^{2-}$) and monobasic phosphate ($H_2PO_4^-$) as the base and acid members.

(*ii*) It has $pK_a$ of about 6.8 which is closer to normal *pH* 7.4 in the body fluids.

(*iii*) It has high buffering capacities.

**Disadvantages :** (*i*) It has lack of buffering capacity in the range 7.5 to 8.0.

(*ii*) They can bind to $Ca^{++}$ and to a lesser extent $Mg^{++}$.

(*iii*) This buffer is known to be toxic to mammalian cells.

**3. Tris buffer : (*Hydroxy Methyl Amino Methane*)**

This buffer is mostly used in biochemistry.

**Advantages :**

(*i*) Since the $pK_a$ is 8, it has a high buffering capacity between *pH* 7.5 and 8.5.

(*ii*) It has very low toxicity.

(*iii*) Does not interfere with most biochemical reactions.

(*iv*) Available in very pure form.

**Disadvantages :**

(*i*) It reacts with a few metallic ions like $Cu^{++}$, $Ca^{++}$, $Ni^{++}$ $Ag^+$ etc.

(*ii*) It reacts with some glass electrodes.

**4. EDTA buffer : (*Ethylene diamine tetraacetate*)**

(*i*) It is a good chelating agent of divalent cations.

(*ii*) EDTA buffers are frequently used when working with nucleic acids. $Mg^{++}$ is a cofactor of nucleases and the use of EDTA abolishes the activity of these enzymes.

(*iii*) EDTA suffers from the disadvantage of absorbing very highly in the U.V. range.

**5. Triethanolamine buffer :**

(*i*) It is favourite for enzymological work.

(*ii*) It is advantageous in the *pH* range 7.5 – 8.0.

(*iii*) It is a volatile buffer and may be chosen for purification work.

- **Application of Buffers:**

**1. In Biological system :**

(*i*) Buffers are important in regulating the *pH* of the body fluids such as blood, lymph, interstitial fluid and intracellular fluid.

(*ii*) Human blood is the best example of a buffer solution of $H_2CO_3$ (**dissolved $CO_2$**) and $NaHCO_3$ and has $pH = 7.35$. When excess $H^+$ ions enter the blood, their effect is neutralised by the reaction

$$H^+ + HCO_3^- \longrightarrow H_2CO_3$$

On the other hand, when excess of $OH^-$ ions are added, they are neutralised by the reaction.

$$OH^- + H_2CO_3 \longrightarrow H_2O + HCO_3^-$$

(*iii*) Thus the pH value of the blood remains constant at $pH = 7.35$ *i.e.* the blood remains slightly alkaline in spite of many acidic and basic food that we take.

(*iv*) A change of $pH = \pm 0.5$ would be fatal.

| Biological system | pH | Nature |
|---|---|---|
| Saliva | 6.35 – 6.85 | Alkaline |
| Gastric juice | 6.9 | Slightly acidic |
| Pure gastric juice | 0.9 | Acidic |
| Intestinal juice | 7.5 – 8.0 | Alkaline |
| Tears (human) | 7.4 | Alkaline |
| Blood (human) | 7.35 – 7.45 | Alkaline |
| Milk | 6.6 – 6.9 | Slightly acidic |
| Urine | 4.8 – 7.5 | Usually alkaline |

**2. In agriculture:**

(*i*) For healthy seedling and growth of crops, the soil needs to be maintained.

(*ii*) Buffers of various *pH* values are of vital importance in the preparation of dairy product.

(*iii*) They are also used in the preservation of foods and fruits.

**3. In industry :**

(*i*) Buffers are highly useful in industrial process.

(*ii*) Manufacture of papers, dyes, inks, paints, drugs, etc and working of latex, tanning of leather, require buffer.

**4. In chemical analysis :**

(*i*) For both qualitative and quantitative chemical analysis buffers of various *pH* values are frequently used.

(*ii*) Many chemical reactions, including those catalysed by enzymes, require *pH* control.

(*iii*) Buffers are used in pathological laboratories to control *pH* of culture media for bacteria and tissues.

- **Titration :**

**It is the operation through which the volume of one solution containing a given substance required to react completely with a known volume of a second solution containing another substance is found out for the purpose of knowing strength or concentration of one solution when that of other is given.**

**Titrant & Titrate :**

The solution which is used for the titration of other solution is known as *titrants* and the solution which is titrated with the help of other solution (*titrant*) is known as titrate.

**Titration Curve :**

When strong base is mixed with a solution of acid and the *pH* recorded, a plot of the base added against *pH* recorded can be obtained and this is known as titration curve.

**Titration curve and choice of indicator :**

By plotting a titration curve, pH change during titration can be studied. The pH titration curve shows the pH change during titration of strong and weak base.

**(*a*) Strong acid and strong base :**

(*i*) By applying strong acid and strong base, rapid ascent in between 4 and 10.

(*ii*) So indicator is applicable for titration.

**(*b*) Weak acid and strong base :**

(*i*) An indicator should be selected showing its neutral colour in the alkaline region.

(*ii*) *Phenolphthalein* whose *pH* of colour change (*pK*) is 8.6.

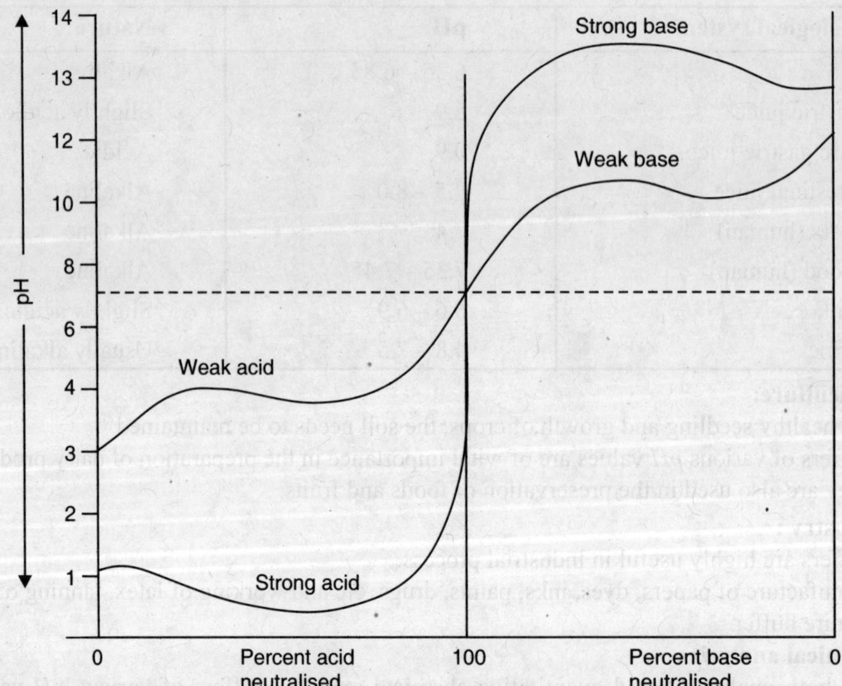


**Fig. 4.5.** *pH* change during titration of acids by alkali (Titration curves)

**(c) Strong acid and weak base :**

(*i*) Here indicator should be used whose pK, *i.e.* neutral colour pH is in the acidic region.

(*ii*) *pH* is 4, and methyl orange is used.

**(d) Weak acid and weak base :**

(*i*) Here no indicator is suitable.

(*ii*) In the solution, by adding alkali or acid during titration its *pH* changes very slowly.

**Suitable pH indicators for different types of titration**

| Type of titration | Suitable indicator | pH of colour change |
|---|---|---|
| (*i*) Strong acid + strong base | Any indicator | 4 to 9 |
| (*ii*) Strong acid + weak base | Methyl orange or methyl red | 3.7 |
| (*iii*) Weak acid + strong base | Phenolphthalein | 8.6 |
| (*iv*) Weak acid + Weak base | No indicator is satisfactory | Slow change of *pH*. |

- **Buffer:**

**Buffer solutions are solutions which resist on appreciable change in their *pH* value on adding small quantity of acid or alkali to this or maintain its *pH* fairly constant even upon the addition of small amount of acid or base (alkali).**

**The resistance of a solution is known as buffer action.**

**Preparation of Buffer solution :**

These solutions are prepared by adding solutions of

(*i*) a weak acid to its salt with strong base. *Example* : $CH_3\ COOH$ (*weak acid*) + $CH_3\ COONa$

(*ii*) a weak base to its salt with a strong acid. *Example* : $NH_4\ OH$ (*weak base*) + $NH_4Cl$.

**Mechanism of Buffer action:**

The capacity of a buffer solution to resist the change of its pH value is called its buffer action.

(*i*) Buffer solutions generally contain a weak acid or base with salts.

(*ii*) Let us consider a buffer solution prepared by mixing equimolar quantities of acetic acid ($CH_3\ COOH$) and sodium acetate ($CH_3\ COONa$) in a common solution.

(*iii*) Sodium acetate ($CH_3COONa$) is completely ionised and due to the presence of common $CH_3\ COO^-$ ion, $CH_3\ COOH$ (acetic acid) is only feebly ionised.

(*iv*) The solution will have relatively large concentration of sodium ($Na^+$), $CH_3COO^-$ ions, and dissolved $CH_3\ COOH$ moles and only small concentration of $H^+$ ions.

(*v*)
$$CH_3\ COONa \rightleftharpoons Na^+ + CH_3\ COO^-$$

Relative concentration : High High High (completely ionized)

$$CH_3\ COOH \rightleftharpoons H^+ + CH_3\ COO^-$$

Relative concentration: High low low (feebly ionized)

**(a) Introduction of Acid :**

(*i*) Now strong acid such as *HCl* is added to this buffer solution. The $H^+$ ions from *HCl* combined with the large reserve $CH_3\ COO^-$ ions produce $CH_3\ COOH$.

$$CH_3\ COO^- + H^+ \rightarrow CH_3\ COOH$$

(*from* $CH_3\ COONa$) (*from HCl*)

(*ii*) Thus $CH_3\ COOH$ is produced. It is weak acid and dissociates slightly and $H^+$ ions are produced.

(*iii*) It hardly affects the $H^+$ concentration or $pH$ value of the solution. Therefore the $pH$ of the solution remains nearly the same.

**(*b*) Introduction of base :**

(*i*) When strong base such as $NaOH$ is added to the buffer solution, $OH^-$ ions are obtained from $NaOH$.

(*ii*) It ($OH^-$) combines with $H^+$ ions to form undissociated moles of water.

$$NaOH \rightleftharpoons Na^+ + OH^-$$
$$OH^- + H^+ \rightarrow H_2O$$

(*iii*) Thus $H_2O$ is produced. It is practically unionised and hence $H^+$ ion concentration or $pH$ value of the solution remains unchanged.

## • pH of Buffer or Henderson-Hasselbach Equation:

Henderson-Hasselbach equation is important for understanding buffer action and acid base balance in the blood and tissues of the mammalian system.

**(*a*) Acidic buffer :**

It consists of a mixture of weak acid and its salt (strong electrolytes).

$$HA \rightleftharpoons H^+ + A^-$$

Applying law of mass action:

$$K_a = \frac{[H^+][A^-]}{[HA]} \quad ......(i)$$

or

$$[H^+] = K_a \frac{[HA]}{[A^-]} \quad .....(ii)$$

Let us denote salt by general formula BA (*B being the metal ion and A$^-$ being the conjugate*). It can be assumed that concentration of A$^-$ ions from complete ionisation of the salt BA is too large to be compared with concentration A$^-$ ions from the acid HA.

$$BA \rightleftharpoons B^+ + A^-$$

$[HA]$ = Initial concentration of the acid as it is feebly ionised in presence of common ion.

$[A^-]$ = Initial concentration of the salt as it is completely ionised.

So $$[H^+] = K_a \frac{[\text{Acid}]}{[\text{Salt}]}$$

Taking negative logarithm of both sides the equation becomes

$$-\text{Log}\,[H^+] = -\log K_a - \log \frac{[HA]}{[A^-]}$$

However $$-\log [H^+] = pH \text{ and } -\log K_a = pK_a$$

Therefore $$pH = pK_a - \log \frac{[HA]}{[A^-]}$$

Changing the negative sign

$$pH = pK_a + \frac{[A^-]}{[HA]}$$

or $$pH = pK_a + \log \frac{[\text{Salt}]}{[\text{Acid}]}$$

This is known as **Henderson-Hasselbach equation.**

**(b) Basic buffer :**

It consists of a mixture of weak base (*BOH*) and its salt (*BA*) or ($B^+$).

$$BOH \rightleftharpoons B^+ + OH^-$$

Applying law of mass action,

$$K_b = \frac{[B^+][OH^-]}{[BOH]} \quad .....(i)$$

or

$$[OH^-] = K_b \cdot \frac{[BOH]}{[B^+]} \quad .....(ii)$$

The salt is fully ionised and BA salt provides $B^+$ ion. The weak base is assumed to be completely unionised.

- **Biological buffer system**

(i) Cells and organisms maintain specific and constant cytosolic pH, keeping biomolecules in their optimal ionic state i.e. pH-7.

(ii) In multicelled organisms, the pH of the extra cellular fluids (e.g. blood) is also specifically regulated.

(iii) Constancy of pH is achieved primarily by biological buffers.

**Body fluids and their Principal buffers**

| Body fluids | Principal buffers |
|---|---|
| • Extra cellular fluids | Bicarbonate buffer |
| | Protein buffer |
| • Intracellular fluids | Phosphate buffer |
| | Protein buffer |
| • Erythrocytes | Haemoglobin buffer |

## PROBLEMS

**1.** *Calculate the pH of 0.02 (M) HCl.*

**Solution:**

HCl is a strong acid and is completely dissociated in aqueous solution in the following manner:

$$HCl \rightleftharpoons H^+ + Cl^-$$

0.02 0.02

For every molecule of HCl, there is one $H^+$

Therefore

$$[H^+] = [HCl]$$

$$[H^+] = 0.02\ M$$

$$pH = -\log(0.02)$$

$$= -\log\left(\frac{0.02}{100}\right)$$

$$= -\log(2 \times 10^{-2})$$

$$= -\log 2 - \log_{10}^{-2}$$

$$= -0.3010 + 2\log 10.$$

$$= -0.3010 + 2 = 1.699 = 1.7.$$

**2.** *Calculate the pH of 0.2 (M) HCl.*

**Solution:**

HCl is a strong acid and is completely dissociated in aqueous solution

$$HCl \rightleftharpoons H^+ + Cl^-$$

For every molecule of HCl, there is one $H^+$, therefore

$$[H^+] = [HCl]$$

$$[H^+] = 0.2 \text{ M}$$

$$p^H = -\log(.2) = 0.69897 \simeq 0.7$$

$$-\log\frac{2}{10} = -\log_2 + \log_{10}$$
$$= 1 - \log_2$$
$$= 1 - 0.30103$$
$$= 0.69897$$

**3.** *Calculate the pH of 0.1 (M) HCl.*

**Solution:**

HCl is strong acid and is completely dissociated in aqueous solution

$$HCl \rightleftharpoons H^+ + Cl^-$$
$$0.1 \quad 0.1$$

For every molecule of HCl, there is one $H^+$, therefore

$$[H^+] = [HCl]$$

$$[H^+] = 0.1 \text{ M}$$

$$pH = -\log(0.1)$$
$$= -\log\left(\frac{1}{10}\right)$$
$$= -\log(1 \times 10^{-1})$$
$$= -\log 1 - \log 10^{-1}$$
$$= 0 - (-1 \log 10)$$
$$= 1.$$

**4.** *The hydrogen ion concentration of fruit juice is 3 × 3 × 10^{-2} (M). What is the pH of the juice? Is it acidic or basic?*

**Solution :**

The definition of *pH* is

$$pH = -\log[H^+]$$

Here, $[H^+] = 3.3 \times 10^{-2}$.

Substituting the value of $H^+$, we get

$$pH = -\log(3.3 \times 10^{-2})$$
$$= -\log(3.3 + \log 10^{-2})$$
$$= -\log(3.3 - 2\log 10)$$
$$= -\log 3.3 + 2$$
$$= 2 - \log 3.3$$
$$= 2 - 0.52$$
$$= 1.48.$$

Since the pH is less than 7.0, the solution is acidic

**5.** *Find the pH of a solution of which $[H^+] = 4 \times 10^{-4}$ molar.*

**Solution :**

The definition of *pH* is

$$pH = -\log[H^+]$$

Here, $[H^+]$ is $4 \times 10^{-4}$ M

Substituting the value of $[H^+]$ we get,

$$\begin{aligned} pH &= -\log(4 \times 10^{-4}) \\ &= -(\log 4 + \log 10^{-4}) \\ &= -(\log 4 - 4 \log 10) \\ &= -(\log 4 - 4) \\ &= 4 - \log 4 \\ &= 4 - 0.602 = 3.398. \end{aligned}$$

**6.** *Calculate the pH of 0.001 (M) HCl.*

**Solution :**

HCl is a strong acid and is completely dissociated in aqueous solution.

$$HCl \rightleftharpoons H^+ + Cl^-$$
$$0.001M \quad 0.001M$$

For every molecule of HCl there is one $H^+$. Therefore

$$[H^+] = [HCl]$$
$$[H^+] = 0.001M$$
$$pH = -\log(0.001)$$

or

$$= -\log\left(\frac{1}{10^3}\right) = -\log_{10} 10^{-3} = +3\log_{10} 10 = 3$$
$$= -(\log 1 - 3\log 10)$$
$$= 3\log 10 - \log 1$$
$$= 3 - 0 = 3.$$

**7.** *Determine $p^H$ of 0.10 M NaOH solution.*

**Solution:**

NaOH is a strong base and is completely dissociated in aqueous solution.

$$NaOH = Na^+ + OH^-$$
$$0.10\ M \quad 0.10\ M$$

Therefore the concentration of $OH^-$ ion is equal to that of the undissociated NaOH.

$[OH^-] = [NaOH] = .10M$ = concentration of NaOH = $10^{-1}$ gram ion/litre.

The pOH of the solution is therefore

$$pOH = -\log[OH^-] = -\log[10^{-1}]$$
$$= 1$$

We know

$$pH + pOH = 14$$
$$pH = 14 - pOH = 14 - 1 = 13.$$

**8.** *Determine the pH of 2(M) of NaOH solution.*

**Solution:**

NaOH is a strong base and is completely dissociated in aqueous solution.

$$NaOH \rightleftharpoons Na^+ + OH^-$$

Therefore the concentration of $OH^-$ ion is equal to that of undissociated NaOH.

The concentration of OH = $C_{OH^-}$ = 2 gram ion/litre.

The pOH of the solution is

$$pOH = -\log C_{OH^-}$$
$$= -\log_{10} 2$$
$$= -0.3$$

We know

$pH + pOH = 14$

$$pH = 14 - (pOH)$$
$$= 14 - (0.3) = 14 + 0.3 = 14.3.$$

**or**

*Determine the pH of 2(M) NaOH solution*

**Solution:**

NaOH is a strong base and N is completely dissociated in aqueous solution.

$$NaOH \rightleftharpoons Na^+ + OH^-$$
$$2M \qquad 2M$$

Therefore concentration of ions is equal to that of the undissociated NaOH.

$$[OH^-] = [NaOH] = 2M.$$

$[H^+]$ can be calculated by applying the expression by substituting the value of $K_w$.

$$= (1 \times 10^{-14}) \text{ and of } [OH^-]$$

$$[H^+] = \frac{K_w}{[OH^-]} = \frac{1\times10^{-14}}{2}$$

By definition

$$pH = -\log\left(\frac{1\times10^{-14}}{2}\right)$$

$$= -\log\frac{10^{-14}}{2} = -(\log 10^{-14} - \log 2)$$

$$= 14 + 0.3 = 14.3.$$

**9.** *If a solution has a pH 7.41, determine its [H⁺] concentration.*

**Solution:**

$$\text{Since } pH = -\log[H^+]$$
$$-pH = \log[H^+]$$

Now, $[H^+]$ can be calculated by taking antilog of $(-pH)$

$$[H^+] = \text{antilog}(-pH)$$

$$\text{Here, } [H^+] = \text{antilog}(-7.41) = 10^{-7.41}$$

$$= 3.9 \times 10^{-8},$$

$$\left[\begin{array}{l} -(7+.41) = -8+1-.41 \\ = \bar{8}.59 \end{array}\right]$$

Therefore, the $[H^+]$ of the given solution is $3.9 \times 10^{-8}$ M.

**10.** *If a solution has a pH 12.4, determine its $H^+$ concentration.*

**Solution:**

Since,

$$pH = -\log[H^+]$$
$$-pH = -\log[H^+]$$

Now,

$[H^+]$ can be calculated by taking antilog of $-pH$

$$\text{Here, } [H^+] = \text{antilog}(-12.4) = 10^{-12.4}$$
$$= 0.398 \times 10^{-12}$$

$$\left[\begin{array}{l} -(12+4) = -12+1-.4 \\ -12.4 = -13\,13+.6 \\ = 13.6 \end{array}\right]$$

Therefore, $[H^+]$ of the given solution is $0.398 \times 10^{-12}$ M.

**11.** *Calculate the pH value of a solution whose hydrogen ion concentration is 0.005 moles/litre.*

**Solution:**

By definition

$$pH = -\log [H^+]$$

Here, $[H^+]$ is 0.005.

Substituting the value of $H^+$ we get.

$$pH = -\log (0.005)$$

$$= -\log \left(\frac{5}{10^3}\right)$$

$$\left[\log 5 = \log \frac{10}{2} = 1 - \log_2 = 1 - .301 = .699 \simeq .7\right]$$

$$= -\log (5 \times 10^{-3}) = -(\log 5 + \log 10^{-3})$$

$$= -(\log 5 - 3 \log 10)$$

$$= 3 \log 10 - \log 5$$

$$= 3 - 0.699$$

$$= 2.301.$$

So, the *pH* value is 2.301.

**12.** *What will be the concentration of $H^+$ ion in a solution having pH = 5.6*

**Solution:**

Since $pH = -\log [H^+]$

$\therefore \quad -pH = -\log [H^+]$

or $\quad [H^+] = \log 5.6 = \log \frac{56}{10}$ 10 log 56 – log 10

$$= 1.748 - 1.00 = 0.748$$

**13.** *Find the pH value of a solution whose hydrogen ion concentration is $2 \times 10^{-6}$ M.*

**Solution:**

By definition

$$pH = -\log [H^+]$$

Here $[H^+]$ is $2 \times 10^{-6}$.

$$pH = -\log (2 \times 10^{-6})$$

$$= -(\log 2 + \log 10^{-6})$$

$$= -(\log 2 - 6 \log 10)$$

$$= 6 - \log 2$$

$$= 6 - 0.3 = 5.7$$

So, the pH value is 5.7.

**14.** *Calculate the pH of a buffer solution that is 0.250 M formic acid (HCOOH) and 0.100 M is sodium formate (HCOONa). $K_a$ for formic acid is $1.8 \times 10^{-4}$.*

**Solution:**

The equilibrium equation for dissociation of formic acid is

$$HCOOH \rightleftharpoons H^+ + HCOO^-$$

$$K_a = \frac{[H^+][HCOO^-]}{[HCOOH]}$$

Since HCOOH is weakly ionized in presence of $HCOO^-$, the [HCOOH] at equilibrium is equal to its initial concentration. Again HCOONa is fully ionized and hence the concentration of $HCOO^-$ ions is the same as that of salt taken.

$$[H^+] = \frac{K_a \times [HCOOH]}{[HCOO^-]}$$

$$= \frac{1.8 \times 10^{-4} \times 0.250}{0.100}$$

$$= 1.8 \times 10^{-4} \times 0.25 = 4.5 \times 10^{-4}\text{ M}$$

$$pH = -\log [H^+]$$

$$= -\log (4.5 \times 10^{-4})$$

$$= -(\log 4.5 + \log 10^{-4})$$

$$= -\log 4.5 + 4 \log 10$$

$$= 4 - \log 4.5$$

$$= 4 - 0.6532$$

$$= 3.3468$$

Therefore *pH* is 3.35.

**15.** *A certain buffer contains equal concentrations of $X^-$ and Hx. $K_b$ for $x^- = 10^{-10}$. Calculate the pH of the buffer.*

**Solution:**

$$K_b \text{ for } X^- = 10^{-10}.$$

For conjugate acid – base pair $K_{HX} \times K_X = 10^{-14}$

$$K_{HX} = \frac{10^{-14}}{10^{-10}} = 10^{-4}$$

Now, [HX] = [X⁻]

Acid Salt.

$$pH = -\log K_a + \log \frac{[\text{salt}]}{[\text{acid}]}$$

$$pH = -\log 10^{-4} + \log 1. \text{ (conc. of salt = conc. of acid)}$$

$$pH = 4.$$

**16.** *What is the pH of a mixture of 5 ml of 0.1 M sodium acetate and 4 ml of 0.1 M acetic acid?*

**Solution:**

$$\text{Concentration of } CH_3COO^- = \frac{5}{3} \times 0.1\text{M}$$

$$\text{Concentration of } CH_3COOH^- = \frac{4}{9} \times 0.1\text{M}$$

$pK_a$ of acetic acid at 25°C = 4.76

Therefore,

$$pH = 4.76 + \log \frac{5}{4}$$

$$pH = 4.76 + (0.097)$$

$$pH = 4.86.$$

**17.** *How is the pH changed on adding 1 ml of 0.1 N HCl to the above mixture?*

**Solution:**

Addition of HCl provides $H^+$ which combines with the acetate ion to give acetic acid. This reduces the amount of acetate ion present and increases the quantity of undissociated acetic acid leading to an alteration in the salt-acid ratio and hence to a change in pH

$$\text{Concentration of } CH_3COO^- = \frac{5}{10} \times 0.1 - \frac{1}{10} \times 0.1$$

$$= 0.04 \text{ M}$$

$$\text{Concentration of } CH_3\,COOH = \frac{4}{10} \times 0.1 + \frac{1}{10} \times 0.1$$

$$= 0.05 \text{ M}$$

$$\left[\begin{aligned} 1 - \log 8 &= 1 - 0903 \\ &= 0.097 \end{aligned}\right]$$

Therefore,

$$pH = 4.76 + \log \frac{0.04}{0.05} = 4.76 - 0.097 = \mathbf{4.663}$$

$$= 4.66.$$

**18.** *Find out pH of 1 (N) HCl.*

**Solution:**

HCl is a strong acid and $H^+$ concentration ($C_{H+}$) in 1(N) HCl is 1 gram ion/litre.

$$p^H = \log C_{H+} = - \log 1 = 0$$

**19.** *The pH of a solution of caustic soda is 9. Calculate the $OH^-$ ion concentration per litre. Assuming complete ionisation of NaOH.*

**Solution:**

We know that

$$pH + pOH = 14$$

or $$9 + pOH = 14$$

or $$pOH = 14 - 9 = 5.$$

Again it is known that

$$[OH^-] = 10^{-pOH}$$

$$= 10^{-5} \text{ moles per litre.}$$

**20.** *Calculate pH of 1(N) NaOH.*

**Solution:**

NaOH is a strong base and $OH^-$ ion concentration. ($C_{OH^-}$) = 1gram ion/litre.

Therefore poH of the solution is -

$$pOH = - \log C_{OH} = \log 1 = 0.$$

$$pH + pOH = 14.$$

$$pH \text{ of the solution is} = 14 - pOH$$

$$pH - 14 - 0 = 14$$

**21.** *Calculate the pH of $10^{-4}$ NaOH solution.*

**Solution:**

NaOH is a strong base and is completely dissociated in aqueous solution.

$$NaOH \rightleftharpoons Na^+ + OH^-$$

Therefore, the concentration $OH^-$ ion is $10^{-4}$ (N) NaOH solution is – $C_{OH-} = 10^{-4}$ gram ion\litre.

$$pOH = - \log {}_{COH^-}$$

$$= -\log 10^{-4}$$

or $$pOH = 4$$

We know $pH + pOH = 14$ $pH = 14 - poH = 14 - 4 = 10$

$\therefore$ $$pH = 10$$

**22.** *Determine the pH of $10^{-5}$ NaOH solution.*

**Solution:**

NaOH is a strong base and is completely dissociated in aqueous solution.

$$NaOH \rightleftharpoons Na^+ + OH^-$$

Therefore the concentration of $OH^-$ ion in $10^{-5}$ (N) NaOH. Solution is.

$$C_{OH^-} = 10^{-5} \text{ gram ion/litre.}$$

$$pOH = -\log C\ OH^-$$

$$= -\log 10^{-5}$$

$$pOH = 5$$

We know

$$pH + pOH = 14$$

$\therefore$ $$pH = 14 - 5$$

$$= 9$$

$\therefore$ $$pH = 9$$

**23.** *Calculate the pH of 0.001 (N) KOH solution.*

**Solution:**

KOH is a strong base and is completely dissociated in aqueous solution.

$$K\ OH \rightleftharpoons K^+ + OH^-$$

Therefore concentration of $OH^-$ ion in 0.001(N) KOH solution is $C_{OH}^- = \frac{1}{10^3} = 10^{-3}$ gram ion/litre.

$$pOH = -\log c_{OH^-}$$

$$= -\log 10^{-3}$$

$$= 3$$

We know

$$pH + pOH = 14$$

$$pH = 14 - 3 = 11$$

$\therefore$ $pH$ of kOH is = 11.

**24.** *pH of $10^{-3}$ (M) HCl is 3 but pH of $10^{-3}$ (M) NaOH is 11. Explain.*

**Solution:**

HCl is a strong acid, and therefore concentration of $H^+$ ion in $10^{-3}$ M HCl is $10^{-3}$ gram ion/litre.

Therefore

$$pH = -\log_e H^+ = -\log 10^{-3} = 3$$

NaOH is a strong base and is completely dissociated in aqueous solution.

Therefore concentration of $OH^-$ ion in $10^{-3}$ (M) NaOH solution is $C_{OH^-} = 10^{-3}$ gram ion/litre

pOH of that solution is

$$pOH = -\log_e OH^-$$

$$= -\log 10^{-3}$$

$$= 3$$

We know

$$pH + \text{pOH} = 14$$

$$pH = 14 - \text{pOH} = 14 - 3 = 11$$

**25.** *A solution of NaOH has pH = 11. What is the strength in gms/litre of NaOH ?*

**Solution:**

$$pH = 11$$

We know $pH + \text{pOH} = 14$

$$\text{pOH} = 14 - 11 = 3.$$

$$[\text{OH}^-] = 10^{-\text{pOH}} = 10^{-3}$$

But one mole $\text{NaOH} = (23 + 16 + 1 = 40) = 40$ gms.

$\therefore$ gms/litre of NaOH $= 40 \times 10^{-3}$

$= 0.04$ gms.

**26.** *Calculate the pH of a buffer solution containing 0.01 (M) $CH_3$ COOH and 0.03 (M) $CH_3$ COONa. ($pK_a = 4.8$)*

**Solution:**

$$pH = \text{pK}_a + \log \frac{[\text{Salt}]}{[\text{Acid}]}$$

(Henderson-Hasselbach equation).

$$= 4.8 + \log \frac{0.03}{0.01}$$

$$= 4.8 + \log 3$$

$$= 4.8 + 0.471 = 5.2\,771$$

**27.** *Find the pH of a buffer solution containing 0.20 mole per litre $CH_3COONa$ and 0.15 mole per litre $CH_3$ COOH $pK_a$ for acetic acid is 4.7447.*

**Solution:**

$$pH = \text{pK}_\text{a} + \log \frac{[\text{Salt}]}{[\text{Acid}]}$$ (Henderson-Hasselbach equation).

$$= 4.7447 + \log \frac{0.20}{0.15}$$

$$= 4.7447 + \log \frac{4}{3}$$

$$= 4.7447 + 0.6021 - 0.477$$

$$= 4.8697.$$

**28.** *Find the pH of a buffer solution containing 0.1 mole acetic acid and 0.1 mole acetate ion, $K_a = 1.8 \times 10^{-5}$.*

**Solution:**

$$\text{p}^\text{H} = \text{pK}_\text{a} + \log \frac{[\text{Salt}]}{[\text{Acid}]}$$ (Henderson-Hasselbach equation).

$\because$ $\text{K}_\text{a} = 1.8 \times 10^{-5}$

$\therefore$ $\text{pK}_\text{a} = -\log (1.8 \times 10^{-5})$

$$= -\log 1.8 - \log 10^{-5}$$

$$= -0.2553 + 5 = 4.7447$$

$$\text{p}^\text{H} = 4.744 + \log \frac{0.1}{0.1}$$

$= 4.7447 + \log 1$

$= 4.7447 + 0 = 4.7447.$

**29.** *Calculate $H^+$ ion concentration of a solution whose pH is2.0*

**Solution:**

We know

$$pH = -\log_{10} [H^+] = -\log_{10} H^+$$

or $2.0 = -\log_{10}{}^{H+}$

or $\log_{10}{}^{H+} = -2$

or $\log_{10}{}^{H+} = \log_{10}{}^{-2}$

$\therefore$ $H^+ = 10^{-2}$ gram ion/litre.

**30.** *Calculate $H^+$ ion concentration of a solution whose pH is 0.2.*

**Solution:**

We know

$$pH = -\log_{10} [H^+]$$

or $0.2 = -\log_{10} H^+$

or $\therefore$ $-0.2 = \log_{10} H^+$

$\therefore$ $H^+ = 10^{-0.2}$

$$= 10^{-\frac{2}{10}} = 10^{-\frac{1}{5}} = \frac{1}{5\sqrt{10}} = \frac{1}{1.584} = 0.63 \text{ gram ion/litre}$$

CHAPTER 5

# Intra- and Intermolecular Interactions in Biological System

## INTRA AND INTERMOLECULAR INTERACTION

The function of biological molecule is decided by its structure. On the other hand, the structure of a biological molecule is determined by the forces between the atoms in a molecule. The interactions between the atoms in the molecule may be weak or strong. Strong interactions are mainly involved in the formation of chemical structure and to some extent also in the molecular structure. On the other hand, weak interactions determine the three-dimensional structure. Therefore, intra and intermolecular interactions are manifested through force (physical or chemical) and result in chemical bond formation.

### Chemical bond:

The attractive force between two or more atoms within a molecule to hold them together as a stable molecule is called **Chemical bond**.

> In the formation of chemical bond atoms interact with each other by losing or gaining or sharing of electrons so as to acquire a stable outer shell of eight electrons.

According to the mode of distribution of valency electron around the nuclei of the combined atoms, there are several types of chemical bonds *viz* **electrovalent** or **ionic bond, covalent bond, co-ordinate bond.** These are also called **interatomic** or **chemical forces**. Particles (atoms, molecules or ions) of substances exert attractive forces on each other when they are brought near to each other. These are **physical forces** known as **van der waals' forces.**

**Electron:** An electron is a fundamental particle which carries one unit negative charge and has a mass nearly equal to $\frac{1}{1837}$th of that hydrogen atom.

I. Electrons are maintained in their orbits by their attraction to the positively charged nucleus.
II. Sometimes their attraction is overcome by other forces and one or more electrons are lost by an atom. In other cases atom may gain additional electrons.

### Chemical Behaviour of Atoms

I. The chemical behaviour of atom lies in the arrangement of the electrons in their orbit.
II. The area around a nucleus where an electron is most likely to be found is called the *orbital* of that electron.

In other words, an atomic orbital represents the most probable space where the electron spends most its time while in constant motion.

| Orbit | Orbital |
|---|---|
| I. It is a definite circular path around the nucleus in which the electron revolves. | It is a three-dimensional region or space around the nucleus within which the probabilities of finding an electron with a certain energy is maximum. |
| II. It indicates an exact position or location of an electron in an atom. | It does not specify the exact position of an electron in an atom. |

- **Energy within Atom**

I. The term energy means the sum of the kinetic and potential energies. The kinetic energy originates from the motion of the electrons and the potential energy of a system when atoms A and B approach each other. Electrons are attracted to the positively charged nucleus and it takes work to keep them in orbit.

(*i*) Nucleus-electron attraction

(*ii*) Electron-electron repulsion

(*iii*) Nucleus-nucleus repulsion

These factors are responsible for potential energy of a system.

II. According to Virial Theorem, the potential energy (P.E) is double the kinetic energy (K.E) and opposite in sign.

$$\text{P.E} = -2\ \text{K.E}$$

The total energy of a system is given as

$$\text{E} = \text{P.E} + \text{K.E}$$

or $$\text{E} = -2\ \text{K.E} + \text{K.E} \qquad [\because \text{P.E.} = -2\ \text{K.E.}]$$

or $$\text{E} = -\ \text{K.E.}$$

III. During the change in energy of the system, the net potential energy of the system must decrease as a result of *chemical bond* formation by electronic rearrangement.

IV. The nucleus-electron attraction is responsible for lowering the potential energy.

V. During the chemical reaction, electrons are transferred from one atom to another. In such reactions, the loss of an electron is called *oxidation* and the gain of electron is called *reduction*.

VI. The amount of energy possessed by an electron is related to its distance from the nucleus. Electrons that are at same distance from the nucleus have the same energy even if they are in different orbitals. Such electrons are said to occupy the same energy level.

- **Chemical Bonding and Atomic Orbital Theory:**

In the light of atomic orbital theory of atom, chemical compound is formed by the chemical bonding between two atoms.

I. The electron wave of the valency orbital of one atom overlaps the electron wave of the other bonding atom and forms **covalent linkage.**

II. But in the case of electrovalent bond, a physical transfer of the electron is involved.

III. Only two electrons are permitted to occupy any one orbital. The electrons of an atom occupy the orbitals of lowest energy.

IV. In molecules, various atomic orbitals of the constituent atoms interact with each other to form a new set of orbitals called *molecular orbitals*. The total number of molecular orbitals is equal to the total number of atomic orbitals.
It allows the electrons greater freedom of movement and this causes lowering of its energy.

V. The amount of energy given off or released per mole at the time of overlapping of orbitals for the formation of bond is known as *bond energy* or *stabilisation energy*.

- **Strong bond:** Bond with high bond energy is known as strong bond. It is more stable. Example: covalent bond, coordinate bond, ionic bond and metallic bond.
- **Weak bond:** Bond with low bond energy is known as weak bond. It is easily dissociable. Example: Hydrogen bond van der Waals interaction.

## • Bond Length:

It represents the equilibrium distance between the centres of two covalently bonded atoms.

I. In a homonuclear diatomic molecule, it is calculated by $d$ (A – A) = 2 × $r$. A.

II. In fact this is the average distance between the nuclei, as they continuously vibrate with respect to each other.

III. In the case of single bond, it is the sum of the covalent radii of two bonded atoms.

$$d\,(A - B) = r\,(A) + r\,(B)$$

IV. Bond lengths are in the order of single bond > double bond > triple bond.

## • Bond Energy:

It is the amount of energy required to separate the units linked by a chemical bond of two covalently bonded atoms in a molecule or the amount of energy released during the formation of bond.

I. When a bond is broken, energy is required and hence bond energy is written with positive sign (+).

$$\text{A} - \text{B} + \text{Energy required (bond energy)} \longrightarrow \text{A} + \text{B (Neutral atom)}$$

II. When bond is formed, energy is evolved and hence energy is written with a negative sign (–).

$$\underset{\text{(AB molecule)}}{\text{A} + \text{B} \longrightarrow \text{A} - \text{B}} + \text{Energy released (called bond energy)}$$

III. The parameter **bond strength** or bond dissociation energy may be assigned to each chemical bond.

IV. The bond energy of a diatomic molecule is obtained from heat formation or heat dissociation of the molecule (Enthalpy change).

## • Bond Angle:

I. It is equal to the angle made of the two bonds between nucleus of the central atom with the neighbouring atoms.

II. It has been observed that with the decrease of electronegativity of the central atom, the bond angle also decreases.

III. It is measured by X-ray diffraction and molecular spectroscopy.

## • Bond Order :

It is defined as the number of electron pairs in bonding molecular orbitals minus the number of electron pairs in antibonding molecular orbitals.

I. Bond order (BO) = [Number of electron pairs in bonding ($N_b$) molecular orbitals (MO's)
– Number of electron pairs in antibonding ($N_a$) molecular orbitals (MO's)]

II. Bond order (BO) = $\frac{1}{2}$ [No. of electrons in bonding MO's ($n_b$) – No. of electrons in antibonding MO's ($n_a$)]

$$\text{Bond order} = \frac{1}{2}\,(n_b - n_a)$$

**Some features:**

(*a*) **Stability of molecule:**

(*i*) A molecule is stable of $n_b > n_a$

(*ii*) A molecule is unstable if $n_b < n_a$

(*b*) **Bond dissociation energy:** Greater the bond order, greater is the bond dissociation energy.

(*c*) **Bond length:** Bond order is inversely proportional to bond length.

(*d*) **Magnetic properties:**

(*i*) A molecular species will be paramagnetic if it has unpaired electrons in its molecular orbitals. Greater the number of unpaired electrons, more will be its paramagnetic character.

(*ii*) The molecular species will be dimagnetic if there are no unpaired electrons in its molecular orbitals.

- **Chemical Bonds:**

I. A chemical bond is responsible for holding the constituent atoms in a chemically distinct species; so, a chemical change is always accompanied by breaking and formation of such bonds.

II. The physical significance of a chemical bond is related in terms of the energy required to break a chemical bond or the same amount of energy is released during the formation of this bond.

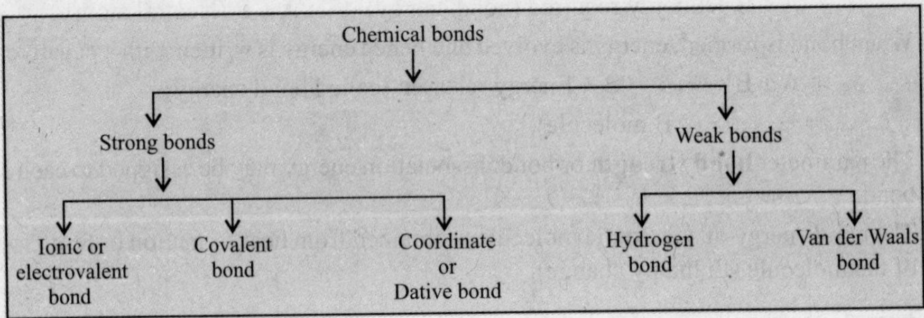


**Ionic Bond or Electrovalent Bond:**

**The chemical bond formed between two atoms by the transfer of one or more valence electrons from one atom to another is called ionic or electrovalent bond or polar bond.**

**Explanation:**

I. Atom P forms an ionic bond with another atom M.

II. Atom P has one electron in its valence shell while atom M has seven electrons.

III. Therefore P transfers on electron to M and in this transaction, both the atoms acquire a stable electron-octet. It results in the formation of positive ion (cation) $P^+$ and negative ion (anion) $M^-$ i.e. the molecule $P^+M^-$ acquire polarity. They are held together by electrostatic force of attraction known as *ionic bond* or *electrovalent* or *electrostatic bond* or, *polar bond.*

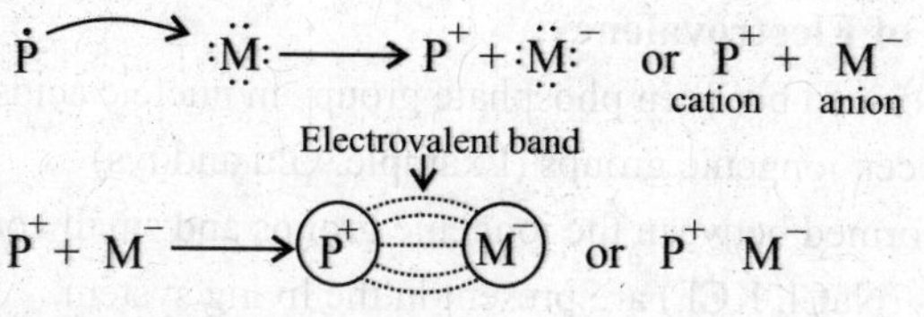

$\dot{P}$ $:\ddot{M}: \longrightarrow P^+ + :\ddot{M}:^-$ or $P^+ + M^-$

cation anion

Electrovalent band

$P^+ + M^- \longrightarrow (P^+)\cdots(M^-)$ or $P^+ M^-$

**Fig. 5.1** Formation of ionic bond between P and M

IV. Ionic bond may be defined as the electrostatic force of attraction between cation and anion which is formed by the transfer of electrons.

**Example:**

**NaCl molecule**

I. Here sodium (Na: 2, 8, 1) transfers its excess one electron to chloride atom (Cl: 2, 8, 7)

II. The electron is lost by Na atom and is accepted by Cl atom and consequently Na atom is converted into a positively charged ion (*i.e.* cation) and Cl atom is converted into a negatively charged ion (*i.e.* anion)

III. Thus two ions are formed and by electrostatic force of attraction, they attract each other. It leads to the formation of ionic or electrovalent bond between Na and Cl.

*Steps :*

(*a*) Na (2, 8, 1) $\longrightarrow Na^+ + e^-$ (2,8)

(*b*) $Cl + e^-$ (From Na atom) $\longrightarrow Cl^-$
(2,8,7) (2,8,1)

(*c*) $Na^+ + Cl^- \longrightarrow NaCl$
(2,8) (2,8,8)

**Nature of Ionic bond:**

I. Here, two oppositely charged ions combine together.

II. They attract each other by electrostatic force of attraction and thus are held together by the lines of force of attraction.

III. Ionic bond is non-directional in nature and extends equally in all directions.

**Characteristics of Ionic Compounds**

1. Ionic compounds consists of three-dimensional solid aggregates of cations and anions which are arranged in a well-defined geometrical pattern.
2. Ionic compounds are freely soluble in polar solvents like water, $NH_3$ and insoluble or slightly soluble in non-polar solvents.
3. Ionic compounds do not conduct electricity when they are in the solid state. However, they conduct electricity when they are in aqueous solution, molten or fused state.
4. They have low volatility and have high melting and boiling points.

- **Electrovalency:**

**Electrovalency of an element is its combining capacity in an ionic compound i.e. when an element forms electrovalent bond its valency is known as electrovalency.**

I. Electrovalency of an element is equal to the number of electrons lost by an atom of that element in forming positive ion or gained by it forming a negative ion, both having the noble gas configuration *i.e.* $S^2P^6$ configuration in their outermost shell.

II. The elements which lose electron show *positive electrovalency* while the elements which gain electrons show *negative electrovalency*.

**Biological application of Electrovalency:**

1. Ionic bonds are formed between phosphate groups in nucleic acids and cations.
2. It is formed between iongenic groups (Example. Glu and lys)
3. Ionic bonds are formed between the iongenic groups and small counter ions.
4. Ionic compounds (NaCl, KCl.) are present in the living system.
5. Antigen and antibody reactions are due to ionic interactions.

## • Covalent Bond

**The chemical bond between two atoms in which the electrons (in pairs) are shared by both the participating atoms is called covalent bond.**

**Explanation:**

I. Here the bonds are formed by sharing of one or more pairs of electrons by two atoms to make a complete and stable electron octet in the outermost orbit of each atom.

II. One electron of each electron pair comes from each atom to make electron orbits of two atoms overlap to hold them together as a molecule.

For stability of the bond, two electrons of the overlapping orbital must spin in opposite direction.

III. For example, there are two atoms P and M each having only seven valency electrons. Both of them are short of one electron to form octet to form stable structure.

IV. The "sharing of a pair of electrons" makes each atom surrounded by an outermost orbit of eight electrons in spite of having total fourteen electrons. This *"shared electron pair"* is called *"covalent bond"* This type of bonding formed by sharing of electron pair is called *Covalency*.

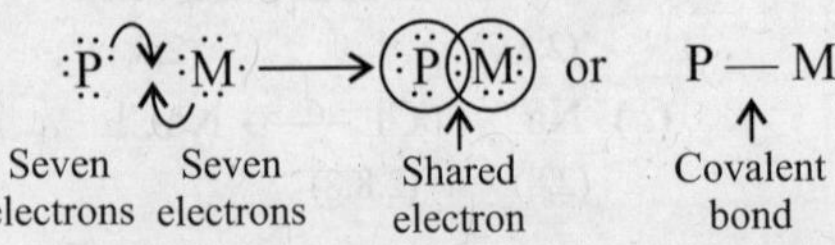


**Fig. 5.2** Covalent bond formation.

**Characteristics:**

1. Covalent linkage may connect two similar atoms when both are a few electrons short of a stable number.
2. In covalent linkage, electrons are not transferred from one atom to the other, the molecule cannot get polarity and no ions are formed. Therefore covalent linkage may be called *non-polar* and *non ionised.*
3. Bonding electron pair between two dissimlar atoms is more under the control of the more electronegative atom.
4. It is exhibited in almost all organic compounds and also in some inorganic compounds.
5. It has a directional character.
6. It does not dissociate spontaneously.

**Types of covalent bond:**

| Type of bond | Represented by | Number of electron pairs involved | Example |
|---|---|---|---|
| Single covalent bonds | Single dash (–) | 1 pair = 1 × 2 = 2 electrons | H – H; Cl – Cl |
| Double covalent bonds | Double dash (=) | 2 pairs = 2 × 2 = 4 electrons | O = O |
| Triple covalent bonds | Triple dash (≡) | 3 pairs = 3 × 2 = 6 electrons | N ≡ N |

**Characteristics of Covalent compound:**

1. Covalent compounds usually consist of discrete molecules and the force of attraction between adjacent covalent molecule is weak.
2. They have low melting and boiling point.
3. They are bad conductors of electricity.
4. Covalent compounds show *isomerism*.

**Example:**

**(*a*) Hydrogen molecule ($H_2$):**

I. It is composed of two H atoms, each having one valence electron.

II. Each contributes an electron to the shared pair and both atoms acquire stable helium configuration. It results in stable $H_2$ molecule.

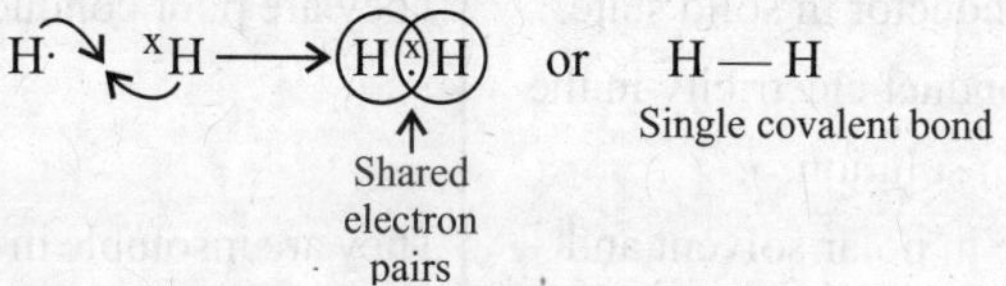


**Fig. 5.3**

**(*b*) Oxygen molecule ($O_2$)**

I. It is composed of two 'O' atoms; each having two valence electrons.

II. Each (2, 6) contributes two electrons to shared pairs and both 'O' atoms achieve the octet. Both 'O' atoms are linked by a double bond.

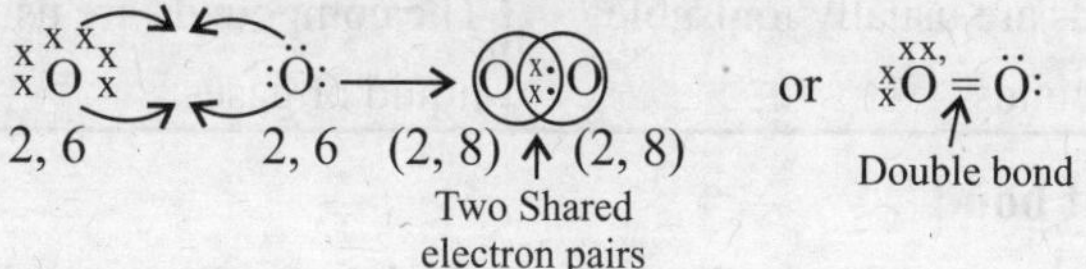


**Fig. 5.4**

**(*c*) Nitrogen molecule ($N_2$)**

I. It is composed of two nitrogen (N=2, 5) atoms each having five electrons in the valence shell.

II. Each contributes three electrons to shared pairs and both atoms (N) achieve the octet.

III. Both atoms are linked by triple bond.

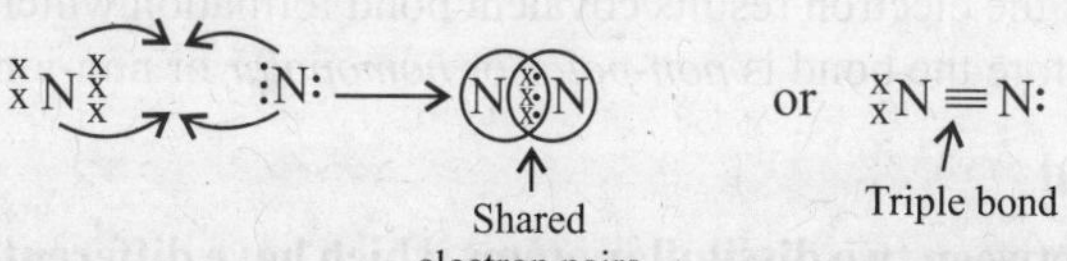


**Fig. 5.5**

## • Application of Covalent Bond

1. Glycosidic bond formation.
2. Peptide bond formation.
3. Ester bond formation.
4. Phosphodiester bond formation
5. Disulphide bond formation

## Difference between Ionic and Covalent Bonds

| | *Ionic bond* | *Covalent bond* |
|---|---|---|
| I. | Bonds are formed by the transfer of electrons from one atom to another. | Bonds are formed by the sharing of electron pair(s) between two atoms. |
| II. | Bond consists of electrostatic force between anions and cations. | Bond consists of shared electrons between atoms. |
| III. | Bond length and bond angle are more variable. | Bond length and bond angle are specific for a particular bond. |
| IV. | They undergo ionic reaction (in solution) which are fast and instantaneous. | They undergo molecular reaction which are slow. |
| V. | They are bad conductor in solid state, however they conduct electricity in the molten state or in solution. | They are poor conductor. |
| VI. | They are soluble in polar solvent and insoluble or slightly soluble in non-polar solvent. | They are insoluble in polar solvent and readily soluble in non-polar solvent. |
| VII. | They have high melting and boiling point. | They have low boiling and melting point. |
| VIII. | These are hard and brittle. | These are soft and waxy. |
| IX. | They cannot exhibit space isomerism. | They can exhibit stereoisomerism |
| X. | The compounds are usually ionisable mineral compounds. | The compounds are usually volatile, solid, liquid or gases. |

### • Non-Polar Covalent bond

**A covalent bond between two similar atoms which have the same electronegativity or zero electronegativity difference is called a *non-polar* or homopolar covalent bond.**

I. Here the shared electron is placed at the centre of the distance between the nuclei of the linked atoms.

II. Due to their same electronegativity or zero electronegativity difference, they have equal tendency to attract the shared electron pair towards them.

III. Equal sharing of the electron results covalent bond formation which has no polarity or ionic character. Therefore the bond is *non-polar* or *homopolar* or non-ionised.

### • Polar Covalent bond

**A covalent bond between two dissimilar atoms which have different electronegativity value is called polar covalent bond.**

I. Take two dissimilar atoms *viz* H and Cl which form covalent bond and have different electronegativity values (H= 2.1, Cl = 3.0), is not equally shared.

II. The electron pair is not placed at the centre of the distance between the nuclei of the two atoms. *viz* H and Cl. It is partially displaced towards the more electronegative Cl atom due to its greater affinity (Cl = 3.0) to electron pair.

III. This type of partial displacement of the shared electron pair towards the more electronegative Cl atoms develops a small negative charge (represented as $\delta^-$) on more electronegative Cl atom and an equal amount of small positive charge (represented by $\delta^+$) on less negative H atom.

IV. As a result, HCl molecule appears to contain two oppositely charged poles namely $H^{\delta+}$ and $Cl^{\delta-}$ at the end of bond and the molecule is depicted as $H^{\delta+}-Cl^{\delta-}$.

V. The covalent bond between H and Cl atoms develops some polarity or partial ionic character and hence called *polar covalent bond.*

VI. Polar covalent bond is not a true or purely covalent bond. It has some polarity or ionic character. Thus it is neither wholly covalent nor wholly ionic but has an intermediate character. The molecules like HCl which have polar covalent bonds are called *Polar molecules*.

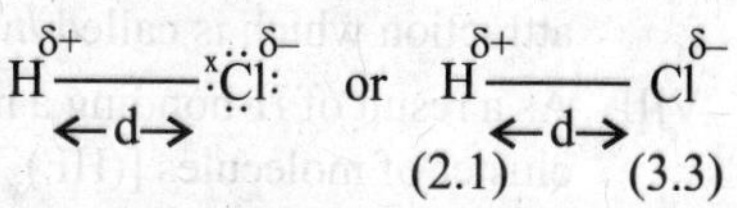

$$\overset{\delta+}{H}\text{———}\overset{\delta-}{\underset{\cdot\cdot}{\overset{\cdot\cdot}{{}^{x}Cl:}}} \quad \text{or} \quad \underset{(2.1)}{\overset{\delta+}{H}}\text{———}\underset{(3.3)}{\overset{\delta-}{Cl}}$$

←d→ ←d→

**Fig. 5.6** HCl molecule.

VII. All covalent bonds between two dissimillar atoms are **polar covalent** bonds.

## Properties of Polar Molecule

I. Polar compounds are intermediate between those of purely covalent and purely ionic compounds.

II. A polar molecule like HCl ($H^{\delta+}-Cl^{\delta-}$) which has a positive and negative charge centre at the end of the covalent bond becomes *dipolar*, hence it is called a *dipole*.

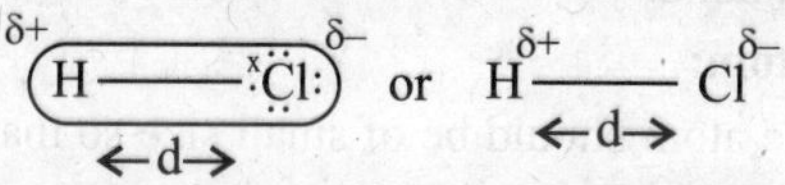

$$\overset{\delta+}{H}\text{———}\overset{\delta-}{{}^{x}\ddot{Cl}:} \quad \text{or} \quad \overset{\delta+}{H}\text{———}\overset{\delta-}{Cl}$$

←d→ ←d→

**Fig. 5.7** A dipole of HCl molecule, Here is the distance between the positive and negative centre of the dipole and is called bond length.

## Dissolution of Polar Solvents:

I. When a polar molecule is dissolved in a polar solvent like water ($H_2O$), the polar molecules of water surround the positive and negative ends of the polar covalent compound (solute) and split it into anions and cations.

II. It is surrounded by a definite and unknown number of water molecules.

*For example:*

III. $HCl^-$ a polar covalent molecule which when dissolved in water, the polar molecules of $H_2O$ break HCl ($H^{\delta+}-Cl^{\delta-}$) molecules into $H^+$ and $Cl^-$ ions which are surrounded by water dipoles.

## Hydrogen Bond

The weak attractive electrostatic force between a hydrogen atom which is already covalently attached with a strongly electronegative atom (N, O, F) of a molecule and another electronegative atom of some other molecule (same molecule or different molecule) is known as hydrogen bond.

## Formation of Hydrogen bond:

I. A hydrogen bond is formed when two electronegative (negative-charged) atoms fulfil their shell requirements of sharing a single hydrogen atom.

II. As an example, hydrogen fluoride (HF) or water ($H_2O$) molecules may be held together by hydrogen bond.

III. Fluorine, Oxygen and Nitrogen are electronegative elements and highly polar.

IV. Due to high electronegativity and small size of fluoride (F) atom, a polar covalent bond (H-F) is formed in H-F molecule. Bonding electron pair in a covalent bond is largely attracted towards electronegative element.

V. Consequently, electronegative element acquires partial negative charge (delta negative $\delta^-$) while the hydrogen atom acquires the same amount of positive charge (delta positive $\delta^+$)

VI. Thus in H-F molecule, a polar covalent bond i.e. $H^{\delta+}–F^{\delta-}$ is formed. It behaves like a dipole.

VII. Now when several such dipoles come nearer to each other, H atom carrying positive charge (i.e.$H^{\delta+}$) in one $H^{\delta+}–F^{\delta-}$ Dipole is attracted towards F atom carrying negative charge (i.e. $F^{\delta-}$) of other $H^{\delta+}–F^{\delta-}$ dipole and gets attached with it (i.e. $F^{\delta-}$ atom) by electronic force of attraction which is called *hydrogen bond.*

VIII. As a result of H-bonding a number of HF molecules get associated together and form a large cluster of molecules $[(HF)_x]$

$$H^{\delta+} - \ddot{\underset{..}{F}}:^{\delta-} - H^{\delta+} - \ddot{\underset{..}{F}}:^{\delta-} - H^{\delta+} - \ddot{\underset{..}{F}}:^{\delta-}$$

Therefore, hydrogen bonding is simply dipole-dipole interaction.

**Conditions for hydrogen bonding:**

Hydrogen bonding comes into existence as a result of dipole-dipole interactions between the molecules in which hydrogen atom is covalently bonded to a highly electronegative atom.

**1. Presence of highly electronegative atom:**

The molecule having hydrogen bonds must have an electronegative atom like N, O, or F directly linked to H atom by a covalent bond.

**2. Presence of small size of atom:**

The highly electronegative atom should be of small size so that the Bond [H-B (N, O, F etc)] may be highly polar ($H^{\delta+}–B^{\delta-}$) and a strong interaction between several such dipoles may occur.

**Strength of H-bonds**

I. It is a weak bond because it is merely an electrostatic force and not a chemical bond.

II. Its strength depends upon the electronegativity of atom to which H atom is covalently bonded. i.e. strength increases with the increase of electronegativity.

III. Since electronegativity of F>0>N, the strength of H bond is in the order H-F..... H(40 KJ $mol^{-1}$) > H– O... H (28 KJ $mol^{-1}$) > H – N...... H (8 KJ $mol^{-1}$).

IV. Hydrogen bonds are much weaker (bond energy (8-40 K J $mol^{-1}$) than covalent bonds (bond energy 40-400 KJ $mol^{-1}$).

**Difference between hydrogen bond and covalent bond**

| | Hydrogen bond | Covalent bond |
|---|---|---|
| I. | It arises due to electrostatic force of attraction | Covalent bond is formed by mutual sharing of electrons |
| II. | Hydrogen bond affects physical properties | Here a change in chemical properties of the constituents of a covalent compound occurs. |
| III. | Hydrogen bond is much weaker bond. | Covalent bond is much stronger than hydrogen bond. |
| IV. | The bond strength of H-F...H is about 40 KJ $mol^{-1}$ | H-H bond is 433 KJ $mol^{-1}$ |

- **Types of Hydrogen bond:**

There are two different types of hydrogen bonds. (*a*) Intramolecular (*b*) Interamolecular

**(*a*) Intramolecular hydrogen bonding**

I. This type of bond is formed between hydrogen atom and N, O or F atom of the same molecule.

II. This type of hydrogen bonding is called *chelation* and is more frequently found in organic compounds.

**(b) Intermolecular hydrogen bonding:**

I. This type of bond is formed between several molecules of the same substance or between several molecules of different substances.

II. In this type of bonding, two or more molecules of the same substance or different substances get polymerised and form a large cluster.

## • Significance of Hydrogen bonding:

**(a) Physical state of water**

Life would have been impossible without liquid water which is the result of intermolecular H-bonding in it.

**(b) Wood fibres:**

I. Hydrogen bonding increases the rigidity and strength of wood fibres.

II. So that it may be used to meet the requirements of housing furnitures etc.

**(c) In clothing:**

The rigidity and tensile strength of the cotton silk or synthetic fibres is due to the bonding in them.

**(d) Food materials:**

Most of our food material such as carbohydrate and proteins also have hydrogen bonding among them.

**(e) In paints and dyes:**

The adhesive action of glue, honey, dyes and paint is also due to the H-bonding in them.

**(f) Structure of compound:**

I. H-bonding is directional and on this account it helps in studying and establishing the structure of many compounds like ice, hydrogen, fluoride, acid, salts etc.

II. H-bonding also accounts for tetrahedral structure of ice, zig-zig arrangement of HF molecules in solid hydrogen fluoride.

**(g) Application in biological investigation:**

I. H-bonding also exists in molecules of living systems, *i.e.*, proteins in various tissues, organs, blood, skin and bones in animals.

II. Fibres like those found in hair, silk and muscle consist of long chains of a large number of amino acids.

III. Long chains are coiled about one another and are joined together by hydrogen bonding.

## • Coordinate Bond

**It is a kind of bond which is formed between two combining atoms by the sharing of two electrons (one electron pair) between them, both electrons being donated entirely by only one atom.**

I. Here one atom donates an electron pair while the other atom accepts it.

II. The atom donating the electron pair is called *donor* while the atom accepting it is called *acceptor*.

III. The atom acting as donor must have one unused pair of electrons called *lone pair of electron* which may be donated by it to the acceptor atom.

IV. The acceptor atom must have an empty orbital to accept the lone pair of electrons.

**Explanation:**

I. A coordinate bond is represented as $P \rightarrow M$ where P acts a donor and M behaves as an acceptor.

II. In the formation of coordinate bond, a partial positive charge is developed on the acceptor atom.

III. Thus the coordinate bond P → M should be represented as $P^{\delta+} \longrightarrow M^{\delta-}$

IV. Due to the development of partial charges, the coordinate bond becomes *slightly polar* and hence it is called *semi-polar* or *dative bond.*

V. Let P and M are two atoms. A coordinate bond is formed between them.

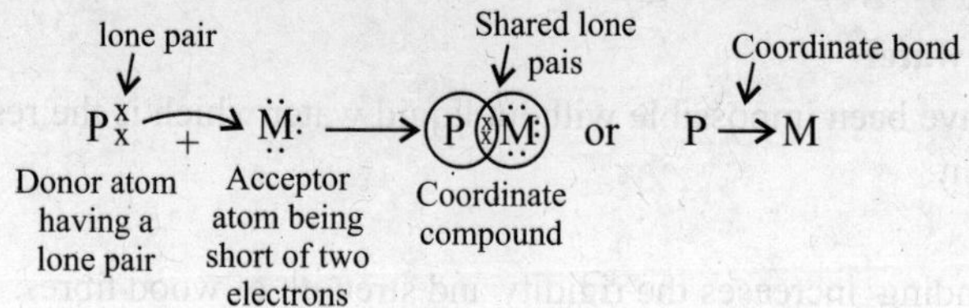


**Fig. 5.8** Coordinate bond formation

VI. The donor atom P has a spare lone pair of electrons on it while the acceptor atom M is short of two electrons.

VII. P donates its lone piar of electrons to M and M accepts it in order to complete its octet. Thus the two electrons of the lone pair which originally belonged to P atom are now shared by both atoms.

VIII. This mutual sharing of electron pair results in the formation of a coordinate bond between P and M (P → M)

IX. Once coordinate bond has been formed, it becomes quite identical with a normal covalent bond. Both are established by a shared electron pair.

X. A coordinate bond has the properties of both covalent bond and ionic bond and therefore it has also been called *Co-ionic bond.*

XI. It is a union of one electrovalent and one covalent bond.

**Steps:**

**1st Step:**

I. The donor atom P transfers one electron of its lone pair to the acceptor atom M.

II. As a result P atom develops unit positive charge (+) and atom M develops unit negative charge (–). This charge is known as *formal charge.*

III. This step is similar to *ionic bond formation.*

**2nd step:**

I. Here two electrons one each with $P^+$ and $M^-$ are shared by both the ions.

II. This step is similar to the formation of *covalent bond.*

Therefore, coordinate bond is equivalent to a combination of electrovalent (polar) bond and a covalent (non-polar) bond. Therefore it is known as *semi-polar bond.*

$$P: + M \longrightarrow \overset{+}{P}\cdot + \cdot M^-$$

$$\overset{+}{P}\cdot + \cdot M^- \longrightarrow P:M \text{ or } P \rightarrow M$$

**Fig. 5.9** Coordinate bond formation.

V. In the light of orbital overlap theory, a coordinate bond is formed when a completely filled orbital (i.e. having lone pair electrons) of an atom overlaps within an empty orbital of the other atom.

**Examples:**

**1. $H_2O_2$ molecule:**

I. It is formed by the combination of $H_2O$ molecule and Oxygen atom.

II. Oxygen atom of $H_2O$ molecule has two lone pairs of electrons on it.

III. One of the two lone pairs on oxygen atom of $H_2O$ is donated to the new oxygen atom and thus a coordinate bond is established between oxygen atom of molecule and new oxygen atom.

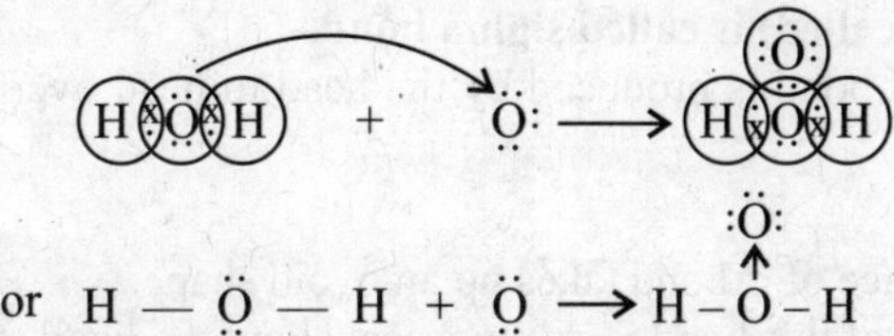

$$\text{or } H - \ddot{O} - H + \ddot{O} \longrightarrow H - \overset{\ddot{O}:}{\overset{\uparrow}{O}} - H$$

**Fig. 5.10** $H_2O_2$ – Coordinate bond formation

**2. Amonium ($NH^+_4$)**

I. It is formed by the combination of $NH_3$ molecule and $H^+$ ion.

II. In $NH_3$ molecule each of the three H atoms are linked to N atom by a covalent bond. Thus N atom in this molecule ($NH_3$) is left with a lone pair of electrons after completing its octet by sharing three of its five valence shell electrons with three H atoms.

III. The electrons of lone pair on N atom are donated to $H^+$ ion and thus a N–H coordinate bond is established in $NH_4^+$ ion.

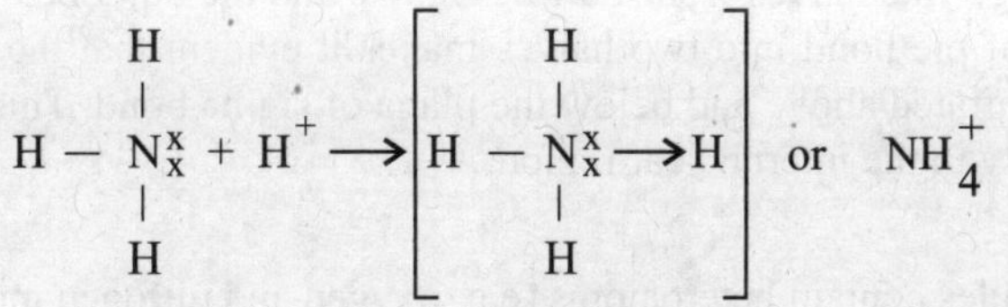

$$H - \underset{H}{\overset{H}{N}}{}^{x}_{x} + H^+ \longrightarrow \left[ H - \underset{H}{\overset{H}{N}}{}^{x}_{x} \longrightarrow H \right] \text{ or } NH_4^+$$

**Fig. 5.11** $NH_4^+$ – Formation of coordinate bond

After the formation of $NH_4^+$ ion, all the four N–H bonds become identical and hence $NH_4^+$ ion is represented as

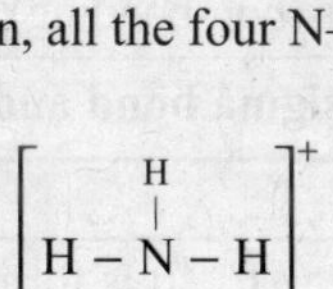

$$\left[ H - \underset{H}{\overset{H}{N}} - H \right]^+$$

**Biological Application**

**1. *Enzyme inhibition***

Some enzymes that require divalent metal ion $Mg^{2+}$, $Mn^{+2}$, $Cl_2^{2+}$ for activity are inhibited non-competitively by chelating agent EDTA **(Ethylenediamine tetraacetate)** which reversibly binds divalent cations by coordinate bonds to form metal chelates.

**2. *Cobalamine or cyanocobalamine or Vit. $B_{12}$***

It is formed by coordinated and covalent bond

**3. *Structure of Haem of haemoglobin***

**4. *Myoglobin***

The covalent bond between the two atoms is formed by the overlap of

(*i*) s-orbital of one atom with s-orbital of the other atom (called s-s overlap).

(*ii*) s-orbital of one atom with p-orbital of the other atom called s-p overlap.

(*iii*) p-orbital of one atom with p-orbital of another atom i.e. p-p overlap.

**Sigma bond : (σ)**

**A covalent bond which is formed between two atoms by the overlap of their half-filled atomic orbitals along the line joining the nuclei of both atoms (i.e. along nuclear axis, bond axis or molecular axis as it is called) is called sigma bond.**

In other words, sigma bond is produced by the head to head overlap of the half-filled atomic orbitals of the two atoms.

**Characteristics:**

(*i*) The boundary surface of σ bond takes up an ovoid shape.

(*ii*) The electron cloud of this bond is symmetrical about the bond axis.

(*iii*) This bond has two electrons which have opposite spin.

**Pie bond : (π)**

A covalent bond which is formed between two atoms by the overlap of their singly-filled p-orbitals along a line perpendicular to their nuclear axis (side to side overlap) is called a pie bond.

In other words, the pie bond is produced by the side to side overlap of half-filled p-orbitals of the two atoms.

**Characteristics:**

I. This bond has one and only one nodal plane which contains the nuclear axis and divides it into two sausage-like halves—one half lies above and the other below the nodal plane.

II. The division of pie bond into two halves makes it evident that the electron density of pie bond is concentrated above and below the plane of sigma bond. This bond has an increased electron density in the internuclear region.

**• n-Orbitals :**

I. Certain molecules contain heteroatoms (e.g. oxygen and nitrogen molecules)

II. The occupied orbitals with highest energy in such molecules are those of lone pairs.

III. These lone pairs are not involved in bonds (non-bonding) and thus retain their atomic character.

**Difference between sigma bond and pie bond**

| | Sigma bond (σ) | Pie bond (π) |
|---|---|---|
| 1. | It is formed by end to end overlap | 1. It is formed by sidewise overlap. |
| 2. | This is formed by overlapping between s-s, s-p and p-p orbitals | 2. This is formed by p-p orbitals. |
| 3. | Overlapping is quite large and hence sigma bond is strong. | 3. Overlapping is to a small extent and hence pie bond is weak. |
| 4. | Electron density of this bond is distributed symmetrically about their nuclear axis. | 4. The electron density of this bond is unsymmetrical about their nuclear axis. |
| 5. | Free rotation about sigma bond is possible | 5. Free rotation about pie bond is not possible. |

**Transition of bonds :**

I. According to molecular orbital theory, when a molecule is excited by the absorption of energy (**UV or visible ray**), its electrons are promoted from a bonding to antibonding (**higher energy state**).

II. Antibonding orbital associated with the σ bond is called $\sigma^*$ orbital and that associated with the π bond is called $\pi^*$ orbital.

III. The major electronic transition within the visible and ultraviolet regions are: $\sigma \rightarrow \sigma^*$, $n \rightarrow \sigma^*$, $\pi \rightarrow \pi^*$ and $n \rightarrow \pi^*$. The energies associated with these transitions are as follows.

$$\sigma \rightarrow \sigma^* > n \rightarrow \sigma^* > \pi \rightarrow \pi^* > n \rightarrow \pi^*$$

**Characteristics:**

(*a*) $\sigma \rightarrow \sigma^*$ (*i*) This transition requires largest energy charge.
(*ii*) Very short wavelengths (190 nm) are absorbed.

(*b*) $n \rightarrow \sigma^*$ (*i*) This transition requires very lower energy in relation to $\sigma \rightarrow \sigma^*$ transition.
(*ii*) This transition usually takes place in saturated compounds.
(*iii*) Absorption band

(*c*) $\pi \rightarrow \pi^*$ (*i*) This transition takes place in the unsaturated centres of molecule.
(*ii*) This transition associated with lower energy.
(*iii*) Longer wavelengths ( ) are generally absorbed.

(*d*) $n \rightarrow \pi^*$ (*i*) It requires the least amount of energy.
(*ii*) Absorption band at 210-220 nm
(*iii*) Shorter wavelengths are absorbed.

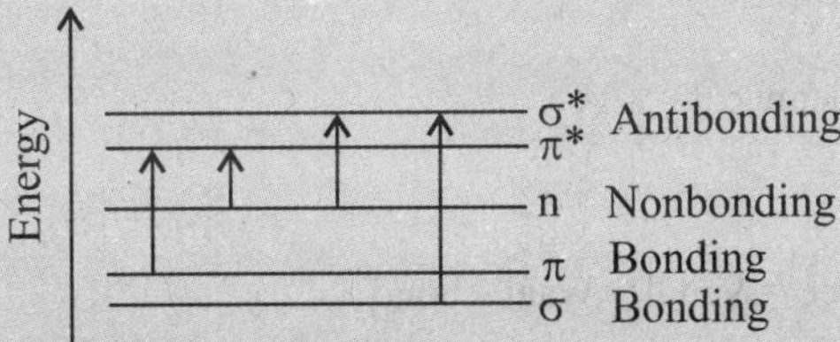


**Fig. 5.12** Schematic molecular orbital energy level. Arrow indicates electronic energy transition.

## Van der Waal's Forces or Bonds

The short-lived intermolecular forces which are believed to exist between all kinds of atoms, molecules and ions when they are sufficiently close to each other are referred to as van der Waal's forces.

- **Characteristics of van der Waal's Forces:**

1. These are very weak attractive forces between groups or atoms of different molecules and weaker than hydrogen bond.
2. These forces are almost absent when the molecules of gas are far apart and are in rapid kinetic motion.
3. Presence of van der Waal's forces in regular and close sequence can generate collective strength and are capable of holding the molecules together to form a stable fold in molecules chain.
4. Bond energies are slightly higher than the average kinetic energy of the molecules.
5. Van der Waals' forces ($F$) are inversely proportional to the seventh power of the distance ($d$) between the dipoles.

$$F \propto \frac{1}{d^7}$$

**Relationship between Distance of Atoms and Force:**

I. When the distance between the atoms or molecules is great, there is neither attraction nor repulsion and the force is zero.

II. When the atoms or molecules are close together at a few Å apart, the intramolecular attraction one to dipole-induced dipole become appreciable.

III. When the atoms or molecules are very close range 3-4 Å apart, there is overlapping of electron clouds and the electrical repulsions between the clouds and the positive nuclei of the two atoms become pronounced.

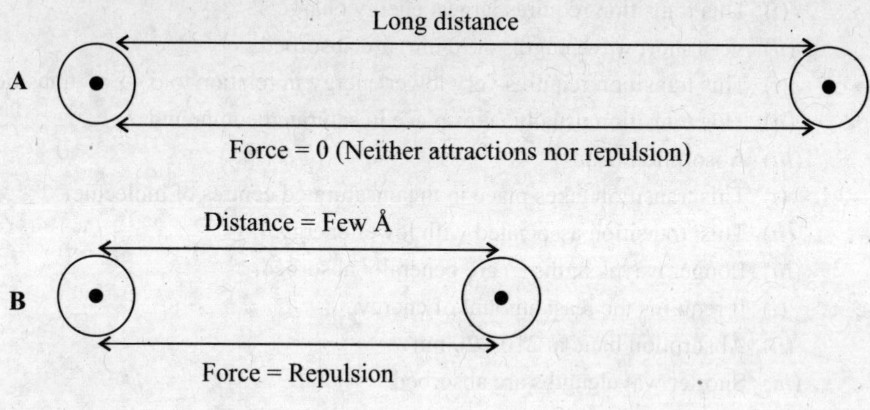


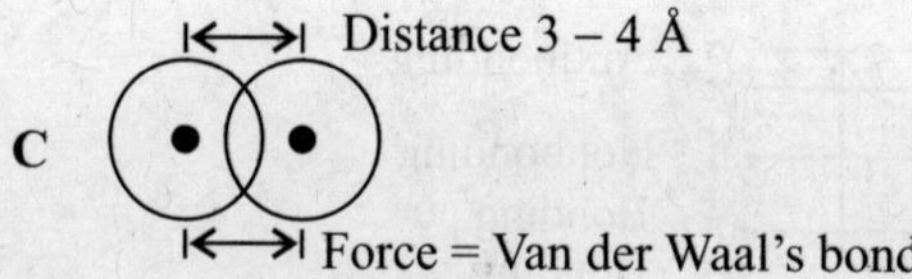


**Fig. 5.13** Distance between atoms (molecules) and the force.

### Origin of van der Waals' forces

I. When two atoms or molecules approach each other, both attractive and repulsive forces operate between their negative electrons and positive protons.

II. **Attractive forces:**

These forces act

(*a*) between the nucleus of a molecule and its own electrons.

(*b*) between the nucleus of one molecule and electrons of other molecule.

III. **Repulsive forces:**

These force act

(*a*) between the electrons of two different molecules.

(*b*) between the nuclei of two different molecules.

**Remark:**

I. The attractive and repulsive forces are in equilibrium at an intermolecular distance which is approximately equal to 4 Å.

II. However, at an intermolecular distance between 4 Å and 10 Å attractive forces predominate. These attractive forces acting between the molecules are called van der Waals' forces.

### Application and Biological Importance

I. These forces are responsible for bringing about condensation and crystallisation (at low temperature) in the noble gases and halogens.

II. Folding of polypeptide chain to form tertiary structure of protein.

III. It helps in binding of an enzyme to substrate to form enzyme substrate (ES) complex.

IV. It also enhances antigen-antibody reaction.

V. Due to van der Waals' forces, lipid and proteins are bonded to form lipoportein.

**Dipole Interaction**

**Diople:**

A polar molecule like HCl ($H^{\delta+}$–$Cl^{\delta-}$) which has positive and negative charge centres or electrical poles, as they are called, and at the ends of the covalent bond becomes dipolar, hence called dipole.

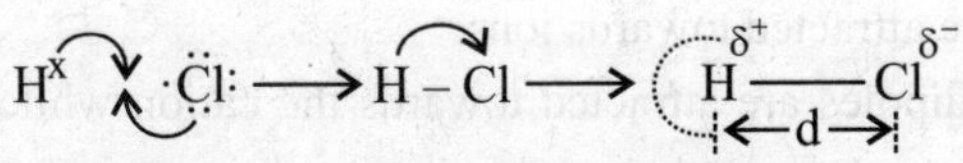

$$H^{x} \; \ddot{\underset{..}{Cl}}: \longrightarrow H - Cl \longrightarrow \overset{\delta^+}{H} \text{———} \overset{\delta^-}{Cl} \quad (\leftarrow d \rightarrow)$$

**Fig. 5.14.** Dipole of HCl, d is the Distance

**Dipole moment**

**The degree of polarity of a polar covalent bond or polar molecule (i.e. having a polar covalent bond) is expressed in terms of its dipole moment (μ) which is equal to the product of the magnitude of electric charge (e) in e.s.u. and distance (d) in Å between the positive and negative centres (i.e. bond length)**

$$\mu = e \times d$$

The greater differences in electronegativities, the greater is the value of dipole moment of a polar molecule, greater is the degree of polarity of the polar covalent bond between linked atoms.

Polar molecules have great attractions for each other and arrange themselves into an endless chain. Polar molecules tend to be oriented in an electric field. The positive ends of dipoles are directed towards the negative electric plate while the negative ends orient towards the positive plate.

**Units of Dipole moment:**

As *e* is of the order $10^{-10}$ e.s.u and *d* of the order of $10^{-8}$ cm.

$$(\mu) = 10^{-10} \text{e.s.u.} \times 10^{-8} \text{cm} = 10^{-10} \times 10^{-8} \text{ esu cm}$$

$$= 10^{-10} \times 10^{-8} \text{ Debye (D)}$$

Thus 1D = $10^{-18}$ e.s.u. cm (C.G.S.) : In S.I. Unit $\mu = 1.6 \times 10^{-19} \text{C} \times 10^{-10}$ meter

$$= 1.6 \times 10^{-29} \text{C meter}$$

**Applications of Dipole moment:**

I. To find the degree of polarity of molecule

II. To distinguish between polar and non polar molecules

III. It can be used for determinining the ionic character of the bond.

IV. It helps in determining the geometry and shape of the molecule.

**Types of Dipole interactions:**

I. Dipole-Dipole interaction.

II. Ion-Dipole interaction.

III. Dipole-Induced dipole interaction.

IV. Instantaneous dipole induced dipole-interaction (**London forces**).

**1. Dipole-Dipole interactions:**

I. These forces are found in polar molecules having permanent polarity as well as permanent dipole moment.

II. When polar molecules are brought nearer to each other, they orient themselves in such a way that positive end of one dipole attracts the negative end of another dipole and *vice versa*.

III. Many molecules are held together by the dipole-dipole interactions.

IV. Dipole-dipole interaction between the polar molecules are also called ***Keesom Forces*** and strongest of all other types of van der Waals forces.

**2. Ion-Dipole Interaction:**

I. Polar molecules are attracted towards ions.

II. Negative ends of dipoles are attracted towards the cation while positive ends towards the anion.

III. This type of interaction is called ion-dipole interaction.

**Example:** Dissolution of NaCl is water ($H_2O$)

**3. Dipole-Induced Dipole Interaction: (Debye forces)**

I. This is found in a mixture containing polar and non-polar molecules.

II. When a non-polar molecule is brought near to the polar molecule, the positive end of the polar molecule attracts the mobile electrons of non-polar molecules; thus polarity is induced in non-polar molecule.

III. Now, both the molecules become dipole. Therefore, the positive of non-polar molecule attracts the displaced electron cloud of non-polar molecule.

IV. Thus two types of molecules are held together.

**Example:** HF and $O_2$, HF and $N_2$ etc

**4. Instantaneous Dipole-Induced Dipole Interaction (London Forces)**

I. These forces are found in non-polar molecules like $H_2$, $O_2$, $Cl_2$ etc as well as monoatomic noble gases like He, Ne, etc.

II. A non-polar atom or molecule may be visualised as a positive centre surrounded by a symmetrical negative electron cloud.

III. Both are in equilibrium. But as the electron cloud oscillates, the electron cloud becomes more dense on one side of the molecule than on the other side and thus the equilibrium gets disturbed for a moment.

IV. The displacement of electron cloud creates an instantaneous dipole temporarily.

V. Thus the non-polar molecule is momentarily self-polarised and becomes temporarily polar.

VI. Now the self, polarise molecule is brought near to a non-polar molecule, it polarises the neighbouring molecule by disturbing its electronic distribution and an induced dipole is created in it *i.e.* non-polar molecule momentarily becomes polar.

**Example:** Diatomic gases like $H_2$, $O_2$, $Cl_2$, $N_2$ etc.

## PROBLEMS

**1.** *Covalent radii of carbon and silicon atoms are 0.77 Å and 11-1 Å respectively. Calculate the bond length of C–Si bond.*

**Solution :**

Since covalent radii have additive character.

$$d\,(C - Si) = r\,(C) + r\,(Si)$$
$$= 0.77 + 1.11 = 1.88 \text{ Å}$$

**2.** *Ionic radius of $Na^+$ ion is 0.95 Å and the internuclear distance between the ion pairs or NaCl ionic crystal is equal to 2.76 Å. Calculate radius of $Cl^-$ ion.*

**Solution :**

$$d\,(Na^+ - Cl^-) = r\,(Na^+) + r\,(Cl^-)$$
$$2.76 \text{ Å} = 0.95 + r\,(Cl^-)$$
$$\therefore \quad r\,(Cl^-) = 2.76 - 0.95$$
$$= 1.81 \text{ Å}$$

**3.** *Covalent radii of chlorine and carbon atoms are 0.99 Å and 0.77 Å respectively. Calculate bond length of Cl–C bond.*

**Solution :**

$$d\,(Cl^\circ - C) = r\,(Cl) + r\,(C)$$
$$= 0.99 + 0.77$$
$$= 1.76 \text{ Å}$$

**4.** *The observed dipole moment for a molecule AB is 1.45 D and its bond length is 1.654 Å. Calculate the percentage of ionic character in the bond.*

**Solution :**

Assuming A-B to be 100% ionic bond and would acquire full unit charge *i.e.* a = 4.8 × $10^{-10}$ esu.

$\mu_{(obs)} = 1.45$ D (given) and $\mu_{ion}$ = can be calculated

$\mu_{ionic} = q \times d$ $\qquad q = 4.8 \times 10^{-10}$ esu

$$d = 1.654 \text{ Å}$$
$$\mu = (4.8 \times 10^{-10} \text{ esu}) \times (1.654 \text{ Å})$$
$$= 4.8 \times 10^{-10} \text{ esu} \times 1.654 \times 10^{-8} \text{ cm}$$
$$= 4.8 \times 1.654 \times 10^{-18} \text{ esu cm} \qquad [10^{-18} \text{ esu cm} = \text{D}]$$
$$= 4.8 \times 1.654 \times \text{D}$$
$$= 7.9392 \text{ D}$$

$$\therefore \quad \% \text{ ionic character} = \frac{\text{Observed dipole moment}}{\text{Dipole moment for 100\% ionic character}} \times 100$$
$$= \frac{1.45 \times \text{D}}{7.9392 \text{ D}} \times 100$$
$$= \frac{1.45}{7.9392} \times 100 = 18.2638\% = 18.264$$
$$= 18.3\%$$

**5.** *The bond length of H–I bond is 1.60 Å and its dipole moment is 0.38 D. Calculate the percentage ionic character of H–I bond.*

**Solution :**

Assuming H–I to be 100% ionic bond and unit charge 4.8 × $10^{-10}$ esu.

Here observed $d = 1.60$ Å and $\mu = 0.38$ D (given)

$\mu_{ionic} = q \times d$ $\qquad q = 4.8 \times 10^{-10}$ esu

$$= 4.8 \times 10^{-10} \times 1.60 \text{ Å}$$

$= 4.8 \times 10^{-10} \times 1.60 \text{ Å} \times 10^{-8}$ cm

$= 4.8 \times 1.60 \times 10^{-18}$ esu cm [$10^{-18}$ = D]

$= 7.68$ D

$$\% \text{ of ionic character} = \frac{\text{Observed dipole moment}}{\text{Expected moment for 100\% ionic character}} \times 100$$

$$= \frac{0.38}{7.68} \times 100 = \frac{0.38}{7.68} = 0.4947 \times 100 = 4.947\%$$

$$= 4.95\%$$

Hence bond in H-I has 4.95% ionic character.

**6.** *The bond length of HF is 0.92 Å and its dipole moment is 2.00 D. Calculate the percentage ionic character of H-F bond.*

**Solution :**

Assuming H–F to be 100% ionic bond and unit charge $q = 4.8 \times 10^{-10}$ esu.

Here observed $d = 0.92$ Å

$\mu = 2.00$ D (given)

$\mu_{ionic} = q \times d = 4.8 \times 10^{-10}$ esu $\times 0.92$ Å

$= 4.8 \times 10^{-10}$ esu $\times 0.92 \times 10^{-8}$ cm

$= 4.8 \times 0.92 \times 10^{-18} = 4.8 \times 9.2$ D [$\because 10^{-18}$ = D]

$= 4.416$

$$\% \text{ of ionic character} = \frac{\text{Observed dipole moment}}{\text{Expected moment for 100\% ionic character}} \times 100$$

$$= \frac{2.00}{4.416} \times 100 = \frac{200}{4.416} = 45.289\%$$

Hence bond in H-F has 45% ionic character.

**7.** *In NaCl ionic crystal, bond length 2°S 2.36 Å and the experimental dipole moment of this molecule is 8.5 D. Calculate the percentage of ionic character in Na-Cl bond in the given molecule.*

**Solution :**

Assuming Na–Cl to be 100% ionic bond and unit charge $q = 4.8 \times 10^{-10}$ esu.

Expected dipole moment $\mu = q \times d$

$= 4.8 \times 10^{-10} \times 2.36$ Å

$= 4.8 \times 10^{-10} \times 2.36 \times 10^{-8}$ cm

$= 4.8 \times 2.36 \times 10^{-18} = 11.328$ D.

Observed dipole moment = 8.5 D

$$\% \text{ ionic character of Na–Cl bond} = \frac{\text{Observed dipole}}{\text{Dipole moment for 100\% ionic character}} \times 100$$

$$= \frac{8.5 \text{ D}}{11.328 \text{ D}} \times 100$$

$$= \frac{8.5}{11.328} \times 100 = 75.035\%$$

Hence bond in NaCl has 75% ionic character.

**8.** *The ionic character in certain A-B bond is 76.81% and the bond length A-B is 159.0 pm. Calculate dipole moment of AB molecule.*

**Solution :**

Ionic character in A-B = 76.81% bond length (d) = 159.0 pm (given)

$= 159 \times 10^{-10}$ cm (1 pm = $10^7$ esu)

$q = 4.8 \times 10^{-10}$ esu.

$\mu_{exp} = ?$

$\mu_{cal} = q \times d = 4.8 \times 10^{-10} \times 159 \times 10^{-10}$

$= 4.8 \times 159 \times 10^{-20}$ esu cm

$$\% \text{ ionic character} = \frac{\text{Observed dipole moment}}{\text{Calculated dipole}} \times 100$$

$$76.81 = \frac{\mu_{exp} \times 100}{4.8 \times 159 \times 10^{-20}}$$

$$\mu_{exp} = \frac{76.81 \times 4.8 \times 159 \times 10^{-20}}{100} = \frac{76.81 \times 4.8 \times 159 \times 10^{-18} \times 10^{-2}}{10^2}$$

$$= \frac{7681 \times 48 \times 159 \times 10^{-2} \times 1\text{D}}{10^5} = 5.86 \text{ D.}$$

Hence dipole moment of A-B molecule is 5.80 D.

**9.** *The bond length of H-I is 1.60 Å. Find the correct value for its dipole moment (Debye unit), if it were present in the completely ionic form (charge on the electron 4.8 × $10^{-10}$ esu): 3.20, 3.00, 7.68, 0.33, 2.77, 6.40, 9.28.*

**Solution :**

For H–I,

$d = 1.60$ Å $= 1.60 \times 10^{-8}$ cm

$q = 4.8 \times 10^{-10}$ esu

$\mu = q \times d = 4.8 \times 10^{-10} \times 1.60 \times 10^{-8} = 4.8 \times 1.60 \times 10^{-18}$

$= 7.68$ D

**10.** *The bond length of HCl is 1.275 Å, and its dipole moment is 1.03 D. Calculate the percentage of ionic character in the bond.*

**Solution :**

Assuming HCl to be 100% ionic bond and each end would acquire unit strength $q = 4.8 \times 10^{-10}$ esu.

$d$ (given) = 1.275 Å = $1.275 \times 10^{-8}$ cm $\mu$ = 1.03 D (Given)

$\mu = q \times d = 4.8 \times 10^{-10} \times 1.275 \times 10^{-8} = 4.8 \times 1.275 \times 10^{-18}$ cm

$= 6.12$ D

$$\% \text{ ionic character} = \frac{\text{Observed dipole moment}}{\text{Dipole moment for 100\% ionic character}} \times 100$$

$$= \frac{1.03}{6.12} \times 100 = 0.1683 \times 100 = 16.83\%$$

Hence bond HCl has 16.83% ionic character.

**11.** *The bond length of HCl is 1.275 Å and dipole moment is 1.07 D. Calculate percentage of ionic character in the bond.*

**Solution :**

Assuming HCl to be 100% ionic bond and each end would acquire unit strength $q = 4.8 \times 10^{-10}$ esu.

d (given) = 1.275 Å = $1.275 \times 10^{-8}$ cm μ = 1.07 D (Given)

$\mu = q \times d = 4.8 \times 10^{-10} \times 1.275 \times 10^{-8} = 4.8 \times 1.275 \times 10^{-18}$ cm

= 6.12 D

$$\% \text{ ionic character} = \frac{\text{Observed dipole moment}}{\text{Dipole moment for 100\% ionic character}} \times 100$$

$$= \frac{1.07}{6.12} \times 100 = 100 = 17.48\%$$

**12.** *The dipole moment of HCl is 1.03 D and its bond length is 1.27 Å. Calculate (i) the charge on the constituent and (ii) the percentage of ionic character.* ***[Pune 1986; B.U.-1987; V.U. 1996]***

**Solution:**

(*i*) Dipole moment of HCl = Charge on the constituent atom ($\delta e$) × bond length

Here observed bond length ($d$) = 1.27 Å

Dipole moment (μ) = 1.03 D (Given) or $1.03 \times 10^{-8}$ esu cm.

∴ 1.03 = $\delta e \times 1.27$ Å

= $\delta e \times 1.27 \times 10^{-8}$ cm

$$\therefore \quad \delta e = \frac{1.03}{1.27 \times 10^{-8}} = \frac{1.03 \times 10^{-18}}{1.27 \times 10^{-8}} = 0.811 \times 10^{-10} = 8.11 \times 10^{-11} \text{ esu.}$$

(*ii*) Observed dipole moment 1.03 D

Calculated dipole moment = $4.8 \times 10^{-10} \times 1.27$ Å

= $4.8 \times 10^{-10} \times 1.27 \times 10^{-8}$ esu cm.

= $4.8 \times 1.27 \times 10^{-18}$

= $6.096 \times 10^{-18}$ esu cm

= 6.096 D

$$\% \text{ ionic character} = \frac{\text{Observed dipole moment}}{\text{Dipole moment for 100\% ionic character}} \times 100$$

$$= \frac{1.03}{6.096} \times 100 = 16.896\%$$

**13.** *A compound AB having partly ionic and partly covalent character has the dipole moment 1.03 debye. Calculate percentage of ionic character. Given : bond length 1.3 Å and charge $4.8 \times 10^{-10}$ esu.* ***[B.U.-1985]***

**Solution:**

If *AB* is completely ionic then $\mu_{cal}$ = $4.8 \times 10^{-10} \times 1.3$ Å

= $4.8 \times 10^{-10} \times 1.3 \times 10^{-8}$

= $6.24 \times 10^{-18}$ esu. cm

= 6.24 D

Observed dipole = 1.03 D

$$\% \text{ ionic character} = \frac{\mu}{\mu_{cal}} \times 100 = \frac{1.03}{6.24} \times 100 = 16.5064\%$$

= 16.51%

**14.** *Deduce the relation between Debye (D) and Coulomb metre (Cm).*

**Solution :**

We know that in CGS units ID = $10^{-18}$ esu cm

Now since 1 esu = $3.335 \times 10^{-10}$ C and 1 cm = $10^{-2}$ m

$$1\text{ D} = 10^{-18} \times (3.335 \times 10^{-10}\text{ C}) \times (10^{-2}\text{ m})$$

$$1\text{ D} = 3.335 \times 10^{-30}\text{ Cm}$$

or

This equation gives the value of 1D in S.I. units.

This equation can also be written as

$$1\text{ Cm} = \frac{1}{3.335 \times 10^{-30}} = \frac{1}{3.335} \times 10^{30}\text{ D}$$

$$1\text{ Cm} = 0.29985 \times 10^{30}\text{ D}$$

This equation gives the value of 1 Cm in C.G.S. units

**15.** *Why covalent bonds are called directional bonds while ionic bonds are called non-directional?*

**Ans.** (*i*) In covalent bond, the shared electron pairs are localized between two atoms or a covalent bond is formed by the overlap of half-filled atomic orbitals which have definite direction. Hence covalent bonds are directional.

(*ii*) In ionic compounds, each ion is surrounded by a number of oppositely charged ions and there is no definite direction.

**16.** *Why HCl is polar while $Cl_2$ molecule is non-polar?*

**Ans.** (*i*) In $Cl_2$ (Cl–Cl) both atoms have same electronegativity. Hence shared pair of electrons are attracted equally by both Cl atoms and remain in the centre.

(*ii*) No ends acquire positive (+) or negative (–) charge.

(*iii*) In HCl, Cl is more electronegative (–) than H. Hence shared pair of electrons is more attracted towards Cl which therefore acquires negative charge while hydrogen acquires positive charge.

CHAPTER 6

# Microscopy

A microscope is a principal tool or optical instrument in biology with the ability to increase the visual size of all the objects and is generally used in laboratories by students. Microscopy serves two independent functions *viz* **magnification** i.e. to enlarge the image of an object and **resolution** *i.e.* the power of distinction between two closely spaced points of the object in its image.

**• Microscope: An instrument which forms a magnified image of a small object placed close to the eye is called microscope.**

**Properties of Light:**

Light is a form of energy emitted from different sources such as sun, lamp, electric bulb etc. in a series of energy pulses.

**Wavelength (λ):**

(*i*) Light consists of a series of successive electromagnetic waves and the path is shown as sine curve, as peaks or crest and ebbs or trough.

(*ii*) The distance between the two successive peaks or two successive ebbs is the wavelength (λ).

(*iii*) Wavelength determines the colours of the visible light.

(*iv*) Our eyes are sensitive to different wavelengths of light in the visible range (380-740 nm).

**Frequency (*f*):**

(*i*) It refers to the number of waves or energy pulses of light emitted per unit time.

(*ii*) Frequency remains constant for a light wave.

(*iii*) Frequency of visible light ranges from $10^{14}$ to $10^{15}$ Hz, each hertz (Hz) amounting to one cycle per minute.

(*iv*) Light rays of identical frequencies radiating from same source and combining or interfering with each other are called **coherent rays**. Light rays of different frequencies radiating from different sources and not combining or interfering with each other are called **non-coherent rays**.

**Amplitude:**

It is the strength of energy of light waves which determines the intensity of brightness of light.

- **Polarization:**

(*i*) It refers to the light wave which vibrates only in one plane at right angles to the direction of wave propagation.

(*ii*) Therefore polarized light differs from ordinary light waves which vibrate in all planes at right angles to the direction of wave propagation.

- **Diffraction:**

The bending of light rays (waves) while emerging through narrow aperture or around edges of opaque obstacles or objects is called diffraction.

- **Refraction:**

The change in the velocity and direction of a light wave passing from one medium to another having different densities is called refraction.

- **Magnification:**

It is the ratio of the apparent size of the object as seen in its image under the microscope and the actual size of the object.

$$\text{Magnification} = \frac{\text{Apparent size of the object (image)}}{\text{Actual size of the object}}$$

or

$$= \frac{\text{Distance of the image from the lens}}{\text{Distance of the object from the lens}} \quad \frac{(v)}{(u)}$$

- **Magnifying power of microscope:**

$$M = \frac{\text{Angle subtended by image at eye}}{\text{Angle subtended by object at eye, when placed at the least distance of distinct vision.}} = \frac{\beta}{\alpha}$$

- **Resolving power or Resolution:**

**The capacity or power of an optical instrument or microscope to produce distinctly separate images of two closely spaced points of the specimen or objects is called resolving power of that instrument.**

(*i*) There is a little difference between resolution and resolving power. **Resolution** stands for actual details obtained in image of a given specimen and **resolving power** is the theoretically calculated capacity of the optical instrument.

(*ii*) A reduction in the resolving power of an optical instrument or microscope by stopping down the lens aperture can result in improved resolution under proper conditions.

The process of separation of two very close objects is called resolution. The ability of an optical instrument to produce separate images of such objects is known as its resolving power.

**Limit of resolution:**

The least or minimum (*d*) distance between two adjacent points in the specimen or objects when they can be seen as two distinct entities or objects by an optical instrument is called "limit of resolution" of that instrument.

Limit of resolution is inversely proportional to the resolving power i.e. smaller the limit of resolution of an optical instrument, greater is said to be its resolving power.

Ernst Abbe (1876) calculated this limit.

$$d = \frac{\lambda}{2\mu \sin \alpha}$$

(*i*) where $\lambda$ is the wavelength of light, $\mu$ is the refractive index between the points and the objective and $2\alpha$ is the angle subtended by the objective in the field of view from the points.

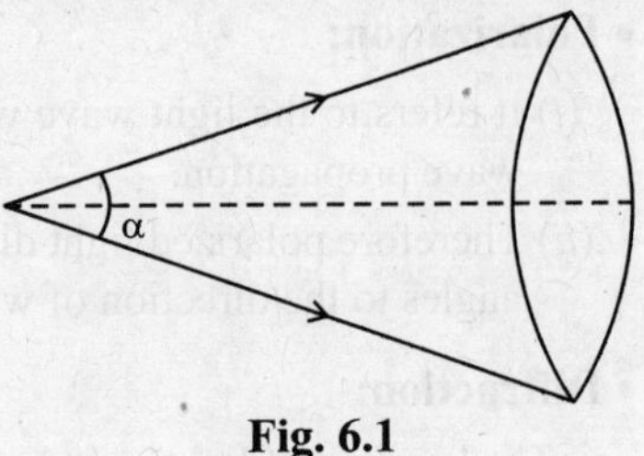


**Fig. 6.1**

(*ii*) The term $\mu \sin \alpha$ is often called numerical aperture (NA) of the objective, *i.e.* NA = $\mu \sin \alpha$. Here $\mu$ is the refractive index of the medium and $\sin \alpha$ is the sine of semiangle of aperture and $\alpha$ is the half angle of the cone of light entering the objective lens from the specimen.

(*iii*) The limit of resolution is directly **proportional** to the wavelength ($\lambda$) of the light and **inversely proportional** to **the semiangle of the light cone** accepted by the objective.

$$\text{Limit of resolution} = \frac{\lambda}{\text{semiangle of the light cone}}$$

(*iv*) If two points are more than 0.1 mm apart, our unaided eyes will be able to distinguish them as two distinct objects.

**The resolving power of human eye is 1' (minute): Explain.**

If two separate distant objects subtend an angle smaller than 1′ on our eye, then the objects will not be seen separated.

- **Power of Accommodation :**

(*i*) Our eye is a natural lens fixed in its place through muscles.

(*ii*) The ability of the eye to change the focal length of the eye lens by adjusting the curvature of the lens, for viewing near and far object, is called the **power of accommodation.**

- **Least distance of distinct vision :**

(*i*) It is the distance of an object from a normal eye, whose image is formed on the retina of the eye when the eye is not exerting its power of accommodation.

(*ii*) For a normal eye, it is about 25 cm.

- **Near point and far point:**

(*i*) The nearest point up to which objects can be seen clearly is called near point.

(*ii*) It is situated at about 25 cm from the eye.

(*iii*) The farthest point up to which objects can be clearly seen without straining the eye is called far point.

(*iv*) For a normal eye the far point is infinity.

- **Visual range:**

(*i*) The range between the near point and the far point is called visual range.

(*ii*) Objects situated anywhere in this visual range will be visible to the eye.

- **Visual angle:**

The angle which an object subtends at our eye is called visual angle.

**Microscope:**

It is an optical instrument consisting of a lens or combination of lenses which forms enlarged, magnified image of a small object.

- **Simple Microscope:**

  (*i*) It consists of a single convex lens of short focal length suitably mounted in a holder.

  (*ii*) Small objects, like small types, small prints which are not clearly visible to the eye, may be seen distinctly with the help of simple microscope.

**Parts of Simple Microscope (Instrumentation)**

**(A) Mechanical parts:**

**(*a*) Base:**

(*i*) It is the basal part of the instrument.

(*ii*) It is made of heavy metal

(*iii*) V or U shaped structure.

**(*b*) Pillar:**

(*i*) It stands vertically on base.

(*ii*) It permits the microscope to be tipped back to any degree desired by the observer.

**(*c*) Draw tube:**

(*i*) It is situated within the pillar.

(*ii*) It may be drawn upward and downward by means of **rack** and **pinion**.

**(*d*) Stage:**

(*i*) This is the part of the microscope on which the object to be examined is placed.

(*ii*) A round hole is present at the middle.

**(*e*) Folded arm:**

(*i*) It is attached with the draw tube.

(*ii*) A ring-shaped structure present at the anterior side which keeps the eyepiece.

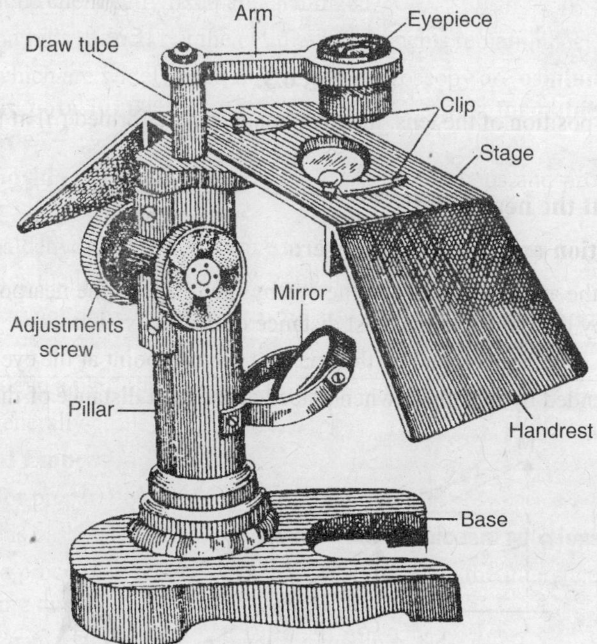


**Fig. 6.2** Diagram of a simple microscope

**(B) Optical parts:**

**(a) Eyepiece:**

(i) It is a hollow tube.

(ii) A single convex lens or a combination of lenses which functions as a convex lens placed in front of the tube.

(iii) It magnifies the object and also helps to produce magnified image of the object.

**(b) Mirror:**

(i) Here a convex lens is mounted in a frame.

(ii) The frame may be of small size.

**Working Procedure:**

(i) A simple microscope is a converging lens of suitable focal length.

(ii) If the object is placed between the principal focus (F) and the optic centre for the convex lens, it gives a virtual, erect and magnified image (IM) on the same side.

(iii) The position of the image depends upon the position of the object with respect to the lens.

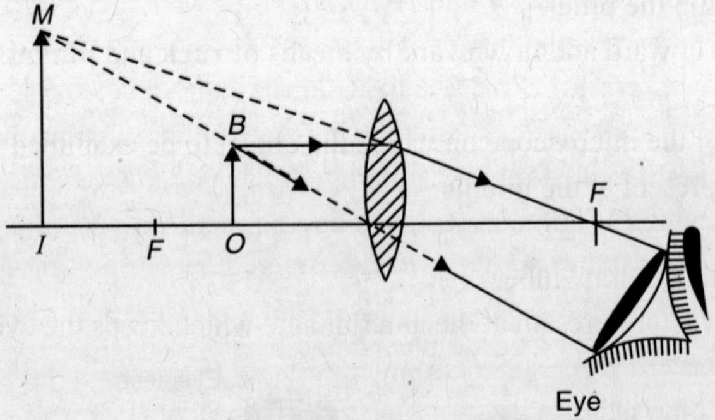


**Fig. 6.3.**

By adjusting the position of the lens, the final image can be formed (a) **at the near point** or (b) **at infinity.**

**(a) Image formed at the near point:**

**Angular magnification or magnifying power:**

It is the ratio of the angle subtended at the eye by the image at the nearpoint, to the angle subtended and unaided by the object, at the least distance of distinct vision.

$$M = \frac{\text{Angle subtended by the image at the nearpoint at the eye}}{\text{Angle subtended by the object when placed at the least distance of distinct vision}} = \frac{\beta}{\alpha}$$

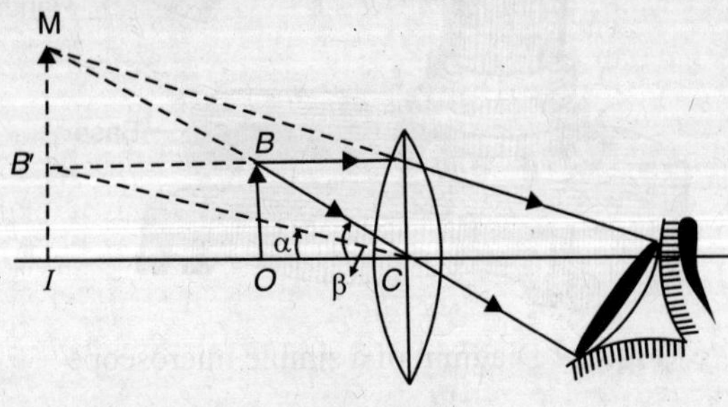


**Fig. 6.4**

In the figure, the image is formed at the nearpoint. Let the angle subtended by the image at the nearpoint at the eye be β. If the object is kept at the nearpoint [marked as IB′ in Fig. 6.4], then the angle subtended by the object at the eye is α.

From the triangle IMC, $\tan\beta = \frac{IM}{IC}$. Since β is small, $\beta = \frac{IM}{IC}$.

From the triangle IB′C $\tan\alpha = \frac{IB'}{IC}$. Since α is small, $\alpha = \frac{IB'}{IC}$.

$$M = \frac{IM}{IC} \times \frac{IC}{IB'}. \text{ But } IB' = OB$$

$$M = \frac{IM}{OB}. \text{ But } \frac{IM}{OB} = \text{Linear magnification}$$

Hence when the simple microscope is adjusted such that the image is formed at the nearpoint, the angular magnification is equal to the linear magnification. Applying new Cartesian sign convention, $u$ is –ve and $f$ is +ve.

The lens equation $\frac{1}{v} - \frac{1}{u} = \frac{1}{f}$

$$\frac{v}{v} - \frac{v}{u} = \frac{v}{f} \quad \text{or} \quad 1 - \frac{v}{u} = \frac{v}{f} \quad \text{or} \quad 1 - M = \frac{D}{f} \qquad [\text{where } v = -D]$$

or $$M = 1 + \frac{D}{f}$$

This adjustment in which the image is formed at the nearpoint of the eye and the eye is accommodated is called the normal arrangement of a microscope.

$$M = 1 + \frac{D}{f}$$

I. The magnifying power depends upon the focal length '$f$' of the lens. The *smaller the focal length, the greater will be the magnifying power*.

II. For a normal eye, the distance of the nearpoint, $D = 25$ cm. $\therefore m = 1 + \frac{25}{f}$.

All persons have not the same nearpoint distance of 25 cm. So, a simple microscope may have different magnification for different persons.

III. If eye is placed behind the lens at a distance '$d$', then $M = \frac{D-d}{f}$.

**(b) Image formed at infinity:**

I. In most cases, the object is placed at the focus and the image is formed at infinity. This adjustment is made, because if the image is formed at $D$, the eye has to exert its maximum power of accommodation.

II. For a relaxed eye,the far point is at infinity, so that the object must be kept at the focus of the lens.

III. When object OB is kept at the principal focus $f$ of the lens, the focal image is formed at the infinity. (Fig. 6.5)

IV. The visual angle subtended at the eye by the image is given by

$$\beta = \frac{OB}{CF} = \frac{OB}{f}; \alpha = \frac{OB}{D}$$

$$\text{Magnifying power} = M = \frac{\beta}{\alpha} = \frac{OB}{f} \times \frac{D}{OB} = \frac{D}{f}$$

$$\boxed{M = \frac{D}{f}}$$

For normal eye, the least distance of distinct vision

$$D = 25 \text{ cm}; \quad \therefore \quad M = \frac{25}{f}$$

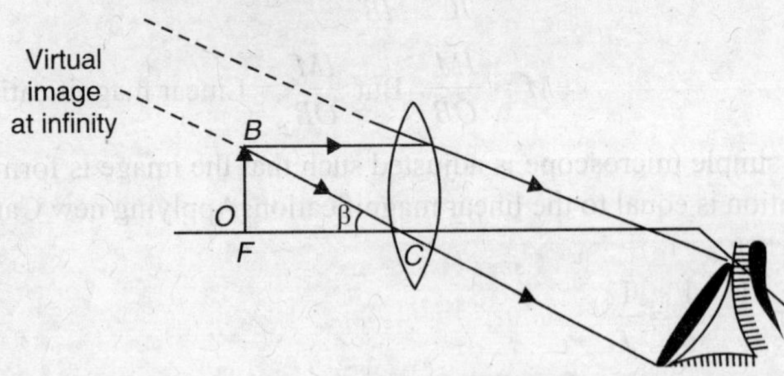


Fig. 6.5

**Uses:**

I. A magnifying glass is used to magnify the **vernier readings** in a microscope or **spectrometer** while performing the experiment.

II. It is also used by the **palmist** to examine the **palm lines.**

III. The magnifying glass (simple microscope) is used by the **watch makers** and **jewellers** to have magnified view of the **tiny components** of the various parts of watches or fine work on jewellery.

IV. A magnifer is used by a **philatelist** to examine the printing details on a stamp.

- **Compound Microscope:**

If the object under examination is very small, the simple microscope can not magnify it sufficiently. Therefore, for getting higher magnification and to get rid of several aberrations of a lens system, a powerful instrument i.e. a compound microscope is used.

**Compound microscope:**

(*i*) It consists of a stand, a stage to hold the specimen, a movable body tube.

(*ii*) It is known as compound microscope because there are two sets of lens, *i.e.* eyepiece and objective.

**Instrumentation:**

**(A) Mechanical parts:**

(*a*) **Base:**

(*i*) It is a supporting stand resting on the table.

(*ii*) It bears weight of the microscope.

(*b*) **Pillar:**

It stands vertically on base.

(*c*) **Arm:**

(*i*) It is a curved, solid piece.

(*ii*) It is movably adjusted with the base at the inclination joint.

(*d*) **Stage:**

(*i*) It is a kind of rectangular platform.

(*ii*) It has a circular hole at the centre.

(*iii*) It is made up of highly polished metal and provided with two spring clips for holding the slide tightly in one position only.

(*iv*) A movable or fixed substage is attached below the stage.

(*e*) **Body tube:**

(*i*) It is a metal tube blackened inside.

(*ii*) A revolving nosepiece carrying objective is attached to the lower end.

(*f*) **Coarse and fine adjustment screw:**

(*i*) The tube is vertically movable with the coarse and fine adjustment screws on the arm.

(*ii*) The coarse adjustment screw moves the tube rapidly while fine adjustment screw does it gradually.

(*g*) **Draw tube:**

(*i*) The upper part of the body tube is called draw tube.

(*ii*) Its length can be adjusted.

**(C) Optical parts:**

(*a*) **Nosepiece:**

(*i*) It is around metal disc like structure.

(*ii*) It is provided with two or three holes.

(*b*) **Objective:**

(*i*) The lens system which is nearest to the specimen is called objective.

(*ii*) It magnifies the specimen a definite number of times.

(*iii*) There are three objectives viz low power (usually 10X), high power (usually 40X) and an oil lens (100X).

(*iv*) There are three types of objectives viz. (*a*) **achromatic,** simplest in construction and adequate for most purposes ; (*b*) **fluorite** which can eliminate aberrations; and (*c*) **apochromatic**—aberrations are corrected by use of it.

(*c*) **Eyepiece:**

(*i*) The second lens system which is nearest to the eye is called eyepiece.

(*ii*) It magnifies the real image of the object as formed by the objective and corrects some of the defects of the objective.

(*iii*) There are three types of eyepieces viz (*a*) Huygenian eyepiece, (*b*) compensating eyepiece, and (*c*) hyperplane eyepiece.

(*d*) **Condenser:**

(*i*) It is fitted below the stage and provided with an iris **diaphragm**.

(*ii*) The opening of the diaphragm can be reduced or even closed completely.

(*iii*) It consists of two or more lenses.

(*iv*) It receives parallel rays of light from mirror and converges and focuses the light into the plane of object (specimen) because the brightness of light is essential for better **resolution**.

(*e*) **Mirror:**

(*i*) A plano concave round mirror is fitted below the stage at some distance from it.

(*ii*) In artificial light plane surface is used, in sunlight or skylight concave surface is used.

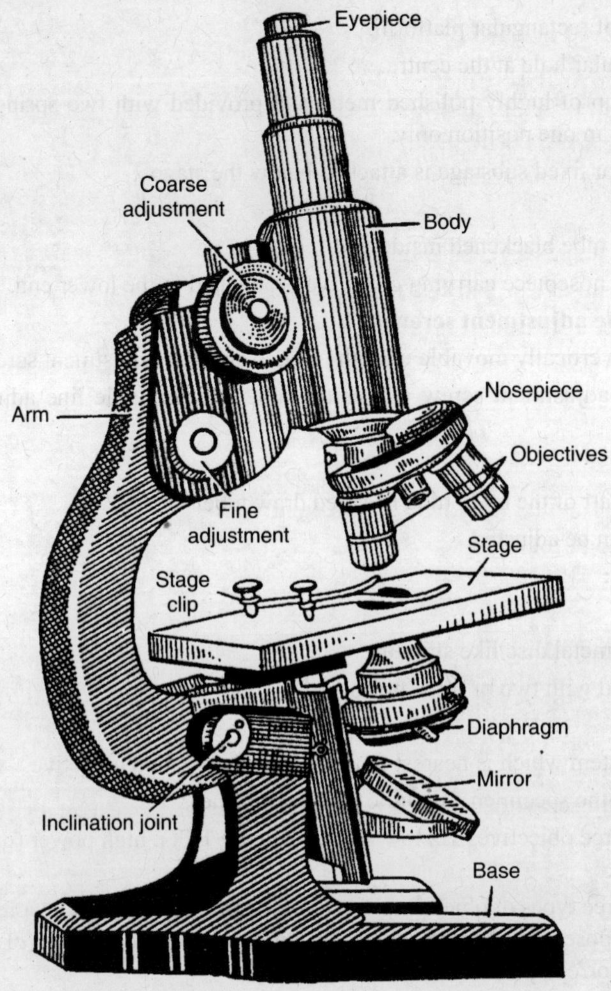


Fig. 6.6 Diagram of a microscope

**Working procedure:**

(*i*) The object (OB) is placed just beyond the focus ($F_o$) of the objective which forms a real inverted and magnified image.

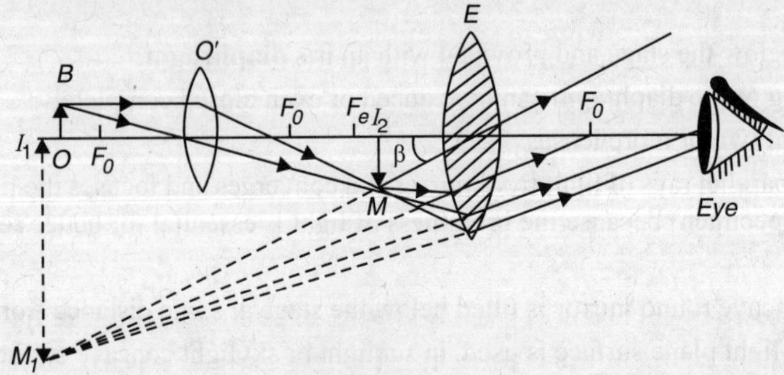


Fig. 6.7

(*ii*) This image is formed within the focus of the eyepiece ($F_e$) which produces a virtual erect and magnified image.

(*iii*) The final image ($I_1M_1$) is adjusted to be formed at the least distance of distinct vision.

(*iv*) The eyepiece acts as a simple microscope or a magnifier.

**Magnifying power or angular magnification:**

**The magnifying power (*M*) of a compound microscope is defined as the ratio of the angle subtended at the eye by the final image ($I_1M_1$) at the least distance of distinct vision (*D*) to the angle (α) subtended at the eye by the object at the same distance *D*.**

$$M = \frac{\beta}{\alpha}$$

- **The final Image is formed at least distance D of distinct vision:**

**Calculation of magnifying power:**

(*i*) When the microscope is in normal use, the final image ($I_1M_1$) is formed at the least distance (*D*) of distinct vision from the eye (Fig. 6.7)

(*ii*) The visual angle (β) subtended by the image ($I_1M_1$) is then given by $\beta = \frac{I_1M_1}{D}$

(*iii*) With the unaided eye, the object subtends a visual angle given by $\alpha = \frac{OB}{D}$

$$\text{Magnification (M)} = \frac{\beta}{\alpha} = \frac{\tan\beta}{\tan\alpha} = \frac{I_1M_1}{D}\Big/\frac{OB}{D} = \frac{I_1M_1}{D} \times \frac{D}{OB}$$

$$= \frac{I_1M_1}{OB}$$

Now $\frac{I_1M_1}{OB}$ can be written as $\frac{I_1M_1}{IM} \times \frac{IM}{OB}$ [where *IM* is the height of intermediate image].

$$\therefore \quad M = \frac{I_1M_1}{IM} \times \frac{IM}{OB} \qquad \text{....(1)}$$

(*iv*) Let the distance of the object *OB* from the objective O′ be $u_0$ and the distance of the image *IM* from the objective $v_0$. Then the linear magnification of the objective $M_0 = \frac{v_0}{u_0} = \frac{IM}{OB}$.

(*v*) The eyepiece is a magnifier. The linear magnification of the eyepiece ($M_e$).

$$M_e = \frac{I_1M_1}{IM} = 1 + \frac{D}{f_e}$$

The magnifying power of the compound microscope,

$$M = \frac{I_1M_1}{IM} \times \frac{IM}{OB} = M_e \times M_o$$

(*vi*) The overall magnifying power of the compound microscope is the product of the magnification produced by the objective $M_o$ and the magnification produced by the eyepiece $M_e$.

$$M = \frac{v_0}{u_0}\left(1 + \frac{D}{f_e}\right)$$

(*vii*) Applying lens equation i.e. $\frac{1}{v}-\frac{1}{u}=\frac{1}{f}$;

By applying new Cartesian sign convention, $u_0$ is –ve, $v_0$ is = +ve

$$\frac{1}{v_0}+\frac{1}{u_0}=\frac{1}{f_0};\ \frac{v_0}{v_0}+\frac{v_0}{u_0}=\frac{v_0}{f}$$

or

$$\frac{v_0}{u_0}=\left(\frac{v_0}{f}-1\right)$$

$$M=\left(1+\frac{D}{f_e}\right)\times\frac{v_0}{u_0}$$

or

$$M=\left(1+\frac{D}{f_e}\right)\times\left(\frac{v_0}{f_0}-1\right)\quad\left(\because\frac{v_0}{u_0}=\frac{v_0}{f_0}-1\right)$$

The object is kept very close to the principal focus of the objective. So $u_0$ is nearly equal to $f_0$. The focal length of the eyepiece is small. So $v_0$ is nearly equal to the distance between the objective and the eyepiece.

$L$ = length of the microscope *i.e.* $u_0 = f_0$

$v_0$ = L

$$M=\frac{v_0}{u_0}\left(1+\frac{D}{f_e}\right)=\frac{L}{f_0}\left(1+\frac{D}{f_e}\right)\quad [f_0 \text{ and } f_e \text{ are small but } f_e > f_0]$$

L is called optical length of the microscope.

The magnifying power $M$ depends upon

**(*a*) the focal length $f_0$ and $f_e$ of the objective**

**(*b*) Eyepiece.**

**(*c*) $M$ is large (*i*) if $f_0$ and $f_e$ are small, (*ii*) when the object is kept very close to the first focus of the objective.**

- **Final image is formed at infinity:**

A compound microscope can be used with the final image formed at infinity which undoubtedly is not the normal use of the microscope.

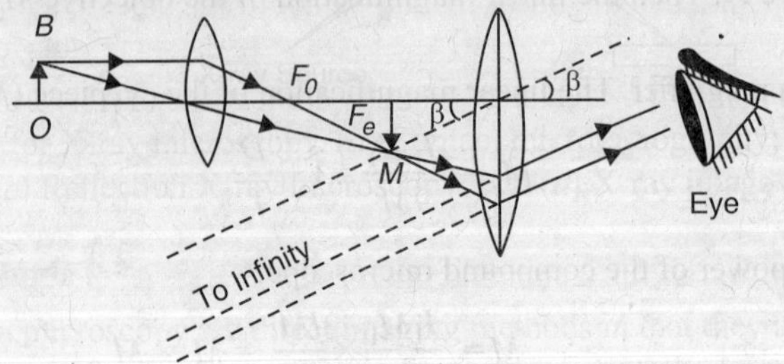


**Fig. 6.8**

I. In this case, the eye is unaccommodated or at rest.

II. The image of the objective must be formed at the focus of the eyepiece.

III. The eyepiece can be so adjusted that the final image is formed at the infinity. Then the magnifying power of the eyepiece becomes $\frac{D}{f_e}$.

IV. The magnifying power of the compound microscope is $M$.

$$M = \frac{v_0}{u_0}\left(\frac{D}{f_e}\right) = \frac{L}{f_0} \times \frac{D}{f_e}$$

**To increase the magnifying power of the microscope:**

(*a*) The focal length of the objective should be small.

(*b*) Object should be placed very near to the first focus of the objective.

(*c*) Focal length of the eyepiece should be small.

- **Light microscope:**

I. The part of the microscope that holds all the components firmly in position is called the stand.

II. There are two basic types of compound light microscope stand—*an upright* or an *inverted microscope*.

**(*a*) Upright microscope:**

(*i*) The light source is below the condenser lens.

(*ii*) The objectives are above the specimen stage.

(*iii*) This is most commonly used format for viewing specimens.

**(*b*) Inverted microscope:**

(*i*) The light source and the condenser lens are above the specimen stage.

(*ii*) The objective lens is beneath the specimen stage.

(*iii*) Moreover, the condenser and light source can often swing out of the light path.

- **Stereomicroscope:**

I. It is a kind of light microscope which is used for the observation of the surfaces of large specimens.

II. This microscope is used when 3D information is required.

III. It is used for routine observation of whole organisms, for screening through vials of fruit flies.

IV. Useful for micromanipulation and dissection.

- **Empty magnification :**

I. Eyepieces are usually magnified by 10X since an eyepiece of higher magnification merely enlarges the image with no improvement in resolution.

II. There is an upper boundary to the useful magnification of the collection of lenses in a microscope.

III. For each objective lens the magnification can be increased above a point where it is impossible to resolve any more detail in the specimen.

IV. Any magnification above this point is called **empty magnification.**

- **Entrance pupil:**

The clear area of the objective is called entrance pupil.

- **Exit pupil:**

The light rays leaving from all the points of the objective after refraction through the eyepiece pass through a small circular area. This is called the **exit pupil** or **eye ring R.**

This is the best position where the eye should be placed so that it collects as much light that passes through the objective as possible.

- **Magnifying power of an optical instrument:**
  It is the ratio of the angle subtended by the final image at the eye to the angle subtended by the distant object at the eye when seen directly.
- **Visual angle:**
  It is the angle subtended by an object at the human eye.

## Aberrations in Lens

The lens system of microscope may suffer from several inherent optical defects called aberrations. Such aberrations cause defective image formations.

(*i*) Light rays from the object which are very close to the principal axis are called **paraxial rays.** The light rays coming from object which fall near the edges of the lens are called **peripheral** or **marginal rays** {Figs. 6.9 ($a$ & $b$)]

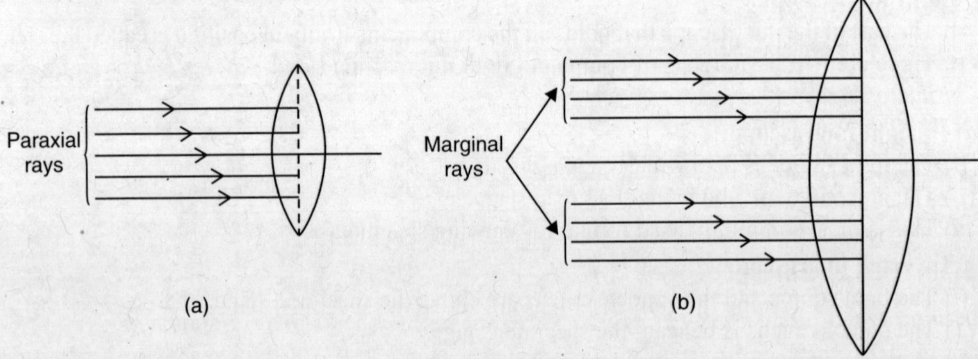


**Fig. 6.9**

(*ii*) In general, non-paraxial rays from an object after refraction through a lens do not converge at a single point. Not only that, the refractive index and focal length of a lens are different for different wavelengths of light.

(*iii*) If the light coming from an object is not monochromatic, different wavelengths corresponding to different colours will form many coloured images of varying sizes.

These deviations from actual size, shape, position and colour of the image produced by a lens are called **aberrations.**

### (A) Spherical aberration

(*i*) It occurs when a wide parallel beam of monochromatic light, parallel to the principal axis, is incident on a lens, the paraxial rays, after refraction, are brought to a focus farthest ($F_p$) while the marginal rays are brought to focus ($F_m$) closer to the lens.

(*ii*) There is no sharp point image of a distant point object. The image is blurred and distorted. This defect is called *lateral spherical aberrations.*

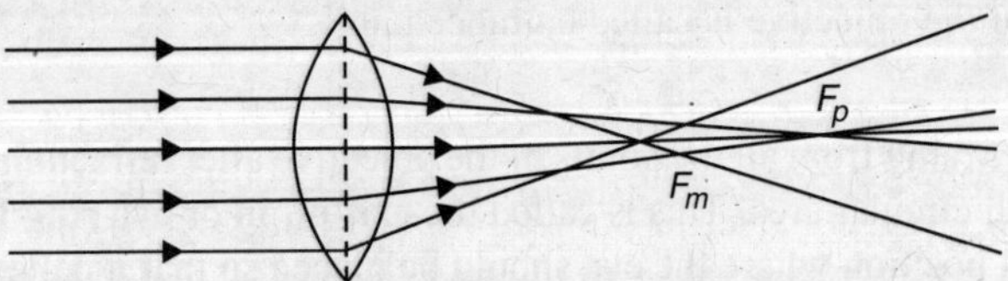


**Fig. 6.10** Spherical aberration

(*iii*) The distance between $F_m$ and $F_p$ is known as *longitudinal spherical aberration.*

(*iv*) If the screen is moved between $F_m$ and $F_P$, a position of the screen is obtained where the image is the best reproduction of object. This is known as circle of least confusion. The radius of circle of least confusion is called *lateral spherical aberration.*

**Correction**

(*i*) By using stops.

(*ii*) By using lenses of large focal length.

(*iii*) By using compound lens systems, each being a combination of concave and convex lens.

(*iv*) By using two planoconvex lens.

**(B) Chromatic aberration:**

The image of a white object (or an object illuminated by white light) formed by a lens is usually coloured and blurred. This defect of the image produced by a lens is called **"Chromatic aberration".**

(*i*) This type of aberration is caused due to the variation in wavelengths of light rays. The focal length of a lens is different for different wavelength of light rays.

(*ii*) For example, the focal lengths of red and violet light are different. The longer wavelength of red light comes to a focus farther from the lens than violet light which has a shorter wavelength.

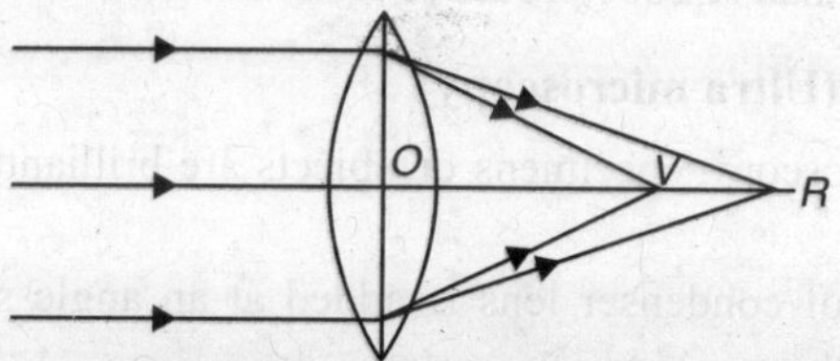


**Fig. 6.11** Chromatic aberration

(*iii*) Therefore we get **coloured blurred image**. This optical defect is called **Chromatic aberration.**

(*iv*) The formation of images of different colours at different positions is called **longitudinal chromatic aberration**.

(*v*) As the different colours are focussed at different places, the magnification produced by the lens also varies from different colours. Therefore the images of different colours are magnified by different amounts. This optical defect is called *lateral chromatic aberration.*

**Corrections.** The method of reducing chromatic aberration is called **achromatisation.**

It is achieved by

(*i*) an achromatic doublet.

(*ii*) using two convex lenses separated by a distance.

(*C*) **Astigmatism:**

If the curvature of the lens is uneven, the image formation becomes defective. This type of optical defect is known as **astigmatism**.

(*i*) This defect is removed by using cylindrical lens.

(*ii*) Sometimes it is associated with myopia or *hypermetropia*, then spherocylindrical glasses are used to correct the defect.

- **What is meant by achromatic combination of lens?**

A combination of two or more lenses which produces images of different colours of the same object at the same position and of the same size is called achromatic combination of lens.

- **What is the cause of chromatic aberration?**
  (*i*) Due to the variation of refractive index with wavelength of light, the focal length of a lens is different for different colours of light.
  (*ii*) Hence for the same position of an object images of different colours are formed at different positions.

- **Bright Field Microscopy:**
  (*i*) Bright field microscopy is obtained in the normal method of illuminating the object.
  (*ii*) Here thin, more or less transparent samples (microorganisms) are viewed.
  (*iii*) In this type of microscopy, the microscope field is bright. The light is transmitted through the sample and the image is formed by the absorption of this light.
  (*iv*) Therefore the image will appear darker than the background which is bright field.
  (*v*) Normally microorganisms do not absorb much light, but staining them with dye greatly increases their absorbing abilities.
  (*vi*) Generally bright field microscopes produce useful magnification of about X 1000 to X 2000. Magnification greater than X 2000 produces fuzzy images.

- **Dark Field Microscopy (Ultra microscopy)**
  (*i*) In the dark field microscopy, specimens or objects are brilliantly illuminated under dark or black background.
  (*ii*) Here, a special type of condenser lens is added at an angle so that the rays of light fall obliquely on the object.
  (*iii*) Only the light refracted (bent) or diffracted (scattered) by the specimen (object) is collected by the objective and sent on to the eyepiece.
  (*iv*) An opaque disc covers the centre of the undersurface of the condenser to prevent any light from entering its central part and allows only bright annularring of light to enter the periphery of the condenser.
  (*v*) These light rays are then focussed in obliquely horizontal directions from the peripheral parts of the condenser lens to the object on the slide above.

- **Polarising Microscopy**

In this microscopy, plane polarised light is used in order to study the birefringent anisotropic structures under dark background produced by the isotropic structure.

**Principle:**

(*i*) The polarising microscope is based on the behaviour of some cell components (**isotropic** or **anisotropic**) when observed under polarised light.

(*ii*) **Isotropic substances** have their molecules randomly oriented at different directions. Polarised light is propagated through this substance with the same velocity irrespective of the impinging direction. It has **identical refractive index**.

(*iii*) **Anisotropic substances** have their well-oriented distribution of molecules. Polarised light is propagated through this substance with varied velocities. They have **different indices of refraction**.

**Birefringence:**

A material which displays two different indices of refraction correspond with two different velocities of transmission is termed birefriengence.

(*i*) Refractive index ($n$) of a substance is defined as the ratio of the velocity of light in vacuum ($c$) to the wave velocity ($v$) of light in that substance (medium). Thus

$$n = \frac{c \text{ (Velocity of light in vacuum)}}{v \text{ (Velocity of light in medium)}}$$

(*ii*) For *ordinary ray* ($o$) its wave velocity ($v_0$) is he same in all directions and for all orientations of the optic axis of the crystal. Hence *refractive index.*

$n_0 = \frac{c}{v_0}$ for ordinary ray is constant for the crystal.

(*iii*) For *extraordinary ray* ($E$) The wave velocity ($v_e$) is different in different directions and consequently the refractive index of the crystal for this $E$-ray will be different in different direction.

$n_e = \frac{c}{v_e}$. It is also known as **principal refractive index** of $E$-ray.

[$v_e$ = wave velocity of $E$-ray along the direction perpendicular to optic axis.]

(*iv*) Along the optic axis both $E$-rays and $O$-rays travel with the same velocity and hence have the same refractive index in this direction.

(*v*) The wave velocity of $E$-ray differs from its ray velocity but they become equal along and perpendicular to the optic axis.

$n_o$ = refractive index for ordinary ($O$) rays.

$n_e$ = Principal refractive index for extraordinary ($E$) rays.

The quantity (birefringence) i.e. $\Delta n = n_e - n_o$.

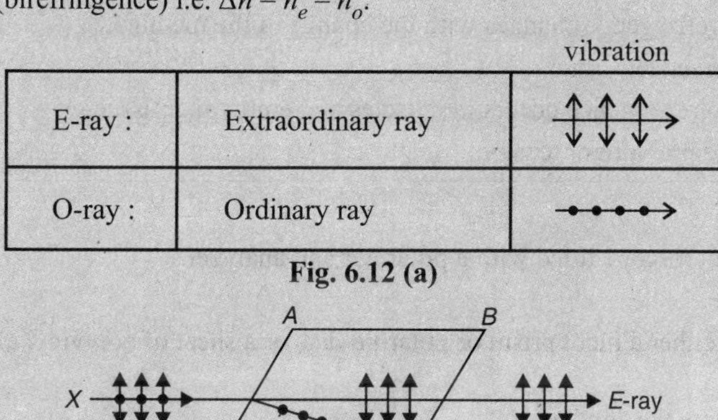

| | | vibration |
|---|---|---|
| E-ray : | Extraordinary ray | |
| O-ray : | Ordinary ray | |

Fig. 6.12 (a)

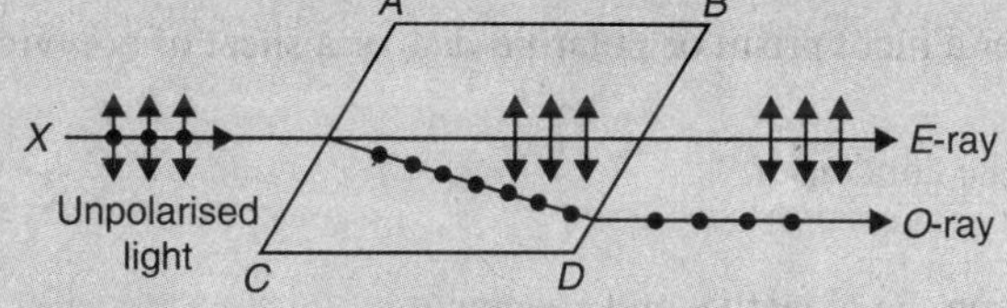


Fig. 6.12 (b)

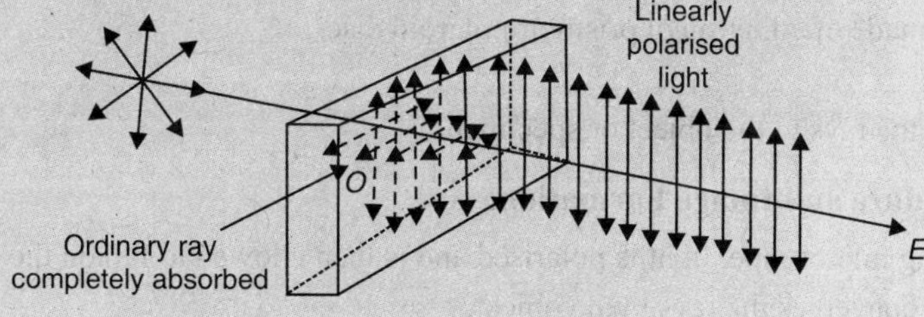


Fig. 6.13

**Types:**

Birefringence → (*a*) Intrinsic (crystalline) birefringence.
→ (*b*) Form birefringence.
→ (*c*) Strain birefringence.

**(A) Intrinsic (Crystalline) Birefringence**:

(*i*) It is the property of material itself.

(*ii*) It is due to the regular asymmetrical arrangement of the ions or molecules.

(*iii*) It is always independent of the refractive index of the medium.

(*a*) **Positive birefringence:**

(*i*) Here the index of refraction is greater along the length of the fibre than the perpendicular plane.

**Example:**

(*ii*) Proteins, carbohydrate and crystals made of long molecules.

(*b*) **Negative birefringence:**

(*i*) Here the index of refraction is greater along the perpendicular plane than along the length of the fibre.

(*ii*) *Example:* Fat molecules, nucleic acids and nucleoproteins.

**(B) Form birefringence:**

(*i*) Here the particles have single (one) refractive index oriented in the medium having different refractive indexes.

(*ii*) Therefore, birefringence changes with the change of the medium.

**(c) Strain birefringence:**

(*i*) This type of birefringence occurs in muscles and embryonic tissue.

(*ii*) It is due to the pressure or tension.

**Instrumentation:**

(*i*) It is a light microscope fitted with a polarizer and analyzer.

(*a*) **Polarizer**

(*i*) It is made of either a nicol prism or polaroid disc or a sheet of polyvinyl alcohol stained with iodine.

(*ii*) It is fitted below the condenser.

(*b*) **Analyzer**

(*i*) It is placed between the objective and eyepiece.

(*ii*) It is used to analyse the direction of polarization of the beam.

(*iii*) It may be made of either nicol prism or polaroid disc.

(*c*) **Condenser:**

It converges the rays to the object or specimen.

**Working Procedure and Image Formation:**

(*i*) In polarising microscope, light is polarised and is then allowed to fall on the object.

(*ii*) Condenser converges the rays to the object.

(*iii*) The polarised light suffers birefringence due to the anisotropic properties of the object (specimen) and gets scattered.

(*iv*) From the object, light passes through the analyzer which is placed just above.

(*v*) If now the object is viewed through the **analyzer** placed perpendicular to the **polarizer**, all the direct light is cut off and only the light from the object will pass through (only the component of the light will pass through) and **birefringent object** is clearly visible.

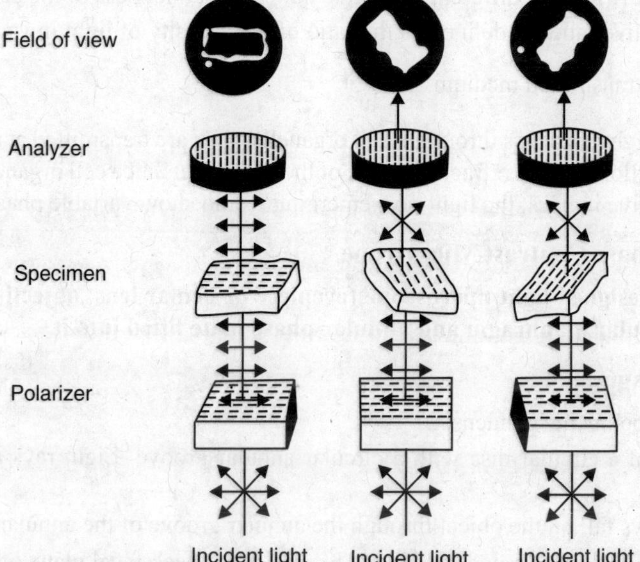


**Fig. 6.14** Diagram to show variations in darkness and brightness of an anisotropic object when placed between crossed polarizer and analyzer and rotated by ± 45°.

(*vi*) The image of the anisotropic object is thus visible on a dark background field.

(*vii*) By rotating the object to a specific direction the change in the intensity of the image can be measured.

## Applications:

(*i*) It is used for both qualitative and quantitative studies.

(*ii*) It is used in histology, anatomy, pathology for the study of qualitative nature of orientation of molecules.

(*iii*) A polarisation microscope is used to study the shape regularity of internal construction and refractive index in cell and tissues.

(*iv*) Applications of polarising microscope in Biology include the study of cell spindle formation and the behaviour of the components of contracting muscle.

## • Phase Contrast Microscope

(*i*) Phase contrast microscope was discovered by Professor Frits Zernicke. He was awarded Nobel Prize for physics in recognition of his discovery of phase contrast.

(*ii*) Phase contrast microscopy is an important instrument for studying living unstained and transparent objects and is widely used in biological and medical research.

## Principle:

(*i*) According to this principle, light waves have variable character for frequency and amplitude. Human eyes cannot perceive a phenomenon when two light rays have similar amplitude and frequency but different phases. (Fig. 6.15) (*ii*) Phase changes are caused by the biological materials through which light rays pass.

(*a*) If material is absorbent, it causes the rays to undergo a change in amplitude. These are perceptible to our eyes.

(*b*) Similarly when light rays pass through a living cell they undergo invisible phase changes (Fig. 6.16) due to different refractive indices and thickness of the cell organelles. (Refractive index is defined as the ratio of the velocity of light in vacuum to its velocity in the transmitted medium. $n = \frac{c}{v}$)

(*c*) When light rays pass through a cell organelle, they are transmitted at a velocity inversely proportional to the refractive index of the organelle. Since cell organelles are of different refractive indices, the light rays emerging would show variable phase changes.

## • Structure of Phase Contrast Microscope

**It is a specially designed light microscope (eyepiece or ocular lens, objective lens, condenser, mirror) with annular diaphragm and annular phase plate fitted into it.**

**Annular diaphragm:**

(*i*) It is placed below the condenser.

(*ii*) It consists of a circular disc with a circular annular groove. Light rays are allowed to pass through it.

(*iii*) The light rays fall on the object through the annular groove of the annular diaphragm.

(*iv*) An image of this illuminated opening is formed at the back focal plane of the objective.

**Annular phase plate:**

(*i*) It is at the back focal plane of the objective.

(*ii*) It consists of a transparent disc with an annular groove.

(*iii*) The groove is made to coincide with the image of the aperture of the annular diaphragm.

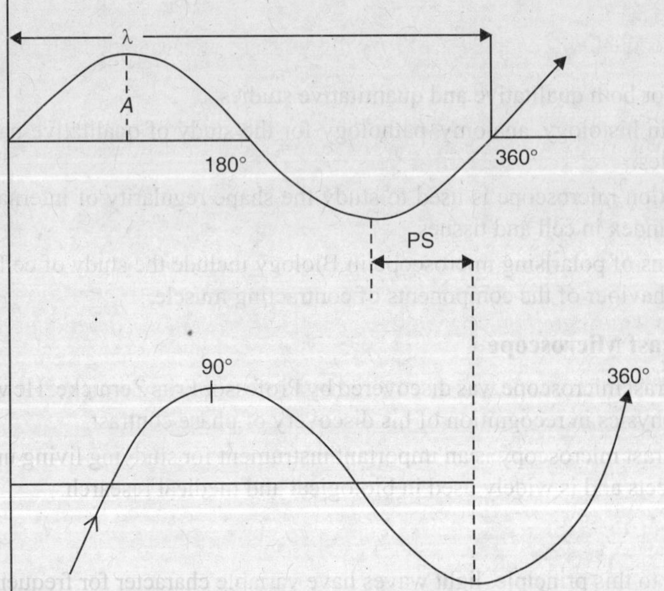


**Fig. 6.15** Light waves exhibiting similar properties of amplitude (A), and wavelength (λ) but differing from one another by being out of phase (P*s*)

(*iv*) All the direct light rays pass through the groove in the phase plate whereas the diffracted light rays pass mainly outside the groove.

(*v*) The proper selection of thickness of groove and refractive index of the phase plate material produces a desired phase difference between the direct light rays to the defracted and retarded light rays.

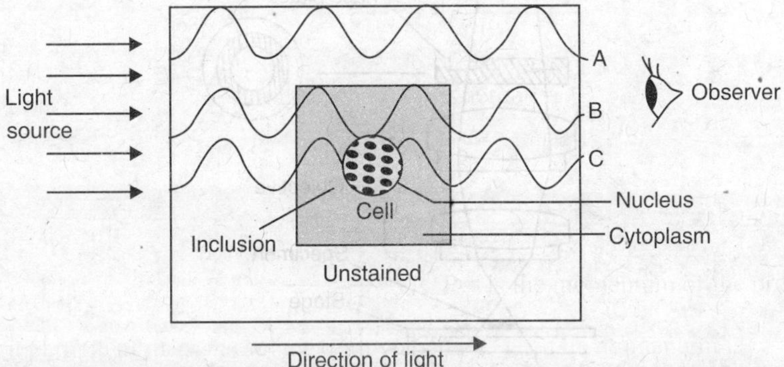


**Fig. 6.16** The effect on light rays when they pass through the various cell components of a living cell. (A) No cell component. (B) Cytoplasm. (C) Nucleus. The light rays get retarded (phase change) while passing through the cell components.

(*vi*) The diffracted light has to pass through thicker part of the phase plate than direct light, hence it is more retarded ($\lambda/4$) than direct light.

(*vii*) The resultant retardation of the diffracted rays from the specimen by $\lambda/2$ may produce phase difference ($\delta$) between the diffracted and direct light rays.

(*viii*) If the crests and troughs of two waves (diffracted light and direct light) having same magnitude coincide with each other, their path difference as well as phase difference become zero.

(*ix*) Thus the two waves and with zero phase difference will reinforce one another (*i.e.* additive superimposition of the diffracted and direct light rays) to give a resultant of **greater amplitude**. The image appears bright in comparison with dark surrounding area. This type of image formation is called **bright contrast** or **negative**. The interference, enhancing the brightness of the resultant waves, is called **constructive resonance**.

(*x*) The diffracted and direct light rays (waves) are in reverse phases with a path difference ($n$) of $\lambda/2$ i.e. peak or crest of one coincides with trough of the other and vice versa (i.e. substractive superimposition of the diffracted and direct light rays) will produce **decreased amplitude**. The image formed in this case is darker than the surrounding area and is termed as **dark** or **positive contrast**. The interference, reducing the brightness of the resultant waves, is called **destructive resonance**.

**Advantages:**

(*i*) It provides a clear image of unstained or living cells without any prior treatment or staining.

(*ii*) It also prevents the possibilities of any damage caused to the cells and altering their morphology during staining.

(*iii*) It introduces higher contrast internal fine details by optical means.

(*iv*) Prolonged observation of the living cells in their natural state can be obtained through this.

(*v*) It utilizes two beams (light) but no beam splitters.

**Disadvantages:**

There is an incomplete separation of the two light rays (direct and diffracted) resulting in the formation of halo around the object.

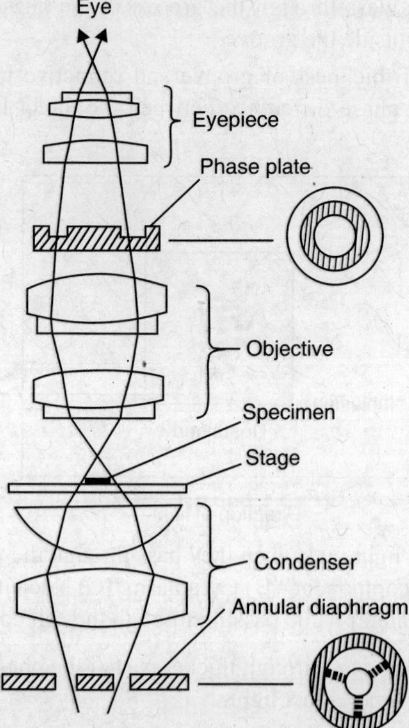


**Fig. 6.17** Path of light through a phase-contrast microscope.

**Example:**

When nucleus is bright, the halo surrounding it is dark and if nucleus is dark, it has bright halo around it.

(*ii*) It cannot create or destroy light energy. It can only redistribute the light energy. Thus one portion of the image is brighter or darker than the rest.

(*iii*) The other related effect of phase contrast microscopy is called **"zone of action"**. It refers to the heightening of the phase contrast effect near edges and discontinuities.

**Applications:**

Phase contrast microscope is an instrument for the examination of living cells without any prior treatment like fixation and staining. The applications of phase contrast microscopy are immense.

(*i*) The process of **mitosis** and **meiosis** and the duration of cell division can be accurately studied in living cells.

(*ii*) Permeability, **phagocytosis** and **pinocytosis** activity of cell membrane can be studied directly.

(*iii*) The behaviour of living protozoa towards physical and chemical factors can be studied.

(*iv*) The **cell cycle** time can be determined using time lapse cinematography.

(*v*) The effects of chemical on living cells can be studied.

- **Interference Microscope**

**The interference microscope is closely allied and improved phase contrast microscope. In this microscope the direct light wave (incident wave) and the diffracted light wave are completely separated while passing through the phase plate. This improvement increases the utility of the**

**interference microscope considerably for visualizing and measuring differences in phase or optical path in transparent or reflecting specimens.**

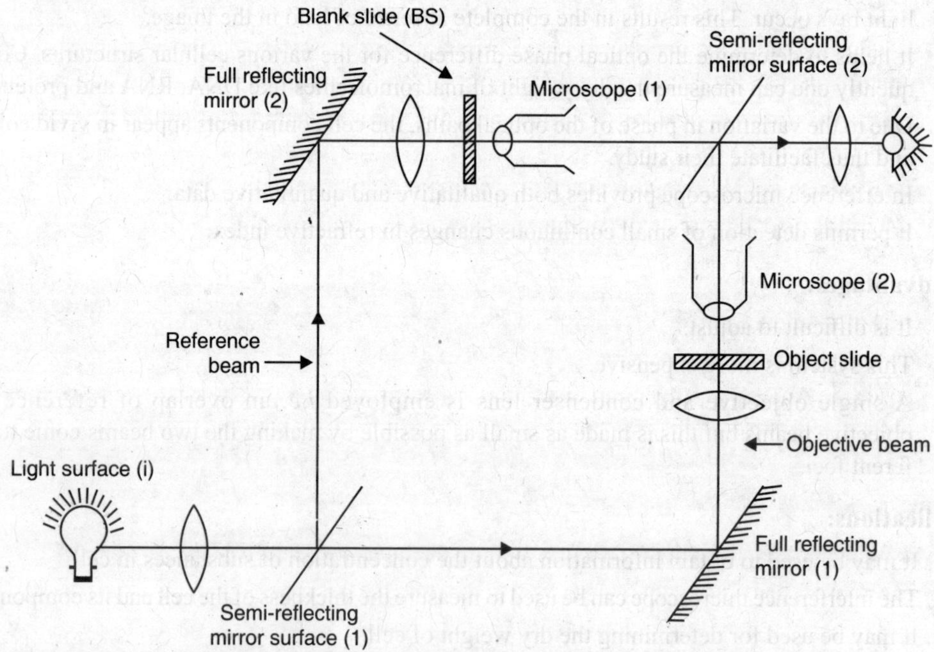


**Fig. 6.18** Schematic diagram of Interference Microscope

**Principle:**

(*i*) The light rays (waves) are splitted into two waves or beams by inserting a beam splitter below the object.

(*ii*) One beam or waves pass through the object known as **objective beam** and undergoes a phase change i.e. **diffracted** wave.

(*iii*) The other beam or waves do not pass through the object known as reference beam and undergoes no change *i.e.* direct light wave or **incident** wave.

(*iv*) The diffracted (objective wave) light waves and the direct (incident) light waves or rays fall separately on the phase plate.

(*v*) The diffracted light wave falls on the phase plate except the groove (annulus). The direct light wave falls exclusively on the groove or annulus.

(*vi*) After passing through the phase plate, the sets of light waves combine and interfere with each other. Since the two sets of light waves combine and interfere with each other in this microscope, it is called **interference microscope**.

(*vii*) The combination and interference of the two sets of light waves form the image with the aid of the ocular lens.

**Instrumentation:**

1. Semireflecting mirror *i.e.* beam splitter ($S_1$ & $S_2$)
2. Fully reflecting mirror (1) & (2).
3. Microscopic lens (1) & (2)
4. Blank slide (BS) and object slide.

**Advantages:**

1. In the interference microscope, there is a complete separation of the diffracted and direct light rays occur. This results in the complete absence of halo in the image.
2. It helps to determine the optical phase difference for the various cellular structures. Consequently one can measure the dry weight of macromolecules like DNA, RNA and proteins.
3. Due to the variation in phase of the optical paths, the cell components appear in vivid colours and thus facilitate their study.
4. Interference microscope provides both qualitative and quantitative data.
5. It permits detection of small continuous changes in refractive index.

**Disadvantages:**

1. It is difficult to adjust.
2. This system is more expensive.
3. A single objective and condenser lens is employed *i.e.* an overlap of reference and objective beams but this is made as small as possible by making the two beams come to different foci.

**Applications:**

1. It may be used to obtain information about the concentration of substances in cell.
2. The interference microscope can be used to measure the thickness of the cell and its components.
3. It may be used for determining the dry weight of cell.
4. It is also used qualitatively to provide contrast in transparent object.

- **Nomarski Interference Contrast Microscope**

It is a variation of Interference microscope.

(*i*) In this optical system a single light beam is used. The beam passes through the object and the objective.

(*ii*) The light beam then passes through a special birefringent prism and gets divided into interfering beams.

(*iii*) This interference microscope has also polarizer and analyzer filter and a compensating prism placed above the condenser.

(*iv*) This optical system makes use of the rate of change of refractive index in the cell. When the rate of change in refractive index is greatest, the contrast is the greatest.

(*v*) The edges of cells and their organelles show excellent contrast. Images produced by this optical system have a "pseudo stereoscopic" or "3 D" effect (three-dimensional).

(*vi*) It is exhibited by a dark shadow on one side of the refractile object.

(*vii*) This image has a great value in interpreting the microanatomy of transparent objects.

**Differences between Phase Contrast and Interference Microscopes**

| | Phase contrast microscope | | Interference microscope |
|---|---|---|---|
| 1. | It utilizes two light beam rays without beam splitters. | 1. | It utilizes two light beam rays with beam splitters. |
| 2. | It produces high contrast images of living cell by incomplete separation of direct and diffracted light rays. | 2. | It also produces high contrast images of living cell by complete separation of direct and diffracted light rays. |

| | | | |
|---|---|---|---|
| 3. | The image suffers from a certain defect *i.e.* a marked 'halo' around the edges. | 3. | Here the image does not suffer from such 'halo' mark. |
| 4. | It does not require beam recombiner. | 4. | This optical system requires recombiner. |
| 5. | This system provides qualitative data. | 5. | It provides both qualitative and quantitative data. |
| 6. | It does not provide dry weight of cellular structure. | 6. | It helps to determine the dry weight of cellular structure. |
| 7. | It helps in detection of sharp changes of refractive index. | 7. | It helps in detection of continuous changes of refractive index. |

- **Fluorescent Microscopy**

It is a microscope used for structural details by observing fluorescence in cells and tissues.

(*i*) Some chemicals are **fluorescent**. If the chemical is irradiated with high energy (short wavelength, *e.g.* uv rays, 290-497 nm) light and become excited to reach higher energy level, after a short interval of excitation (less than $10^{-5}$ sec), the excited electrons come back to their original low energy state *i.e.* "ground state" with simultaneous remission of specific visible light wave length. Such emission of light by excited molecule is called **fluorescence.**

(*ii*) Here the absorbed light is called **exciting light** and the emitted light is called **fluorescent light.** Fluorescent light is always of longer wavelength than exciting light.

(*iii*) The phenomenon of fluorescence is thought to involve an electronic rearrangement in the irradiated substance.

(*iv*) If the delay between absorption and emission is less than 10 seconds, the effect is called **fluorescence** and more than 10 seconds it leads to the phenomenon of **phosphorescence**.

Both the phenomena (fluorescence and phosphorescence) go under the general name of *luminescence i.e.* emission of visible light wavelength owing to the reason other than heat.

(*v*) If the emitted light comes from atoms, it has the same wavelength and identical frequency as the exciting (absorbed) light and is distinctively called **resonance fluorescence** or **resonance radiation.**

(*vi*) **Chemiluminescence** is the emission of light from a substance raised to a transient electronically "excited" state by a chemical reaction.

**Bioluminescence** is the emission of visible light from living organisms (protozoa, sponge beetle etc.).

(*vii*) Naturally inherent fluorescence properties are termed as **primary fluorescence**. Artificially induced fluorescence is called **secondary fluorescence**.

It is a kind of microscope used for structural details of cells and tissues by fluorescence.

Kohler (1904) was the first to detect fluorescence while observing tissues irradiated with ultra violet light in a microscope. **Coons et al** (1941) used fluorescent antibody technique. Nowadays fluorescent microscopy is widely used in detecting fluorescent dyes bound to cell organelles.

**Principles:**

**Fluorescence is the phenomenon by which certain substances (fluorescent) absorb light of short wavelength (uv ray) and after a very short interval of time the visible light of long wavelength is re-emitted.**

(*i*) The substance which emits fluorescence is termed **fluorophore.**

(*ii*) Ultraviolet light is actually absorbed by a fluorophore and simultaneously re-emitted as rays of longer wavelength visible range.

(*iii*) The change in wavelength from uv to the visible range is termed **Stoke's effect.**

(*iv*) The photon of uv is absorbed by the molecules of fluorophore, they get excited, absorb energy and change from ground state to an excited state.

(*v*) In this condition the energy of each molecule is higher than its normal or ground state. The excess energy either gets dissipated as heat or gets emitted as **fluorescence.**

(*vi*) Each fluorophore has characteristic spectra of excitation and emission.

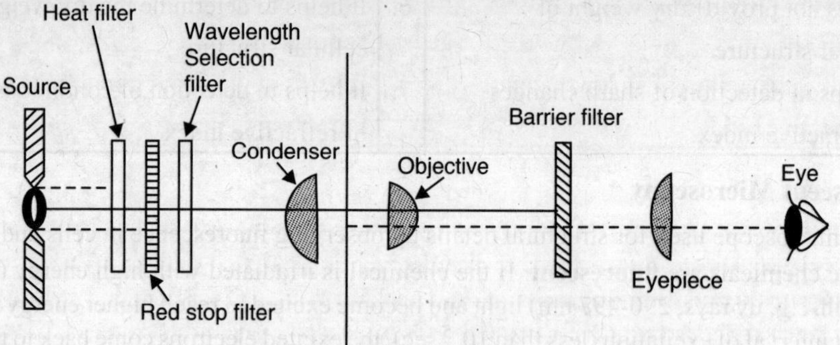


**Fig. 6.19.** Components of the fluorescent microscope.

**Instrumentation:**

The fluorescent microscope consists of the following :

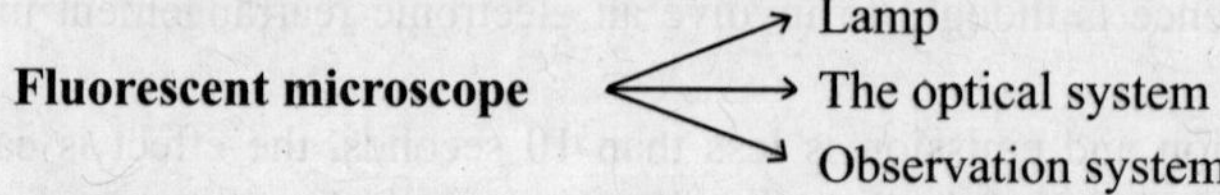


**(A) Lamp:**

(*i*) There are three types of lamps *viz* mercury arc lamps, xenon arc lamps and tungsten-halogen lamps.

(*ii*) A high pressure mercury vapour lamp or halogen lamp is most commonly used as the source of uv light and blue light respectively.

(*iii*) They give their light at a specific wavelength so that selection of fluorescent excitation is convenient.

**(B) The Optical system:**

The uv near the visible range excites the fluorophore and emits light of a longer wavelength in the visible range.

**1. Primary or Exciter Filter:**

(*i*) It is placed between the uv source and the specimen.

(*ii*) It absorbs light of a longer wavelength emitted from the uv source.

(*iii*) It gives nominally monochromatic illumination.

**2. Secondary Filter of Barrier:**

(*i*) It is located between the objective and eyepiece lenses.

(*ii*) It absorbs all short wavelength rays to darken the background field.

(*iii*) It transmits only the long wavelength fluorescent light to the eyeplece lens system.

**3. Dichronic mirror:**

(*i*) Located between the objective lens system and barrier filter.

(*ii*) It reflects short wavelength (below 500 nm) light but transmits larger wavelength (above 500 nm) light in single direction.

(*iii*) It helps to produce dark background field.

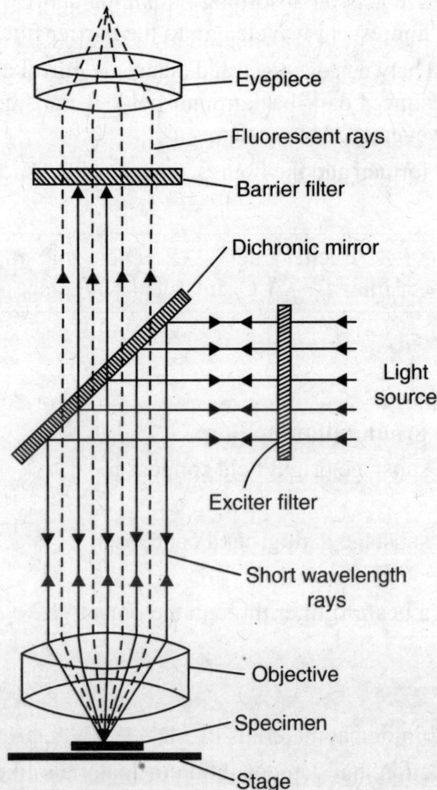


**Fig. 6.20** Path of light through an incident-light fluorescence microscope.

**(C) Observation system:**

(*i*) The objectives and oculars are used for the formation of image of fluorescence.

(*ii*) The lens system consists of quartz or fluorite ($CaF_2$) or lithium fluorite condensers. All these materials are transparent to the electromagnetic rays.

(*iii*) Strong fluorescence can be easily observed but weak fluorescence requires optimal choice of objectives of high numerical aperture.

(*iv*) Oculars of low to moderate magnification are ideal for observation in fluorescent microscopy.

(*v*) In this system, material (fluorophore) is observed against a dark-background.

**Working procedure and image formation:**

(i) Collecting lenses of the microscope transmit UV or blue light from the source to the exciter filter placed between the light source and stage bearing specimen.

(*ii*) The exciter filter pass on the UV or blue wavelength as are near the absorption maximum of fluorescence to fall on the **dichronic mirror**.

(*iii*) The dichronic mirror located between the objective lens system and barrier filter reflects back the shorter wavelength (below 500 nm) and transmits the comparatively longer wavelengths to the condenser.

(*iv*) Here the condenser converges the light rays and focusses it to the specimen which get excited by the U.V. light (short wavelength-500 nm).

(*v*) The specimen thus emits electron to the higher energy level and fluorescence light (visible area) are produced to reach the dichronic mirror again.

(*vi*) The dichronic mirror here acts on absorbing remaining short wavelength but transmits only the characteristic long fluorescent wavelength to the barrier filter.

(*vii*) The barrier filter placed between eyepiece and dichronic mirror absorbs only short wavelength harmful to eye and produces dark background field. It transmits only the long wavelength flourescent light to the eyepiece lens system.

(*viii*) Thus the final image is formed and the flourescent part of the specimen looks coloured against a dark background.

| Lamp ⟶ collecting lens ⟶ Exciter filter ⟶ Dichronic mirror ⟶ Condenser ⟶ Stage specimen ⟶ Barrier filter ⟶ Occular lens (Eyepiece) & image |
|---|

**(Pathway of light)**

**I. Illumination System:**

**1. Dia-illumination** or **dark ground illumination:**

It is obtained by using a substage bright field condenser.

**2. Oblique illumination:**

It is developed by using substage dark ground condenser.

**3. Epi-illumination:**

It is developed by using a beam splitter through the objective.

**Application**

**(A) Detection of materials:**

(*i*) It is used for studying biological materials like DNA, RNA, proteins etc.

(*ii*) It is also used for detection and demonstration of materials like chlorophyll, carbohydrates, proteins and nucleic acids.

(*iii*) Proteins can be detected with dye called **fluorescein isocyanate** or **rhodamine**.

(*iv*) Biological materials stained with dyes for conventional microscopy can also be studied with fluorescent microscopy.

(*v*) Study of distribution of Vitamin A porphyrins in the cell by primary fluorescence.

(*vi*) Distribution of noradrenaline in the terminals of sympathetic nerve endings.

**(B) Banding Pattern of Chromosome:**

(*i*) Using fluorescent dyes, banding pattern on both sister chromatids of chromosome can be studied.

(*ii*) Fluorescent dyes like **quinacrine** and **quinacrine mustard** are used. The chromosomal bands obtained with **quinacrine dye** are called Q bands and with quinacrine mustard are called **QM bands**. Bands are obtained when fluorescent dyes attach to specific regions of the chromosome.

(*iii*) Y chromosome of human gives strong fluorescence. It can be clearly observed as fluorescent body in interphase nuclei.

(*iv*) It also helps in studying and identifying homologous chromosomal pairs, and detection of deletion or alternation of chromosome.

**(C) Detection of cell components with labelled Antibodies:**

Fluorescent dyes such as **fluorescein isothiocyanate** and **lissamine rhodamine** B are employed to chemically labelled blood serum proteins called antibodies.

## • Electron Microscopy

The electron microscope is a practical application of **de Broglie waves**. The electron microscope (EM) uses beam of electrons for magnification.

The basic principle on which the electron microscope works is :

(*a*) **The electron behaves like light rays having shorter wavelength.**

(*b*) **An electron beam can be focussed by suitable electric and magnetic fields.**

(*i*) Electrons are negatively charged subatomic particles. Their use in imaging can be explained in terms of their quantum mechanical wave like properties.

(*ii*) Since electrons are easily absorbed in air at atmospheric pressure, the EM has to operate in vacuum.

(*iii*) The wavelength of the electron beam depends on its energy which is in turn dependent on the voltage used to accelerate the electrons. In modern electron microscopes, imaging electrons are usually accelerated between 30 and 100 kilovolts. The most popular EM uses about 100 kilovolts.

$$\lambda = \frac{h}{\sqrt{2meV}} = \frac{6.62\times10^{-34}}{\sqrt{2\times9.1\times10^{-31}\times1.6\times16^{-19}V}} = \frac{12.27\times10^{-10}}{\sqrt{V}}$$

If $$V = 100\,\text{kV} = 100{,}000\,\text{V},\ \lambda = \frac{12.27\times10^{-10}\,m}{10\times10\sqrt{10}} = 0.0388\times10^{-10}\ \text{m} = 0.04\ \text{Å}$$

(*v*) According to de Broglie equation $\left(\lambda = \frac{h}{\sqrt{2\,me\,V}}\right)$, this voltage (100 kV) will correspond to a wavelength of about 0.0388A° or 0.04 A° (approx).

(*vi*) Electron microscope makes use of the wave properties of the moving electron.

$$\lambda = \frac{h}{m\,v}$$

## Working Procedure

(*a*) **Image formation:**

(*i*) In electron microscope, image formation occurs by electron scattering. Electrons strike the atomic nuclei and get dispersed and dispersed electrons form the image.

(*ii*) The electron image is converted into a visible form by projecting on fluorescent screen.

(*iii*) Here a narrow beam of electron is obtained from the electron gun. Electrons in the form of a collimated beam pass through the condenser lens or coil and fall on the object.

(*iv*) They get scattered and transmitted through the object and pass through the objective lens (coil). It magnifies the image of the object.

(*v*) The **projector lens** (coil) further magnifies the image and projects it on the fluorescent screen or the photographic film.

(*vi*) The image formation occurs when the energy of the electrons is transformed into visible light through excitation of chemical coating of the screen.

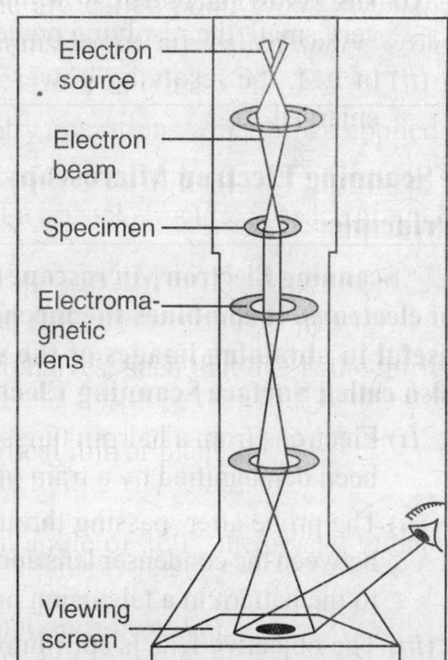


**Fig. 6.21** Diagram depicts the basic position and arrangement of components of electron microscope in a sequence from top downwards: at the lower end the fluorescent display plate or viewing screen displays the image indirectly to the viewer, as an electron beam cannot be directly seen by eye.

(*vii*) The object must be very thin so that electron can pass through it. More electrons will be passing through the transparent parts of the objects and less electrons will be passing through the comparatively denser portions.

(*viii*) The electrons which reach the fluorescent screen form the **bright spot** while the areas where the electrons do not reach the screen form **dark spot**. The areas which scatter electrons are termed as **electron dens**.

(*ix*) The varying degree of intensity of electrons forms the image with varying degrees of grey.

(*x*) The lighter the atomic number, the greater the dispersion. Biological material generally have a low atomic number, the dispersion of electrons is poor. Very poor dispersion leads to very poor contrast in the image formation. Therefore a number of salts with high atomic number are used to increase the contrast of the image.

Such salts are used during fixation and staining.

(*b*) **Magnification:**

(*i*) The objective lens and projector lens play active role in the image formation in electron microscope.

(*ii*) Intermediate lens may be fitted between objective and projector lens in order to get maximum magnification.

(*iii*) If the magnification of the objective lens or coil is 100 and the projector lens or coils is 200, the net magnification will be $100 \times 200 = 20{,}000$.

(*iv*) With the help of intermediate lens, we can achieve a magnification as high as 1,60,000.

(*c*) **Resolving Power:**

(*i*) The resolving power of a microscope is inversely proportional to wavelength. Since $\lambda$ is very small, the resolving power is very high.

(*ii*) In EM, the resolving power is so high that the image from the objective can be greatly enlarged.

## • Scanning Electron Microscopy (SEM)

**Principle:**

**Scanning Electron Microscope (SEM) generates image by scanning the specimen with a beam of electrons. It combines the mechanism of electron microscopy and television. A SEM is very useful in obtaining images of the surface of the thick specimen. This method of examining is also called Surface Scanning Electron Microscope (SCEM).**

(*i*) Electrons from a hairpin tungsten filament are formed into fine electron probe. This probe has been demagnified by a train of lenses.

(*ii*) The probe after passing through the condenser lens is deflected by **beam deflectors** located between the condenser lens and the specimen in a raster pattern over the specimen stages similar to the pattern in a television picture tube.

(*iii*) The objective lens is split into two parts. One part is placed between the condenser lens and the specimen (regarded as additional condenser lens). It focuses the electron beam onto small spot on the specimen. As the electron beam sweeps rapidly over the specimen, molecules in the specimen are excited to high energy levels and emit **secondary electrons**. These electrons are used to form an image of the specimen surface.

(*iv*) The image signal can be collected by the detector. Secondary electrons knocked out of the specimen by the primary beam, as well as the back scattered electron reflected from the surface of the sample are also captured by detector.

(*v*) The essential component of the detector is **scintillator** which emits protons of light when excited by the electron incident upon it.

(*vi*) The protons are used to generate an electronic signal to a video screen. The image then develops point by point, line by line on the screen as the primary electron sweeps over the specimen.

**Instrumentation:**

Scanning Electron Microscope (SEM) consists of four components:

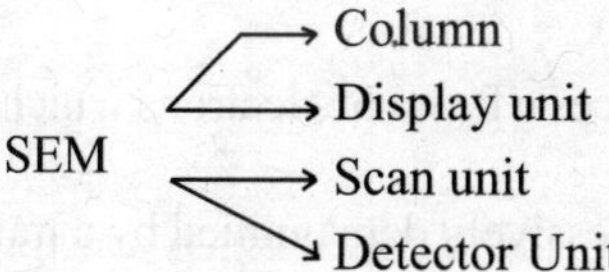


**(A) Column:**

(*i*) It is an evacuated tube of metal.

(*ii*) Other components are sequentially arranged from top to bottom.

(*iii*) It contains a tungsten filament enclosed in a cathode shield, an anode plate.

(*iv*) Three electromagnetic lenses are present. The first two lenses act as condenser and demagnify (reduce the width and increase the narrowness) the incident electron probe. The **third lens** called **objective** lens further reduces the probe-diameter and focusses it onto the object surface. It has two sets of vertical and horizontal scanning coils. One set near the top of the lens defecctings the beam away from the axis while the second set deflects it in the opposite direction towards the axis so that the beam crosses the axis in the exit aperture of the lens. **Stigmators** are used to correct the imperfection of the lens.

(*v*) In SEM, accelerating voltage is generally between 5kV and 50kV.

(*vi*) The vacuum pump, stabilized electric supply, water supply for cooling and air locks are built-in accessories in SEM.

**(B) Display Unit:**

(*i*) In SEM, secondary electrons form a 3D image and they do not have sufficient energy to excite a fluorescent screen and form an image.

(*ii*) In SEM, the secondary electrons are first collected and then amplified to form an image on the phosphor screen of cathode ray tube (CRT).

(*iii*) In most models there are two tubes, one to observe the visual image of the specimen and the other is used for photography. The long after-glow of CRT gives a persistent image for selection of the sites for observation and focussing. The tube for photography has short after-glow.

**(C) Scan Unit:**

(*i*) In SEM, images are formed line by line or point by point in square raster pattern.

(*ii*) The scanning action is controlled by a scan generator that sends signal to deflection coils (two) placed between last two lenses or in the final lens.

(*iii*) The coils deflect the probe in a regular manner on the surface of the specimen.

(*iv*) Scan generator also operates deflection coils that centre the electron beam of C.R.T. (Cathode Ray Tube) of display unit.

(*v*) The magnification of SEM is expressed as the ratio of the scan length on the screen which remains constant to the scan length on the specimen surface, which is variable.

(*vi*) If the beam scans a 1mm square area of the specimen and the resultant image is displayed on a $10nm^2$ screen, the magnification achieved is 100 X. If the scan area is reduced to $0.1mm^2$ the magnification is increased to 1000 X.

**(D) Detector Unit:**

(*i*) The collisions of the electrons of beams with the surface of the specimen give out various types of radiations.

(*ii*) Secondary electrons emitted by the specimen are captured by detector. It should be noted that secondary electron forms 3 D image.

(*iii*) The essential component of detector is **scintillator**. A plastic scintillator coated with evaporated film of aluminum is the main component of secondary electron detector.

(*iv*) Collision of electrons (10-15KV energy) with the scintillator emits photons of light, the photons are used to generate an electronic signal to a video screen.

**Image formation:**

(*i*) Electron gun i.e. hairpin tungsten filament produces a beam of electrons which are accelerated by microscope column.

(*ii*) The beam passes through condenser and is progressively demagnified by a train of lenses into fine probe.

(*iii*) The objective lens reduces the probe and focusses it on the object surface.

(*iv*) The deflection coils move the focussed probe across the specimen in square raster pattern. The interaction of the electron beam with the atom of the specimen (object) surface emits **secondary electrons**. The electron fall on the specimen surface is called **primary electron**.

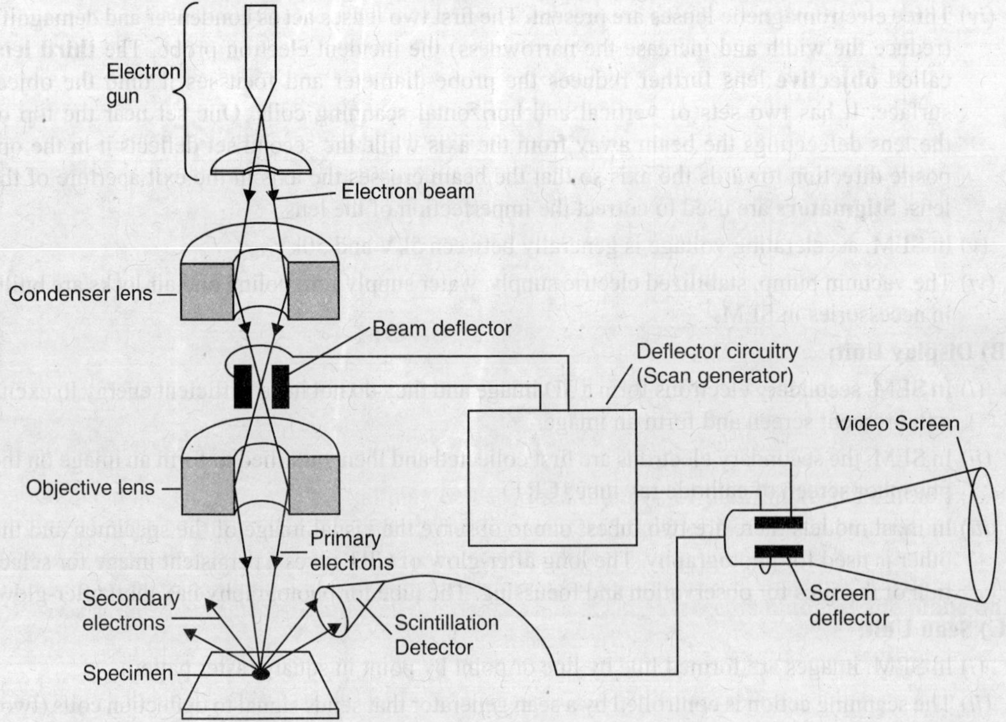


**Fig. 6.22** Schematic diagram of ad SEM.

(*v*) The secondary electrons are then captured by the positively charged detector. The collison of electron with the scintillator present in detector emits photon of light which generates electromagnetic signal to a video screen. The image is then developed point by point, line by line on video screen.

(*vi*) The secondary electrons are first collected, amplified and then used for the formation of 3D image on the phosphor screen of cathode ray tube (CRT).

**Application:**

1. Study of the surface of the cell.
2. Study of hard materials. (cartilage, bone and dentine etc.)

**Advantages:**

(*i*) It can separately detect the scattered electrons, unscattered electrons and the combination of two sets of electrons.

(*ii*) It is free of chromatic aberration and provides a clear image of thick specimen.

(*iii*) It provides a picture of better contrast with the use of electrons.

(*iv*) It can operate at low magnification.

(*v*) It has a great depth of field which helps to obtain 3D image.

## • Transmission Electron Microscopy (TEM)

**Electron microscopes in which electrons are allowed to be transmitted through the objects are known as Transmission Electron Microscope (TEM).**

**Instrumentation:**

The transmission electron microscope consists of an electron gun, central column, electromagnetic lenses and a fluorescent screen.

**(A) Electron gun:**

(*i*) The source of electron is the electron gun and is located at the top of the microscope body.

(*ii*) It consists of 'cathode' a hot hairpin tungsten filament (2mm long) or lanthanum hexaboride, enclosed in a cup-shaped electrode at cathode.

(*iii*) The electrons boil away from the white hot tungsten and are shaped into a **conical beam** by an electrode system called **electron gun**.

(*iv*) The tungsten filament is surrounded by a negatively biased grid, shield or wehnelt cylinder with an aperture.

A negative grid biased against the cathode determines the beam current that passes through the grid.

(*v*) The voltage applied to cathode as a negative potential (50-100 KV) so that anode facing the grid is at ground potential.

(*vi*) This provides a narrow beam of electrons of uniform velocity.

**(B) Microscope column:**

(*i*) It consists of an evacuated metal tube.

(*ii*) It is housed at the top of the electron gun, a series of electromagnetic lenses, viewing screen and photographic plate. These components are aligned one above the other.

(*iii*) A vacuum is present inside the microscope column.

**(C) Electromagnetic lenses or coils:**

(*i*) The electrons are negatively charged particles and their path can be deflected by electromagnetic field i.e. trajectory of an electron can be controlled by using electromagnets.

(*ii*) Each electromagnetic lens or coil has coils of electric wire and wound on a hollow metal cylinder designed in such a way that an electric current passing through the magnetic coils produces an axillary symmetrical magnetic field in the centre of the lens.

(*iii*) The magnetic field forces the electrons to spiral around a central axis.

(*iv*) The electron beam passes through the microscope column and gets deflected by the current flowing through the coils of the lens.

(*v*) The electromagnetic lens are of different types. *viz.* condensor lens, objective lens and projector or image lens.

(*vi*) These are used to retract and focus the electron beam properly.

**(a) Condenser lens:**

(*i*) It is the first lens to affect the electron beam.

(*ii*) It increases the electron density by focussing the beam on the specimen, i.e. it acts as collimator.

(*iii*) It has an aperture (0.25 to 0.5 mm diameter) which reduces the amount of stray electrons reaching the specimen.

(iv) For better focussing of electron beam, two lenses may be used.

**(b) Objective lens:**

(*i*) It is the most important and sophisticated part of the electron microscope.

(*ii*) It is a strong, thin lens with high refractory power.

(*iii*) The specimen is positioned on the specimen stage within the objective lens.

**(c) Projector lens/image lens:**

(*i*) It selects small portion of the intermediate image produced by the objective lens and magnifies it again.

(*ii*) In many cases second projector lens may be used.

(*iii*) Final images on viewing screen are produced.

**(d) Chambers:**

**(a) Specimen Chamber:**

(*i*) It is located above the objective lens.

(*ii*) It can be opened to outside for inserting new specimens.

(*iii*) It is pumped out before reconnecting it to the main column.

**(b) The Viewing Chamber:**

(*i*) It is situated at the bottom of the microscope column.

(*ii*) It contains fine grain fluorescent screen. It is coated with a chemical which by excitation forms the image.

(*iii*) The image can be observed through glass windows either directly or through a telescope.

**(E) Ancillary equipment:**

It includes:

(*i*) Power supplies for high voltage cathode heater voltage and focussing current.

(*ii*) Vacuum systems.

**Comparison between Light Microscope and Electron Microscope**

| | | Light microscope | Electron microscope |
|---|---|---|---|
| 1. | **Specimen:** (Objects) | Both living and dead, hydrated and dehydrated objects or specimens are examined as light can easily pass through water. | Dead, and dehydrated specimens (objects) are examined. as electrons are stopped by water molecules. |
| 2. | **Fixation:** | Specimens or tissues to be studied under light or optical microscopes are fixed with ordinary fixatives like acetoalcohol, formalin alcohol etc. | Fixation is very critical process in case of electron microscope. Fixation is done with osmium tetraoxide ($OSO_4$) or glutaral dehyde potasium permanganate etc. |

(*Contd.*)

| | | Light microscope | Electron microscope |
|---|---|---|---|
| 3. | **Embedding** | Paraffin: | Araldite, vesto plan, Epan 812 Maraglas, Durocopan |
| 4. | **Sectioning** | (*i*) Very thin sections (6 μm thick) cut out with a razor blade on a microtome are usually examined under this microscope.<br>(*ii*) Light can pass though thin sections | (*i*) Very thin slices of tissues (0.05 μm) cut out with a glass or diamond knife on an ultra microtome are usually examined under TEM.<br>(*ii*) Electron cannot pass through thick sections. |
| 5. | **Mounting** | (*i*) On glass-slide with an egg albumin.<br>(*ii*) Deparaffinized in xylol for staining, | (*i*) On a perforated metal disc (grid) usually covered with fomvar or paralodian. |
| 6. | **Staining** | Usually ordinary dyes like eosine, haemotoxylin etc are used for staining. | Usually salts of heavy metals such as lead acetate, uranyl acetate, phosphotungstic acid etc. are used for staining. |
| 7. | **Viewing** | Image is viewed in the occular lens. | Image is viewed on a phosphorescent screen. |
| 8. | **Source of radiation** (light) | (*i*) Light beam (i.e.) Sun Light gathered by a planoconcave mirror is used as source of light.<br>(*ii*) Vissible light rays illuminate the objects under study. | Electron beam is generated by an electron gun (consist of tungsten filament. i.e. *cathode*, cathode shield, metalic plate, with central hole i.e. **anode**, high voltage current (tungsten filament emits electron) are used as source of light.<br>(*ii*) Invisible electrons are passed through the objects under study. |
| 9. | **Wavelength of used radiation** (light) | The wavelength of visible light is about 5000A° | The wavelength electron beam is very short i.e. 0.05A° |
| 10. | **Lens system**<br>(*a*) Condenser lens: | (*a*) A glass made condenser lens (to concentrate the light beam and focus it on to the object) is used. | (*a*) More than one electromagnetic condenser lenses (to concentrate the electron beam and focus it on to the object) are used. |
| | (*b*) Objective and eyepiece lenses: | (*b*) Here objective and eyepiece lens are used to magnify the image of the object. | (*b*) Here electromagnetic objective and projector lenses are used to magnify the image of the object. |
| | (*c*) Intermediate lens: | A single objective lens is only used. | (*c*) A few objective lenses are used in succession to give a very highly magnified image of the object. These are intermediate object. |

(*Contd.*)

| | Light microscope | Electron microscope |
|---|---|---|
| **11. Vacuum setup** | (*i*) Light easily passes through the air. | (*i*) Electrons cannot pass through air and are stopped by the gas molecule. |
| | (*ii*) Therefore no vacuum has to be created in the column of the microscope. | (*ii*) Therefore, the air pressure in the column of the microscope is decreased to $10^{-4}$ mm of Hg by using a suction pump. |
| **12. Working procedure and image formation** | (*i*) Light beam comes from sunlight or lamps are condensed by condenser lens. | (*i*) Electron beam emited by tungsten filament passes through anode → high voltage electromagnetic field → electromagnetic coil acts as condenser lens. |
| (*a*) Mechanism and image formation | (*ii*) Light rays reflected from the surface of the object forms the image of the object. The image is called light image. | (*ii*) Electrons scattered by the atoms of the object form the image of the object. The image is called electron image. |
| (*b*) Visibility of the image | (*i*) Light image of an object is visible to human eye. | (*i*) Electron image is not visible to human eye. |
| | (*ii*) It shows all the colours present in the object. | (*ii*) It can be cast on a zinc sulphide coated fluorescent screen for taking black and white photograph. |
| (*c*) Contrast on image | The recognition of different parts of an object in the image is called 'contrast'. It results from bitterness of intensity and wavelength of light reflected from different parts of the object. | Contrast is due to differential scattering of electrons by different parts of an object. Areas looking dark are electron dense and scatter electrons more strongly than white or electron lucent regions that have weak scattering power. |
| **13. Magnifying power** | The magnifying power is low and does not exceed 1500X, since the focal length of glass lenses is high. | The magnifying power is very very high and may exceed 100,000 X because the focal length of electromagnetic lenses is very small. |
| **14. Resolving power and resolution** | (*i*) Resolving power is very low.<br>(*ii*) Resolution. | (*i*) Resolving power is high. |
| **15. Inverted set-up** | (*i*) It is about 1½ feet in height. | (*i*) It is about 6 feet in height |
| | (*ii*) The light is passed from below and object is viewed from the top. | (*ii*) The light i.e. electron gun is placed above and viewing screen is kept at the bottom. |

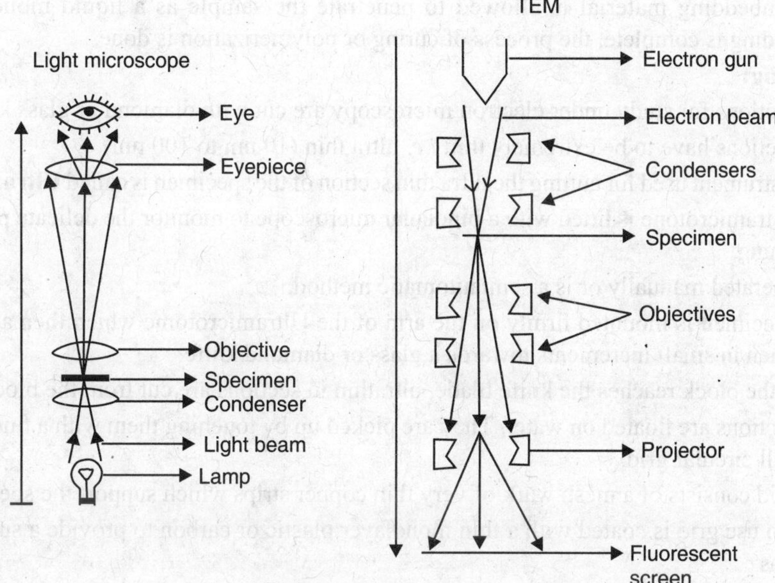


**Fig. 6.23** Light & Electron microscopes

## • Tissue Processing for Electron Microscopy

The preparation of specimen for electron microscopy includes fixation, dehydration, staining and sectioning.

**(A) Fixation:**

(*i*) Specimen must be chemically fixed and stabilized.

(*ii*) Primary fixation kills the cell but the cellular components remain intact.

(*iii*) The fixatives which are widely used in electron microscopy are **osmium tetraoxide ($OsO_4$), formaldehyde, potasium permanganate ($KMnO_4$) paraformaldehyde, aerolein and glutaral- dehyde.**

(*iv*) **Osmium tetraoxide** is an additive fixative. It preserves lipids and proteins against damage during freezing.

(*v*) **Buffered** formaldehyde is a good fixative in the presence of calcium ions. It is a non-coagulant fixative. It makes the lipid insoluble for good fixation.

(*vi*) **Potasium permanganate** ($KMnO_4$) is used as fixative for the membranous structures (Endo plasmic reticulum & Golgi complex)

(*vii*) **Glutaraldehyde** in a phosphate buffer is widely used in the study of enzyme histo-chemistry.

(*viii*) Fixatives are generally chilled before use.

**(B) Dehydration and Embedding:**

(*i*) The next step for preparation of specimen is dehydration.

(*ii*) Here the tissue is passed through a successive grade of alcohol solutions.

(*iii*) Embedding is a process of adding a supporting medium to the interstices of a block of tissue for easy handling during section cutting.

(*iv*) The supporting material is called embedding medium.

(*v*) The specimen is then placed into a fluid such an acetone or propylene oxide to prepare it for embedding.

(*vi*) The various embedding materials are **epoxy resins, methacrylate, epon 812 etc.**

(*vii*) The embedding material is allowed to penetrate the sample as a liquid monomer. After embedding is complete, the process of curing or polymerization is done.

**(C) Sectioning:**

(*i*) The sections for study under electron microscopy are cut with diamond or glass knives.

(*ii*) The sections have to be extremely thin *i.e.* ultra thin (10 nm to 100 nm).

(*iii*) The instrument used for cutting the ultra thin section of the specimen is called **Ultramicrotome**.

(*iv*) The Ultramicrotone is fitted with a binocular microscope to monitor the delicate procedure of sectioning.

(*v*) It is operated manually or is a semiautomatic method.

(*vi*) The specimen is mounted firmly on the arm of the Ultramicrotome which then advances the specimen in small increments towards a glass or diamond knife.

(*vii*) When the block reaches the knife blade, ultrathin in sections are cut from the block face.

(*viii*) The sections are floated on water. They are picked up by touching them with a fine wire mesh or small circular grid.

(*ix*) The grid consists of a mesh wark of very thin copper strips which support the specimen.

(*x*) Prior to use grid is coated with a thin monolayer plastic or carbon to provide a support to the sections.

(*xi*). The specimen is visualised through the holes of screen.

**(D) Post staining:**

(*i*) Sections to be examined with the electron microscope are generally not stained.

(*ii*) However to the contrast, may be improved by "post staining" with electron dense or electron stain materials **(Uranyl acetate, lead citrate, osmium tetraoxide).**

(*iii*) Post staining, the specimen is ready for viewing or photographed with TEM.

**Negative Staining:**

(*i*) It is one of the simplest techniques for the preparation of ultrathin section used for transmission electron microscopy.

(*ii*) Negative staining involves treatment of electron dense embedding materials such as **phospho tungstenic acid** ($H_3PW_{12}O_{40}$).

(*iii*) It penetrates all the empty spaces (opening cervices) of the cells between the macromolecules.

(*iv*) When the material (specimen) is washed and studied under electron microscope, it shows negative contrast *i.e.* the background is dark and heavily stained (filled with phosphotungstic acid) whereas the specimen (material) itself is lightly stained.

(*v*) This technique is usually used for examimng very small objects such as virus or isolated organalles.

**• Shadow casting (shadowing) :**

**It is a special technique for increasing contrast as well as to study three-dimensional appearence of virus and certain macromolecules such as DNA and collagen fibres.**

It involves the deposition of a thin layer of an electron dense metal such as **gold, chromium, palladium or uranium** on a biological specimen.

(*i*) The specimen (material) is first spread on a clean mica surface and dried. It is then kept in a vacuum **evaporator**.

(*ii*) Within the evaporator two electrodes *viz* one carbon electrode and other metal electrode (Chromium, Palladium, uranium etc) placed at an angle 10° – 45° relative to the specimen.

(*iii*) After a vacuum is created in the evaporator, current is applied to the electrode. Metal is evaporated from the electrode (a filament of incandescent tungsten) and sprayed over the surface of the specimen.

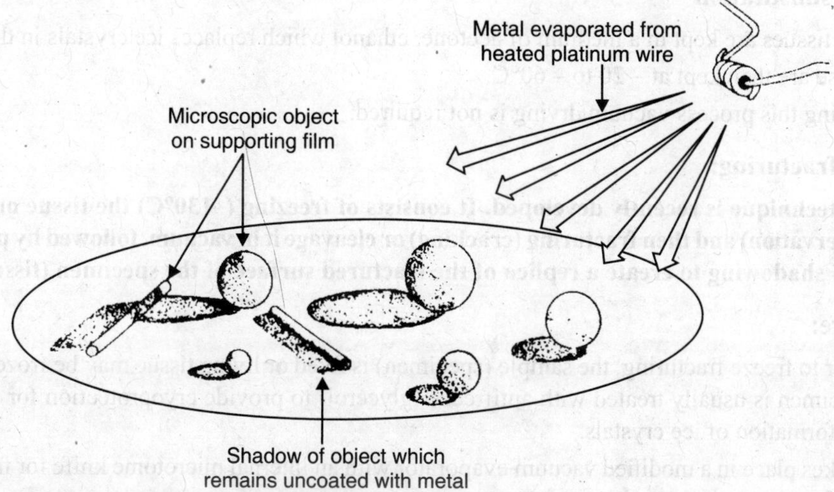


**Fig. 6.24** Shadow casting preparation is made for the electron mircoscopy by exposing the materials like viruses to metal evaporated from heated platinum wire from one side.

(*iv*) The vapour of the heavy metal is deposited at an angle on one side of the surface of the elevated particles as a thin coating and other side remains clear of it.

(*v*) The carbon electrode is then fired and the evaporated carbon renders stability and support to metal replicas. The mica support containing the specimen is then removed from the vacuum evaporator and lowered gently onto a water causing float away from the mica surface.

(*vi*) The replica is then transferred to an acid bath and then again returned into water surface. Then clean replica of the specimen is retrieved on a standard copper grid.

(*vii*) It can be observed in the TEM. Thus by shadow casting, shape and profile of a particle can be discerned. It permits determination of the height of the particle.

- **Freezing:**

(*i*) It is a method of preparation of tissues for microscopical studies.

(*ii*) It is used instead of usual method of chemical fixation.

(*iii*) It has an advantage of quick fixation without any loss of time.

- **Freeze drying:**

**The principle of freeze drying methods are rapid freezing of tissue at about 160°C and their subsequent desiccation in vacuum at a higher temperature. Finally the preparation of sections by normal procedure.**

This technique can be processed in the following steps:

(*a*) Initial freezing or quenching

(*b*) Subsequent drying

(*c*) Embedding

(*d*) Treatment before examination or observation.

**Advantages:**

(*i*) The tissue gets fixed preventing any changes.

(*ii*) Enzymes are arrested to their original locations.

(*iii*) The fixation is quick and homogeneous.

(*iv*) All the components of the cells remain unchanged.

- **Freeze substitution**

(*i*) The tissues are kept in a medium of acetone, ethanol which replaces ice crystals in the tissue.

(*ii*) These are then kept at – 20 to – 60°C.

(*iii*) During this process vacuum drying is not required.

- **Freeze fracturing:**

**This technique is recently developed. It consists of freezing (–130ºC) the tissue or sample (cryopreservation) and then fracturing (cracking) or cleavage it in vacuum, followed by platinum or carbon shadowing to create a replica of the fractured surface of the specimen (tissue).**

**Procedure:**

(*i*) Prior to freeze fracturing, the sample (specimen) is fixed or living tissue may be frozen. Fixed specimen is usually treated with antifreeze (glycerol) to provide cryoprotection for reducing the formation of ice crystals.

(*ii*) It takes place in a modified vacuum evaporator with an internal microtome knife for fracturing the frozen specimen.

(*iii*) The cryoprotected specimen is mounted on a metal specimen support and immersed in cooled liquid nitrogen. It occurs within the vacuum evaporator along with the temperature –100ºC.

(*iv*) The frozen specimen (sample) is fractured with a blow from the microtome knife. After fracture the specimen is left in vacuum for evaporation of water from the exposed surfaces. This process is called **freeze etching.**

(*v*) At the cut surface, a coating of mixed platinum and carbon is made at an angle, so that replica of the fractured specimen is made on the material according to its contour.

(*vi*) The coating along with the tissue is removed from the chamber and floated on the acid which dissolves tissue. The replica of the carbon-platinum deposition of the tissue is removed.

(*vii*) The replica can be trimmed and cut to shape to study under TEM.

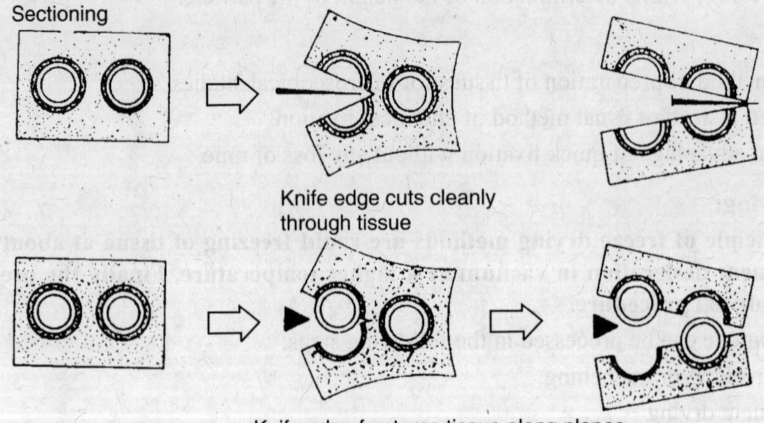


**Fig. 6.25.** Diagram showing the difference between sectioning and freeze fracturing for electron microscopy.

**Observation:**

(*i*) The electron micrograph shows the two faces of the specimen *viz* 'P' face is the interior face of the inner monolayer (*i.e.* Protoplasmic side) and 'E' face is the interior of the outer monolayer

(Exterior side).

(*ii*) Electron can easily pass through carbon but the platinum is electron dense and therefore appears black in the electron micrograph.

(*iii*) With the help of this technique, three-dimensional impact of the sample (specimen) is observed through micrograph.

- **Freeze etching:**

**Process by which water is evaporated from the exposed surfaces of the sample which is left in the vacuum after freeze fracturing is called freeze etching.**

- **Fixation**

**It may be defined as the process by which tissues or their components are fixed selectively to a desired extent.**

(*i*) The purpose of fixation is to kill the tissue without causing any distortion of the components to be studied as far as practicable.

(*ii*) It is required for the following reasons:

(*a*) For sectioning of the tissues.

(*b*) For staining of counterparts.

(*c*) For histochemical techniques applied to living cells.

(*d*) For permanent preparation of tissues.

- **Fixatives**

**The chemicals which are used for fixation are called fixative.**

**Characteristics:**

(*i*) It should stop post-mortem changes like autolysis and bacterial decay.

(*ii*) It should result in no or minimum changes like shrinkage or distortion. It also prevents autolysis.

(*iii*) It should render the cell components in a state which would resist further processing.

(*iv*) It should increase the visibility and keep reality of the components by making them soluble.

- **Additive fixative :**

(*i*) They can remove water and also bind with the protein by means of cross bridges and remain as the integral part of the protein.

(*ii*) Its atoms combine with some part of cellular proteins.

(*iii*) It has precipitating property.

**Example:** Formaldehyde, Ethyl alcohol.

- **Non-additive fixative:**

(*i*) They can remove bound water but do not combine with the protein part of the cell.

(*ii*) They can change the protein by the process of denaturation .

**Example:** Osmium tetraoxide, Potasium dichromate.

- **Dye:**

**Dyes are complex organic molecules or compound and imparts its colouration to specific organelle or components.**

**It has two active chemical groups. *viz.*, chromophore and auxochrome.**

- **Chromophore:**

  (*i*) It imparts colour to the compound.

  (*ii*) They are isolated covalently bonded groups and show a characteristic absorption in the ultra-violet and visible region.

  (*iii*) They are associated with some histological dyes—quinoid (O=⬭=O), Azo (—N=N—) and Nitro (—N(→O)=O) groups.

- **Auxochrome:**

  (*i*) It is clinging property of dyes and enables the dye to attach with the tissue protein.

  (*ii*) These are ionic groups of the dye and have tendency to increase the intensity of colour.

  (*iii*) It is of two types, viz., cataionic and anionic.

  (*iv*) Cataionic anxochrome– $NH_2^+$ histological dye.

  Anionic anxochrome– $COOH^-$, $OH^-$, $SO_3H^-$ histological dye.

- **Acid dye:**

  (*i*) Chromophoric groups are anions.

  (*ii*) They have greater combining capacity at acidic pH.

  (*iii*) They are also known as *anionic dye*.

  (*iv*) They are used for cytoplasmic organelles. Substances which show affinity with this dye are termed as *acidophilic*.

  **Example:** Acid fuchsin, methyl blue etc.

- **Basic dye:**

  (*i*) Chromophoric groups are *cations*.

  (*ii*) Cataions are associated with chromophore.

  (*iii*) Dyes are usually chloride, sulphate, acetate etc.

  (*iv*) Substances which show affinity with this dye, are termed as *basophilic*.

  **Example:** Basic fuchsin, methyl green, haematoxylene.

| Name of the Dye | Acid/Basic | Colouration |
|---|---|---|
| 1. Haematoxylene | Basic | Blue |
| 2. Eosin | Acid | Red |
| 3. Basic fuchsin | Basic | Red |
| 4. Acid fuchsin | Acid | Violet red |
| 5. Toluidin | Basic | Blue |
| 6. Erythrocene | Acid | Violet |
| 7. Methyl blue | Acid | Blue |

- **Neutral dye:**

  (*i*) Chromophoric group consists of both anionic and cataionic groups.

  (*ii*) They have both properties of acidic and basic stain i.e. they are amphoteric dyes.

  **Example:** Janus green B, methylene blue.

- **Metachromatic dyes:**

  (*i*) These groups of dyes stain the cell component in colour different from its own colour.

(*ii*) The action of the stain is due to the substances which bind with it as **dimer** or **polymers.**

(*iii*) Azure-B is a blue stain. Stains the nucleus blue but stain the cytoplasm *i.e.* pink and it is due to the RNA in cytoplasm.

- **Vital stain:**

(*i*) These dyes are used for staining living cells without damaging its normal physiology.

(*ii*) It includes Janus green B, Methylene blue, neutral red, etc.

(*iii*) Janus green B is used to stain mitochondria. It is *green* in oxidized state but at reduction state it becomes **colourless.**

Neutral red is used for staining vacuoles.

Methylene blue is used for staining nucleus as well as chromosome in the dividing cell.

- **Mordants :**

The chemicals which binds with the cell components on one side and the stain on the other. Thus it facilitates absorption of the stain to the cell components for better staining.

Example: Iron-alum (Ferric amonium sulphate).

Lake = The mordant – dye complex.

Lake formation: the phenomenon by which mordant binds to the cell component on one side and stain on the other forming tissue-mordant-dye complex:

- **X-ray Microscope**

**In this type of instrument, X-ray radiation is utilized for chemical analysis and for magnification of 100-1000 diameter. It is an ultramicrochemical analytical technique by which small samples ($10^{-12}$ to $10^{-14}$g) can be analysed.**

Image formation in X-ray microscopy occurs on the basis of four general principles:

(A) Contact microradiography

(B) Projection X-ray microscopy

(C) Reflection X-ray microscopy

(D) X-ray image spectrography.

**(A) Contact microradiography:**

(*i*) Here thin specimen is placed in contact with an extremely fine grained photographic emulsion.

(*ii*) The emulsion has a resolution of more than 25,000 lines/cm or inch and radiographed with X rays of suitable wavelength.

(*iii*) Therefore an absorption image in scale 1:1 is obtained and viewed in a light microscope.

(*iv*) The maximal resolution is that of the optical microscope (0.25 vm). The image has more information. It can be obtained by examining microradiogram in electron microscope.

**(B) Projection X-ray microscopy:**

(*i*) It is based on the production of fine X-ray focal spot.

(*ii*) X-ray focal spot is obtained by an electronic lens system.

(*iii*) The final focal spot is produced on their metal foil which serves as transmission target.

(*iv*) The X-rays are generated on the target by the impact of electrons. The sample is placed near the target.

(*v*) The primary magnification depends on the ratio of the distances from focal spot to sample and sample to film.

(*vi*) The resolution is about 0.1 μ. It is of the same order as the size of the focal spot.

**(C) Reflection X-ray microscopy:**

(*i*) It is based on the refractive index of the X-rays in solids. It is very small (less than 1).

(*ii*) The X-rays are totally reflected at grazing incidence (incidence and very small angle).

(*iii*) If the reflecting surface is made cylindrical, there will be a focussing action in one dimension.

(*iv*) By crossing two such surfaces, a true image can be formed.

(*v*) The resolution is about 0.5 - 1 μm.

**(D) X-ray image spectrography:**

(*i*) It utilizes Bragg reflections in a cylindrically bent crystal.

(*ii*) It produces slightly enlarged emission images.

(*iii*) Here, the resolution is about 50 μm.

**Application:**

It has been utilized for the quantitative determination of dry weight, water content, elementary composition of tissues.

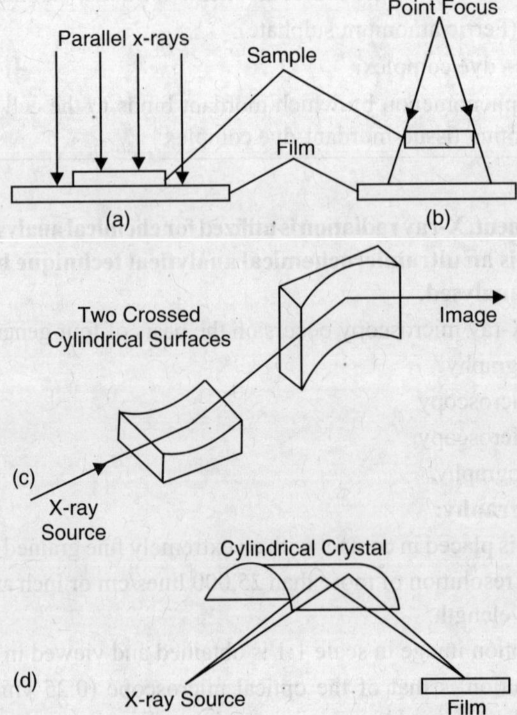


**Fig. 6.26** Principles for X-ray microscopy. (*a*) Contact microradiography; (*b*) Projection X-ray microscopy: (*c*) Reflection X-ray microscopy; and (*d*) X-ray image spectrography.

**Other Imaging Methods :**

Light and electron microscopy are direct imaging methods in that they use protons or electrons to produce actual images of a specimen. The indirect imaging methods are **scanning tunnelling microscopy** (STM), **atomic force microscopy** etc. Each method has the potential for showing molecular structure at near atomic resolution, ten times better than the best electron microscope.

## • Scanning Tunnelling Microscope (STM):

### Principle :

**It is based on the principle of electron tunnelling. At the quantum mechanical level, an electron has both wave-like and particle-like properties. These properties allow the eletron to cross barrier that it cannot penetrate as a particle, but it can penetrate in the form of wave. This penetration is called *tunnelling*.**

### Instrumentation :

(*i*) In this microscope a sharp needle is brought close to the surface to be imaged.

(*ii*) The distance is almost the order of a few angstroms.

(*iii*) At such a distance electron from the surface (electrically conductive) will tend to tunnel across the gap and set up a measurable tunnelling current in the needle.

(*iv*) This current is a precise indicator measure of the distance between the tip and the surface.

(*v*) The needle is then moved in a raster pattern across the surface. At each point its height is adjusted for maintaining the tunnelling current at a constant value.

(*vi*) Thus a record of the vertical portion of the needle at each position becomes equivalent to a record of the surface topography and can be converted into an easily observable picture.

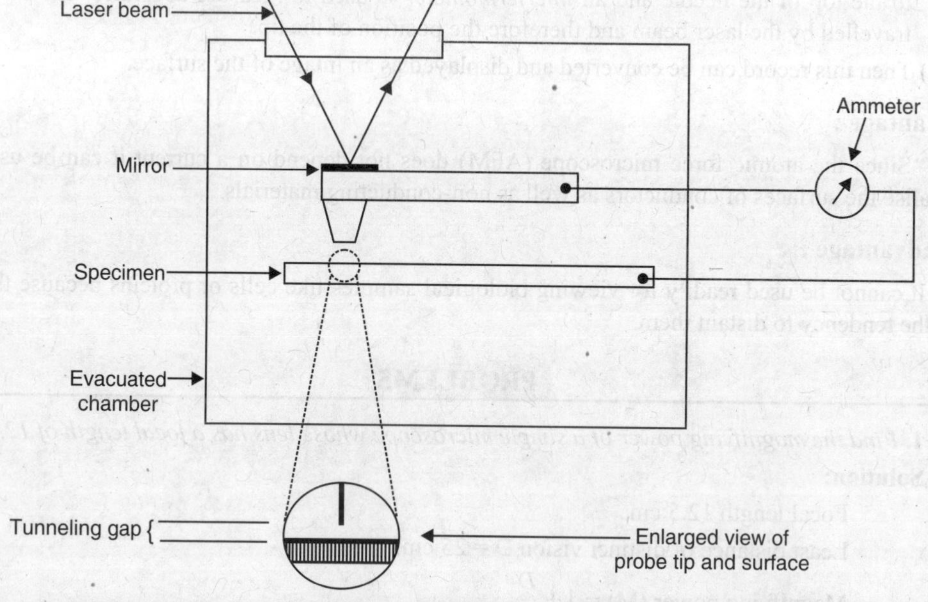


**Fig. 6.27.** A schematic diagram of a scanning tunnelling electron microscope.

(*vii*) The tip of the needle can be made so fine that only one single atom is present at the tip or very end.

(*viii*) The modern electrical techniques are readily available to measure nano-amperes of tunnelling current.

(*ix*) This makes possible the visualisation of individual atoms on the surface of the materials.

### Advantages:

(*i*) It is used to produce image of DNA molecules which enable precise measurement of the pitch of the double helix of DNA, proteins and macromolecular assemblies.

(*ii*) The technique is better studied in to producing images of physical surfaces rather than biological specimen.

**Disadvantages :**

(*i*) An important limitation of STM is that the specimen must be electrically conductive.

(*ii*) The non-conducting or semiconducting biological specimen or samples or materials must be stained with heavy metals before they are imaged. This causes a loss of resolution.

(*iii*) A precise control has to be made in the preparation of the tip and its positioning above the surface.

**• Atomic Force Microscopy (AFM) :**

(*i*) It is a modified form of scanning tunnelling microscope (STM).

(*ii*) In this modified system the needle tip is not kept at a small distance from the sample but it (needle) is pushed right against the surface.

(*iii*) The force present in the tip is maintained constant and the tip is scanned across the surface like ponograph needle running the groove of a record.

(*iv*) A record of the vertical motion of the tip is obtained by reflecting a laser from a mirror fixed to the top of the needle and an *interfermometer* is used to measure accurately the distance travelled by the laser beam and therefore the position of the tip.

(*v*) Then this record can be converted and displayed as an image of the surface.

**Advantage :**

Since the atomic force microscope (AFM) does not depend on a current it can be used to visualise the surfaces of conductors as well as non-conducting materials.

**Disadvantage :**

It cannot be used readily for viewing biological samples like cells or proteins because the tip has the tendency to distant them.

## PROBLEMS

**1.** *Find the magnifying power of a simple microscope whose lens has a focal length of 12.5 cm.*

**Solution:**

Focal length 12.5 cm.

Least distance of distinct vision D = 25 cm.

$$\text{Magnifying power (M)} = 1 + \frac{D}{f}$$

$$= 1 + \frac{25}{12.5} = 1 + 2 = 3$$

**2.** *The focal length of a convex lens used as a magnifier is 5 cm. Calculate the magnifying power when the image is formed at the least distance of distinct vision (25cm) from the lens.*

**Solution:**

When the image is formed at the nearpoint

$$M = 1 + \frac{D}{f}$$

$$= 1 + \frac{25}{5} = 1 + 5 = 6$$

**3.** *A converging lens of focal length 6.25 cm is used as a magnifying glass. If the nearpoint of the observer is 25 cm from the eyes and the lens is close to the eye, calculate the angular magnification.*

**Solution:**

Here the image is formed at the nearpoint.

$$M = 1 + \frac{D}{f}$$

$$= 1 + \frac{25}{6.25} = 1 + \frac{2500}{625} = 1 + 4 = 5$$

∴ Magnification is 5

**4.** *The focal lengths of the objective and eyepiece of a compound microscope are 4 cm and 6 cm respectively. If an object is placed at a distance 6 cm from the object, what is the magnification produced by the microscope? Distance of distinct vision is 25 cm.*

**Solution:**

The focal length ($f_0$) = 4 cm

$u_0 = -6$ cm $\quad v_0 = ?\quad f_e = 6$ cm.

$$\frac{1}{v_0} - \frac{1}{u_0} = \frac{1}{f}$$

or
$$\frac{1}{v_0} = \frac{1}{f_0} + \frac{1}{u_0} = \frac{u_0 + f_0}{u_0 f_0}$$

or
$$\frac{1}{v_0} = \frac{u_0 + f_0}{u_0 f_0}$$

or
$$v_0 = \frac{u_0 f_0}{u_0 + f_0} = \frac{-6 \times 4}{-6 + 4} = \frac{-24}{-2} = 12 \text{ cm}$$

$$\text{Magnifying power} = \frac{v_0}{u_0}\left(1 + \frac{D}{f_e}\right) = \frac{12}{-6}\left(1 + \frac{25}{6}\right)$$

$$= -2\left(\frac{6+25}{6}\right) = -2 \times \frac{31}{6} = 10.33$$

**5.** *A convex lens of focal length 5 cm is to be used as a simple microscope. Where should an object be kept so that the image formed by the lens lies at least distance D of distinct vision (D = 25 cm)? Also calculate the magnifying power of this microscope in this set-up.*

**Solution:**

Let $u_0$ be the distance of the object from the lens. The convex lens forms the virtual image of the object at a distance of 25 cm (nearpoint) infront of it. Thus for the lens we have

$v_0 = -25$ cm and $f = +5$ cm

From the lens formula

$$\frac{1}{v_0} - \frac{1}{u_0} = \frac{1}{f}$$

or
$$\frac{1}{u_0} = \frac{1}{v_0} - \frac{1}{f} = -\frac{1}{25} - \frac{1}{5}$$

or $$\frac{1}{u_0} = \frac{-1-5}{25} = \frac{-6}{25}$$

or $$u_0 = -\frac{25}{6}\text{ cm} = -4\frac{1}{6}\text{ cm}$$

This is the least distance at which observer can see clearly through the lens.

The magnifying power of the instrument when the image is formed at the least distance of distinct vision $D$,

$$M = \frac{D}{u_0} = \frac{25}{\frac{25}{6}} = \frac{25}{25} \times 6 = 6.$$

**6.** *A compound microscope consists of an objective of lens of 1 cm focal length and an eyepiece of 2.5 cm focal length. What is the distance between the lenses? What is the magnification if the object is the sharp focus when it is 1.05 cm from the objective?*

**Solution:**

Here objective $u_0 = 1.05$ cm; focal length $(f_0) = 1$ cm

From the lens formula

$$\frac{1}{v_0} - \frac{1}{u_0} = \frac{1}{f}$$

or $$\frac{1}{v_0} - \frac{1}{1.05} = -1$$

or $$\frac{1}{v_0} = \frac{1}{1.05} - 1 = \frac{1-1.05}{1.05} = \frac{-.05}{1.05} = -\frac{1}{21}$$

or $$\frac{1}{v_0} = -\frac{1}{21}$$

or $$v_0 = -21\text{ cm}$$

The image is formed on the other side at distance 21 cm from the objective.

For the eyepiece $v_0 = 25$ cm (nearpoint); $f_0 = -2.5$ cm

Again $$\frac{1}{v_0} - \frac{1}{u_0} = \frac{1}{f_0}$$

or $$\frac{1}{25} - \frac{1}{u_0} = -\frac{1}{2.5}$$

or $$\frac{1}{u_0} = \frac{1}{25} + \frac{1}{2.5} = \frac{1}{25} + \frac{10}{25} = \frac{11}{25}$$

or $$u_0 = \frac{25}{11} = 2.27\text{ cm}$$

So, the distance between the lenses: 21 + 2.27 = 23.27 cm. Again magnification produced by the objective *m*, *i.e.* $m_1 = \frac{v_0}{u_0} = \frac{21}{1.05}$ and the magnification produced by the eyepiece *i.e.*

$$m_2 = 1 + \frac{D}{f_e} = 1 + \frac{25}{2.5}$$

$$\therefore \quad \text{Total magnification} = \frac{21}{1.05}\left(1 + \frac{25}{2.5}\right) = 220$$

**7.** *In a compound microscope consisting of two convex lenses of focal length 1.5 cm and 10 cm, the object is 2 cm from the objective and the final image is formed at the least distance of distinct vision (D = 25 cm).*

**Solution:**

For the objective, we have $u_0 = -2$ cm

$f_0 = 1.5$ cm $\quad v_0 = ?$

From lens formula $\frac{1}{v_0} - \frac{1}{u_0} = \frac{1}{f_0}$

or $\frac{1}{v_0} = \frac{1}{f_0} - \frac{1}{u_0} = -\frac{1}{2} + \frac{1}{1.5} = \frac{1}{1.5} - \frac{1}{2} = \frac{20 - 15}{30}$

$\frac{1}{v_0} = \frac{5}{30}$ or $v_0 = \frac{30}{5} = 6$ cm

The image of the object is formed by the objective at 6 cm behind the objective.

The final image is formed at the least distance of distinct vision i.e. 25 cm in front of the eyepiece –ve = – 25 cm $f_e = 10$ cm $u_e = ?$

From lens formula

$$\frac{1}{v_0} - \frac{1}{u_0} = \frac{1}{f_0}$$

$\therefore \frac{-1}{25} - \frac{1}{u_e} = \frac{1}{10}$ or $\frac{1}{u_e} = -\frac{1}{25} - \frac{1}{10} = \frac{-2-5}{50} = \frac{-7}{50}$

$$u_e = -\frac{50}{7} = -7.14 \text{ cm}$$

The image of the object formed by the objective serves as an object for the eyepiece and it is at distance of 7.14 cm in front the eyepiece. Hence the distance between objective and eyepiece is 6 + 7.14 = 13.14 cm

**8.** *A compound microscope with an objective of 1.0 cm focal length and an eyepiece of 2.0 cm focal length has a tube length of 20 cm. Calculate the magnifying power of the microscope if the final image is formed at the near point of the eye.*

**Solution:**

The final image is formed at the nearpoint i.e. 25 cm (least distance of distinct vision) in front the eyepiece. Hence for the eye lens

$$v_e = -25 \text{ cm}, \quad f_e = +2.0 \text{ cm} \quad u_e = ?$$

From the lens formula $\frac{1}{v_0} - \frac{1}{u_0} = \frac{1}{f_o}$

$$\frac{1}{u_e} = \frac{1}{v_e} - \frac{1}{f_e} \quad \text{or} \quad \frac{1}{u_e} = -\frac{1}{25} - \frac{1}{2} = \frac{-2-25}{50} = \frac{-27}{50} \text{ cm}$$

I. The image of an object formed by the objective serves as an object for the eyepiece and it is at a distance of 50/27 cm in front of the eyepiece.

II. Since the distance between the objective and the eyepiece is 20 cm, the distance of this image from the objective is

$$20 \text{ cm} - \left(\frac{50}{27}\right) \text{cm} = \frac{540-50}{27} = \frac{490}{27} \text{ cm}$$

Now for the objective: $v_0 = +\left(\frac{490}{27}\right) \text{cm}, f_0 = 1 \text{ cm}, u_0 = ?$

From lens formula $\frac{1}{v_0} - \frac{1}{u_0} = \frac{1}{f_0}$

Substituting the value $\frac{1}{\frac{490}{27}} - \frac{1}{u_0} = \frac{1}{1.0}$

$$\text{or} \quad \frac{1}{u_0} = \frac{27}{490} - \frac{1}{1.0 \text{ cm}} = \frac{27.490}{490} = \frac{-463}{490 \text{ cm}} \quad \therefore u_0 = -\left(\frac{490}{463}\right) \text{cm}.$$

III. Magnifying power of the microscope for final image at the nearpoint $M = -\frac{v_0}{u_0}\left(1 + \frac{D}{f_e}\right)$

$$M = -\left(\frac{490}{27} \Big/ \frac{490}{463}\right) \text{cm} \left(1 + \frac{25}{2.0}\right) = -\frac{463}{27} \times 13.5 = -231.5$$

**9.** *The total magnification produced by a compound microscope is 20, while that produced by the eyepiece is 5 when the microscope is focussed on a certain object, the distance between the objective and eyepiece being 14 cm. Find the focal length of the objective and the eyepiece, if the distance of distinct vision is 20 cm.*

**Solution:**

Total magnification $= M = M_0 \times M_e = 20$

$M_0 \times 5 = 20$ $\quad [\because M_e = 5, D = 20]$

$\therefore \quad M_0 = 4$

The focal length of the eyepiece $= f_e = ?$

$$1 + \frac{D}{f_e} = 5$$

$$\frac{D}{f_e} = 5 - 1 = 4 \quad [D = 20]$$

$$20 = 4 f_e \quad \text{or} \quad f_e = \frac{20}{4} = 5 \text{ cm}$$

For the eyepiece, $v_e = -20$ cm $\quad f_e = 5$ cm, $u_e = ?$

$$u_e = \frac{v_e f_e}{f_e - v_e} = \frac{-20 \times 5}{5-(-20)} = \frac{-100}{5+20} = \frac{-100}{25} = -4 \text{ cm}$$

Distance between the objective and the eyepiece = 14 cm

For the objective $v_0 = 14 - 4 = 10$ cm

Magnification of the objective

$$-4 = \frac{v_0}{u_0}$$

$$-4 = \frac{10}{u_0} \quad \text{or} \quad u_0 = \frac{10}{-4} = 2.5 \text{ cm}$$

$$\text{Focal length of the objective} = f_0 = \frac{u_0\, v_0}{u_0 - v_0} = \frac{-2.5 \times 10}{-2.5 - 10} = \frac{-25}{-12.5} = \frac{25}{12.5} = 2 \text{ cm}$$

**10.** *The length of a microscope is 14 cm and for relaxed eye the magnifying power is 25. The focal length of the eyepiece is 5 cm. Calculate the distance of the object from the objective and the focal length of the objective.*

**Solution:**

Length of the microscope is $L = v_0 + f_e$

$$14 = v_0 + 5 \qquad [\because f_e = 5]$$

$$v_0 = 9 \text{ cm}$$

Now $M = -\dfrac{v_0}{u_0} \times \dfrac{D}{f_e}$

$$25 = \frac{-9}{u_0} \times \frac{25}{5} = \frac{-9 \times 5}{u_0} = \frac{-45}{u_0}$$

$$u_0 = -\frac{45}{25} = \frac{-9}{5} = -1.8 \text{ cm}$$

Now $\dfrac{1}{f_0} = \dfrac{1}{v_0} - \dfrac{1}{u_0}$ but $u_0$ is negative. Therefore

$$\frac{1}{f_0} = \frac{1}{9} - \left(-\frac{5}{9}\right)$$

$$\frac{1}{f_0} = \frac{1}{9} + \frac{5}{9} = \frac{6}{9} = \frac{2}{3}$$

$$\frac{1}{f_0} = \frac{2}{3} \quad \text{or} \quad f_0 = \frac{3}{2} = 1.5 \text{ cm}$$

**11.** *An object is seen first in red light and then in violet light through a simple microscope. In which case will the magnifying power be larger?*

**Answer:**

(*i*) The magnifying power of a simple microscope is $\dfrac{D}{f}$ (normal use).

(*ii*) As $f_V < f_R$, the magnifying power will be larger for violet light.

**12.** *Why is, in a microscope, objective lens of a small aperture taken?*

**Answer :**

(*i*) Microscope is used to observe very small objects placed very close to it.

(*ii*) Therefore, if the aperture of the objective is large, then light coming from the object will spread in large aperture and the object will be seen less bright.

(*iii*) On the other hand, if aperture is small; then light will spread in small aperture and image formed will be more bright.

**13.** *Magnifying power of a simple microscope is inversely proportional to the focal length of the lens. Then,why can we not obtain a very high magnifying power by using a lens of very small focal length?*

**Answer:**

(*i*) For a very high magnifying power, the focal length of the lens would have to be very small.

(*ii*) Lenses of very small focal length are not easy to manufacture.

(*iii*) Such lenses would be very thick in the middle and cause appreciable dispersion of light due to their prismatic action.
Hence they will produce coloured image.

**14.** *In a compound microscope, both the objective and the eyepiece should be of small focal lengths. Why?*

**Answer:**

(*i*) This is to achieve a high magnifying power.

(*ii*) When a compound microscope forms the final image of an object at the least distance of distinct vision *D*, its magnifying power is given by

$$M = -\frac{v_0}{u_0}\left(1+\frac{D}{f_e}\right)$$

$u_0$ = distance of the object from objective.
$v_0$ = distance of the image formed by the objective.

(*iii*) For a high magnifying power, the object should be placed very near the objective ($u_0$, very small) and for a real distance (large $v_0$) image, it should be just beyond the focus of the objective.

(*iv*) For this, the focal length $f_0$, of the objective should be small. Also, for high magnification $\left(1+\frac{D}{f_e}\right)$ by the eyepieces the focal length $f_e$ of the eyepiece should also be small.

**15.** *When viewing through a compound microscope, our eye should be positioned not on the eyepiece, but a short distance away from it. Why? How much should be short distance between the eye and eyepiece?*

**Answer:**

(*i*) The eyepiece produces an image of the objective lens at a short distance on the outer side of it. This image is called "eye ring".

(*ii*) An important property of the eye ring is that all the rays emerging from the microscope pass through the eye ring.

(*iii*) Therefore, eye ring is the best position for the eye to see the image.

(*iv*) So, our eye should be placed a few mm away from the eyepiece. Then it would receive all the rays from the object passing through objective.

(*v*) If eye is placed very close to the eyepiece, then it would not collect all the rays from the object and see a fainter image. Moreover, the field of view will be smaller.

**de Broglie Waves**

The waves associated with material particles are called **matter waves** or **de Broglie waves.** The wavelength of matter waves is called **De Broglie wavelength.**

(*i*) The energy of a proton is E = *h* (Planck's equation) where *h* = planck's constant and $v$ = frequency of radiation.

(*ii*) The energy of same proton is E = mc$^2$ (Einstein's equation) where *m* = mass of the proton and *c* = velocity of light.

(*iii*) From the above

$$E = h\nu = \frac{hc}{\lambda} \quad .....(1)$$

$$E = mc^2 \quad .....(2)$$

$$\frac{hc}{\lambda} = mc^2 \qquad \text{where } v = \frac{c}{\lambda}$$

$$\therefore \quad mc^2 = h\frac{c}{\lambda}$$

or $$\lambda = \frac{h}{mc} = \frac{h}{P}$$ P = is the momentum of the proton.

Consider a particle of mass *m* moving with velocity *v*. Then its momentum is $p = mv$.

de Broglie wavelength $\lambda = \frac{h}{mv}$. Hence, de Broglie wavelength depends upon the mass of the particle and its velocity.

**de Broglie wavelength of an electron:**

Consider an electron, under a potential difference of V volt. Its kinetic energy is $\frac{1}{2}mv^2$.

But $\frac{1}{2}mv^2 = eV$ where $m$ = mass of the electron

$$2m \times \frac{1}{2}mv^2 = 2\,me\,V$$ $e$ = charge of the electron

or $$m^2v^2 = 2\,me\,V$$

Momentum of the electron $= mv = \sqrt{2meV}$

de Broglie wavelength $= \lambda = \frac{h}{mv} = \frac{h}{\sqrt{2meV}}$

Substituting the values of *e*, *m*, and *h* we get

$$\lambda = (12.27)/\sqrt{V}\ \text{Å}$$ where $h = 6.62 \times 10^{-34}$ Js, $m = 9.1 \times 10^{-31}$ Kg, $e = 1.6 \times 10^{-19}$C

$$\therefore \quad \lambda = \frac{h}{\sqrt{2mE}}$$ where E = eV = Kinetic energy of the particle

**16.** *An electron microscope uses electrons, accelerated by a voltage of 50 KV. Determine de Broglie wavelength associated with the electrons.*

**Solution:**

$$\lambda = \frac{h}{\sqrt{2emV}} = \frac{6.62\times10^{-34}}{\sqrt{2\times1.6\times10^{-19}\times9.1\times10^{-31}\times50\times10^{3}}}$$

$$= 5.47 \times 10^{-12}\text{m}$$

Wavelength of electron = $5.47 \times 10^{-12}$m.

**17.** *What is the potential difference required in electron microscope to give electrons of wave-length 0.5Å.*

**Solution:**

$$\lambda = \frac{12.25\,\text{Å}}{\sqrt{V}}$$

Here $\lambda = 0.5$ Å

$$0.5 = \frac{12.25}{\sqrt{V}}$$

or $$\sqrt{V} = \frac{12.25}{0.5} = 24.5$$

or $V = 24.5 \times 24.5 = 600.25$ or 601 volt.

**18.** *Electron beams are accelerated to potential 40 KV and 150 V. Estimate the wavelengths of the electrons in the two beams. Given:*

$h = 6.6 \times 10^{-34}$ Js, m $= 9.04 \times 10^{-31}$Kg and 1 eV $= 1.6 \times 10^{-19}$ J

**Solution:**

$$\lambda = \frac{h}{mv} = \frac{h}{\sqrt{2meV}}$$ Substituting the values, we obtain

$$\lambda = \frac{6.6 \times 10^{-34}}{\sqrt{(2 \times 9.04 \times 10^{-31} \times 1.6 \times 10^{-19} \times 40 \times 10^{3}}}$$

$$= 0.613 \times 10^{-11}\,\text{m}$$

Similarly in case of 150 V,

$$\lambda = \frac{6.6 \times 10^{-34}}{\sqrt{(2 \times 9.04 \times 10^{-31} \times 1.6 \times 10^{-19} \times 150}}$$

$$= 10^{-10}\,\text{m}.$$

**19.** *The velocity of a body of mass 1 kg is 1m/s. Calculate de Broglie wavelength.*

**Solution:**

$$m = 1\text{kg}, \; v = 1\text{m/s}, \qquad \lambda = \frac{h}{mv}$$

$$\lambda = \frac{6.62 \times 10^{-34}}{1 \times 1} = 6.62 \times 10^{-34}\,\text{m}.$$

**20.** *Calculate de Broglie wavelengths for electron and proton moving with same speed of $10^5$ ms$^{-1}$. ($m_e = 9.1 \times 10^{-31}$ kg, $m_p = 1.67 \times 10^{-27}$ kg and $h = 6.63 \times 10^{-34}$ Js)*

**Solution:**

The de Broglie wavelength of a particle of mass 'm' with speed 'v' is given by

$$\lambda_e = \frac{\lambda}{mv}$$

(*a*) For electron $$\lambda_e = \frac{6.63 \times 10^{-34}\,\text{Js}}{9.1 \times 10^{-31}\,\text{kg} \times 10^{-5}\,\text{ms}^{-1}}$$

$$= \frac{6.63 \times 10^{-34} \times 10^{31} \times 10^{-5}\ \text{Js}}{9.1\ \text{kg ms}^{-1}} = 0.728 \times 10^{-8}\ \text{m}$$

$$= 7.28 \times 10^{-9}\ \text{m}$$

(*b*) For proton $\quad \lambda_p = \dfrac{6.63 \times 10^{-34}}{1.67 \times 10^{-27}\ \text{kg} \times 10^{5}\ \text{ms}^{-1}}$

$$= \frac{6.63 \times 10^{-34} \times 10^{22}\ \text{Js}}{1.67\ \text{kg ms}^{-1}} = 3.97 \times 10^{-12}\ \text{m}$$

**21.** *What is de Broglie wavelength of a 2 kg object moving at speed of 1 ms$^{-1}$?*

**Solution:**

Applying de Broglie's equation

$\lambda = \dfrac{\lambda}{mv}$ $\qquad m = 2$ kg $\qquad v = I$ ms$^{-1}$

$h = 6.63 \times 10^{-34}$ Js (Planck's constant)

$$\therefore \quad \lambda = \frac{6.63 \times 10^{-34}}{2\ \text{kg} \times 1\ \text{ms}^{-1}} = 3.315 \times 10^{-34} = 3.3 \times 10^{-34}\ \text{m}$$

**22.** *Calculate the wavelength of a body of mass 1 mg moving with a velocity of 3 ms$^{-1}$.*

***(Delhi Univ. 1975)***

**Solution:**

According to de Broglie's equation

$\lambda = \dfrac{\lambda}{mv}$ $\qquad h =$ (Planck's constant) $= 6.624 \times 10^{-34}$ Js

$m = I$ mg $= 10^{-6}$ kg

$v = 3$ ms$^{-1}$

$$\therefore \quad \lambda = \frac{6.624 \times 10^{-34}\ \text{Js}}{(1 \times 10^{-6})\ \text{kg} \times 3\ \text{ms}^{-1}}$$

$$= \frac{6.624 \times 10^{-34} \times 10^{6}\ \text{Js}}{3\ \text{kg ms}^{-1}}$$

$$= \frac{2.208 \times 10^{-28}\ \text{kg m}^{-2}\text{s}^{2}}{\text{kg ms}^{-1}}$$

**23.** *Calculate the wavelength of an electron moving with a velocity of 2.5 × 10$^{7}$ ms$^{-1}$ (h = 6.63 × 10$^{-34}$ Js, mass of an electron = 9.11 × 10$^{-31}$ kg).*

**Solution:**

Applying de Broglie's equation $\lambda = \dfrac{h}{mv}$

$$\lambda = \frac{6.63 \times 10^{-34}}{9.11 \times 10^{-31} \times 2.5 \times 10^{7}}$$

$$= \frac{6.63 \times 10^{-34} \times 10^{+31} \times 10^{-7}}{9.11 \times 2.5}$$

$$= \frac{6.63\times10^{-10}\times10^{2}}{9.11\times2.5\times10^{2}} = \frac{663\times10^{-10}}{9.11\times2.5}$$

$$= \frac{663\times10^{-10}}{2277.5} = 0.2911\times10^{-10}\text{ m}$$

**24.** *Calculate de Broglie wavelength of electrons of speed 0.5 km s$^{-1}$ [$m_e$ = 9.1 × 10$^{-31}$ kg, h = 6.63 × 10$^{-34}$ Js]*

**Solution:**

According to de Broglie's equation $\lambda = \frac{h}{mv}$

Here mass of electrons = $m_e$ = 9.1 × 10$^{-31}$ kg
= 9.1 × 10$^{-31}$ × 10$^3$ g = 9.1 × 10$^{-28}$

Speed $v$ = 0.5 km s$^{-1}$ = 0.5 × 10$^3$ m s$^{-1}$.
$h$ = 6.63 × 10$^{-34}$ Js

$$\therefore \quad \lambda = \frac{6.63\times10^{-34}}{9.1\times10^{-28}\text{ g}\times0.5\times10^{3}\text{ ms}^{-1}} = \frac{6.63\times10^{-34}\text{ Js}}{4.55\times10^{-25}\text{ g m s}^{-1}} = \frac{6.63\times10^{-34+28}\times10}{4.55}$$

$$= 1.457\times10^{-6}\text{ m} = 1.46\times10^{-6}\text{ m}$$

**25.** *Calculate de Broglie wavelength of a proton of momentum 2.55 × 10$^{-22}$ kg ms$^{-1}$. [h = 6.63 × 10$^{-34}$ Js]*

**Solution:**

According to de Broglie's equation $\lambda = \frac{h}{mv}$

$$\therefore \quad \lambda = \frac{6.63\times10^{-34}}{2.55\times10^{-22}} \qquad [mv = 2.55\times10^{-22}]$$

$$= \frac{6.63\times10^{-12}}{2.55} = 2.6\times10^{-12}\text{ m}$$

**26.** *A moving particle has a de Broglie wavelength of 1.0 Å. What is the momentum of the particle?*

**Solution:**

According to de Broglie's equation $\lambda = \frac{h}{mv}$

Momentum = $mv$ $\qquad \lambda$ = 6.63 × 10$^{-34}$ Js

$\therefore \quad mv = \frac{h}{\lambda} \qquad \lambda$ = 1.0 Å = 10$^{-10}$ m

$$\therefore \quad mv = \frac{6.63}{10^{-10}}\times10^{-34} = 6.63\times10^{-34+10}\times10\text{ kg ms}^{-1} = 6.63\times10^{-24}\text{ kg ms}^{-1}.$$

**27.** *Calculate the wavelength associated with the motion of the earth, a stone and an electron. The masses and the velocities of the objects are given below:*

| *Mass* | *Velocity* |
|---|---|
| *Earth = 6 × 10$^{24}$ kg* | *3 × 10$^4$ ms$^{-1}$* |
| *Stone = 0.1 kg* | *1.0 ms$^{-1}$* |
| *Electron = 10$^{-30}$ kg* | *6 × 10$^5$ ms$^{-1}$* |

*Planck's constant = 6.6 × 10$^{-34}$ Js. Which of these objects will have the measureable wavelength?*

**Solution:**

(*a*) For earth:

$$\lambda = \frac{h}{mv} = \frac{6.6\times10^{-34}}{6\times10^{24}\times3\times10^{4}} = \frac{1.1\times10^{-34-24}\times10^{-4}}{3\times10} = \frac{11\times10^{-63}}{3} = 3.67\times10^{-63}$$

$$= 3.67\times10^{-63}\text{ m}$$

(*b*) For stone

$$\lambda = \frac{h}{mv} = \frac{6.6\times10^{-34}}{0.1\times1.0} = \frac{66\times10^{-34}\times10}{10\times1} = 66\times10^{-34}\text{ m}$$

(*c*) For electron

$$\lambda = \frac{h}{mv}$$

$$\therefore \quad \lambda = \frac{6.6\times10^{-34}}{10^{-30}\times6\times10^{5}} = 1.1\times10^{-34}\times10^{30}\times10^{-5} = 1.1\times10^{-9}\text{ m}$$

Obviously the wavelength of electron is maximum and hence is a measurable quantity.

**28.** *Calculate the wavelength of matter-wave associated with an electron moving with a velocity of 1.20 × 10⁷ cm s⁻¹. (Mass of electron = 9.1 × 10⁻²⁸ g; Planck's constant h = 6.626 × 10⁻²⁷ ergs).*

**Solution:**

$$\lambda = \frac{h}{mv}$$

Here $\lambda = 6.626\times10^{-27}$ ergs

$m = 9.1\times10^{-28}$

$v = 1.20\times10^{7}\text{ cm s}^{-1}$.

$$\therefore \quad \lambda = \frac{6.626\times10^{-27}\text{ erg.s}}{(9.1\times10^{-28}\text{ g})\times(1.20\times10^{7})\text{ cm s}^{-1}} = \frac{6.626\times10^{-27}\times10^{21}}{9.1\times1.20} = \frac{6.626\times10^{-6}\text{ g cm}^{2}\text{s}^{2}\text{.s}}{9.1\times1.2\text{ g cm s}^{-1}}$$

$$= \frac{6.626\text{ g.cm}^{2}\text{.s}^{2}\text{.s}}{10.92\text{ g.cm.s}^{-1}} = 0.6067\times10^{-6}\text{ cm}$$

**29.** *Calculate the de Broglie wavelength of a proton of kinetic energy 500 eV. The mass of proton is 1.67 × 10⁻²⁷ kg (h = 6.63 × 10⁻³⁴ Js and I eV = 1.6 × 10⁻¹⁹ J)*

**Solution:**

The de Broglie wavelength of a particle of mass $m$ and kinetic energy $K$ is given by

$$\lambda = \frac{h}{\sqrt{2\,mK}}$$

$m = 1.67\times10^{-27}$ kg and $K = 500\text{ eV} = 500\times1.6\times10^{-19}$ J

$$\therefore \quad \lambda = \frac{6.63\times10^{-34}}{\sqrt{(2\times1.67\times10^{-27}\text{ kg})\times(500\times1.6\times10^{-19})}}$$

$$= \frac{6.63\times10^{-34}}{\sqrt{1000\times2.67\times10^{46}}} = \frac{6.63\times10^{-34}}{\sqrt{2670\times10^{46}}}$$

$$= \frac{6.63\times10^{-34}}{51.67\times10^{-23}} = 0.1283\times10^{-11} = 1.283\times10^{-12}\text{ m}$$

**30.** *What kinetic energy of a neutron will be associated if de Broglie wavelength be 1.40 × $10^{-10}$ m? The mass of neutron is 1.675 × $10^{-27}$ kg and h = 6.63 × $10^{-34}$ J.*

**Solution:**

Applying de Broglie's equation:

$$\lambda = \frac{h}{\sqrt{2\,mK}} \qquad h = \text{Planck's constant}$$

Here $m = 1.675 \times 10^{-27}$ kg $\quad h = 6.63 \times 10^{-34}$

$\lambda = 1.40 \times 10^{-10}$ m

$$1.4 \times 10^{-10} = \frac{6.63 \times 10^{-34}}{\sqrt{2 \times 1.675 \times 10^{-27} \times K}}$$

$$(1.4 \times 10^{-10}) = \frac{6.63 \times 10^{-34} \times 6.63 \times 10^{-34}}{2 \times 1.675 \times 10^{-27} \times K}$$

$$\therefore \quad K = \frac{6.63 \times 6.63 \times 10^{-68}}{2 \times 1.675 \times 10^{-23} \times 1.4 \times 1.4 \times 10^{-20}}$$

$$= \frac{43.9569 \times 10^{-68} \times 10^{+47}}{6.566}$$

$$= 6.69 \times 10^{-21} \text{ J.}$$

CHAPTER 7

# X-ray Crystallography

The name crystal comes from Greek word **"Kyestallas"** meaning clear ice. But later the name was extended to manifest. The science which deals with the study of geometrical forms and physical properties of crystalline solids is known as **crystallography**. The study of crystallography is necessary to understand the strong correlation between the structure of a material and its physical properties. Although, biological materials are rarely found in crystalline state in natural condition. But for the investigation of three-dimensional structure of biomolecules, the study of crystallography is essential.

- **Crystal:**

**A crystal is a solid of definite shape with a regular, repetitious, three-dimensional, array of atoms or molecular groups.**

(*i*) The outer surface of a crystal is made up of some plane surface called face.

(*ii*) When two faces meet, an edge is formed.

(*iii*) The angle between two faces is called interfacial angle.

- **Crystalline solid:**

The solid in which atoms or molecules are arranged in periodic, regular, repeated geometrical and three-dimensional pattern is called crystalline solid.

- **Lattice:**

The regular periodic arrangement of points in a space (or the atoms in a crystal) and looks net like structure is called lattice.

- **Crystal lattice:**

The periodic arrangement of atoms in a crystal is called crystal lattice or it is an infinite array of discrete points with an arrangement and orientations that appears exactly same, from whichever of the point the array is viewed.

- **Basis:**

It is defined as an assembly of atoms, ions or molecules identical in composition and orientation which when repeated in three dimensions generates a crystal structure.

- **Plane lattice:**

The periodic arrangement of atoms in a plane is called plane lattice.

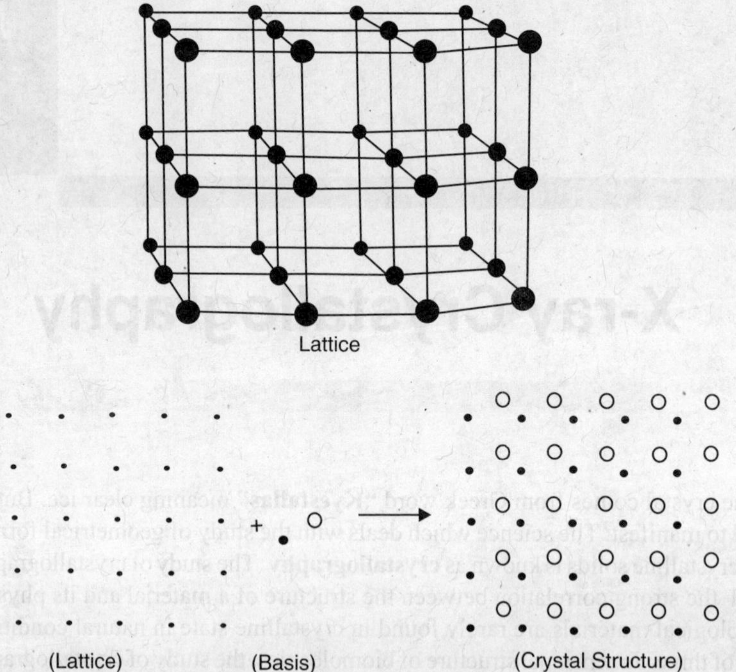


**Fig. 7.1** Generation of crystal structure from lattice and basis.

- **Space lattice:**

To study the details of crystal, each structural unit is represented by a geometrical point, such geometrical point is known as *space lattice* or an infinite array of points (atoms or molecules) in a three dimensions in which every point has an identical surroundings.

- **Lattice array:**

The regular periodic arrangement of points in the space position of the atoms in a crystal is called *lattice array*.

- **Crystal Structure:**

(*i*) The crystal structure is always described in terms of atoms rather than points.

(*ii*) Thus in order to obtain a crystal structure, an atom or a group of atoms must be placed on each lattice point in a regular fashion. Such an atom or groups of atom identical in composition, arrangement and orientation is called *basis* which when repeated in three-dimensions generate a *crystal structure*.

(*iii*) Thus a crystal structure is formed by the addition of a basis to every lattice point i.e. lattice combined with a basis generates a crystal structure.

Lattice + Basis = Crystal Structure.

(*iv*) Lattice is a mathematical concept and crystal structure, a physical concept.

- **Unit cell:** The smallest repeating unit in a space lattice in a crystal which can generate the complete crystal by repeating its own dimensions in various directions is called unit cell.
- **Primitive cell:** It is a minimum-volume unit cell and which contains only one lattice point at corners.
- **One dimentional lattice:** A set of points repeated at regular interval along a straight line is called one dimentional lattice.

- **Two dimentional lattice:** A set of points on a plane along two coordinate axes is called two dimentional lattice.
- **Crystal (three dimentional) lattice:** A set of points arranged in a regular ways along three dimentional axes is called three dimentional or crystal lattice.

**Liquid crystal:**

- **Liquid crystals are substances exhibiting the properties of liquid and crystals.**

(i) They possess surface tension and have the property to flow (fluidty).

(ii) They form drops but not spherical and resemble jelly to some extent.

(iii) They are anisotropic and have much of the long range order of solids.

**Example** Soap solution used in gas-liquid chromatography.

**Crystal symmetry:**

The definite ordered arrangement of the faces and edges of a crystal is known as crystal symmetry.

- **Symmetry Operation:**

**A symmetry operation is that which transforms the crystal to itself i.e. crystal remain invariant under a symmetry operation.**

1. A symmetry operation is one that leaves the crystal and its environment invariant.
2. A sense of symmetry is a powerful tool for the study of the internal structure of crystal.
3. It is simplyfying key to the endless arrays of atoms which make up crystalline solids enabling us to think of them in terms of a familiar pattern.
4. Symmetry operations performed about a point or a line are called *point group symmetry* operations and symmetry operation is performed by translations are called *space group symmetry*.

- **Point group:** It is defined as the collection of the symmetry operations which when applied about a lattice point leaves the lattice invariant.
- **Space group:** The group of all the symmetry elements of a crystal structure is space group.

- **Elements of Symmetry:**
  - **Plane of symmetry:**

(*i*) Here crystal is divided by an imaginary plane into two halves, such that one is the mirror image of the other.

(*ii*) It is of two types: *viz*. Horizontal mirror plane and Vertical mirror plane.

- **Axis symmetry:**

(*i*) Here crystal is rotated around its own axis through an angle of 360°, the crystal remains invariant.

(*ii*) If the original is rotated anticlockwise direction the rotation is positive)

- **Centre of symmetry:**

(*i*) It is a point such that any line drawn through it will meet the surface of the crystal at equal distance on either side.

(*ii*) A crystal can have one or more planes and one or more axis symmetry but only have one centre of symmetry.

**Types of symmetry operation:**

**(*a*) Translation operation:**

(*i*) It follows from the orderly arrangement of a lattice.

(*ii*) It means a lattice point $\vec{r}$ under lattice translation vector operation $\vec{T}$ gives another point $r'$ which is exactly similar to *r, i.e.*

$$\vec{r^1} = \vec{r} + \vec{T}$$

**(*b*) Rotation operation:**

(*i*) A lattice is said to possess the rotation symmetry of its rotation by an angle θ about an axis transform the lattice itself.

(*ii*) As the lattice remains invariant under a rotation of 2π, the angle θ must be an intregral multiple of *i.e.*

$$n\theta = 2\pi$$

$$\theta = \frac{2\pi}{n}$$

(*iii*) The factor *n* takes integral values and is known multiplicity of rotation axis.

**(*c*) Reflection:**

(*i*) A lattice is said to possess reflection symmetry if there exists a plane in the lattice which divides it into two identical halves which are mirror image of each other.

(*ii*) Reflection across a line or plane changes the character of lattice from left-handed to right-handed or *vice versa.*

**(*d*) Inversion:**

(*i*) It is a point operation which is applicable to three-dimensional lattices only.

(*ii*) This symmetry element implies that each point located at $\vec{r}$ relative to a lattice point was an identical point located at $-\vec{r}$ relative to the same lattice.

(*iii*) It also changes the character of lattice from left-handed to right-handed.

- **Cubic crystal system:**

  (*i*) The crystal axes are perpendicular to each other *i.e.,* $\alpha = \beta = \gamma = 90°$.

  (*ii*) Lengths are equal *i.e.,* $a = b = c$.

  (*iii*) Cubic lattices may be simple (*sc*) face centred (*fcc*) or body centred.

  **Example:** NaCl crystal.

- **Tetragonal crystal system:**

  (*i*) The crystal axes are perpendicular to each other *i.e.,* $\alpha = \beta = \gamma = 90°$.

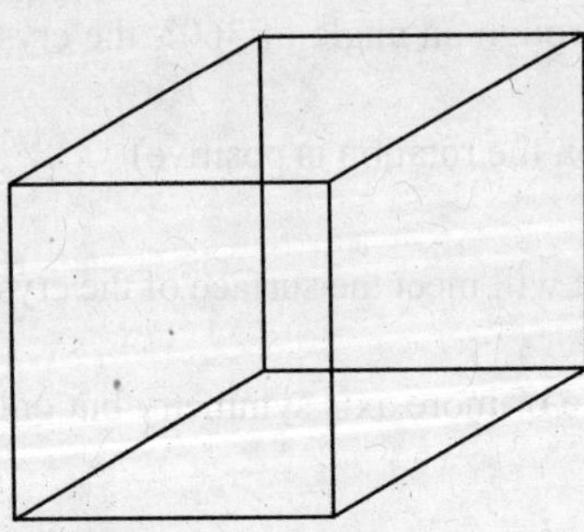

**Fig. 7.2** Cubic crystal

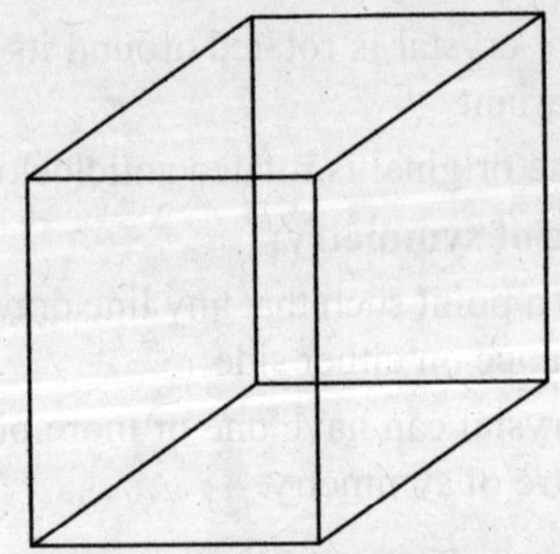

**Fig. 7.3** Tetragonal crystal

(*ii*) Lengths along the two axes are same but along the third axis is different *i.e.,* $a = b \neq c$.

(*iii*) Tetragonal crystal may be simple (*P*) or body centred.

**Example:** *Sn* crystal.

- **Hexagonal crystal system:**

(*i*) Two of the crystal axes are 120° apart i.e. $\gamma = 120°$ while the third axis is perpendicular to both of them *i.e.,* $\alpha = \beta = 90°$.

(*ii*) Lengths are same along the axes that are 120° apart but length along the third axis is different *i.e.,* $a = b \neq c$.

**Example:** AgCl crystal.

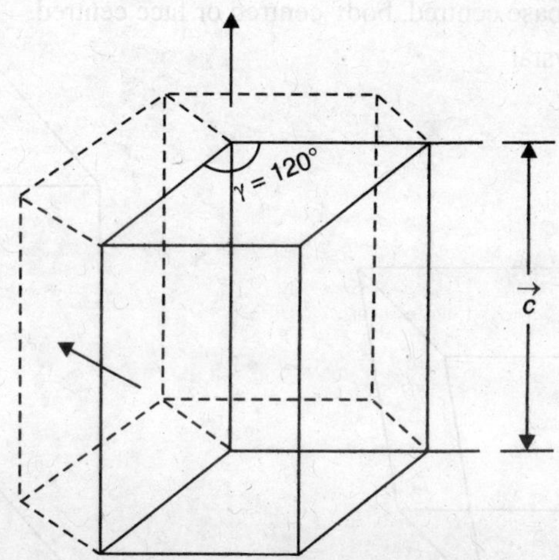


**Fig. 7.4** Hexagonal crystal

- **Trigonal (rhombohedran) crystal system.**

(*i*) The angle between each pair of crystal axes are the same but are not 90° *i.e.,* $\alpha = \beta = \gamma \neq 90°$.

(*ii*) Lenths are same along the three axes *i.e.,* $a = b = c$.

**Example:** *Sb* crystal.

- **Triclinic crystal system:**

(*i*) Three axes are not perpendicular to each other *i.e.,* $\alpha \neq \beta \neq \gamma \neq 90°$.

(*ii*) Lengths and angles are unequal *i.e.,* $a \neq b \neq c$.

**Example:** $CuSO_4$, $5H_2O$ crystal.

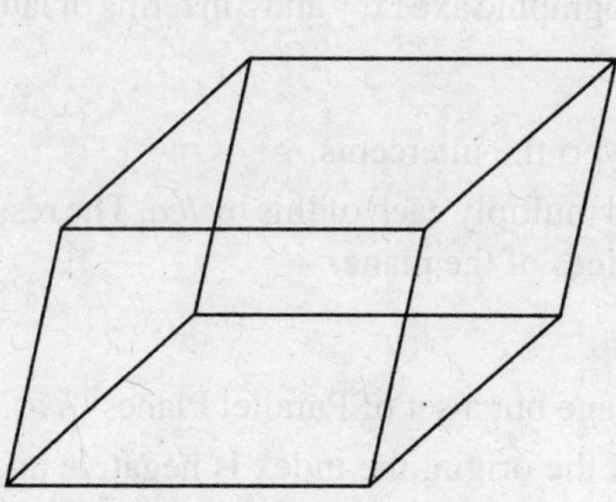

**Fig. 7.5** Trigonal crystal

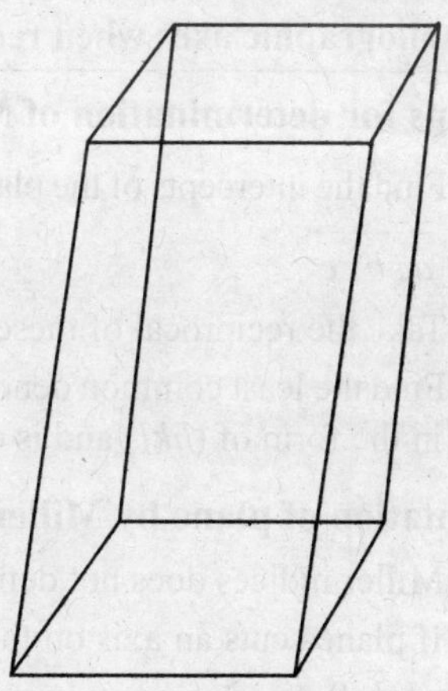

**Fig. 7.6** Trichlinic crystal

- **Monoclinic crystal system:**

  (*i*) Two of the crystal axes are perpendicular to each other *i.e.*, $\alpha = \gamma = 90° \neq \beta$, but third is obliquely inclined.

  (*ii*) Lengths are unequal *i.e.*, $a \neq b \neq c$.

  **Example:** $FeSO_4$ crystal.

- **Orthorhombic crystal system:**

  (*i*) Crystal axes are perpendicular to each other *i.e.*, $\alpha = \beta = \gamma = 90°$.

  (*ii*) Lengths are unequal *i.e.*, $a \neq b \neq c$.

  (*iii*) They may be simple, base centred, body centred or face centred.

  **Example:** $MgSO_4$ crystal.

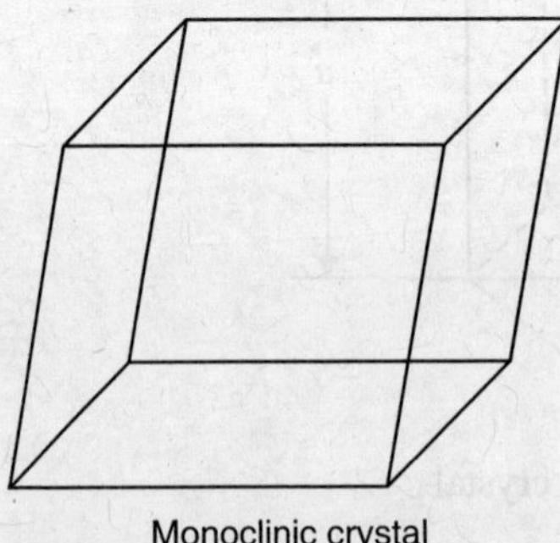


**Fig. 7.7** Monoclinic crystal

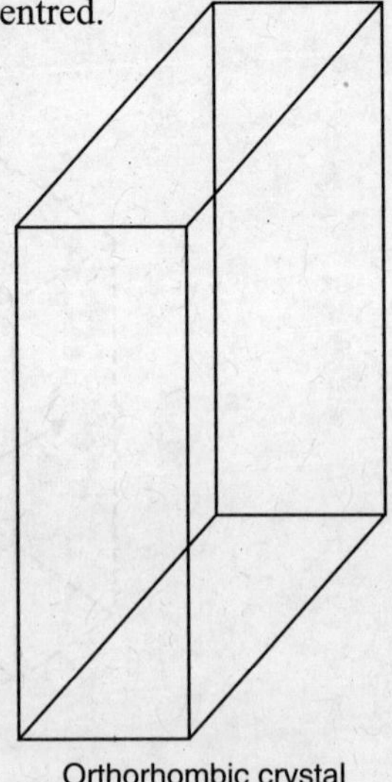


**Fig. 7.8** Orthorhombic crystal

## MILLER INDICES:

Orientation of planes in a crystal may be described in terms of their intercepts on the axis of the crystal. W.H. Miller evolved a method to designate a set of planes in a crystal by three numbers $(h\ k\ l)$ known as Miller indices.

**Miller indices may be defined as the reciprocals of the intercepts made by the plane on crystallographic axes when reduced to smallest number.**

- **Steps for determination of Miller Indices:**

  (*i*) Find the intercepts of the plane on the crystallographic axes *x*, *y* and *z* in terms of lattice constant $\vec{a}, \vec{b}, \vec{c}$.

  (*ii*) Take the reciprocal of these intercepts i.e. invert the intercepts.

  (*iii*) Find the least common denominator (*lcd*) and multiply each of this by *lcd*. The result is written in the form of (*hkl*) and is called **Miller Indices** of the plane.

**Orientation of plane by Miller Indices:**

(*i*) Miller indices does not define a Particular Plane but a set of Parallel Planes (*hkl*).

(*ii*) If planes cuts an axis on the *negative* side of the origin, the index is negative and denotes as $(h\ \bar{k}\ l)$ by placing a minus (–) sign above the corresponding index (here $\bar{k}$)

(*iii*) The zero index means that the plane considered is parallel to one of the axis so that the corresponding intercept become infinity.

(*iv*) For the crystal plane having intercepts $2a$, $3b$ and $c$ on crystal axes $\vec{a}, \vec{b}$ and $\vec{c}$ the reciprocal of the number 2, 3 and 1 are $\frac{1}{2}, \frac{1}{3}$ and 1.

The smallest three integers having the same ratio 3, 2, 6, the plane is therefore referred to as (3 2 6) plane.

- **Important features of Miller Indices of Crystal Planes**

(*i*) If a plane is parallel to a co-ordinate axis, then the intercept of this plane is infinity and hence is **Miller index is zero.**

(*ii*) A plane passing through the origin is defined interms of a parallel plane having non zero intercept.

(*iii*) Parallel plane equally spaced have the same index numbers.

(*iv*) It is the ratio of the indices which is important and determines a plane. For example (246) and (123) represent the same plane. If the Miller indices of two planes have the same ratio *i.e.,* (844) and (422) or (211), then the planes are parallel to each other.

(*v*) If (*hkl*) are the Miller indices of a plane, then the plane cuts the axes into *h*, *k* and *l* equal segments respectively.

(*vi*) The direction (*hkl*) is perpendicular to the plane (*hkl*).

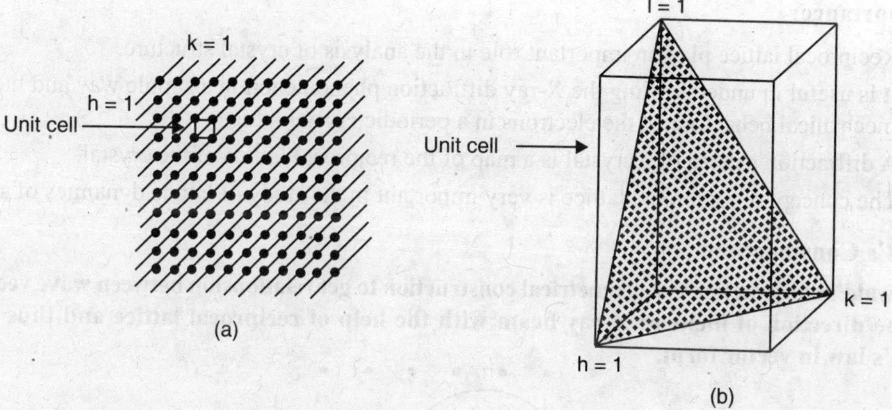


**Fig. 7.9** Lattice planes and miller indices. (*a*) A two-dimensional lattice showing the set of 'planes' (lines) with the Miller indices $h = 1$, $k = 1$ or (11) (*b*) A three-dimensional unit cell showing the plane (111)

Three crystallographic axes are X, Y, Z.

| | $x$ | $y$ | $z$ |
|---|---|---|---|
| | $2a$ | $3b$ | $c$ |
| Reciprocal : | $\frac{1}{2}$ | $\frac{1}{3}$ | 1 |
| Simplification | $6\times\frac{1}{2}$ | $6\times\frac{1}{3}$ | $6 \times 1$ |
| | 3 | 2 | 6 |

So Miller indices of the Plane is (3 2 6).

**Fig. 7.10** Miller indices of planes

**Miller indices of a direction:**

(*i*) The indices of direction in a crystal are expressed as the set of the smallest integers which have the same ratio as the components of a vector in the desired direction referred to the axes vectors.

(*ii*) The integers are written between square brackets [*h k l*].

- **Reciprocal Lattices:**

(*i*) Every crystal has two lattices associated with it. *viz,* the crystal lattice (direct space) and the reciprocal lattices.

(*ii*) This concept was devised for the purpose of tabulating two important properties of crystal planes like slope and interplanar spacing.

(*iii*) Each set of parallel planes in a direct lattice can be represented by a normal to these planes having length equal to the reciprocal of the interplanar spacing.

(*iv*) The normals are drawn with reference to any arbitary origin and points are marked at their ends. These points form a regular arrangement which is called reciprocal lattice.

(*v*) Thus each point in a reciprocal lattice is a representative point of a particular parallel set of planes.

(*vi*) The direct lattice is the reciprocal of its own reciprocal lattice.

(*vii*) The reciprocal lattice is so named because the length assigned to each normal representing the reciprocal lattice is proportional of the interplanar spacing of that plane in ordinary space.

- **Importance:**

1. Reciprocal lattice play an important role in the analysis of crystal structure.
2. It is useful in understanding the X-ray diffraction phenomenon in a simple way and the wave mechanical behaviour of the electrons in a periodic crystal lattice.
3. A diffraction pattern of a crystal is a map of the reciprocal lattice of the crystal.
4. The concept of reciprocal lattice is very important in the study of lattice dynamics of solids.

**Ewald's Construction:**

**Ewald's construction is a geometrical construction to get relationship between wave vector $\vec{K}$ and the direction of incident X-ray beam with the help of reciprocal lattice and thus derive Bragg's law in vector form.**

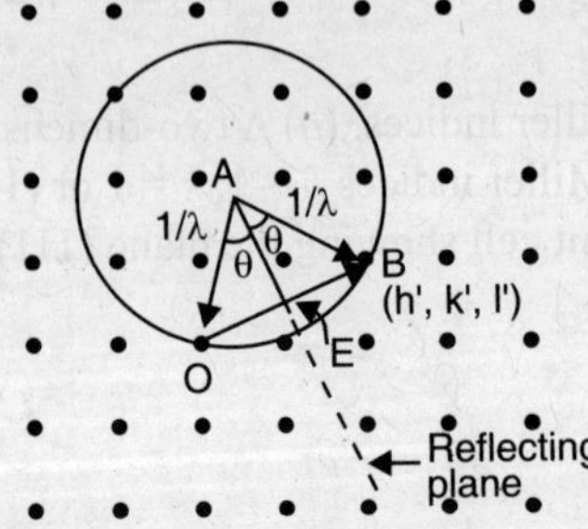


**Fig. 7.11** Ewald Construction in the reciprocal lattice.

(*i*) Draw a vector $\vec{AO}$ of length $\frac{1}{\lambda}$ in the direction of incident X-ray beam originating at *A* and terminating any reciprocal lattice point *O*.

(*ii*) Taking A as the centre, draw a sphere of radius *AO* $\left(\frac{2\pi}{\lambda}\right)$ which may intersect at some point *B* of the reciprocal lattice.

(*iii*) Let the coordinate point *B* be ($h'$, $k'$, $l'$) which may have a highest common factor *n i.e.,* the coordinates are of the type ($nh$, $nk$, $nl$) where *h*, *k* and *l* do not have a common factor other than units.

(*iv*) Apparently $\vec{OB}$ represent the reciprocal lattice vector and is normal to some set of direct lattice planes.

(*v*) $|\vec{OB}| = \frac{2\pi n}{d}\left[\begin{array}{l} n = \text{is the largest factor common to number } (h'\ k'\ l') \\ d = \text{interplanar spacing)} \end{array}\right].$

## X-Rays:

A German scientist **W. K. Roentgen** in 1895 observed that when fast moving cathode rays strike a metal piece of high atomic weight and high melting point, a new kind of rays are produced. Since nothing was known about these rays, or radiation Roentgen called these rays as *X-rays.*

**Short wave length electromagnetic waves (0.5Å – 15Å) that are produced when high speed electrons are suddenly stopped are known as X-rays.**

## Production of X-rays:

### Principle:

X-rays are produced whenever electrons with energies in the region of kV strike heavy metallic target. The impinging of electrons lose their energy on collision with the target and at each collision, a part or extremely rarely the whole of the lost energy appears as X-rays.

### Instrumentation:

**Dr. W.D.Coolidge,** 1913 designed the X-ray tube. The coolidge tube consists of the following parts.

(*i*) It consists of a bulb 'B' with two coaxial projecting tubes $T_1$ and $T_2$.

(*ii*) In $T_1$, there is a tungsten filament (spiral) F which act as cathode. A cylindrical shield 'S' made of molybdenum cover it. When it is heated at 2000 °C the thermoelectrons are emitted.

(*iii*) The anode A is made of copper with a target tungsten or molybdenum having high atomic number and high melting point housed in other projecting tube $T_2$.

(*iv*) High potential difference is maintained between cathode and anode by the high tension source.

### Mechanism:

(*i*) High potential is maintained between 50-200 kV between the cathode and anode by means of a step up transformer.

(*ii*) This large potential difference across the tube causes the emission of electrons from the heated tungsten filament F.

(*iii*) Electrons emitted from the tungsten filament are focussed by cylindrical shield 'S' to the target with a high speed.

(*iv*) When the electrons strike the target, the electron's kinetic energy will be converted into X-rays. A part of the Kinetic energy will be converted into heat (97.8%).

(*v*) Due to continuous striking of electrons, the target T becomes very hot. The target is kept cool by running ice cold water in the copper tube around it.

(*vi*) The electrons emitted from the filament strike the target only during one half cycle of the alternating potential difference i.e. only when the target is positive with respect to the filament. During other half cycle, target becomes negative with respect to the filament and the electrons are repelled by the target. Thus the tube act as its own rectifier.

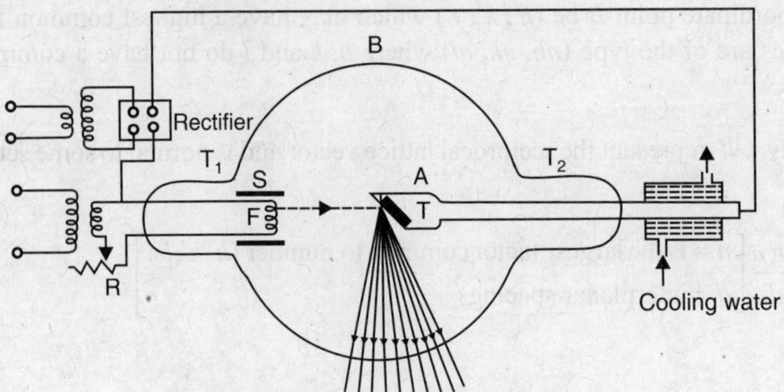


Fig. 7.12 Coolidge' tube and X-ray production

- **Types of X-rays:**

**(*a*) Soft X-rays:**

(*i*) X-rays having wavelength of 4Å or above longer wavelength, smaller frequency and hence smaller energy.

(*ii*) Due to their low penetrating power, they are called soft X-rays.

(*iii*) They are produced at comparatively low potential difference and high pressure.

**(*b*) Hard X-rays:**

(*i*) X-rays having smaller wavelength (1Å), high frequency and hence high energy.

(*ii*) Their penetrating power is large therefore they are called hard X-rays.

(*iii*) They are produced at comparatively low pressure and high potential difference.

- **Properties of X-rays:**

(*i*) X-rays like light are **electromagnetic waves** extremely short wavelength (0.01 nm to 10 nm). They show reflection, refraction, interference, diffraction and also polarisation under suitable condition.

(*ii*) X-rays do not contain any charged particles. So they are not deflected in electric and magnetic field. They travel in **straight lines** with the speed of light.

(*iii*) X-rays are highly **penetrating**. The shorter the wavelength, the greater is the penetrating power. They penetrate more in less density substance and less in high density one.

(*iv*) X-rays affect photographic plate even when wrapped in black papers by several fields.

(*v*) They produce **illumination** on falling on fluorescent material such as barium platinocyanide, zinc sulphide etc.

(*vi*) X-rays ionise gases through which they pass.

(*vii*) X-rays produce **photoelectric effect.**

(*viii*) They are capable of destroying living cell and cause **genetic mutation.**

- **Application of X-rays:**

**(*a*) In surgery:**

(*i*) X-rays easily penetrate into low-density substances like wood, paper, skin, flesh and blood but do not pass through high density substances like bone, iron etc.

(*ii*) This property is used to detect fractures of bones, tuberculosis of lungs and presence of bullet or stone in the body.

(*iii*) It also helps to detect bone fracture, bone decay or the presence of bullet or stone.

**(b) In radio therapy:**

(*i*) In cancer like disease, X-rays are allowed to fall on the affected part of the body to destroy cancer cell.

(*ii*) Heavy atoms diffract X-rays sufficiently, therefore to get information about the patient stomach, the patient is administered some heavy-atom substance (like barium sulphate solution).

**(c) In Detective Department:**

X-rays are used to detect contraband goods like gold hidden in sealed parcels, leather cases and human body.

**(d) In Engineering Department:**

X-rays are used to detect cracks, flaws and air bubles in iron girders fitted in buildings and bridges.

**(e) In Trade:**

X-rays are used to distinguish artificial diamonds from real diamonds.

**(f) In Laboratory:**

(*i*) It is used to study the internal structure of crystals and molecules. Example, the arrangement of sodium and chlorine atoms in a sodium-chloride crystal and their separation can be studied by X-rays.

- **X-ray diffraction:**

**X-ray diffraction constructs images from the diffraction pattern of X-rays passing through a crystalline specimen. This method is used to deduce the molecular structure at the atomic level of resolution and to analyze the structure of proteins, nuclic acids and other biological molecules at this resolution.**

**Features of X-ray diffraction:**

(*i*) Diffraction occurs when two or more waves are combined.

(*ii*) The peaks of one wave fall on the peaks of the other, the total effect constructive i.e. to enhance the intensity.

(*iii*) On the other hand, if the peaks of one wave falls on the trough of the other, the effect is destructive i.e. one of cancelling each other, i.e. combined intensity is zero.

(*iv*) If an incoming wave is split into two by slits and the two parts of the wave are allowed to interfere each other, then a pattern is produced on the screen placed on the other side of the screens.

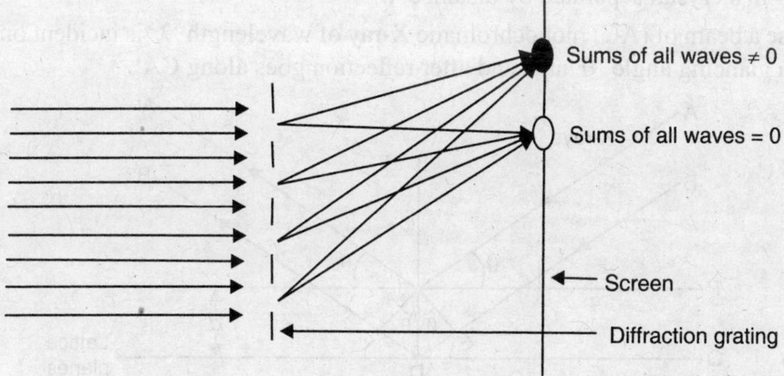


**Fig. 7.13** A schematic sketch illustrating diffraction from a grating

(*v*) Whether the two beams interfere *constructively* or *destructively* at a particular point on the screen is decided by the difference in length of the paths that each of the waves take from the slit to the point on the screen.

(*vi*) If the difference corresponding to the whole number of wavelengths, the combination of two waves is perfectly constructive and the intensity is greatly enhanced.

(*vii*) A regular spaced series of slits have the same type of effect and is called a *diffraction grating*.

(*viii*) It is possible to have two or three-dimensional gratings such as *crystall lattice* (three dimensional grating).

(*ix*) It cannot diffract visible light because the diffraction effect is noticeable where the spacing between the slits in the grating should be of the same order of magnitude as the wavelength of the incident radiation.

(*x*) In a crystal, the distance between the lattice points corresponds to the dimensions of atoms and molecules is a few angstrom, X-rays are radiation with wavelength of this size hence diffraction from crystal is observed with X-rays.

**Importance of X-ray for Crystal Structure Studies :**

I. The wavelength of X-rays (5Å to 15Å) is comparable for the interatomic distances in actual crystal.

*Example*: NaCl crystal, atomic spacing is $2.8 \times 10^{-10}$ m = 2.8Å

(*ii*) X-rays scattered elastically without change of wavelength by the charged particles of the atom.

• **Bragg' Law of Crystal Diffraction**

**W.L. Bragg discovered that a X-rays can be regularly reflected by the cleavage planes of the crystals when the rays are incident on their surface nearly at glancing angle. In a crystal, certain special successive atomic planes called cleavage planes, along which the atoms are more densely packed.**

The X-ray that penetrate more deeply into crystal are reflected from the lower planes are thus several beams of X-rays reflected from various planes are obtained.

The intensity of reflected beam at certain angles will be maximum where the two reflected waves from two different planes have a path difference equal to an integral multiple of the wave- length of rays, while at some other angle the intensity will be minimum.

**Procedure:**

(*i*) Let the horizontal lines represent the consecutive parallel lattice planes equidistant from one another in a crystal separated by distance '*d*'.

(*ii*) Suppose a beam of (AC) monochromatic X-ray of wavelength 'λ' is incident on the first plane (P) at a glancing angle 'θ' at C and after reflection goes along CA′

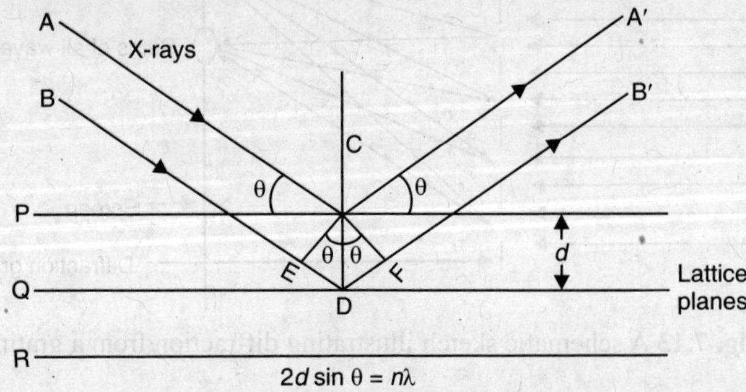


**Fig. 7.14** Braggs' Law

(*iii*) The other rays 'BD' is incident on cleavage plane 'Q' at the same glancing angle 'θ' After reflection it goes along DB′ making the angle 'θ' with the Plane Q.

(*iv*) The resultant beam arises from the reinforced reflection from all planes reached the incident beam such reinforcement occurs only if the path difference between successive rays such as AA′, BB′ is nλ where n is an integers.

**Observations:**

(*i*) Draw CE perpendicular on BD and CF per perpendicular on DB′, then CE is the incident wave front and CF is the reflected wavefront.

(*ii*) The path difference between two reflected rays CA′ and DB′ or two wavefronts is Δ *i.e.,*

$$\Delta = ED + DF$$

Now in Δ CED

$$ED = CD \sin\theta$$

and in Δ CDF, $DF = CD \sin\theta$

Path Difference $= ED + DF$

$$= CD \sin\theta + CD \sin\theta$$

$$= 2\, CD \sin\theta$$

$$= 2\, d \sin\theta \qquad [\because CD = PQ = d]$$

(*iii*) If the path difference is an integral multiple of wavelength λ, constructive interference will take place between the reflected beams.

(*iv*) Thus the intensity will be maximum if $2\, d \sin\theta = n\lambda$.

where $n = 1, 2, 3 \ldots.$

If $n = 1$ it is called first order reflection.

If $n = 2$ is is called second order reflection.

(*v*) This phenomenon is similar to that of the optical diffraction with diffraction grating with only this difference that have various parallel lattice planes of the crystal act as lines in the grating.

**Inference:**

(*i*) Bragg' law reveal that the diffraction intensities can be stronger only at certain values of θ for specific values of λ and d.

(*ii*) The knowledge of diffraction angle and wavelength of X-rays can provide information about the interplanar spacing of crystal.

**Bragg's Law and Equation**

**When X-rays of wavelength 'λ'are incident on a crystal surface making an angle 'θ' with the surface of the crystal, these suffer reflection and constructive interference takes place between the beam reflected between two consecutive lattice planes of the crystal separated by a distance 'd' then the relation**

$$2d \sin\theta = n\lambda$$

The law represented by an equation *i.e.,* called **Bragg's equation.**

$2d \sin\theta = n\lambda$ $\quad\lambda$ = wavelength

$d$ = consecutive lattice plane distance

θ = angle (Glancing Angle)

$n$ = the order of diffraction spectrum maxima.

- **Characteristic features of Bragg's Law:**

1. It is a consequence of periodicity of the space lattice.

2. The law does not refer to the arrangement or basis of atoms associated with its lattice point.
3. The composition of the basis determines the relative intensity of the various orders of diffraction from a given set of parallel planes.
4. Bragg's reflection can occur only for wavelength $\lambda \le 2d$. **[This is the reason for that visible wavelength can not be used in diffraction]**
5. For same order and spacing the angle θ decreases as the wavelength decreases. Consequently for the gamma rays angle must be used.

**Significance:**

(*i*) The diffraction intensity can be stronger only at certain values of θ for specific values of λ and *d*.

(*ii*) Diffraction angle and wavelength of X-ray can provide the information about inter planar spacing of the crystal.

(*iii*) By using monochromatic X-ray in any diffraction experiment, the space, shape, orientation of the unit cell can be determined.

- **Difference between Bragg's reflection and ordinary reflection**

(*i*) In ordinary reflection all reflected waves are in phase and there are no maxima and minima.

(*ii*) The phenomenon of Bragg's reflection is similar to that of diffraction of light from a diffraction grating with only this difference that in Bragg's reflection various parallel planes of the crystal serve the same purpose as the ruled lines on the grating which produce opacities and transparencies.

- **Experimental methods of X-ray Diffraction:**

The Bragg's law $2d \sin \theta = n\lambda$ for X-ray diffraction requires that θ and λ be matched *i.e.*, X-rays of wavelength λ striking a three dimensional crystal at an arbiratory angle of incidence in general will not be reflected. To satisfy the Bragg's law it is necessary to scan in either wavelength λ or angle θ.

The standard experimental methods are

(*a*) Laue's method

(*b*) Powder method

(*c*) Rotating crystal method

- **Laue's method of diffraction:**

(*i*) In this method a single crystal is held stationary, in a beam of X-rays having continuous wavelength.

(*ii*) The crystal selects and produces diffraction corresponding to discrete values of λ for which planes of spacing '*d*' exist and angle of θ satisfies Bragg's relation.

**Procedure:**

(*i*) A narrow beam of continuous wavelength over a wide range (perhaps 0.2Å to 2Å) X-rays from an X-ray tube were used.

(*ii*) It is collimated by the fine slits $S_1$ and $S_2$ and was allowed to pass through a single thin crystal specimen. Zinc sulphide (ZnS) not more than one mm in size held stationary in a stand.

(*iii*) The transmitted radiation was read on the other side on a photographic plate 'P'.

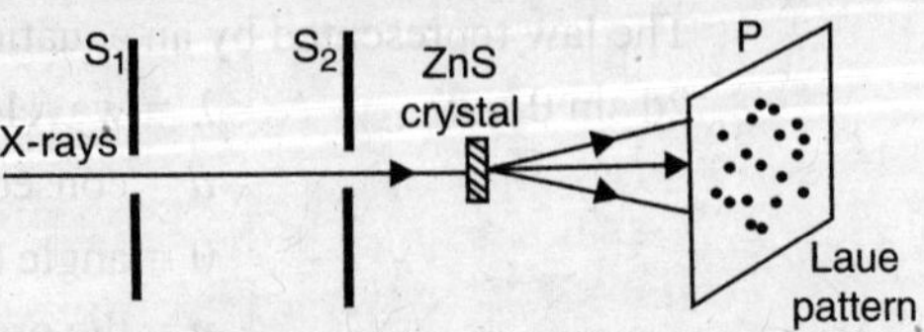


**Fig. 7.15** Laue diffraction

**Observation :**

(*i*) After several hours of exposure, the plate was developed.

(*ii*) It was observed that (*a*) a central intense black spot and (*b*) a series of fainter black spot arranged in a definite regular pattern around the central black spot.

**Inference :**

(*i*) This pattern is known as *Laue diffraction pattern* or the *laue photograph* and spots are known as *Laue spots*.

(*ii*) It establishes that X-rays are electromagnetic waves and crystals behave as three-dimensional gratings.

**Usefulness/Significance :**

(*i*) It is suitable for rapid determination of crystal orientation and symmetry. For example, If a crystal has four-fold axial symmetry and is oriented with the axis paralled to the X-ray beam, the laue pattern will show four-fold symmetry. The symmetry of pattern also help to determine the shape of the unit cell.

(*ii*) An imperfect or strained crystal has atomic plane which are not exactly plane and but are slightly curved, we get streak's in laue's pattern instead of sharp diffraction spot. This spot is known as *asterism*.

(*iii*) It is important for studying the extent of crystalline imperfection under mechanical or thermal treatment.

(*iv*) This is not suitable for determining crystal structure.

- **Powder Method of Crystal Diffraction:**

(*i*) There are many substances that are basically crystalline but cannot be obtained as large single crystal.

(*ii*) The powder sample is actually a aggregate of small crystals or crystallites. The studies of this sample is called powdered technique.

(*iii*) This technique was developed by Debye-Scherrer in Germany 1916 for the determination of the crystal structure of small grained crystallites or finely powdered polycrystallines.

**Procedure:**

(*i*) In this method a powdered sample of crystalline material is placed in a fixed position in a monochromatic X-ray beam.

(*ii*) Here the substance is ground and finely powdered so that it could be assumed that all the minute crystals have completely random orientation.

(*iii*) A pencil of monochromatic X-ray is incident on the powdered specimen S. Some of the microcrystal are so oriented that the Bragg's condition or equation of reflection, $2d \sin\theta = n\lambda$ is satisfied.

(*iv*) The diffracted rays leave the specimen along the direction which generate a number of cones concentric with direction of the original beam and make an angle $2\theta$ with it, where '$\theta$' is Bragg's angle.

**Observation:**

(*i*) The scattered beams has the maximum intensity at an angle '$2\theta$' with the incident X-ray beam. Since the crystal have random orientations, the locus of these maxima would be a cone of semivertical angle $2\theta$.

(*ii*) If the photographic film PP be kept at right angle to incident beam, the above cone would intersect the plane of the film in a ring.

(*iii*) A number of concentric ring corresponding to different orders of reflection would thus be obtained in the film.

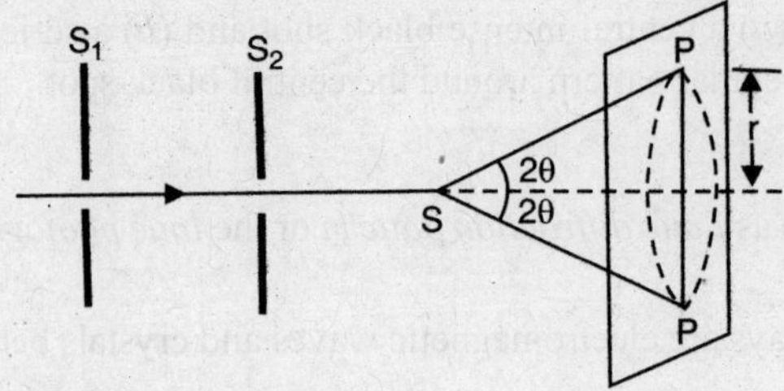


(*a*) Debye-Scherrer method of powder crystal diffraction

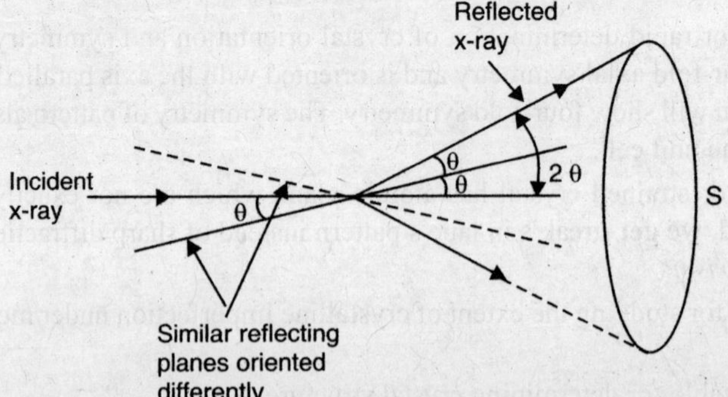


(*b*) A cone produced by reflection of x-ray from identical planes having different orientations.

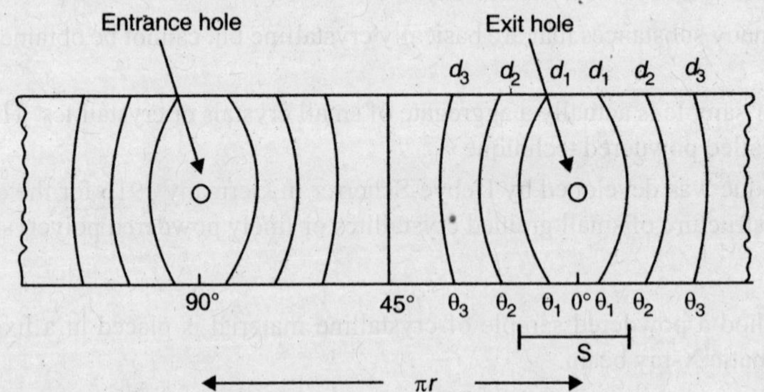


(*c*) Flattened photographic film after developing and indexing of diffraction lines.

**Fig. 7.16** Powder Method of Crystal diffraction

**Inference:**

(*i*) From the distances between the lines and the dimensions of the film, the angle θ can be calculated for each line.

(*ii*) It is most useful for crystal with only one or two parameters.

- **Rotating crystal:**

(*i*) In this method a small and well formed single crystal is mounted perpendicular to the beam.

(*ii*) The film is generally held in a cylindrical camera.

(*iii*) Now when the crystal is rotated slowly, and during the time of exposure, a successive planes pass through orientation, which are necessary for Bragg's reflection, each produceds a spot on the film.

(*iv*) The position on the film when developed indicates orientation of the crystal at which the spot was formed.

(*v*) The data obtained from these spots give information about the structures of ordinary and complex molecules.

- **Atomic radius ($r$):**

  It is defined as the half distance between the two nearest neighbouring atoms in a crystal of a pure element.

- **Nearest neighbour distance ($2r$):**

  (*i*) The distance between the centres between two nearest neighbouring atoms is called nearest neighbour distance.

  (*ii*) If $r$ is the radius of the atoms, nearest neighbour distance is $2r$.

- **Co-ordination number ($N$):**

  (*i*) The co-ordination number in a crystal is defined as the number of equivalent nearest neighbours to an atom in a given crystal, structure.

  (*ii*) The coordination number is calculated by taking nearest neighbouring atoms at a distance $2r$.

- **Electron volt:**

  An electron-volt is the kinetic energy acquired by an electron when it is made to move in an electric field and 1 volt potential gradient.

  According to the definition of potential difference, the work done in accelerating charge $e$ through a potential $V$ is eV.

  Charge of electron, $e = 1.6 \times 10^{-19}$ coulomb.

  Potential difference, $V = 1$ volt.

  $\therefore$ 1 electron volt, eV $= 1.6 \times 10^{-19} \times 1$ volt $= 1.6 \times 10^{-19}$ joule.

## Problems on Miller Indices:

**1.** *Find the Miller indices for planes in each of the following sets which intercept $\vec{a}, \vec{b}$ and $\vec{c}$ at (i) 3a, 3b, 2c (ii) a, 2b, α (iii) a, b/2, c (iv) 3a, 2b, 2c* ***(P.U. 2003)***

**Solution :**

(*i*) Set of intercepts = $3a, 3b, 2c$

Reciprocal number representing the intercepts are $\frac{1}{3}, \frac{1}{3}, \frac{1}{2}$. Smallest three integers having the same ratio are $\frac{1}{3}\times 6, \frac{1}{3}\times 6, \frac{1}{2}\times 6$

$= 2, \quad 2, \quad 3$

$\therefore$ Miller indices are (223)

(*ii*) Set of intercepts $a, 2b, \alpha$

Reciprocal of numbers are $1, \frac{1}{2}, 0$

Smallest three integers having the same ratio 2, 1, 0

$\therefore$ Miller indices are (210)

(*iii*) Set of intercepts $a, \frac{b}{2}, c$

Reciprocal of numbers are = 1, 2, 1

Smallest three integers having the same ratio 1, 2, 1

∴ Miller indices are (121)

(*iv*) Set of intercepts $3a, 2b, 2c$

Reciprocal of numbers are = $\frac{1}{3}, \frac{1}{2}, \frac{1}{2}$

Smallest three integers having the same ratio $\frac{1}{3} \times 6, \frac{1}{2} \times 6, \frac{1}{2} \times 6$

$= 2, \quad 3, \quad 3$

∴ Miller indices are (233)

**2.** *Explain the meaning of (322) and (111) planes.*

**Solution.**

(*i*) (3, 2, 2) plane has miller indices 3, 2, 2

Its intercepts on $\vec{a}, \vec{b}$ and $\vec{c}$ axes are $\frac{1}{3}, \frac{1}{2}$ and $\frac{1}{2}$.

Taking the whole numbers, intercepts are 2, 3, 3 *i.e.* $2a$, $3b$ and $3c$.

(*ii*) (111) plane has Miller indices 1, 1, 1.

Its intercepts on $\vec{a}, \vec{b}$ and $\vec{c}$ axes are 1, 1 and 1

*i.e.*, $a$, $b$ and $c$.

**3.** *In a crystal, a lattice plane cuts intercepts of 2a, 3b, and 6c along the axes, where a, b and c are primitive vectors of the unit cell. Determine the Miller indices of the given plane.*

**Solution:**

Set of intercepts are $2a$, $3b$ and $6c$.

Reciprocal number representing the intercepts are $\frac{1}{2}, \frac{1}{3}, \frac{1}{6}$. Smallest three integers having the same ratio are $\frac{1}{2} \times 6, \frac{1}{3} \times 6, \frac{1}{6} \times 6$ *i.e.*, 3, 2, 1.

∴ Miller indices are = (3 2 1)

**4.** *Find the Miller indices of the plane which intercepts $\vec{a}, \vec{b}$ and $\vec{c}$ at 4, 1, 2.*

**Solution:**

Set of intercepts are a 4, 1, 2.

Reciprocal number representing the intercepts are $\frac{1}{4}, \frac{1}{1}, \frac{1}{2}$.

Smallest three integers having the same ratio are $\frac{1}{4} \times 4, 1 \times 4, \frac{1}{2} \times 4$ i.e. 1, 4, 2.

∴ Miller indices are (1 4 2).

**5.** *Miller indices are (316). Find the intercepts made by the plane on the three crystallographic axes.*

**Solution:**

Intercepts on the $\vec{a}, \vec{b}$ and $\vec{c}$ axes are $\frac{1}{3}$, 1 and $\frac{1}{6}$.

Taking the whole numbers; intercepts are

$\frac{1}{3} \times 6 = 2,\ 1 \times 6 = 6,\ \frac{1}{6} \times 6 = 1$ i.e. 2, 6, 1 i.e. $2a$, $6b$ and $c$.

**6.** *In a triclinic crystal, a lattice plane makes intercepts of lengths a, 2b and* $\frac{3}{2}c$. *Find the Miller indices of the plane.*

**Solution.**

Set of intercepts are $a$, $2b$, $\frac{-3}{2}c$, i.e. $1, 2, \frac{-3}{2}$. Reciprocal number representing the intercepts are $1, \frac{1}{2}, \frac{-2}{3}$.

Smallest three integers having he same ratio are $1 \times 6, \frac{1}{2} \times 6, \frac{-2}{3} \times 6$ *i.e.,* 6, 3, – 4.

Hence the Miller indices of the plane are $(6\ 3\ \bar{4})$.

**7.** *A plane intercepts at a* $\frac{b}{2}$, *3c in a simple cubic unit cell. Find the Miller indices of the plane.*

**Solution:**

Intercepts are $a, \frac{b}{2}, 3c$ *i.e.,* $1, \frac{1}{2}, 3$.

Reciprocal number representing the intercepts are $1, 2, \frac{1}{3}$.

Smallest integers having the same ratio are $1 \times 3, 2 \times 3, \frac{1}{3} \times 3$ *i.e.,* 3 6 1. Hence Miller indices of the plane are (3 6 1).

**8.** *Miller indices of the planes are (3 2 6). Find its intercepts of the plane.*

**Solution:**

Its Miller indices are (3 2 6) intercepts on $\vec{a}, \vec{b}$, and $\vec{c}$ axes are $\frac{1}{3}, \frac{1}{2}$ and $\frac{1}{6}$.

Taking the whole numbers, the intercepts are 2, 3, 1 *i.e.,* $2a$, $3b$ and $c$.

**9.** *In an orthorhombic crystal, a lattice plane cuts intercepts of lengths 3a, 2b and* $\frac{3c}{2}$ *along three axes. Deduce the Miller indices of the plane a, b, c are the primitive vectors of the unit cell.*

**Solution:**

Set of intercepts are $3a, -2b, \frac{3c}{2}$. *i.e.,* $3, -2, \frac{3}{2}$. Reciprocal number representing the intercepts are $\frac{1}{3}, -\frac{1}{2}, \frac{2}{3}$. Smallest three integers having the same ratio are $\frac{1}{3} \times 6, -\frac{1}{2} \times 6, \frac{2}{3} \times 6$ *i.e.,* 2, –3, 4.

Hence the Miller indices of the plane are $(2\ \bar{3}\ 4)$.

**10.** *Find the Miller indices of a set of parallel planes which make intercepts in the ratio of 2a : 3b on the X and Y axis and are parallel to Z axis;* $\vec{a}, \vec{b}, \vec{c}$ *being primitive vectors of the lattice.*

**Solution:**

Intercept on the $X$ axis = $2a$

Intercept on the $Y$ axis = $3b$

Intercept on the $Z$ axis = $\alpha$ [as the set of planes parallel to $Z$ axis]

Reciprocal number representing the intercepts are $\frac{1}{2}, \frac{1}{3}, \frac{1}{\alpha}$ or zero.

Smallest three integers having the same ratio are $\frac{1}{2} \times 6, \frac{1}{3} \times 6, \frac{1}{0} \times 6$ *i.e.*, 3, 2, 0.

Hence the Miller indices ($hkl$) are (3 2 0).

**11.** *Find the Miller indices of a set of parallel planes which make intercepts in the ratio 3a : 4b on the X and Y axes and are parallel to the Z axis a, b, c being primitive vectors of the lattice?*

Solution:

Intercept on the $X$ axis = $3a$

Intercept on the $Y$ axis = $4b$

Intercept on the $Z$ axis = $\alpha$ [as the set of planes parallel to $Z$ axis]

Reciprocal number representing the intercepts are $\frac{1}{3}, \frac{1}{4}, \frac{1}{\alpha}$ or zero.

Smallest three integers having the same ratio are $\frac{1}{3} \times 12, \frac{1}{4} \times 12, \frac{1}{\alpha} \times 12 = 4\ 3\ 0$

Hence the Miller indices ($hkl$) are (4 3 0).

**12.** *Calculate interplanar spacing for a (3 2 1) plane in a simple cubic lattice whose lattice constant is $4.2 \times 10^{-10}$ m.*

**Solution:**

$$d_{hkl} = \frac{a}{\sqrt{h^2+k^2+l^2}} = \frac{4.2 \times 10^{-10}}{\sqrt{3^2+2^2+1^2}} = \frac{4.2 \times 10^{-10}}{\sqrt{9+4+1}} = \frac{4.2 \times 10^{-10}}{\sqrt{14}}$$

$$= \frac{4.2 \times 10^{-10}}{3.74} = 1.123 \times 10^{-10}.$$

**13.** *Show that for a simple cubic lattice*

$$d_{100} : d_{110} : d_{111} = \sqrt{6} : \sqrt{3} : \sqrt{2}.$$

*where $d_{hkl}$ is the separation between adjacent (hkl) parallel planes.* **(C.U. 2005)**

**Solution:**

The spacing $d_{hkl}$ of plane ($hkl$) in simple cubic lattice of a side a is $d_{hkl} = \frac{a}{\sqrt{h^2+k^2+l^2}}$.

For (100) plane $d_{100} = \frac{a}{\sqrt{1^2+0^2+0^2}} = a$

For (110) plane $d_{110} = \frac{a}{\sqrt{1^2+1^2+0^2}} = \frac{a}{\sqrt{2}}$

For (111) plane $$d_{111} = \frac{a}{\sqrt{1^2+1^2+1^2}} = \frac{a}{\sqrt{3}}$$

$$d_{100} : d_{110} : d_{111} = a : \frac{a}{\sqrt{2}} : \frac{a}{\sqrt{3}} = \sqrt{6} : \sqrt{3} : \sqrt{2} \text{ (proved)}$$

**14.** *For a cubic lattice calculate the distance of (1 2 3) and (2 3 4) planes from a plane passing through the origin.* ***(HPU 1993)***

**Solution :**

For (1 2 3) plane $$d_{123} = \frac{a}{\sqrt{1^2+2^2+3^2}} = \frac{a}{\sqrt{1+4+9}} = \frac{a}{\sqrt{14}}$$

For (2 3 4) plane $$d_{234} = \frac{a}{\sqrt{2^2+3^2+4^2}} = \frac{a}{\sqrt{4+9+16}} = \frac{a}{\sqrt{29}}$$

**15.** *Lattice constant of a cubic lattice is a calculate the spacing between (211), (112), (001), (110), (011), (100), (111) and (101) planes.* ***(Meerut Univ. 2003).***

**Solution:**

The spacing $d$ of plane $(hkl)$ in a simple cubic lattice of side $a$ is $d = \frac{a}{\sqrt{h^2+k^2+l^2}}$

$$= \frac{a}{(h^2+k^2+l^2)^{\frac{1}{2}}}$$

(*i*) For (211) plane $$d = \frac{a}{\sqrt{2^2+1^2+0^2}} = \frac{a}{\sqrt{6}}$$

(*ii*) For (112) plane $$d = \frac{a}{\sqrt{1^2+1^2+2^2}} = \frac{a}{\sqrt{6}}$$

(*iii*) For (001) plane $$d = \frac{a}{\sqrt{0^2+0^2+1^2}} = \frac{a}{\sqrt{(1)^2}} = a$$

(*iv*) For (110) plane $$d = \frac{a}{\sqrt{1^2+1^2+0}} = \frac{a}{\sqrt{2}}$$

(*v*) For (011) plane $$d = \frac{a}{\sqrt{0^2+1^2+1^2}} = \frac{a}{\sqrt{2}}$$

(*vi*) For (100) plane $$d = \frac{a}{\sqrt{1^2+0^2+0^2}} = a$$

(*vii*) For (111) plane $$d = \frac{a}{\sqrt{1^2+1^2+1^2}} = \frac{a}{\sqrt{3}}$$

(*viii*) For (101) plane $$d = \frac{a}{\sqrt{1^2+0^2+1^2}} = \frac{a}{\sqrt{2}}$$

**16.** *Calculate the interplaner distance d of the planes (3 2 0) taking the lattice to be a cube with a = b = 3 Å.*

**Solution:**

Interplaner distance $d = \dfrac{a}{\sqrt{h^2 + k^2 + l^2}}$

Here $a = 3$ Å

$$\therefore \quad d = \frac{3\text{ Å}}{\sqrt{3^2 + 2^2 + 0^2}} = \frac{3\text{ Å}}{\sqrt{9 + 4 + 0}} = \frac{3}{\sqrt{13}}\text{ Å}$$

**17.** *The lattice constant of a cubic crystal is 2.25 Å. Find the interplanar spacing or the set of crystallographic plane having Miller indices (100).* ***(C.U. 2001)***

**Solution:**

The lattice constant $a = 2.25$ Å

We know $d_{hkl} = \dfrac{a}{\sqrt{[h^2 + k^2 + l^2]}} = \dfrac{a}{[h^2 + k^2 + l^2]^{\frac{1}{2}}}$

$$\therefore \text{ Interplanar spacing } d_1 = d_{100} = \frac{2.25}{\sqrt{1^2 + 0^2 + 0^2}} = \frac{2.25}{1} = 2.25\text{ Å}$$

Let $d_2$ be the perpendicular distance from next plane parallel to (100) plane.

$$\therefore \quad d_2 = \frac{2\,a}{\sqrt{h^2 + k^2 + l^2}} = \frac{2 \times 2.25}{\sqrt{1^2 + 0^2 + 0^2}} = 4.50\text{ Å} = 4.5\text{ Å}$$

Thus interplanar spacing between two adjacent planes of Miller indices (100) is $d = d_2 - d_1$

$= 4.5 - 2.25 = 2.25$ Å

**18.** *Calculate interplanar spacing 'd' of the plane (iii) which is a simple cubic lattice of side a.* ***(W.B.U.T 2003)***

**Solution:**

Interplanar spacing '$d$' is given by $d_{111} = \dfrac{a}{\sqrt{(h^2 + k^2 + l^2)}}$.

$$d_{111} = \frac{a}{\sqrt{(1^2 + 1^2 + 1^2)}} = \frac{a}{\sqrt{3}}$$

**19.** *The orthorhombic crystal has axial units in the ratio of 0.424 : 1 : 0.367. Find the Miller indices of a crystal face whose intercepts are in the ratio 0.212 : 1 : 0.183.*

**Solution:**

The axial units are $a : b : c :: 0.424 : 1 : 0.367$

$pa = 0.212$ or $p \times 0.424 = 0.212$

$$\therefore \quad p = \frac{0.212}{0.424} = \frac{1}{2}$$

Similarly $qb = 1$ or $q \times 1 = 1$

$\therefore q = 1$

and $rc = 0.183$

$r \times 0.367 = 0.183$ or $r = \dfrac{0.183}{0.367} = \dfrac{1}{2}$

Hence parameter of the plane is $\dfrac{1}{2}, 1, \dfrac{1}{2}$

Miller indices $= \dfrac{1}{1/2} : \dfrac{1}{1} : \dfrac{1}{1/2} = 2 : 1 : 2$

$\therefore$ (2 1 2)

**20.** *To pay an orthorhombic semi-precious gem has a ratio of a : b : c of 0.529 : 1 : 0.477. Find the Miller indices of faces whose intercepts are as follows:*

*0.264 : 1 : 0.238*

*1.057 : 1 : 0.954*

*0.529 : α : 0.159*

**Solution:**

The axial units are $a : b : c = 0.529 : 1 : 0.477$.

(*a*) $pa = 0.264$ or $p \times 0.529 = 0.264$

$\therefore p = \dfrac{0.264}{0.529} = \dfrac{1}{2}$

Similarly

$qb = 1$ or $q \times 1 = 1$ $\therefore q = 1.$

and

$rc = 0.238$ or $r \times 0.477 = 0.238$

$\therefore r = \dfrac{0.238}{0.477} = \dfrac{1}{2}$

Hence parameter of the plane is $\dfrac{1}{2} : 1 : \dfrac{1}{2}$

(*b*) $pa = 1.057$ $p \times 0.529 = 1.057$

$\therefore p = \dfrac{1.057}{529} = 2$

$qb = 1$ or $q \times 1 = 1$ $\therefore q = 1$

and $rc = 0.954$ $r \times .477 = 0.954$

$\therefore r = \dfrac{0.954}{0.477} = 2$

$\therefore$ Here parameter of the plane is 2 1 2.

(*c*) $pa = 0.529$ or $p \times 0.529 = 0.529$ $\therefore p = \dfrac{0.529}{0.529} = 1$

$qb = \infty$ or $q \times 0 = \infty$ $\therefore q = \dfrac{\infty}{0} = 0$

$rc = 0.159$ or $r \times 0.477 = 0.159$ $\therefore r = \dfrac{0.159}{0.477} = \dfrac{1}{3}$

Here parameter of the plane is 1 0 1.

Parameters are $\frac{1}{2}:1:\frac{1}{2};\ 2:1:2;\ 1:0:\frac{1}{3}$.

Taking the reciprocal of the number and reducing to the smallest set of whole numbers, the following Miller indices are obtained.

(*a*) $\frac{1}{1/2}=2\ \frac{1}{1}=1\ \frac{1}{1/2}=2$ i.e. (2 1 2)

(*b*) $\frac{1}{2}:1:\frac{1}{2}=1\ 2\ 1$ i.e. (1 2 1)

(*c*) $\frac{1}{1}=\frac{0}{0}=\frac{1}{1/3}=3$ i.e. (1 0 3)

**21.** *A certain orthorhombic crystal has a ratio a : b : c = 0.429 : 1 : 0.377. Find the Miller indices of the faces whose intercepts are*

(*i*) *0.214 : 1 : 0.188*

(*ii*) *0.8585 : 1 : 0.754*

(*iii*) *0.429 : ∝ : 0.126*

**Solution:**

The axial units are $a : b : c = 0.429 : 1 : 0.377$

(*i*) $pa = 0.214$ or $p \times 0.429 = 0.214$ $\therefore p = \frac{0.214}{0.429} = \frac{1}{2}$

$qb = 1$ or $q \times 1 = 1$ $\therefore q = 1$

$rc = 0.188$ or $r \times 0.377 = 0.188$ is $r = \frac{0.188}{0.377} = \frac{1}{2}$

Parameter of the plane is $\frac{1}{2}, 1, \frac{1}{2}$.

(*ii*) $pa = 0.858$ or $p \times 0.429 = 0.858$ $\therefore p = \frac{0.858}{0.429} = 2$

$qb = 1$ or $q \times 1 = 1$ $\therefore q = 1$

$rc = 0.754$ or $r \times 0.377 = 0.754$ $\therefore r = \frac{0.754}{0.377} = 2$

Parameter is 2, 1, 2.

(*iii*) $pa = 0.429$ or $p \times 0.429 = 0.429$ $\therefore p = \frac{0.429}{0.429} = 1$

$qb = \propto$ or $q \times 1 = \propto$ $\therefore q = \frac{\propto}{1}$

$rc = 0.126$ or $r \times 0.377 = 0.126$ or $r = \frac{0.126}{0.377} = \frac{1}{3}$

Parameter is $1 \propto \frac{1}{3}$.

Taking reciprocal of the number and reducing smallest whole numbers the following Miller indices are obtained.

(*i*) $\frac{1}{2}, \frac{1}{1/2} = 2, 1, \frac{1}{1/2} = 2$ *i.e.*, (2 1 2)

(*ii*) $\frac{1}{2}, \frac{1}{2} 1 = 1, 1, 2$ *i.e.*, (1 1 2)

(*iii*) $1, \frac{1}{3}, 3 = 1, 0, 3$ *i.e.*, (1 0 3)

**22.** *Lead is a face centred cubic with an atomic radius of 1.746 Å. Find the spacing of 200 planes are 220 planes.[Lattice constant of lead (FCC) i.e.* $a = \frac{4r}{\sqrt{2}}$]

**Solution:**

$$d_{hkl} = \frac{a}{\sqrt{h^2 + k^2 + l^2}}$$

The lattice contant $a = \frac{4r}{\sqrt{2}} = \frac{4 \times 1.746}{\sqrt{2}} = \frac{6.984}{1.414} = 4.93$

(*a*) For (200) plane $= \frac{4.93}{\sqrt{(2)^2 + (0)^2 + (0)^2}} = \frac{4.93}{\sqrt{4}} = \frac{4.93}{2} = 2.465$ Å

(*b*) For (220) plane $= \frac{4.93}{\sqrt{(2)^2 + (2)^2 + (0)^2}} = \frac{4.93}{\sqrt{4+4}} = \frac{4.93}{\sqrt{2 \times 4}} = \frac{4.93}{2\sqrt{2}}$

$$= \frac{4.93}{2 \times 1.414} = \frac{2.465}{1.414} = 1.74 \text{ Å}$$

**23.** *An X-ray bulb is operated on 50 kV. Find the shortest wave length of X-rays given out by the bulb.*

**Solution:**

If $\lambda_{min}$ be the minimum wave length of X-ray photon, $v_{max} = \frac{C}{\lambda_{min}}$ [$C$ = Velocity of light]

$v_{max} = \frac{eV}{h}$ or $\lambda_{min} = \frac{C}{v_{max}} = \frac{ch}{eV}$ (Duane Hunt Law)

Given $V = 50$ kV $= 50 \times 10^3$ GV.

Known $\left[\begin{array}{l} c \text{ (velocity of light) } 3 \times 10^{10} \text{ cm/sec} = 3 \times 10^8 \text{ m s}^{-1} \\ h \text{ (Plank's constant)} = 6.624 \times 10^{-34} \text{ J-s} \\ e = 1.6 \times 10^{-19} \text{ coulomb} \end{array}\right]$

$$\lambda_{min} = \frac{3 \times 10^8 \times 6.624 \times 10^{-34}}{1.6 \times 10^{-19} \times 50}$$

$$= \frac{3 \times 6.624 \times 10^{-26} \times 10^{19}}{1.6 \times 50 \times 10} = \frac{3 \times 6.624 \times 10^{-7}}{1.6 \times 50} = \frac{3 \times 6624 \times 10^{-7}}{1.6 \times 10^3 \times 50}$$

$$= \frac{3 \times 6624 \times 10^{-10}}{1.6 \times 50 \times 10^{3}} = \frac{19872 \times 10^{-10}}{1.6 \times 50 \times 10^{3}} = \frac{12420 \times 10^{-10}}{50 \times 10^{3}}$$

$$= \frac{12420}{50000} \times 10^{-10} = 0.2484 \times 10^{-10} = 0.248 \text{ Å}$$

**24.** *An X-ray bulb is working at 30 kV. Find the shortest wavelength of X-rays produced by the bulb.*

**Solution:**

$$\lambda_{min} = \frac{ch}{eV} = \frac{3 \times 10^{8} \times 6.624 \times 10^{-34}}{1.6 \times 10^{-19} \times V} = \frac{12420 \times 10^{-10}}{V} = \frac{12420 \times 10^{-10}}{30 \times 10^{3}}$$

$$= \frac{12420}{30000} \times 10^{-10} = 0.414 \times 10^{-10} = 0.414 \text{ Å}$$

**25.** *Calculate the maximum frequency and minimum wavelength of X-rays produced in a tube maintained at 13.26 kV. ($h = 6.63 \times 10^{-34}$ Js, $c = 3.0 \times 10^{8}$ m s$^{-1}$, $e = 1.6 \times 10^{-19}$ C).*

**Solution:**

Maximum frequency of the X-ray produced in the tube at $V$ volt is given by $v_{max} = \frac{eV}{h}$

Here given $V = 13.26$ kV $= 13.26 \times 10^{3}$ V

$$\therefore v_{max} = \frac{1.6 \times 10^{-19} \times 13.26 \times 10^{3}}{6.63 \times 10^{-34}}$$

$$= \frac{1.6 \times 13.26 \times 10^{-16}}{6.63 \times 10^{-34}} = \frac{1.6 \times 13.26 \times 10^{-16} \times 10^{34}}{6.63}$$

$$= \frac{21.216 \times 10^{18}}{6.63} = 3.2 \times 10^{18} \text{ Hz}$$

$$\lambda_{min} = \frac{C}{v_{max}} = \frac{3 \times 10^{8}}{3.2 \times 10^{18}} = \frac{3 \times 10^{-9}}{32} = 0.09375 \times 10^{-9} = 0.9375 \times 10^{-10}$$

$$= 0.9375 \text{ Å}$$

**26.** *Find the maximum frequency of X-rays emitted by an X-ray tube operating at 30 kV.*

**Solution:**

We know

$$v_{max} = \frac{eV}{h} = \frac{1.6 \times 10^{-19} \times 30 \times 10^{3}}{6.624 \times 10^{-34}} = \frac{1.6 \times 30 \times 10^{-16} \times 10^{34}}{6.624}$$

$$= 0.2415 \times 30 \times 10^{18}$$

$$= 7.246 \times 10^{18} \text{ Hz.}$$

**27.** *Calculate maximum frequency of X-rays operating at a voltage of 50 kV. $h = 6.6 \times 10^{-3}$ Js, $e = 1.6 \times 10^{-19}$ C.*

**Solution:**

We know

$$v_{max} = \frac{eV}{h} = \frac{1.6 \times 10^{-19} \times 50 \times 10^{3}}{6.6 \times 10^{-34}}$$

$$= \frac{1.6 \times 50 \times 10^{-16} \times 10^{34}}{6.6} = \frac{1.6 \times 50 \times 10^{18}}{6.6} = 0.242 \times 50 \times 10^{18}$$

$$= 12.12 \times 10^{18} \text{ Hz}$$

**28.** *An X-ray bulb operates at 50 kV. Calculate the shortest wavelength of the X-ray produced. Given: h = 6.62 × 10^{–34} Js, e = 1.6 × 10^{–19} coulomb.* ***(Vid. U. 2005)***

**Solution:**

If all the energy of the electron is converted into X-radiation the X-ray produced will have the shortest wavelength which is given by

$$\lambda_{min} = \frac{ch}{eV}$$

Here $h = 6.62 \times 10^{-34}$ Js  $e = 1.6 \times 10^{-19}$ C

$V = 50$ kV $= 50 \times 10^3$ volt  $c = 3 \times 10^8$ m/s

$$\therefore \lambda_{min} = \frac{6.62 \times 10^{-34} \times 3 \times 10^8}{1.6 \times 10^{-19} \times 50 \times 10^3} \text{ metre}$$

$$= 0.25 \times 10^{-10} \text{ m}$$

$= 0.25$ Å nearly.

**29.** *X-rays penetrate through the flesh, but not through the bones. Why*

**Ans.**

(*i*) The penetrating power of the X-rays depends upon the voltage applied across the tube producing X-rays.

(*ii*) Doctor's apply a potential difference of about 60,000 volt for taking X-ray of the patient body.

(*iii*) These X-rays penetrate through the lighter elements like the flesh (which is composed of carbon, oxygen and hydrogen) but can not pass through heaviour elements like bone (which are made of phosphorus and calcium).

**30.** *X-rays are obtained by electrons and electrons can be produced by X-rays. Discuss.*

**Ans.**

(*i*) In X-ray tube, when energetic electrons fall on the heavy target, they are stopped after striking one or more atoms of the target.

(*ii*) Their kinetic energy is obtained in the form of X-ray photons of a continuous frequency range.

(*iii*) If these X-ray photons are passed through some gas then they separate electrons from gas atoms (ionisation).

**31.** *Why is the structure of crystals studied by X-rays?*

**Ans.**

(*i*) Atoms in crystal are arranged in a regular order.

(*ii*) The distance between the atoms of crystals is of the order of wave length of X-rays.

(*iii*) When X-rays fall on a crystal, they are diffracted. The diffraction is helpful in the study of crystal.

**32.** *The phenomenon of X-ray production is also called "inverse photoelectric effect" Why?*

**Ans.**

(*i*) In photoelectric effect, the photon falling on some matter is absorbed by the matter and is energy is transferred to an electron in the matter which emits photoelectron.

(*ii*) In X-ray production photons are produced by getting energy from high speed electrons which ultimately come to the rest.

(*iii*) Thus this phenomenon is also called "inverse photoelectric effect".

**33.** *X-ray from a tube undergo first order reflection at a glancing angle of 12° from the face of a calcite crystal. The Lattic Plane spacing of the calcite crystal is $3.04 \times 10^{-8}$cm. Calculate the wave-length of X-rays.*

**Solution:**

$$\theta = 12° \quad n = 1 \quad d = 3.04 \times 10^{-8} \text{ cm} = 3.04 \times 10^{-10} \text{ m.}$$

We know from Bragg's equation, $2d \sin\theta = n\lambda$

$$\text{or} \qquad \lambda = \frac{2d \sin\theta}{n} = \frac{2 \times 3.04 \times 10^{-10} \times \sin 12}{1} = 1.26 \times 10^{-10} \text{ m}$$

**34.** *X-ray from a tube undergo first order reflection at a glancing angle 60° from the face of a crystal. The wavelength of X-ray is $0.066 \times 10^{-9}$ meter. Determine the lattice plane spacing of the crystal.*

**Solution:**

$$n = 1$$
$$\lambda = 0.066 \times 10^{-9} \text{ meter}$$
$$d = ?$$
$$\theta = 60°$$

We know $2\, d \sin\theta = n\lambda$

$$2d = \frac{n\lambda}{\sin\theta} = \frac{1 \times 0.066 \times 10^{-9}}{\sin 60}$$

$$\text{or} \qquad d = \frac{1 \times 0.066 \times 10^{-9}}{2 \times \sin 60°} = 0.038 \text{ mn.}$$

**35.** *At what angles may a ray of wavelength 0.440 Å be reflected from a cube face of a rock salt crystal (d = 2.814 Å).*

**Solution:**

We know from Bragg's law $n\lambda = 2d \sin\theta$.

$$\text{or} \qquad \sin\theta = \frac{n\lambda}{2d} \qquad d = 2.814 \text{ Å}$$
$$\lambda = 0.440 \text{ Å}$$

For first order $n = 1$

$$\sin\theta_1 = \frac{1 \times 0.440}{2 \times 2.814} = 0.7818 = 0.782$$

$$\theta_1 = \sin^{-1} 0.782 = 4.48°$$

For second order $n = 2$

$$\sin\theta_2 = \frac{2 \times 0.440}{2 \times 2.814} = 0.15636$$

$$\theta_2 = \sin^{-1} 0.15636 = 8.99°$$

For third order $n = 3$

$$\sin\theta_3 = \frac{3 \times 0.440}{2 \times 2.814} = \frac{1.320}{5.628} = 0.2345$$

$$\theta_3 = \sin^{-1} 0.2345 = 13.56°$$

For fourth order $n = 4$

$$\sin \theta_4 = \frac{4 \times 0.440}{2 \times 2.814} = \frac{0.88}{2.814} = 0.3127$$

$$\theta_4 = \sin^{-1} 0.3127 = 18.22°$$

**36.** *The spacing between principal planes of NaCl crystal is 2.82 Å. It is found that first order Brag reflection occurs at glancing angle 10°. Calculate the wavelength of X-rays.*

**Solution:**

We know from Bragg's law

$$2d \sin \theta = n\lambda$$

Given $d = 2.82$ Å $= 2.82 \times 10^{-10}$ m

$Q = 10°$ $\sin 10 = 0.173648 = 0.17365$

For first order Bragg's reflection, $n = 1$

$$n\lambda = 2d \sin \theta$$

$$1.\lambda = 2 \times 2.82 \times 10^{-10} \times \sin 10°$$

$$= 2 \times 2.82 \times 10^{-10} \times 0.17365$$

$$= \frac{2 \times 282 \times 10^{-10}}{10^2 \times 10^5} \times 17365 = \frac{9793860 \times 10^{-10}}{10^7} = 0.979386 \times 10^{-10} \text{ m.}$$

**37.** *The spacing between successive (100) plane in NaCl is 2.82 Å. X-rays incident on the surface of the crystal is found to give rise to first order Bragg's reflection at glancing angle 8.8°. Calculate the wavelength of X-rays.* ***[H.P.U. 1996]***

**Solution:**

We know from Bragg's law,

$$2d \sin \theta = n\lambda$$

Given $d = 2.82$ Å $= 2.82 \times 10^{-10}$ m

$\theta = 8.8°$ $\sin 8.8° = 0.1529$

For first order Bragg's reflection, $n = 1$.

$$n\lambda = 2d \sin \theta$$

$$1.\lambda = 2 \times 2.82 \times 10^{-10} \times 0.1529 = \frac{2 \times 282 \times 1529 \times 10^{-10}}{10^6}$$

$$= \frac{862356 \times 10^{-10}}{10^6} = 0.862356 \times 10^{-10} \text{ m}$$

$$= 0.8624 \times 10^{-10} \text{ m}$$

$$= 0.8624 \text{ Å}$$

**38.** *Calculate wave length of X-ray if lattice constant of a crystal $d = 2.8 \times 10^{-10}$ metre and first order glancing angle is 12°.*

**Solution:**

We know from Bragg's law

$$n\lambda = 2d \sin \theta$$

Given $d = 2.8 \times 10^{-10}$ m

$\theta = 12°$ $\sin 12° = 0.2079$

For first order Bragg's reflection $n = 1$

$$n\lambda = 2d \sin \theta$$

$$1.\lambda = 2 \times 2.8 \times 10^{-10} \times \sin 12 = 2 \times 2.8 \times 10^{-10} \times 0.2079$$

$$= \frac{2 \times 28 \times 2079 \times 10^{-10}}{10^5} = \frac{116424 \times 10^{-10}}{10^5} = 1.164 \times 10^{-10} \text{ m}$$

$$= 1.164 \text{ Å}$$

**39.** *The Brag's angle for first order reflection from the face of a crystal is 60°. The wave length of X ray is 1.8 × $10^{-10}$ m. Calculate lattice plane spacing crystal.*

**Solution:**

According to Bragg's law. $n\lambda = 2d \sin \theta$

Given $\theta = 60°$ $\sin 60° = 0.86602$ $d = ?$ $n = 1$

$\lambda = 1.8 \times 10^{-10}$ m

$$1 \times 1.8 \times 10^{-10} = 2d \times 0.86602$$

$$2d = \frac{1.8 \times 10^{-10}}{0.86602}$$

$$d = \frac{1.8 \times 10^{-10}}{2 \times 0.86602} = \frac{0.9 \times 10^{-10}}{.86602} = 1.0392 \times 10^{-10} \text{ m}$$

**40.** *NaCl has its principal planes spaced at 2.820 Å. The first order of Bragg reflection is located at 10°. Calculate (a) wave length of X rays and (b) the angle for the second order Bragg reflection.*

**Solution:**

(*a*) We know from Bragg's law. $n\lambda = 2d \sin \theta$.

Given $d = 2.820$ Å, $\theta = 10°$ $\sin 10° = 0.173648,$ $n = 1$

$\lambda = ?$ $= 0.17365$

$$\lambda = \frac{2d \sin \theta}{n} = \frac{2 \times 2.820 \times 0.1736}{1} = 979386 = 0.9794 \text{ Å}$$

(*b*) Given $\lambda = 0.9794$ Å $d = 2.820$ $n = 2$

$\theta = ?$

We know from Bragg's law. $n\lambda = 2d \sin \theta$

or $$\sin \theta = \frac{n\lambda}{2d}$$

$$\sin \theta = \frac{2 \times 0.9794}{2 \times 2.820} = \frac{0.9794}{2.820} = 0.347$$

$$\therefore \theta = \sin 0.347 = 20.30°$$

**41.** *Calculate the galancing angle on the cube face (100) of a rock salt crystal (a = 2.814 Å) corresponding to second order reflection of X-rays of wavelength 0.710 Å.* ***(M.D.U. 2001)***

**Solution:**

We know from Bragg's law, $n\lambda = 2d \sin \theta$.

Given $a = 2.814$ Å $\lambda = 0.710 \text{ Å} = 0.710 \times 10^{-10}$ m

or $d_{100} = 2.814$ Å $n = 2$

$= 2.814 \times 10^{-10}$m

$$d_{100} = \frac{2.814}{\sqrt{1^2 + 0^2 + 0^2}} = \frac{2.814}{1} = 2.814$$

$2 \times 0.710 \times 10^{-10} = 2 \times 2.814 \times 10^{-10} \times \sin\theta$

or $$\sin\theta = \frac{2\times 0.710\times 10^{-10}}{2\times 2.814\times 10^{-10}} = \frac{0.710}{2.814} = \frac{710}{2814} = 0.2523$$

$\therefore \sin\theta = 0.2523$

$\theta = \sin^{-1} 0.2523 = 14.61° = 14.6°$

**42.** *X-ray's of wavelength 1.5 Å make a galancing angle of 16° in the first order when diffracted from NaCl crysal. Find lattice constant of NaCl.* ***(Osmania Univ. 2004)***

**Solution:**

We know from Bragg's law $n\lambda = 2d \sin\theta$

Given $\lambda = 1.5$ Å $= 1.5 \times 10^{-10}$ m

$\theta = 16°$, $\sin 16° = 0.2756$ $d = ?$ $n = 1$

$1 \times 1.5 \times 10^{-10} = 2 \times d \sin 16° = 2 \times d \times 0.2756$

$$2d = \frac{1.5\times 10^{-10}}{0.2756}$$

or $$d = \frac{1.5\times 10^{-10}}{2\times 0.2756} = \frac{1.5\times 10^{-10}}{0.5512} = 2.72 \times 10^{-10}\text{ m} = 2.72\text{ Å}$$

**43.** *If X-rays of wave length 0.5 Å are diffracted at an angle 5° in the first order. What is the spacing between the adjacent planes of the crystal? At what angle will second maximum occur?*

**Solution:**

According to Bragg's law $n\lambda = 2d \sin\theta$

Given $\lambda = 0.5$ Å $= 0.5 \times 10^{-10}$ m $\sin\theta = \sin 5° = 0.087155$

$= 0.08716$

$d = ?$ $n = 1$

$1 \times 0.5 \times 10^{-10} = 2\text{d} \sin 5°$

$$2d = \frac{1\times 0.5\times 10^{-10}}{\sin 50} = \frac{0.5\times 10^{-10}}{0.08716} = 5.7365 \times 10^{-10}$$

$2d = 5.7365 \times 10^{-10}$, $2d = 5.7365$ Å

or $2d = 5.7365$

$$d = \frac{5.736}{2} = 2.868 = 2.87\text{ Å}$$

2nd part:

According to Bragg's law $n\lambda = 2d \sin\theta$

Given $n = 2$ $d = 2.87$ Å $= 2.87 \times 10^{-10}$ $\sin\theta = ?$

$2d \sin\theta = 2 \times 0.5 \times 10^{-10}$

$$\sin\theta = \frac{2\times 0.5\times 10^{-10}}{2\times 2.87\times 10^{-10}} = \frac{0.5}{2.87} = 0.1742$$

$\theta = \sin^{-1} 0.1742 = 10.032°$

**44.** *Calculate Bragg angle for X-rays having wave length 1.54 Å in different order 1, 2, 3. If intraplanar distance is 2.67 Å.* ***(A.U. 1994)***

**Solution:**

According to Bragg's law $n\lambda = 2d \sin \theta$

Given $\lambda = 1.54$ Å $d = 2.67$ Å $n = 1$ $\sin \theta = ?$

$1 \times 1.54 = 2 \times 2.67 \times \sin \theta$

or $\sin \theta = \dfrac{1.54}{2 \times 2.67} = \dfrac{1.54}{5.34} = 0.28838 = 0.2884$

$\sin \theta = 0.2884$

$\theta = \sin^{-1} 0.2884 = 16.76°$

Second order, $n = 2$

$$\sin \theta = \frac{n\lambda}{2d} = \frac{2 \times 1.54}{2 \times 2.67} = \frac{1.54}{2.67} = 0.5767 = 0.577$$

$\sin \theta = 0.577 = 35.22°$

Third order, $n = 3$

$$\sin \theta = \frac{n\lambda}{2d} = \frac{3 \times 1.54}{2 \times 2.67} = \frac{4.62}{5.34} = 0.865$$

$\sin \theta = 0.865$ or $\theta = \sin^{-1} 0.865 = 59.9°$

**45.** *A certain crystal reflects monochromatic X-rays strongly when Bragg glancing angle (first order) is 15°. What are the glancing angles for second and third order spectrum.* **(H.P.U. 1993)**

**Solution:**

We know from Bragg's law $n\lambda = 2d \sin \theta$

Given $\theta = 15°$ $\sin 15° = 0.2588$

$n = 1$

$1.\lambda = 2d \times 0.2588$

or $\dfrac{\lambda}{2d} = 0.2588$

For second order, Given $n = 2$ $\theta_2 = ?$

$\therefore 2 \times \lambda = 2d \sin \theta$

$$\sin \theta_2 = \frac{2\lambda}{2d}$$

$$\sin \theta_2 = \frac{\lambda \times 2}{2d}$$

$$\sin \theta_2 = 2 \times 0.2588 \qquad \left[\because \frac{\lambda}{2d} = 0.2588\right]$$

$\theta_2 = \sin^{-1} 0.5176$

$\theta_2 = 31.17°$

For third order, Given $n = 2$ $\sin \theta_3 = ?$

$n\lambda = 2d \sin \theta_3$

$$\sin \theta_3 = \frac{3.\lambda}{2d} = 3 \times 0.2588 \qquad \left[\because \frac{\lambda}{2d} = 0.2588\right]$$

$$\sin \theta_3 = 0.7764$$
$$\theta_3 = \sin^{-1} 0.7764 = 50.93°$$

**46.** *X-rays of wave length 0.71 Å are reflected from the plane (1,1,0) of a rock salt crystal (a = 2.82 Å). Calculate the galancing angle corresponding to second order reflection.*

**Solution:**

Given $a = 2.82$ Å

$$d_{110} = \frac{a}{\sqrt{1^2 + 1^2 + 0^2}} = \frac{a}{\sqrt{2}} = \frac{2.82 \text{ Å}}{\sqrt{2}} = \frac{2.82}{1.41} = 2 \text{ Å}$$

We know from Bragg's law, $n\lambda = 2d \sin \theta$

$d = 2$ Å $\quad n = 2$ ($\because$ second order) $\quad \lambda = 0.71$ Å

$$2d \sin \theta = n\lambda$$

$$\sin \theta = \frac{n\lambda}{2d} = \frac{2 \times 0.71}{2 \times 2} = \frac{0.71}{2} = 0.355$$

$$\therefore \sin \theta = 0.355$$
$$\theta = \sin^{-1}$$
$$= 20.79°$$

**47.** *Bragg's angle corresponding to the first order reflection from (111) planes in a crystal is 30°, when X-rays of wavelength 1.75 Å are used. Calculate the interatomic spacing.*

**Solution:**

Since $d_{hkl} = \dfrac{a}{\sqrt{h^2 + k^2 + l^2}}$, $\quad d_{111} = \dfrac{a}{\sqrt{1^2 + 1^2 + 1^2}} = \dfrac{a}{\sqrt{3}}$

From Bragg's law $n\lambda = 2d \sin \theta$

Given $\lambda = 1.75$ Å $\quad n = 1 \quad d = \dfrac{a}{\sqrt{3}} \quad \sin \theta = \sin 30$

$$\sin \theta = \frac{1}{2}$$

$$2 \times \frac{a}{\sqrt{3}} \times \sin 30 = 1.75 \times 1$$

or $\dfrac{2a}{\sqrt{3}} \times \dfrac{1}{2} = 1.75 \times 1$ $\quad$ or $\quad a = \dfrac{1.75 \times \sqrt{3} \times 2}{2} = 1.75 \times 1.73205 = 3.031$ Å

CHAPTER 8

# Radioactivity

**Elementary particles:**

According to **Dalton's atomic theory** (1803-1808) every substance is composed of minute particles known as atoms.

Atom of all elements is composed of the smallest and elementary particles like *neutron*, *proton* and *electron*.

In addition to these observative 'particles' and 'quanta', there are other elementary particles like *positron* (anti electron), *anti proton*, *antineutron, neutrino, antineutrino*.

- **Salient features of the atomic nucleus:**

(*i*) *Nucleus* is the central part of an atom and is spherical or prolate shaped.

(*ii*) It is rigid and dense. It consists of *protons* and *neutrons* and carries positive charge in the entire mass of the atom. *Positive* charge is due to the presence of positively charged protons in it. Neutrons are neutral particles.

(*iii*) *Neutrons* and *protons* present in the nucleus are collectively called *nucleons*.

(*iv*) The number of electrons ($e$) (outside the nucleus) is equal to the number of protons ($p$) in the nucleus. i.e. $p = e$

(*v*) The sum of the number of protons ($p$) and neutrons ($n$) in an atom is called *mass number* (A) of the atom i.e. $\text{A} = p + n$

(*vi*) The number of protons in the nucleus of an atom is called *atomic number* ($Z$) or *proton number*. $\text{Z} = p = e$

(*vii*) Protons and neutrons are made up of still smaller particles called *quark*. A proton contains two up quarks ($u$) and a down quark ($d$). A neutron contains two down quarks and one up quark.

$p = uud \quad n = udd$

- **Neutrino:**

(*i*) It is a particle with zero mass and zero charge.

(*ii*) It is formed during neutron decay.

(*iii*) Fermi named like the particle as *neutrino*.

- **Antineutrino:**

It is a particle identical to neutrino but has opposite spin.

$$\underset{\text{Neurtron}}{{}_0n^1} \longrightarrow \underset{\text{Proton}}{{}_1p^1} + \underset{\text{Electron}}{{}_{-1}e^0} + \underset{\text{Antineutrino}\downarrow}{\text{neutrino}\uparrow}$$

- **Anti-electron or positron ($+\ {}_1e^0$)**

 (*i*) It has same negligible mass and amount of charge as electron.

 (*ii*) However the charge is positive.

- **Antiproton (${}_{-1}p^1$)**

 (*i*) It has same mass and spin as protons.

 (*ii*) But it has opposite (negative) charge.

 (*iii*) Its symbol is $p^-$.

- **Antineutron.**

 (*i*) It is a particle identical with neutron but has opposite spin to neutron.

 (*ii*) It is produced when a proton strikes an antiproton.

$$\underset{\text{Proton}}{{}_1p^1} + \underset{\text{Antiproton}}{{}_{-1}p^1} \longrightarrow \underset{\text{Neutron}}{{}_0n^1\uparrow} + \underset{\text{Antineutron}}{{}_0n^1\downarrow}$$

- **Mesons:** These are particles having mass intermediate between that of electron and proton.
- **Isobars:**

 (*i*) Elements with the same mass number but different atomic number (z) are called isobars.

 (*ii*) They have different places in periodic table and differ in chemical properties.

 **Example:** ${}^{16}_{8}O \quad {}^{16}_{7}N$

- **Isotones:** Atoms whose nuclei have same number of neutrons but different mass number and atomic number.

 **Example :** ${}_6C^{14(8+6)} \quad {}_8O^{16(8+8)}$ [Each has 8 neutrons]

- **Isotopes:** Atoms with same atomic number but different mass number are known as isotope.

 **Example :** ${}_1H^1$, ${}_1H^{2(1+1)}$ and ${}_1H^{3(1+2)}$

- **Isomers:** Elements with same atomic number and same mass number but different half life for radioactivity is called isomers.
- **Nuclear forces:** In nucleus, the positively charged protons and the uncharged neutrons are held together. There are some strong attractive forces (protons to protons, neutrons to neutrons and protons to neutrons) operating in the nucleus which keeps the nucleus intact. These are called nuclear forces.
- **Magic number:**

 (*i*) The nuclei which contain 2, 8, 20, 50, 82 or 126 protons (*p*) or neutrons (*n*) have been to be extra stable, have large number of isotopes and have their nuclear shells completely filled (i..e closed shells).

 (*ii*) The number 2, 8, 20, 50, 82 or 126 are called **magic number** for the nuclear shells.

 **Example:** ${}_{18}Ar^{38}$ ($n = 38 - 18 = 20$) ${}_{20}Ca^{40}$ ($n = 40 - 20 = 20$)

- **Relation between the value of *n*/*p* ratio.**

The nuclei having higher value of $n/p$ ratio (or lower value of $p/n$ ratio) are more unstable and hence are radioactive.

- **Radioactivity:**

**The phenomenon of spontaneaous emission of highly penetrating ionising radiations (*α-rays, β-rays and γ-rays*) from the unstable-atomic nuclei of the heavy elements (*atomic weight greater*

***than 206*) like uranium is called radioactivity. The nucleus thereby loses energy and changes into the more stable nucleus of another element. It is a self-disintegrating and irreversible process.**

**The substances which exhibit this behaviour (*radioactivity*) are said to be radioactive.**

## • Types of radioactivity:

**(A) Natural Radioactivity :**

The naturally occuring elements (*viz.* uranium, radium, polonium etc) disintegrate continously and emit alpha ($\alpha$), beta ($\beta$) and gamma ($\gamma$) rays. They change into other elements. This process of spontaneous change is called natural radioactivity.

**(B) Artificial Radioactivity :**

It is a process by which element is converted into a new radioactive isotope of a known element by artificial means.

## • Units of Radioactivity :

The units of radioactivity of an element is the measure of the rate at which it changes to the daughter element. It has been derived on the scale of disintegration of radioactive substance per second.

- **Curie (*Ci*):** The quantity of any radioactive substance which undergoes $3.7 \times 10^{10}$ disintegrations per second.
- **Millicurie (*mCi*):** Represents $3.7 \times 10^7$ radioactive disintegrations per second.
- **Becquerel (*Bq*):** It is the quantity of radioactive material in which one nuclear disintegration occurs in one second.

$$1 \text{ Bq} = 2.7 \times 10^{-11} \text{ Ci.}$$

- **Rutherford (*rd*):** One rutherford is the amount of radioactive substance which undergoes $10^6$ disintegration per second.

$$1 \text{ Bq} < 1 \text{ Rd} < 1 \text{ Ci}$$

## • Radiations from Radioactive materials:

(*i*) A small hole is made in a lead block L and a piece of radium is placed in it.

(*ii*) Radiation emitted from radium (radioactive material) escape only through the hole and others are absorbed by wall.

(*iii*) This arrangement is placed in an evacuated enclosure.

(*iv*) The nature of radiations emitted by the radioactive element can be studied (*a*) by applying electric field, (*b*) applying magnetic field.

(*a*) When electric field is applied, the ray which deflects towards the negative plate is $\alpha$ (alpha) particle or ray, the ray which deflects towards the positive plate is the $\beta$ (beta) particle and the ray which goes undeflected is the $\gamma$ (gamma) ray.

(*b*) When a uniform magnetic field is applied perpendicular to the plane of the paper and directed top to bottom, into it, by applying **Flemming's left hand rule**, it can be seen that the ray which bend towards left is $\alpha$ particle, the ray which bends towards right in a semicircle is $\beta$ particle and the ray which goes straight undeflected is the $\gamma$ ray.

Thus the radiation emitted by radium (radioactive element) through strong electric field between parallel plates or through magnetic field separates the radiation into three kinds ($\alpha$, $\beta$ and $\gamma$) of rays.

**(*a*) Alpha ($\alpha$) rays :**

(*i*) They are positively charged and deflected towards the negative plate.

(*ii*) These rays consists merely of helium nuclei.

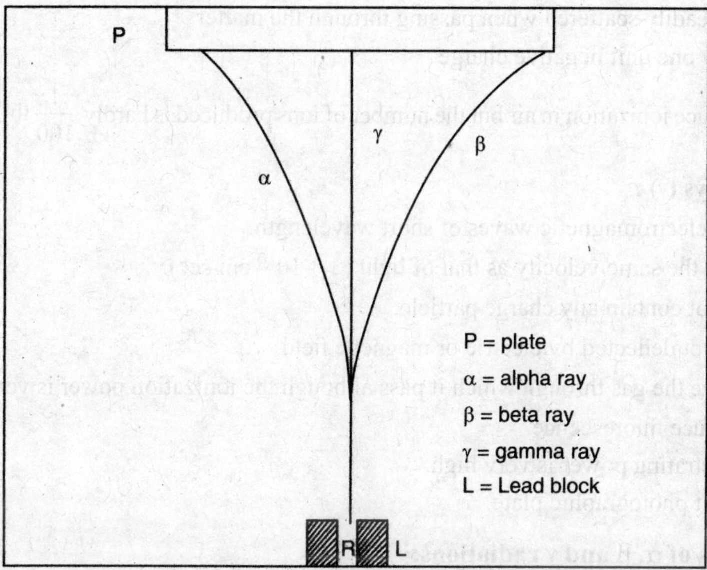


**Fig. 8.1** Diagramatical representation of α, β and γ rays

**(*b*) Beta (β) rays :**

(*i*) They are negatively charged and deflected towards the positive plate.

(*ii*) These rays merely consists of electrons.

**(*c*) Gamma (γ) rays:**

They are not deflected and are neutral.

## • Properties of Radioactive Radiation:

### (A) Alpha Particles (α rays) :

(*i*) These are positively charged particle.

(*ii*) It is identified as nuclei of helium atom with two (2) protons and two (2) neutrons.

(*iii*) They move with high velocities ranging from $1.4 \times 10^9$ to $1.7 \times 10^9$ cm per second.

(*iv*) They have greatest ionizing power than β particle.

(*v*) They have shortest penetrating power.

(*vi*) They affect photographic plate.

(*vii*) They get scattered while passing through thin metal foil.

(*viii*) They are deflected by electric and magnetic field.

(*xi*) They are doubly charged helium, so they are called “Helions”.

### (B) Beta Particles (β Rays) :

(*i*) They are identified as electrons.

(*ii*) They are negatively charged particle.

(*iii*) It can penetrate through large thickness of matter.

(*iv*) They are deflected by electric and magnetic field.

(*v*) They affect photographic plate.

(*vi*) They move with high velocity ($2.36 \times 10^8 - 2.83 \times 10^8$ cm/sec)

(*vii*) They produce fluorescence and phosphorescence.

(*viii*) They are readily scattered when passing through the matter

(*ix*) They carry one unit negative charge.

(*x*) They produce ionization in air but the number of ions produced is hardly $\frac{1}{100}$th those produced by α rays.

**(C) Gamma Rays (γ) :**

(*i*) These are electromagnetic waves of short wavelength.

(*ii*) They have the same velocity as that of light ($3 \times 10^{10}$ cm/sec).

(*iii*) They do not contain any charge particle.

(*iv*) They are not deflected by electric or magnetic field.

(*v*) They ionize the gas through which it pass although the ionization power is very low.

(*vi*) They produce fluorescence.

(*vii*) Their penetrating power is very high.

(*viii*) They affect photographic plate.

**Characteristics of α, β and γ radiations:**

| Sl. No. | Property | α Rays | β Rays | γ Rays |
|---|---|---|---|---|
| 1. | **Symbol** | α : $_2He^4$ | β : $_{-1}\beta^0$; $_{-1}e^0$ | γ: $_0e^0$ ; $_{-1}{}^0e$ |
| 2. | **Mass** | $6.65 \times 10^{-24}$g | $9.11 \times 10^{-28}$g | negligible |
| 3. | **Charge** | +2 | –1 | 0 |
| 4. | **Speed & Nature** | High speed Helium nuclei | High speed electrons | High energy radiations |
| 5. | **Velocity** | Nearly $\frac{1}{10}$ to $\frac{1}{20}$th that of light ($1.4 \times 10^9$ – $1.7 \times 10^9$ cm/sec) | Nearly same as that of light ($2.36 \times 10^{10}$ – $2.83 \times 10^{10}$ cm/sec) | Same as that of light ($3.0\times10^{10}$ cm/sec) |
| 6. | **Effect of electric & magnetic fields** | Deflected towards negative plate | Deflected towards positive plate | No deflection |
| 7. | **Penetrating power** | Low | More than α particles i.e. 100 times more | More than β particles i.e. 100 times more. |
| 8. | **Ionizing power** | Greatest ionizing power than β particle | low nearly 100 times that of γ rays | very low |
| 9. | **Luminescence** | Positive i.e. Produce phosphorescence and fluorescence on ZnS. | Produce phosphorescence on barrium platino cyanide, calcium, tungsten. | Produce phosphorescence. |

(*Contd.*)

| Sl. No. | Property | α Rays | β Rays | γ Rays |
|---|---|---|---|---|
| 10. | **Effect of photographic plate** | They affect photographic plate | The effect on photographic plate is less than α rays | Effect on photographic plate is least. |

- **Radioactive decay:**

**It is the phenomenon of spontaneous breaking up of the high energy level of unstable radiactive nuclei to low energy, more or less stable nuclei with simultaneous emission of particles.**

- **Features of atomic nucleus:**

(*i*) The atomic nucleus consists of neutrons and protons known together as *nucleus*.

(*ii*) The nucleus is stable when neutron-proton ratio (n:p ratio) lies within a specific narrow range.

(*iii*) The stable n:p ratio rises slowly but progressively with rise of the total number of nucleus.

(*iv*) Deviation from stable n:p ratio makes the nucleus unstable and radioactive.

**1. Alpha decay:**

(*i*) Alpha (α) decay leads to the emission of alpha rays from the nucleus of an atom (called parent element) which are streams of high velocity *alpha particles*.

(*ii*) Each alpha particle consists of two neutrons and two protons and corresponds to the stable helium nucleus ($He^{2+}$).

(*iii*) Thus it lowers the mass number by 4 and the atomic number by 2 and the new element (daughter element) is formed raises the $n:p$ ratio slightly and thereby increases the stability.

(*iv*) (*i*) $^{238}_{92}U \xrightarrow{\alpha\ decay} {}^{234}_{90}U$ $n:p$ ratio form 1.587 to 1.600

(*ii*) $^{226}_{88}Ra \xrightarrow{\alpha\ decay} {}^{222}_{86}Rn + {}_2He^4$

Radium Radon

**2. Beta (β) decay:**

(*i*) Negatron decay leads to the emission of beta (β) rays, which are streams of very light, highly accelerated and negatively charged beta particles.

(*ii*) A β particle is just an electron and hence its symbol is $_{-1}e^0$. Since ejection of a β particle is also followed by the emission of gamma (γ) ray.

(*iii*) When a β particle is ($_1e^0$) is emitted from the nucleus of an element, the nucleus of the new element formed possesses the same mass number but the positive charge i.e. the atomic number will be increased by one unit. Example: lead (Pb) transform into Bismuth (Bi)

**Example:** $_{82}Pb^{210} \longrightarrow {}_{83}Bi^{210} + {}_{-1}e^0 + \gamma$ rays

(*iv*) Radioactive elements possessing surplus neutrons and consequently too high $n:p$ ratio emits β particles. A neutron breaks into a proton, a beta particle and an uncharged and massless *neutrino*. The proton retained in the nucleus, the beta particles and neutrino are emitted.

Neutron ⟶ Proton + Electron + Neutrino.

$$_0n^1 \longrightarrow P^1 + {}_{-1}e^0 + {}_0\nu^0$$

(*v*) Neutron-deficient radioisotopes with too low $n:p$ ratio may raise the ratio through positron emission.

(*vi*) During positive beta (β) decay, a proton changes into a neutron by emitting a positron ($\beta^+$) which is highly accelerated and very transient beta particle with positive charge.

(*vii*) Positron emission replaces a proton by a neutron and consequently raises the n : p ratio. Here the atomic number is lowered by 1 with no change in the mass number.

$$_6C^{11} \longrightarrow {}_5B^{11} + {}_{+1}e^0 + \mho \text{ (Neutrino)}$$

**3. K-capture or electron capture.**

**The change of a nuclear proton into a neutron by capturing an extranuclear orbital electron of the innermost electron shell is known as K-capture.**

(*i*) It occurs to neutron deficient radioisotopes with the too low n : p ratios provided neither positron emission nor alpha (α) decay is possible.

(*ii*) An extranuclear orbital electron for the *k* shell of the atom may be taken into the nucleus.

(*iii*) This electron converts a proton into a neutron and produces more stable condition.

(*iv*) Thus is raises the $n : p$ ratio. The atomic number falls by 1 without any change in mass number.

(*v*) K capture immediately followed by the fall of an electron from an outer orbit to the k-shell to occupy the position vacated by the captured electron. This is accompanied by the emission of a low energy–X ray photon.

(*vi*) K-capture may also leave the daughter nucleus in an "excited" state from which it reverts to the "ground" state by emitting gamma photons.

**4. Gamma emission or decay:**

Gamma rays are short wave electromagnetic radiation with no charge, no mass. Thus the emission of gamma rays from a radioactive element will not produce a new element.

(*i*) When radioactive element emits α particles or β particles, after emission, the nucleus will be in excited state.

(*ii*) The nucleus can return to its ground state by emitting gamma rays.

**• Law of Radioactive Decay**

**(*i*) Atoms of all radioactive elements are undergoing spontaneous disintegration i.e. constantly breaking to form fresh radioactive products with the emission of α, β and γ rays.**

**(*ii*) The rate of breaking is not affected by external factors like temperature, pressure, chemical combination but depends entirely on the law of chance i.e. the quantity of radioactive element which disintegrates in unit time is directly proportional to the amount of the radioactive element present.**

**Rate of Radioactive Decay:**

It is a spontaneous process and occurs at definite rate which is the characteristic of the source. The rate of radioactive decay follows an exponential law.

The number of atoms disintegrating at any time is proportional to the number of atoms of the isotope (N) present at that time ($t$).

Suppose an element A disintegrate into another element B. Let $N_o$ is the number of atoms of the radioactive element A at the begining i.e. $t = 0$.

Now as the time passes, the element A undergoes continuous disintegration into B and hence the number atoms of A goes on decreasing. N is the number of atoms present of the substance A at the time $t = t$.

$$A \longrightarrow B$$

At $t = 0$, $N_0$

At $t = t$, N

**dN** a small number of atoms which disintegrate during small interval $dt$ after $t$.

Then $\frac{dN}{dt}$ = rate of disintegration at that instant. *i.e.,* A into B.

The rate of disintegration is proportional to the number of atoms present at that time because more the number of active nuclei present at '$t$', more will decay in the next interval $dt$.

Hence

$$\frac{dN}{dt} \propto N$$

or $\frac{dN}{dt} = -\lambda N$ .............. (*i*) where λ is a constant, called *decay constant* or disintegration constant or **radioactive constant**.

Since the intensity and the number of atoms of A decreases with time, the rate of disintegration has been indicated by negative sign.

The differential equation (1) can be put in the form

$$\frac{dN}{dt} = -\lambda N$$

Integrating,

$$\frac{dN}{N} = -\lambda dt$$

$$\log_e N = -\lambda t + C$$

when $t = 0$, $N = N_0$, So $\text{Log}_e N_0 = 0 + C$

$$\therefore \quad \log_e N = -\lambda t + \text{Log}_e N_0$$

$$\log_e \frac{N}{N_o} = -\lambda t$$

$$\therefore \quad \lambda = N_0 e^{-\lambda t}$$

## • Activity:

**In the process of radioactive decay, the activity of radioactive sample is measured by the number of atoms that disintegrate per second.**

If dN is the number of atoms disintegrating in time $dt$ second then

activity $$R = \frac{-dN}{dt}$$

$$\frac{dN}{dt} = -\lambda N = -\lambda N_0 e^{-\lambda t} \qquad [\because N = N_0 e^{-\lambda t}]$$

$$\therefore R = \lambda N_0 e^{-\lambda t}$$

S.I. unit of radioactivity is called Becquerel.

1 Bq = 1 disintegration per second.

## • Mean life

(*i*) Since atoms of a radioactive element are constantly disintegrating one after another, the life of every atom is different.

(*ii*) The atoms which disintegrate first has zero half life and that disintegrate last has a very long life. Thus the life of every atom is different.

(*iii*) The mean ages of all the atoms in a radioactive element is called the average life or mean life. It is denoted by tau τ.

**The mean life of a radioactive element is measured by the ratio of the total life time of all the radioactive atoms to the total number of such atoms in it.**

$$\text{Mean life } (\tau) = \frac{\text{Sum of the lives of all the atoms}}{\text{Total number of atoms}}$$

(*iv*) It can be shown that the average life or mean life of a radioactive element is equal to the reciprocal of decay constant.

$$\boxed{T_{av} = \text{tau } (\tau) = \frac{1}{\lambda}}$$

- **Radioactive constant:**

**It may be defined as the ratio of the amount of the substance which disintegrate in a unit time to the amount of the substance present**

$$\lambda = \frac{\frac{-dN}{dt}}{N}$$

We know that $N = N_0 e^{-\lambda t}$ Putting $t = \frac{1}{\lambda}$

The number of the original atoms present after this time is

$$N = N_0 e^{-\lambda \frac{1}{\lambda}} = N_0 e^{-1}$$

$$N = \frac{No}{e} = \frac{N_0}{2.718}$$

$$N = 0.368\, N_0 = 0.37\, N_0.$$

- **Half life:**

**The half life of a radioactive substance is defined as the time interval in which one half quantity (mass/number of atoms) of the sample (radioactive substance) disintegrate from the parent nuclei to daughter nuclei.**

Suppose after time $T$ (or $t_{1/2}$) half of the atoms of the radioactive element have disintegrated *i.e.*, initial quantity of a radioactive element be $N_0$ and $t = T$, the quantity left after $T$ time *i.e.*, $N = \frac{N_0}{2}$.

We know $\qquad N = N_0\, e^{-\lambda t}$

on putting $t = T$ and $N = \frac{N_0}{2}$ we get

$$\frac{N_0}{2} = N_0\, e^{-\lambda t}$$

or $$\frac{1}{2} = e^{-\lambda T} = \frac{1}{e^{\lambda T}}$$

or $$2 = e^{\lambda T}$$

or $$\lambda T = \log_e 2$$

or $$T = \frac{\log_e 2}{\lambda} = \frac{0.693}{\lambda}$$

Hence $$\boxed{t_{1/2} \text{ or } T = \frac{0.693}{\lambda}}$$

So, half life of a radioactive element is inversely proportional to its decay constant. If the initial quantity of a radioactive element be $N_0$, then the quantity $N$ of the element left after '$n$' half lives is given by $N = N_0\left(\frac{1}{2}\right)^n$.

It should be noted that at the end of half life period, the number of nuclei persent is $N = N_0\left(\frac{1}{2}\right)^n$

**Isotopes:**

**The atoms of an element whose nuclei have the same number of protons but different number of neutrons are called the isotopes of that element.**

**Special features of Isotopes:**

(*i*) All isotopes of an element have the same number of electrons (atom has equal number of proton and electron). Therefore *chemical properties* of different isotopes of an element *are same.*

(*ii*) The mass numbers (number of nucleons) of different isotopes of an element are different. Hence their *physical properties* are *not the same.*

(*iii*) As chemical properties are same, two isotopes of the same element can not be separated by any chemical process. To separate them, physical processes based on atomic mass like gaseous diffusion are used.

(*iv*) Among isotopes of the same element, some may be stable and some radioactive. This is so because of the difference in their nuclear structure.
For example: ${}_6C^{12}$ is stable while ${}_6C^{14}$ is radioactive.

• **Applications of Radioisotope in Bio-medical Science:**

Radioisotopes are used to study various problems in Biology. It is used due to the following properties.

(*i*) They are chemically identical.

(*ii*) They are easily detectable and measurable. Even small amounts of radioisotopes may be accurately detected.

(*iii*) The radiations pass through the living cells and tissues and be detected *in vivo.*

(*iv*) They may be incorporated into the chemical substances of the tissues of living beings.

**Isotopic tracer-tagging:**

(*i*) The process of adding very small amount of radioisotopes to an element or compound is called *tagging* or *labelling*.

(*ii*) The radioactive isotope is called *tracer* or *tracer* element because it can be traced and observed by suitable instrument.

(*iii*) The method of studying the dynamics of a particular system by observing the pathway of radioisotope added to it, is called **tracer technique**.
Sometimes when suitable radioisotope is available, molecules are tagged with non-radioactive isotope.
Example: deuterium (${}_1^2H$) for Hydrogen and $O^{18}$ for Oxygen. The tracing is done with **Mass Spectrometer.**

• **Criteria of radioisotope for tracer studies:**

(*i*) The isotope should be highly specific.

(*ii*) The isotope should be incorporated into the biological system.

(*iii*) The isotope should not leave the molecule easily after chemical incorporation.

(*iv*) Isotopic effect should be negligible.

(*v*) The isotope should not alter the biological system.

**1. Application of radioisotope in medicine for diagnosis as well as therapy.**

(*i*) **Detection and location of brain tumor:**

The exact position of brain tumor in the brain is detected by using two isotopes *viz* $^{131}I$ and $^{32}P$.

(*ii*) **Diagnosis and treatment of disorder in thyroids:**

Both the disorders in thyroid *i.e.* hyperthyroidism (overactive) and hypothyroidism (underactive) can be diagnosed by administering *NaI* tagged with $^{131}I$.

(*iii*) **In the treatment of leukemia and Polycythemia Vera:**

- In the treatment of *leukemia* (overproduction of W.B.C) an intravenous injection of sodium phosphate tagged with $^{32}P$ is used.
- In the treatment of *polycythemia vera* (bone marrow produces excessive amount of RBCS) an intravenous injection of sodium phosphate tagged with is $^{32}P$ used. Here β rays emitted from $^{32}P$ to destroy malignancy.

(*iv*) **Circulatory disorder (viz. thrombosis) detection:**

Here, sodium chloride tagged with $^{24}Na$ is injected into a vein of a forearm of the patient and the counter is placed at the foot. Moving the Geiger counter in the different parts of the body, the exact position of the circulatory impairment is detected.

(*v*) **Determination of abnormalities in the pumping action of the heart:**

Sodium iodide tagged either by $^{24}Na$ or $^{131}I$ is administered into the blood stream. The rhythmic pumping action of the chambers of the heart get recorded by the Geiger counter with an automatic pen recorder.

(*vi*) **Estimation of total blood content of a patient:**

For the detection of anemia, a suitable compound tagged with $^{59}Fe$ (radioiron) is injected. After sometime, the blood is tested by measuring its activities, thus anamia can be detected.

(*vii*) **Radio therapy in the treatment of cancer:**

Radiocobalt ($^{60}$Co) emits high energy gamma (γ) rays which destroy the malignancy of cancerous growth. $^{60}$Co is known as **"Poor Man's Radium"**.

(*ix*) **Treatment of skin disease:**

For the treatment of skin disease radiophosphorus ($^{32}P$), $T = 14.3$ days, β emitter is found good for skin disease.

**2. Uses of radioisotope in Agriculture:**

(*i*) **Uptake of fertilizers by plant:**

The plant takes phosphorus from two sources *i.e.,* from the soil and from the added fertilizer. If the fertilizer tagged with radioactive phosphorus ($^{32}P$) is used, the exact proportion taken up by the plant can be determined accurately.

(*ii*) **Production of new plant varieties**

Irradiation of maize with radioactive cobalt ($^{60}$Co) produces 15% increase in the quantity of green plants growing in the same field but not irradiated.

(*iii*) **The process of photosynthesis** can be studied by using radioisotopes.

(*iv*) **Transportation** of minerals from roots to leaves and its subsequent redistribution within the plants can be followed through use of radioisotopes.

**3. Use of radioisotope in Microbiology**

For sterilization radioisotopes are used

(*i*) Bacteria are killed when suitable radiations are allowed to fall on them.

(*ii*) If perishable articles are exposed to radiations, they remain fresh beyond their normal life-span.

(*iii*) Microorganisms are very much useful for metabolic studies with isotope. The studies with isotopically labelled orotic acid on mutant strains of *Neurospora* yielded labelled pyrimidines. It indicates that orotate is an intermediate precursor of pyrimidine.

**4. Use of radioisotope in Cellular and Molecular Biology:**

(*i*) Location of synthesis and storage by different organelles *e.g.* Golgi Body, Endoplasmic reticulum (ER) are detected by radioisotopic studies.

(*ii*) DNA replication analysis by using $^{14}N$ & $^{15}N$ known as Menselson Stahl (1950) experiment. It proved semiconservative mode of DNA replication.

(*iii*) Radioisotopic compounds such $^{3}H$ thymidine & $^{3}H$ uridine are used for studying replication & transcription.

- **Detection and Measurement of radioactivity**

Apart from half life, radioactive decay can also be expressed as disintegration per minute (dpm) or per second (dps). The unit of radioactivity used is curie (Ci) millicurie (mCi) and micro curie (μCi)

Radioactivity can be measured by counting devices for photographic methods, Scintillation and gas ionization and solid state.

- **Photographic method:**

(*i*) Here radioactive emission interact with matter.

(*ii*) Here ionizing radiation from a radioactive source interact with photographic emulsion and leaves an image.

(*iii*) In this method permanent record has been made.

- **Scintillation:**

It is based on its ability to excite an atom leading to emission of visible of light.

- **Gas ionization:**

It is based upon the ability of the radioactive emission to ionize matter.

- **Radiation Detection and Measurement:**

Detection instruments are used to detect subatomic particles or nuclear charged particles like α, β etc or ionizing radiations like γ rays or high energy photons. Subatomic particles are too small to be detected directly. Therefore, suitable detecting instruments are required. Radiations are detected and measured from their characteristic properties like **ionization, fluorescene** and **photographic effect**.

- **Ionization**

  - The process of creating ions in a gas assembly is called *ionization* and the external agents that bring about the process are called **ionizing agents**.
  - The migration of the ions through the gas constitutes an electrical current, is known as *ionization current.*

**A • Ionization chamber:**

It works on the principle that charged subatomic particles can ionize a gas. The number of ion pairs formed gives information with regard to the nature of the incident particle as well as it energy.

(*a*) **Pulse ionization chamber:**

(*i*) In this type of ionization chamber, ionization produced by charged particles traversing the gas is detected as a single voltage pulse.

(*ii*) Here, electronic pulse is proportional to the number of electrons released by the ionizing radiation. It is known as non-integrating type.

(*b*) **Current ionization chamber:**

(*i*) In this type of ionization chamber, pulses are not separated but continuous accumulation of charge leads to flow of electric current.

(*ii*) It integrates the event and is known as **integrating type**.

## Principle:

**The ionization chamber works in the principle of measurement of the number of ions and electrons produced by the radiation in a gas filled chamber. It is achieved by creating an electrical potential across the chamber by means of two electrodes. The electrons then rush towards the anode and the positive ions towards the cathodes. The migration of the ions (charged particles) give rise to a current which can be measured directly with the help of galvanometer or can be amplified electronically.**

- **Description of the instrument :**

**1. Ionization chamber:**

(*i*) It consists of a closed metallic cylinder (*C*) having a a suitable gas (*e.g.* argon).

(*ii*) A thin axial wire (ω) enclosed in a glass envelope is present within the cylinder.

(*iii*) A high potential difference is established between the cylinder and the wire electrode.

(iv) The wire electrode act as anode *i.e.,* positive potential and cylinder act as cathode.

**2. Other components:**

A resistance (R), a capacitor ($C_1$) and an amplifier cum pulse counter are connected.

## Working mechanism:

(*i*) Radiation emitted by the radioactive element enters the counter tube.

(*ii*) It ionizes the gas molecules and produces a large number of ion pairs along its path.

(*iii*) The positive ions move towards the cylindrical electrode and the negative ions (electrons) towards the positive wire electrode (anode).

(*iv*) The positive ions being heavier will move slowly.

(*v*) As the electrons get collected by the central anode, the wire potential suffers a drop *i.e.,* $\Delta V$ is the drop potential in volts, $C_1$ is the capacitance of the capacitor formed by the cylinder wire system. $n$ = the number of electrons produced and collected by the axial wire, $e$ is the charge of electrons in columb; Q is the magnitude of columbs of the total collected charge.

$$\Delta V = \frac{Q}{C} = \frac{ne}{C}$$

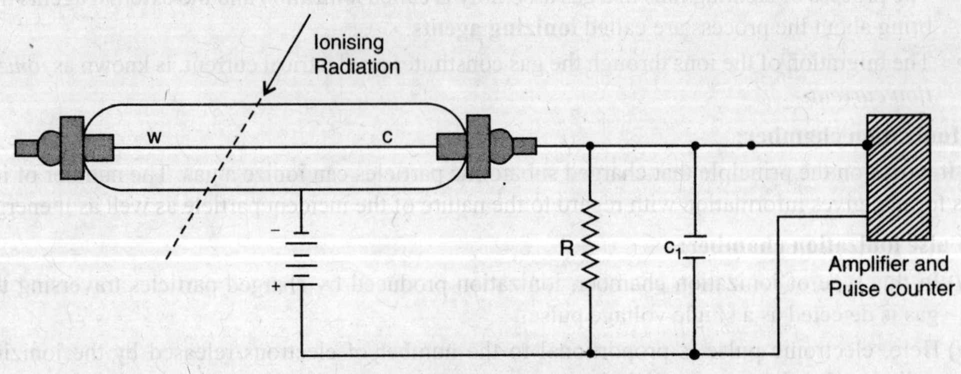


**Fig. 8.2** Ionization chamber

**Application:**

(*i*) Ionization chambers have been used to study a particles, protons and nuclei of lighter elements. These particles have large specific ionization produce strong pulses and are easily counted by the ionization chamber.

(*ii*) For neutron detection the chamber is filled with **Boron trifluoride** (Boron = $_5B^{10}$)

(*iii*) The ionization chambers are not used the electrons because of there low primary ionization. A β-rays & γ-rays do not produce enough ionization to give pulses detectable above the background noise, these cannot be detected by ionization chamber.

## (B) Radiation Detection by Proportional Counter

The proportional counter is basically a gas filled ionization based nuclear detector.

Low energy ionizing particles can not be detected by an ionization chamber as the voltage pulse produced by them is very small amplitude. In proportional counter, the voltage pulse is sufficiently high.

**Principle:**

(*i*) When the potential difference between the central wire and the metallic cylinder increases past a certain value, the electrons produced due to ionization are accelarated towards the central wire which is positive potential.
In this process, the electrons gain large energy and create further ion pairs along their path. The resulting 'avalanche' of secondary electrons that reaches the anode may represent a gas multiplication factor as high as 1000.

(*ii*) The size of the output pulse is proportional to the number of primary electrons produced by the ionizing particle and the number of primaries is proportional to the energy of the ionising particle. Thus the height of the output pulse is proportional to the energy of the ionizing particle.

(*iii*) If *w* is the average energy for generation of one pair and ΔE the energy lost to the incident charged particle in the gas between the chamber electrodes, then the number of ion pairs produced

$$n = \frac{\Delta E}{W}$$

(*iv*) If '*m*' is multiplication factor, then each primary ion will produce 'm' ion pairs. Hence charge carried by central electrode $Q = \frac{m\Delta E}{W} e$ [∵ Q α ΔE] and pulse height is proportional to energy of ionizing particle. This is why it is called a proportional counter.

**Description of the instrument:**

**1. Ionization chamber :**

(*i*) It consists of hollow copper cylindrical tube about 20 cm long and 2 cm in diameter.

(*ii*) It contains a mixture of 10% methane and 90% argon at a pressure of about one atmosphere.

(*iii*) *A tungsten electrode* consists of a fine tungsten wire of radium 0.001 cm and fixed along the axis of the tube. It is well-insulated.

(*iv*) A positive potential of about 700 volts is applied to the central electrode with respect to the cylindrical tube.

**2. Other components :**

(*i*) The electrical pulses generated in the counter are amplified with the help of amplifier and then fed to a counting device known as *scalar cum counter.*

(*ii*) A *discriminator* is used which allow pulses of required voltage height & cut off low voltage pulses.

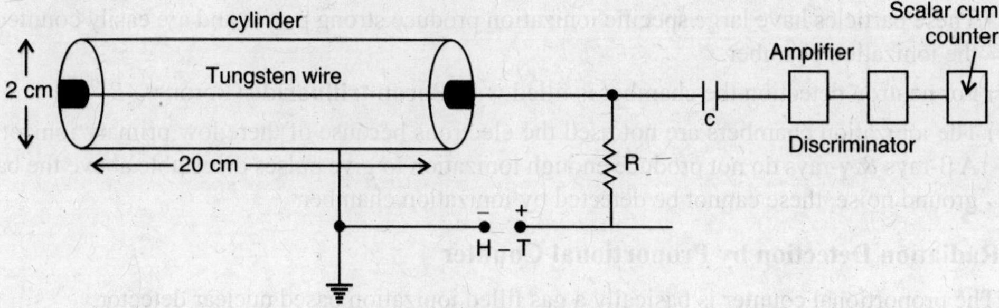


**Fig. 8.3** Proportional counter

**Working mechanism:**

(*i*) Since the central wire is very thin and the potential difference is very high, the electric field from the central wire is very high.

(*ii*) The electrons rushing towards the electrodes (wire) are involved in several collisions with the intervening neutral gas molecules or atoms. As a result, additional electrons are produced.

(*iii*) At each collision, the primary ion pair produced due to radiation ionizes more gas molecules to produce secondary ions which are also accelerated.

(*iv*) As a result a whole torrent of ions reach the electrode and is known as *Townsend Avalanche effect*.

(*v*) As a consequence of gas amplification, the current flow increases. In the proportional counter region the number of ion pairs is directly proportional to the number of primary electrons produced by the radiation until a certain voltage reached.

(*vi*) The voltage pulses filtered by the capacitor are passed through the discriminator circuits.

(*vii*) The discriminator level is adjusted in such a way that it does not allow the passage of low voltage pulses.

The discriminator sends to the counter only genuine pulses which are registered counted.

**Application:**

(*i*) Detection of low energy X-rays, β-rays and γ-rays.

(*ii*) It has been found very useful for detection and measurement of soft radiations like X-rays γ rays, β rays, mesons & protons.

(*iii*) The counter not only counts the incoming particles but is also capable of measuring the energy of the particles.

(*iv*) The particles of different energies can be distinguished by pulse height analysis because the size of the current pulse is proportional to the energy of the original charged particles causing ionization.

(*v*) If a sample is labelled with two different isotopes, both the isotopes can be measured by these counters.

(*vi*) Proportional counters can be used to detect neutrons although neutrons do not ionize the gas they pass through.

**(C) Geiger Muller Counter :**

It is based upon the ability of the radioactivity emission to ionize matter. Here, the voltage applied to a gas filled detector is increased beyond the proportional counter region, the pulse height does not remain proportional to the energy of the incident ionizing radiation or charged particle. The counter which operates in this region of higher voltage is known as a G.M. counter.

- **Principle:**

**Ionizing particles produce ions in a gases filled tube which are accelerated towards a high voltage electrode and collide with gas molecules to form other ions. These new ions are also accelerated and may produce more ions. These shower of ions appear as a voltage pulse across a resistor placed in the high voltage circuit. Then this voltage pulse is fed to an amplifier through a capacitor, the amplified pulse is finally passed on a scalar and detected and counted by a counter.**

- **Description of the Instrument:**

**1. Ionization chamber:**

(*i*) It is a cylindrical copper tube one end of which is closed by a thin mica window. The other end of the tube is closed by insulated plug.

(*ii*) A thin metal wire (tungsten) serves as *anode* enters the tube and runs along the axis.

(*iii*) The wall of the tube acts as *cathode*.

(*iv*) A potential difference is maintained between anode (axial wire) and cathode (metal wall) by connecting them to a source of voltage.

(*v*) The cylinder is filled with low pressure gas mixture of helium (He), argon (Ar) etc. A small volume of easily dissociable "quencing" substance like alcohol vapour is also present within the cylinder.

**Other components :**

(*i*) *Resistor capacitor* and pulse amplifier:

Through *capacitor* voltage pulse is fed to amplifier for *amplification* of pulse.

(*ii*) **Scalar & counter :**

Scalar records the arrival of each individual pulse separetly and finally detected and counted by counter.

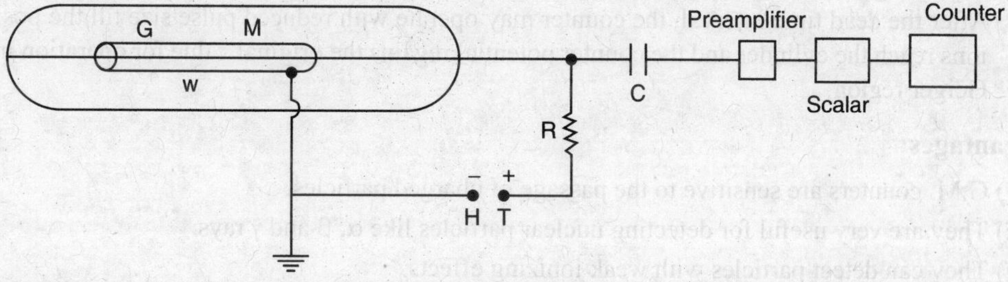


**Fig. 8.4**

- **Working :**

(*i*) The central thin wire acts as the *anode* and is mounted at high positive potential with respect to the metallic cylinder i.e. counter tube that acts as *cathode*.

(*ii*) Radiation emitted by the radioactive substance enters the counter tube through the mica window and ionize the gas molecules into ions which triggers a pulse of current.

(*iii*) The current pulse flows and produces voltage pulse. This voltage pulse is fed to a pulse amplifier through a capacitor.

(*iv*) The amplified pulse is finally passed to scalar and finally detected and counted by a counter.

(*v*) Let an ionizing particle enter the G.M. tube and produce one single ion pair in the enclosed tube.

(*vi*) The resulting electron would be rapidly accelerated towards the positive central wire and will reach relatively high velocity.

(*vii*) The electrons strike the gas molecule with the sufficient kinetic energy to knock out electrons out of the atoms and the process is called ionization by *collisions*.

(*viii*) The new electrons are also accelerated and may in their turn produce more ion pairs. The process is cumulative and as a result an avalanche occurs and a very large number of electrons reaches the anode, leaving behind a space charge due to massive slow moving positive ions. In this way, the initial formation of a single ion pairs results in a very *large pulse* of *current* to the anode.

(*ix*) Some of the free electrons on collisions with argon atoms merely excite them, which is their turn return to the normal state emitting photons. If a photon is absorbed by another excited atom, it may get ionized on releasing electrons which produce further *avalanches*. Thus the avalanche spreads throughout entire volume of the counter. It leads to a amplification as high as $10^8$.

(*x*) The total number of ions i.e. the whole length of the wire output pulse no longer depends upon the number of primary ions initially produced by the ionising particles.

(*xi*) **Dead time:**

(*i*) The wire potential changes with the drift of the positive ions. The rate of change of wire potential is most rapid as the positive ions move away from the immediate neighbourhood of the wire.

(*ii*) The counter remain dead bill the positive ions have moved away from the avalanche site to put the wire back to Geiger threshold potential.

The time interval between the production of initial pulse and initiation of second Geiger discharge is called *dead time* and during this period the counter is insensitive (dead) to further ionization pulses.

(*xii*) **Recovery time:**

After the dead time interval, the counter may operate with reduced pulse size till the positive ions reach the cylinder and the counter potential regains the original value for operation in the Geiger region.

**Advantages:**

(*i*) G.M. counters are sensitive to the passage of charged particles.

(*ii*) They are very useful for detecting nuclear particles like $\alpha$, $\beta$ and $\gamma$ rays.

(*iii*) They can detect particles with weak ionizing effects.

(*iv*) The pulse height is sufficiently high which requires little amplification.

(*v*) The pulse is constant over a wide range of applied voltage (known as plateu width)

(*vi*) They are very efficient in detecting and counting high energy $\beta$ particles.

**Disadvantage:**

(*i*) It is used for comparative measurement only.

(*ii*) G.M. counters have very low intrinsic efficiency.

(*iii*) It cannot detect uncharged particles like neutrons.

(*iv*) Less efficient than scintilation counter in detecting $\gamma$ rays.

- **Scintillator Counter:**

Rutherford and his contemporaries observed that when a screen of zinc sulphide (ZnS) is hit by an ionizing particle, the emission of a small flash of light (scintillation) is detected and converted into amplified voltage pulses by a photomultiplier. This property was utilized by Rutherford and others for counting the number of alpha (α) particles emitted from a radioactive source and formed the basis of scintillation.

- **Scintillation:**

**The phenomenon of fluorescence due to excitation by radioactivity is known as scintillation.** The light emitted in scintillation can be detected by coupling it to a photomultiplier which converts the photon energy into an electrical pulse whose magnitude remains proportional to the energy of the original radioactive event.

- **Types of Scintillation counting:**

Scintillation can be counted by two different techniques *viz* solid or external scintillation counting and liquid or internal scintillation counting.

**(a) *External* or *solid scintillation counting*:**

(*i*) Here the sample is placed close to the fluor crystal which in turn is placed adjacent to photomultiplier.

(*ii*) *Fluor crystal* — *Emitters*

- Crystallized Zinc sulphide (ZnS) — α emitter
- sodium iodide (NaI) — γ emitter
- anthracene — β emitter

(*iii*) Photomultiplier is connected to a high voltage supply and scalar.

(*iv*) It is particularly useful for measurement of gamma (γ) emitting isotopes and detection & and analysis of X-rays.

**(b) *Internal or liquid scintillation counting:***

(*i*) Here, the radioactive sample is suspended in scintillation system compossed of the solvent and an appropriate scintillator (Primary and Secondary).

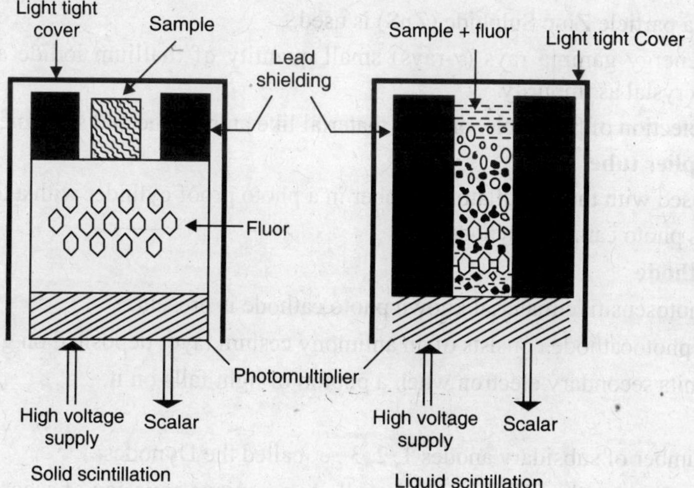


**Fig. 8.5** Solid and liquid scintillation counting, a schematic illustration.

(*ii*) The radiation from the suspended samples molecules collids with a solvent molecules imparting a discrete amount of the energy to the solvent molecule.

(*iii*) Aromatic hydrocarbons are very effective as scintillator solvents, toulene, xylene, Ethanol, Acetone are also used as solvent.

(*iv*) Liquid scintillation are employed largely for measurements on pure megatron emitters especially where the particles energy is low (< 1 MeV).

**Principle:**

**It is based on the principle that certain crystals produce flashes of light called scintillation when a photon of ionizing radiation is absorbed by crystals of scintillators or phosphors which are detected and converted into amplified voltage pulses by photomultiplier.**

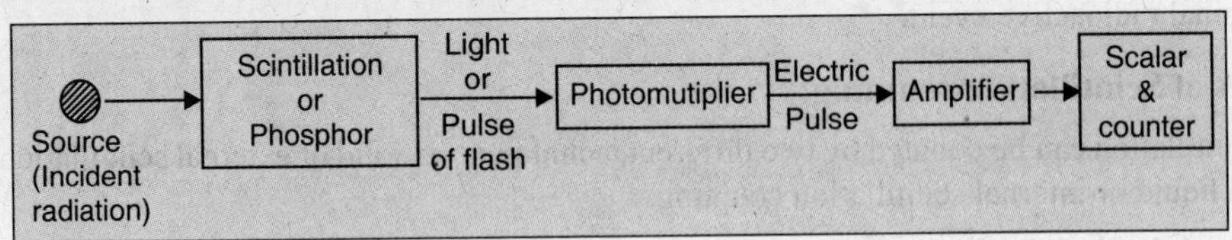


**Fig. 8.6** Flow chart of scintillation.

**Instrumentation :**

**(A) Source** (*i*) It emits ionizing radiations.

(*ii*) It falls on scintillator or phosphor.

**(B) Scintillation counter:**

(*i*) It is an energy transducer which converts the energy of an ionizing radiation or of a gamma photon into scintillation flashes.

(*ii*) It consists of scintilation chamber and photomultiplier tube.

**(*a*) Scintillaton chamber:**

***Scintillator:***

(*i*) The scintillator crystal or phosphor is housed in the scintillator chamber.

(*ii*) A wide variety of phosphor is used for different ionizing particles.

(*iii*) To detect a particle Zinc Sulphide (ZnS) is used.

(*iv*) For low energy gamma rays ($\gamma$-rays) small quantity of thallium iodide activator is used with NaI crystal as impurity.

(*v*) For the detection of $\beta$ particle, organic material like anthracene and stilbeme are used.

**(*b*) Photomultiplier tube:**

(*i*) It is enclosed with the scintillator chamber in a photo proof cylinder with a terminal cap.

(*ii*) It consists photo cathode, dynode

**(1) Photo cathode**

(*i*) A photosensitive material known photo cathode is present.

(*ii*) The photocathode consists of an antimony cesium layer deposited on glass.

(*iii*) It emits secondary electron when a photon of light falls on it.

**(2) Dynodes**

(*i*) A number of subsidary anodes 1, 2, 3 . . . called the Dynodes.

(*ii*) They are coated with materials that allows a copious emission of secondary electrons.

(*iii*) The voltages on the Dynodes are arranged to increase successively in positive steps.

(*iv*) In a typical photomultiplier there are 9 *dynodes* and 10 stages of electron amplification.

**(3) Collector plate:**

(*i*) The electrons emitted by final dynode (9) are collected at a solid metal disc electrode called *anode* or collecting electrode or collector plate.

(*ii*) The enormous number of electrons strike the collector plate to generate a pulse of current which is amplified further.

**(c) Amplifier and scalar counter**

(*i*) The output pulse of current from the photo multiplier is fed to a pulse amplifier.

(*ii*) After amplification, the current is recorded by an electronic pulse counter.

**Mechanism or Operation.**

(*i*) The incident radiation is absorbed by the scintillator (phosphor) and the molecules (atoms) are excited.

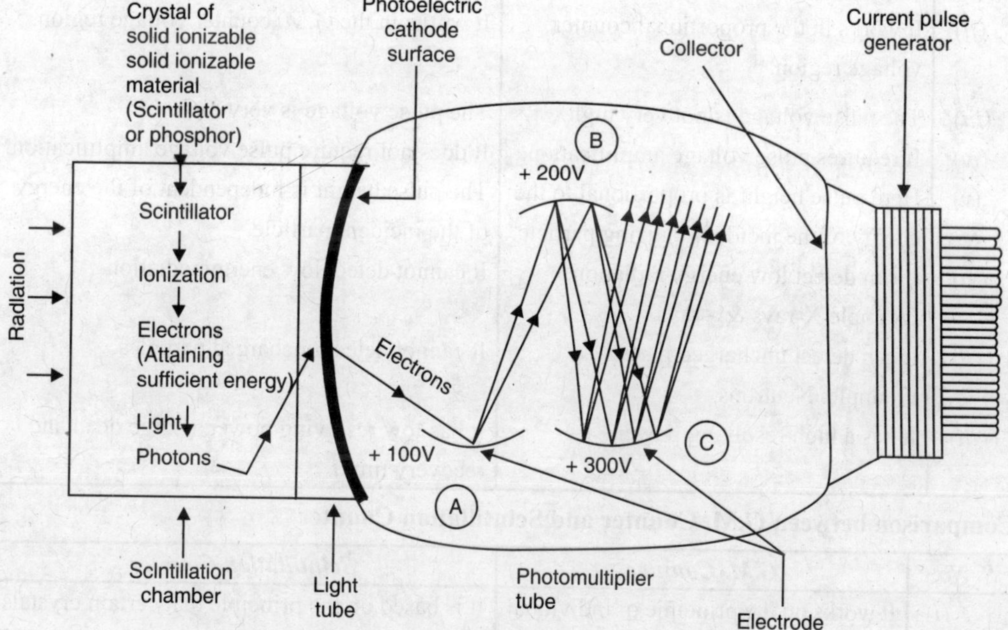


**Fig. 8.7** Schematic diagram of Scintillation counter

(*ii*) The excited atoms (molecules) of the flurorescent material of the phosphor de-excite and produce a small flash of light (scintillation) in a very short time.

(*iii*) The emitted light photons are transmitted to photo cathode of the photomultiplier tube.

(*iv*) Photo electrons are emitted due to absorption of light photons at the photo cathode.

(*v*) Electron multiplication takes place quickly at the photomultiplier dynodes. A single photoelectron emitted by photo cathode produces approximately $10^{-6}$ electrons.

(*vi*) The final voltage pulse is amplified by the amplifier, detected, measured and counted by the scalar-counter assembly.

- **Advantages of Scintillation counting:**

(*i*) It can detect α, β particles and γ rays.

(*ii*) Liquids, solids, suspensions, emulsions, gels etc can be accomodated in a scintillation counter and its radioactivity can be determined accurately.

(*iii*) Dual or triple labelled samples can be measured in a scintillation chamber.

(*iv*) This method is capable of rapid counting since the pulses given by scintillation counters are of extremely short duration.

- **Disadvantages of Scintillation counting:**

(*i*) It require higher voltages for operation.

(*ii*) The most disadvantages of scintillation counting is *quenching* (reduction in the efficiency of transferring energy from the β particles to the photomultiplier).

(*iii*) The cost per sample is higher in scintillation counting.

**Comparison between Proportional Counter and G.M. Counter**

| *S. No.* | *Proportional Counter* | *G M Counter* |
|---|---|---|
| (*i*) | It works on the principle of gas amplification and/or multiplication. | It works on the principle of individual gas ionization. |
| (*ii*) | It works in the proportional counter voltage region. | It works in the G.M. counter voltage region. |
| (*iii*) | The pulse voltage is not very high. | The pulse voltage is very high. |
| (*iv*) | It requires pulse voltage amplification. | It does not require pulse voltage amplification. |
| (*v*) | Here pulse height is proportional to the energy of the incident ionizing particle. | The pulse height is independent of the energy of the incident particle. |
| (*vi*) | It can detect low energy radiation. Example X-rays & γ rays. | It cannot detect low energy radiation. |
| (*vii*) | It can detect uncharged particles. Example: Neutrons | It cannot detect uncharged particles. |
| (*viii*) | It has a high resolving power. | It has low resolving power. (Large dead and recovery time). |

**Comparison between G.M. Counter and Scintillation Counter**

| *S. No.* | *G.M. Counter* | *Scintillation Counter* |
|---|---|---|
| (*i*) | It works on the principle of individual gas ionization. | It is based on the principle that certain crystals produces light called scintillation when photon ionizing radiation is absorbed by the crystals. |
| (*ii*) | By isonisation of gas electrons and ions are produced. | Only electrons are produced and no ions. |
| (*iii*) | It is operated with lower voltage. | It is operated with higher voltage. |
| (*iv*) | Here two electrodes like anode and cathode are used for the production of secondary electrons. | Here a series of electrodes are used for the production of secondary electrons. |
| (*v*) | It can not detect very minute quantities of radiation. | It can detect minute quantities of radiation. |
| (*vi*) | It is less efficient and sensitive to γ rays. | It is more efficient and sensitive to γ rays. |
| (*vii*) | Smaller than scintillation counter. Example: Neutrons | Bigger than G M Counter. |

## PROBLEMS

**1.** *What are the changes in the mass number A and atomic number Z of a nucleus due to radioactive disintegration?*

**Solution:**

(*i*) On emission of α particle from the nucleus, the mass number *A* decreases by 4 while the atomic number *Z* decreases by 2.

(*ii*) On emission β particle, the atomic number *Z* increases by 1.

(*iii*) On emission γ rays, both *A* and *Z* remain unchanged.

**2.** *Why are all radioactive elements ultimately converted into lead?*

**Solution:**

(*i*) All those elements which are heavier than lead is radioactive.

(*ii*) This is because in the nuclei of heavy atoms, besides the nuclear attractive forces repulsive forces between the protons are also effective and these forces reduce the stability of the nucleus.

(*iii*) Hence the nuclei of heavier elements are being converted into lighter and lighter elements by emission of radioactive radiation.

When they are converted into lead, the emission stopped because the nucleus of lead is stable.

**3.** *If the decay series of thorium $_{90}Th^{232}$, 6 α particles and 4 β particles are emitted. What will be the mass number and atomic number of the final decay product?*

**Solution:**

(*i*) One α decay results the decrease of mass number by 4 and decrease of atomic number by 2.

(*ii*) Therefore, the emission 6 α particle result, of mass number *i.e.,* 6 × 4 = 24. So the mass number will be 232 – 24 = 208.

Consequently the atomic number will be 6 × 2 = 12 *i.e.,* 90 – 12 = 78.

(*iii*) Again emission of one β particle result the increase of atomic number by 1. Here emission of 4 β particle result increase of atomic number by 4 × 1 = 4. So the atomic number will be 78 + 4 = 82.

$$_{90}Th^{232} \xrightarrow{6\alpha} {}_{78}Th^{208} \xrightarrow{4\beta} {}_{82}Th^{208}$$

**4.** *How many a particles will be emitted by the decay of $_{90}Th^{228}$ into $_{82}Pb^{212}$?*

**Solution:**

Let the *x* number of particle be emitted.

Therefore the mass number will be reduced by 4*x* and atomic number will be reduced by 2*x*.

$\therefore \quad 228 - 4x = 212 \qquad \therefore \quad -4x = -16 \qquad \therefore \quad x = 4$

Again $\quad 90 - 2x = 82 \qquad$ or $\quad -2x = -8 \qquad \therefore \quad x = 4$

So the number of emission α particle is 4.

**5.** *Calculate the number of alpha (α) and beta (β) particles emitted when $_{92}U^{238}$* and *$_{82}Pb^{206}$*.

**Solution:**

Let the element formed between $_{92}U^{238}$ and $_{82}Pb^{206}$ be *X*. The mass of *X* be same as that of *Pb* because there is no loss in the mass number of the daughter element formed as a result of beta (β) emission. Now let the number of α and β particles emitted be *x* and *y*. The disintegration sequence will be as follows.

$$_{92}U^{238} \xrightarrow{-x\alpha} X^{206} \xrightarrow{-y\beta} {}_{82}Pb^{206}$$

Thus

$$238 - 4x = 206$$

$$x = \frac{238-206}{4} = \frac{32}{4} = 8$$

The atomic number of X = 92 – 2 × 8 = 92 – 16 = 76

Consequently:

$$y = 82 - 76 = 6$$

Thus the number of various particles emitted is

$$\alpha = 8$$
$$\beta = 6$$

**6.** *How many α and β particles will be emitted when $_{90}Th^{234}$ changes into $_{84}Po^{218}$*

**Solution:**

Let the α and β particles emitted from the radioactive element be *x* and *y*.

$$_{90}Th^{234} \xrightarrow{-x\alpha} X \xrightarrow{-y\beta} {}_{84}Po^{218}$$

Thus $234 - 4x = 218$

$$4x = 234 - 218 = 16$$

or $x = 4$

Thus the number of α particle emitted = 4 Decrease the atomic number due to loss of 4 α particle.

Number of β particle emitted : The atomic number $x = 90 - 2 \times 4 = 82$ consequently $y = 84 - 82 = 2$

**7.** *What will be the mass number and atomic number of nucleide formed when $_4Be^7$ undergoes K electron capture ?*

**Solution:**

When the nucleus of an element captures a K-electron, a proton is converted into a neutron

$$_{+1}p^1 - {}_{-1}e^0 \rightarrow {}_0n^1$$

Therefore, mass number remains the same but the atomic number decreases by one unit.

Hence the mass number of the nucleide = 7 and the atomic number = 4 – 1 = 3

**8.** *$_{15}P^{32}$ undergoes β emission. What are the mass number and atomic number of the product ?*

**Solution:**

$$_{15}P^{32} \xrightarrow{\beta} X^{32}_{15+1} = X^{32}_{16}$$

The mass number 32 and atomic number 16

**9.** $_{90}Th^{232} \longrightarrow {}_{82}Pb^{208}$

*Mention the number of α and β particles emitted during the above reaction.*

**Solution:**

The number of α particle emitted

$$4x = 232 - 208 = 24$$

$$\therefore \quad x = 6 \quad i.e. \quad 6\,\alpha.$$

The number of β particle emitted: The atomic number decreases by loss of 6α; atomic number: $x = 90 - 2 \times 6 = 90 - 12 = 78$.

$$y = 82 - 78 = 4 \quad i.e. \quad 4\beta.$$

**10.** $_{90}U^{238}$ *emits α particle, what are the atomic and mass number of the product ?*

**Solution:**

$$_{92}U^{238} \xrightarrow{\alpha} {}_{90}X^{234}$$

$$238 - 4 = 234$$
$$92 - 2 = 90$$

Mass number = 234

Atomic number = 90

**11.** $_{90}Th^{228}$ *disintegrate to* $_{83}Bi^{212}$ *, what is the number of α and β particles emitted ?*

**Solution:**

The number of α particle emitted:

$$4x = 228–212 = 16$$

or $x = 4$ Thus the number of α particle is 4.

The number of β particle emitted:

Decrease the atomic number due to loss of 4 α particle

The atomic number $x = 90 - 2 \times 4 = 82$

Consequently $y = 83 - 82 = 1$

$$\alpha = 4$$
$$\beta = 1$$

**12.** *The radionucleide* $_{90}Th^{234}$ *undergoes two successive β decays followed by one α decay. What is the atomic number and mass number of the resultant product.*

**Solution:**

$$_{90}Th^{234} \xrightarrow{\beta} {}_{91}X^{234} \xrightarrow{\beta} {}_{92}Y^{234} \xrightarrow{\alpha} {}_{94}Z^{230}$$

mass number = 230

atomic number = 94

**13.** *In the following radioactive decay* $_{92}X^{232} \longrightarrow {}_{89}Y^{220}$, *how many α and β particles are ejected from X and Y.*

**Solution:**

The number of α particle emitted

$$4x = 232 - 220 = 12$$

$x = \frac{12}{4} = 3$ Thus the number of α particle is 3

The number of β particle emitted

Decrease the atomic number due to loss of α particle *i.e.*

3α the atomic number $x = 92 - 3 \times 2 = 86$

Consequently, $y(\beta) = 89 - 86 = 3$

3α and 3β are ejected.

**14.** *In the following radioactive decay* $_{90}Th^{234} \longrightarrow {}_{82}Pb^{206}$, *how many α and β particles are emitted?*

**Solution:**

(*i*) Here mass number change (234 – 206) = 28 unit.

Let $x$ be the number of α particle emitted emission of one a particle result decreases the mass number by 4.

Therefore
$$234 - 4x = 206$$
$$-4x = 206 - 234$$
$$-4x = -28 \quad \therefore x = 7$$

Therefore the number of α particle is 7.

The atomic number will be decreases by 7 × 2 i.e. 14 i.e. 90 – 14 = 76.

(*ii*) Emission of one β particle result, the increase of atomic number by 1. Let $y$ be the number of β particle be emitted. Therefore
$$76 + 1.y = 82$$
$$y = 82 - 76 = 6$$
$$_{90}Th^{234} \xrightarrow{7\alpha} {}_{76}Th^{206} \xrightarrow{6\beta} {}_{82}Th^{206}.$$

**15.** *From an element $_{84}X^{202}$ an α particle is liberated and then a β particle. If the ultimate thus formed is $_aY^b$, then find of a and b.*

**Solution:**

(*i*) Emission of α particle result decrease of mass number by 4 and atomic number by 2. Therefore after emission of α particle the mass number will be 202 – 4 = 198 and the atomic number will be 84 – 2 = 82. So element will be $_{82}X^{198}$.

(*ii*) Emission of β particle result only increase of atomic number by 1. Here after β emission mass number will remain as 198 and the atomic number will be 82 + 1 = 83.

Therefore the element will be $_aY^b = {}_{83}Y^{198}$.
$$a = 83 \text{ and } b = 198.$$

**16.** $^{227}_{89}AC \xrightarrow{-\beta} Th \xrightarrow{-\alpha} Ra \xrightarrow{-\alpha} Rn$

*Calculate mass number and atomic number of Th and Rn.*

**Solution:**

(*i*) After β emission, the mass number of thorium will be 227 and atomic number will be 89 + 1 = 90.

(*ii*) Then after α radiation in two steps, the mass number of *Rn* will decrease to 227 – (4 + 4) = 219. The atomic number will decrease to 90 – (2 + 2) = 86.

**17.** *The half life time period of a radioactives element 6930 days. What is the average life of the element?*

**Solution:**

The average life period of a radioactive element is reciprocal of the disintegration constant (λ).

Thus average life ($l$) = $\frac{1}{\lambda}$

$$\text{Here } \lambda = \frac{0.693}{T} = \frac{0.693}{6930} = \frac{693}{6930 \times 10^3} = 1 \times 10^{-4} \text{ day.}$$

$$\lambda = \frac{1}{\lambda} = \frac{1}{1 \times 10^{-4}} = 1 \times 10^4 \text{ day}$$

$$\therefore \quad \lambda = 1 \times 10^{-4}/\text{day}$$

**18.** *The half life of radium is 1580 years. What is the average life of the element?*

**Solution:**

The half life $t_{1/2}$ or $T$ = 1580 days.

The average life ($l$) = $\frac{1}{\lambda}$

We know $\lambda = \frac{0.693}{T} = \frac{0.693}{1580} = \frac{693}{158 \times 10^4} = 4.386 \times 10^{-4}$

$\therefore \quad l = \frac{1}{4.386 \times 10^{-4}} = \frac{1 \times 10^7}{4386} = \frac{10000}{4386} \times 10^3 = 2.279 \times 10^3/\text{day}.$

**19.** *Calculate half life time and mean life time of the radioactive substance whose decay constant is 4.28 × 10$^{-4}$ per year.*

**Solution:**

Decay constant ($\lambda$) = 4.28 × 10$^{-4}$ per year.

Average or mean life ($T_a$) = $\frac{1}{\lambda} = \frac{1}{4.28 \times 10^{-4}} = \frac{10^4}{4.28} = \frac{10^6}{428}$ = 2336 years.

Half life $t_{1/2}$ or $T = \frac{0.6931}{\lambda} = \frac{0.6931}{4.28 \times 10^{-4}} = \frac{6931 \times 10^4 \times 10^2}{428 \times 10^4}$

$= \frac{6931 \times 10^2}{428}$ = 16.1939 × 100

= 1619 years.

**20.** *The decay constant of radioactive element is 2 × 10$^{-4}$ per year. What is its half period?*

**Solution:**

Here decay constant ($\lambda$) = 2 × 10$^{-4}$

Half period $t_{1/2}$ or $T = \frac{0.693}{2 \times 10^{-4}} = \frac{0.693 \times 10^4}{2 \times 10^3}$ = 346.5 × 10

= 3465 years.

**21.** *The disintegration constant of a radioactive sample is 0.113 per minute. Find its half period?*

**Solution:**

Here disintegration constant ($\lambda$) = 0.113

Half life $T = \frac{0.693}{\lambda} = \frac{0.693}{0.113}$ = 6.13 minute.

**22.** *Calculate half life of $I^{131}$. The decay constant ($\lambda$) for $I^{131}$ is 0.0866/day.*

**Solution:**

Here decay constant $\lambda$ = 0.0866/day

Half life ($T$) = ?

$T = \frac{0.693}{0.0866}$ = 8.002 day = 8 days.

**23.** *Find half life and average life of a radioactive sample whose disintegration constant is 2.31 × 10$^{-3}$ per day.*

**Solution:**

Disintegration constant ($\lambda$) = 2.31 × 10$^{-3}$ /day

Half life ($T$) = ? Mean life ($Tav$) = ?

$$T = \frac{0.693}{2.31\times10^{-3}} = \frac{0.693\times10^3}{2.31} = \frac{693\times10^2}{231} = 300 \text{ days}$$

$$T_{av} = \frac{1}{\lambda} = \frac{1}{2.31\times10^{-3}} = \frac{1\times10^3}{2.31} = \frac{1\times10^5}{231} = 4.329 \times 10^2 = 4.329 = 432.9 \text{ days.}$$

**24.** *Calculate the decay constant of $^{32}P$ whose half life is 14 days.*

**Solution:**

Half life ($T$) = 14 days

$$\text{Decay constant } (\lambda) = \frac{0.693}{T} = \frac{0.693}{14} = \frac{693}{14\times10^3} = 49.5 \times 10^{-3}\text{/day.}$$

or 0.0495/day

$$\text{or} \quad \frac{0.0495}{24} = 0.00206\text{/hr} = \frac{0.00206}{60} = 0.000034\text{/sec.}$$

$$\text{or} \quad \frac{0.000034}{60} = 0.0000005\text{/sec.}$$

**25.** *$Ca^{45}$ has a half life of 163 days. Calculate decay constant ($\lambda$) interms $day^{-1}$ and $sec^{-1}$.*

**Solution:**

Here $T$ = 163 days

$$\text{Decay constant } (\lambda) = \frac{0.693}{T} = \frac{0.693}{163} = 0.00425 = \frac{425}{10^5}$$

$$= 4.25 \times 10^{-3}\text{/day}$$

$$\frac{4.25\times10^{-3}}{24\times60\times60} = \frac{425}{10^3\times10^2\times864\times10^2} = \frac{425}{10^5\times864\times10^2} = \frac{425\times10^{-7}}{864}$$

$$= 0.4918 \times 10^{-7} = 4.92 \times 10^{-8}\text{/sec.}$$

**26.** *Potassium 40 has a half life of $4 \times 10^8$ years. Calculate decay constant.*

**Solution:**

Half life ($T$) = $4 \times 10^8$

Decay constant ($\lambda$) = ?

$$\lambda = \frac{0.693}{T} = \frac{0.6931}{4\times10^8} = 0.173275 \times 10^{-8} \text{ yr}^{-1}$$

$$\frac{0.693}{4\times10^8} = \frac{693}{4\times10^{11}} = \frac{693\times10^{-11}}{4\times365\times24\times60\times60} = \frac{693\times10^{-13}}{1261440} = \frac{693\times10^{-14}}{126144} = 0.0054937\times10^{-14}$$

$$= \frac{5.4937\times10^{-14}}{10^3} = 5.49 \times 10^{-17} = 5.5 \times 10^{-17}\text{/sec.}$$

**27.** *A radioactive substance has a half life of 30 days. Calculate radioactive disintegration constant.*

**Solution:**

Half life ($T$) = 30

$$\text{Decay constant } (\lambda) = \frac{0.693}{T} = \frac{0.693}{30} = \frac{693}{3\times10^4} = 231 \times 10^{-4}\text{/day}$$

**28.** *A certain radioactive substance has a half life of 20 days. Calculate disintegration constant and the average life.*

**Solution:**

Half life ($T$) = 20 days

Disintegration constant ($\lambda$) = ?

$$\lambda = \frac{0.693}{T} = \frac{0.693}{20} = 0.34655 \text{ day}^{-1}$$

$$\text{Average life } (T_{av}) = \frac{1}{\lambda} = \frac{1}{0.34655} = 28.855 \text{ days}$$

$$= 28.86 \text{ days.}$$

**29.** *The half life of a certain radioactive substance is 13.86 days. What is the value of disintegration constant?*

**Solution:**

$$\text{Half life } (T) = 13.80 \text{ days}$$

$$\text{Disintegration constant } (\lambda) = \frac{0.693}{13.86}$$

$$= 0.05 \text{ per days}$$

**30.** *What percentage of given mass of a radioactive substance will be left undecayed after four half-lives.*

**Solution:**

Initial mass be $N_0$, then quantity $N$ of the element left after '$n$' half-lives *i.e.,*

$$N = N_0 \left(\frac{1}{2}\right)^n$$

Therefore mass of element remaining after 4 half lives is

$$N = N_0 \left(\frac{1}{2}\right)^4 = \frac{N_0}{16} = 6.25\%$$

**31.** *The half life of radon is 3.8 days. Calculate how much radon will be left out of 1024 milligram after 38 days.*

**Solution:**

The half life of radon is 3.8 days i.e. $T = 3.8$ days. The number of half life ($n$) in 38 days ($t$) *i.e.,*

$n = \frac{t}{T}$ or $n = \frac{38}{3.8} = 10.$

The initial quantity of radon is 1024 milligram *i.e.,* 1024 mg. Therefore the mass of radon left after 10 half lives is $N = 1024 \times \left(\frac{1}{2}\right)^{10} = \frac{1024}{1024} = 1.0$ mg.

**or**

| **Time ($t$)** | **0** | ***T*** | **2*T*** | **3*T*** | **4*T*** | **5*T*** | **6*T*** |
|---|---|---|---|---|---|---|---|
| Amount left (mg) | 1024 | $\frac{1024}{2}$ | $1024 \times \frac{1}{4}$ | $\frac{1}{8} \times 1024$ | $\frac{1}{16} \times 1024$ | $\frac{1}{32} \times 1024$ | $\frac{1}{64} \times 1024$ |

*(Contd.)*

| Time (t) | | | | 7T | 8T | 9T | 10T |
|---|---|---|---|---|---|---|---|
| Amount left (mg) | | | | $\frac{1}{128}\times1024$ | $\frac{1}{250}\times1024$ | $\frac{1}{512}\times1024$ | $\frac{1}{1024}\times1024=1$ |

**32.** *Half life of a radioactive element is 2 years. Initial weight of the element is Ig. After 6 years what fraction of the element will remain?*

**Solution:**

Here half life is 2 years *i.e.,* $T = 2$ years

Initial weight of the element i.e. $N_0 = 1$ gram

∴ Number of half lives ($n$) for 45 years *i.e.,*

$$t = 6 \qquad \therefore \qquad n = \frac{t}{T} = \frac{6}{2} = 3$$

Amount of element left after 6 years *i.e.,* $N = N_0\left(\frac{1}{2}\right)^n$

$$= 1\left(\frac{1}{2}\right)^3 = \frac{1}{8} = 0.125g$$

**33.** *The half life of radioactive element is 2 hours. Initial weight was 32 gram. After 10 hours what fraction will remain?*

**Solution:**

Here half life ($T$) = 2 hours. Initial weight was 32 grams.

Therefore, the number of half lives ($n$) after 10 hours ($t$) will be

$$n = \frac{t}{T} = \frac{10}{2} = 5$$

The amount of element will remain after 10 hours

i.e. $N = N_0\left(\frac{1}{2}\right)^n = 32\times\left(\frac{1}{2}\right)^5 = 32\times\frac{1}{32} = 1$ gram.

**or**

| Time | 0 | T | 2T | 3T | 4T | 5T |
|---|---|---|---|---|---|---|
| Amount left | 32 | $32\times\frac{1}{2}=16$ | $32\times\frac{1}{4}=8$ | $32\times\frac{1}{8}=4$ | $32\times\frac{1}{16}=2$ | $32\times\frac{1}{32}=1$ |

**34.** *Half life of a radioactive element is 2d what fraction of 1 gram sample of this element will remain radioactive after 16d?*

**Solution:**

Here half life ($T$) = $2d$. Initial weight ($N_0$) = 1 gram.

Therefore, the number of half lives ($n$) after $16d$ ($t$) will be

$$n = \frac{t}{T} = \frac{16d}{2d} = 8$$

∴ The amount will remain after $16d$ will be $N = N_0\left(\frac{1}{2}\right)^n$

$$\therefore \quad N = 1\times\left(\frac{1}{2}\right)^8 = 1\times\frac{1}{256} = 0.0039 \text{ gram.}$$

**35.** *A sample of radioactive substance has $10^6$ radioactive nuclei. It has half time is 20 seconds. How many nuclei will remain after 10 seconds?*

**Solution:**

Initial number = $N_0 = 10^6$. Half life ($T$) = 20 second.

Therefore, the number of half lives ($n$) in 10 seconds ($t$) will be

$$n = \frac{t}{T} = \frac{10}{20} = \frac{1}{2}$$

$$\therefore \text{ The amount will remain } N = N_0\left(\frac{1}{2}\right)^{1/2} = 10^6\times\frac{1}{\sqrt{2}} = 10^6\times\frac{1}{1.41}$$

$$= \frac{1000}{141}\times 10^5 = 7\times 10^5 \text{ (approx.)}$$

**36.** *The half life period of $_{53}I^{125}$ is 60 days. What percent of radioactivity would be present after 180 days?*

**Solution:**

Here half life ($T$) or $I$ is 60 days.

Therefore, the number of half lives ($n$) after 180 days ($t$)

i.e. $$n = \frac{t}{T} = \frac{180}{60} = 3$$

$\therefore$ Amount of the element ($I$) would be present after 180 days *i.e.*

$$N = N_0\left(\frac{1}{2}\right)^n = 100\times\left(\frac{1}{2}\right)^3 = \frac{100}{8} = 12.5\%$$

**37.** *A radioactive element has a half life of 8 years. Calculate how much of the radioactive element will be left out of I milligram after 24 years.*

**Solution:**

Here half life ($T$) of the radio active element = 8 years.

Initial weight ($N_0$) = 1 milligram.

Therefore, the number of half lives ($n$) in 24 years ($t$) will be

$$n = \frac{t}{T} = \frac{24}{8} = 3$$

$$\therefore \text{ The amount left after 24 years i.e. } N = N_0\times\left(\frac{1}{2}\right)^n$$

$$N = 1\times\left(\frac{1}{2}\right)^3 = \frac{1}{8}\, mg$$

**38.** *$I^{131}$ has half life period of 13.3 hours. After 79.8 hours what fraction of $I^{131}$ will remain?*

**Solution:**

Here half life of radioactive element ($T$) is 13.3 hours.

Therefore the number of half lives ($n$) in 79.8 hours will be

$$n = \frac{t}{T} = \frac{79.8}{13.3} = 6$$

∴ The amount of the element will remain *i.e.*

$$N = N_0 \times \left(\frac{1}{2}\right)^n = \left(\frac{1}{2}\right)^6 = \frac{1}{64}$$

**39.** *The half life of a radioactive substance is 2d. After how many days $\frac{1}{64}$ of the sample will remain behind?*

**Solution:**

Here half life of the substance is $2d$ *i.e.* $T = 2d$.

| Time | 0 | $T$ | $2T$ | $3T$ | $4T$ | $5T$ | $6T$ |
|---|---|---|---|---|---|---|---|
| Residue left behind | 1 | $\frac{1}{2}$ | $\frac{1}{4}$ | $\frac{1}{8}$ | $\frac{1}{16}$ | $\frac{1}{32}$ | $\frac{1}{64}$ |

∴ Required time is $6T$ i.e. $6 \times 2d = 12d$.

**40.** *The half life of a radioactive element (RaA) is 30 minutes. Find many days the quantity of RaA is reduced to $\frac{1}{4}$.*

**Solution:**

| Time | 0 | $T$ | $2T$ | $3T$ |
|---|---|---|---|---|
| Residual mass | 1 | $\frac{1}{2}$ | $\frac{1}{4}$ | $\frac{1}{8}$ |

∴ The required time is $2T$ i.e. $3.0 \times 2 = 6.0$ minutes.

**41.** *The half life of radium is 1600 years. After how many years 25% of the radium block undecayed?*

**Solution:**

Here half life i.e. $T = 1600$ years, $25\% = \frac{25}{100} = \frac{1}{4}$.

| Time | 0 | $T$ | $2T$ | $3T$ | $4T$ | $5T$ | $6T$ |
|---|---|---|---|---|---|---|---|
| Residual mass | 1 | $\frac{1}{2}$ | $\frac{1}{4}$ | $\frac{1}{8}$ | $\frac{1}{16}$ | $\frac{1}{32}$ | $\frac{1}{64}$ |

∴ Required time is $2T$ *i.e.* $2 \times 1600 = 32$ years.

**42.** *The half life of radio active substance is 30 days. What time will it take for $(3/4)^{th}$ of its original mass to disintegrate.*

**Solution:**

Here half life *i.e.* $T = 30$ days, the mass remain undecayed $1 - \frac{3}{4} = \frac{1}{4}$.

| Time | 0 | $T$ | $2T$ | $3T$ | $4T$ |
|---|---|---|---|---|---|
| Residual mass | 1 | $\frac{1}{2}$ | $\frac{1}{4}$ | $\frac{1}{8}$ | $\frac{1}{16}$ |

Therefore required time $2T = 30 \times 2 = 60$ days.

**43.** *Calculate the time required for 90% of a radioactive sample of thorium to disintegrate. Assume the half life of thorium to be 1.4 × $10^{10}$ years.*

**Solution:**

Here half life ($T$) = 1.4 × $10^{10}$

$$\text{The decay constant } \lambda = \frac{0.6931}{T} = \frac{0.6931}{1.4\times10^{10}} = 0.495\times10^{-10}$$

If initially (at time $t = 0$), the amount or atoms of radioactive sample thorium be $N_0$ and then the amount left at time $t$ be $N$.

$$\therefore \quad N = N_0\, e^{-\lambda t}$$

$$\text{Here 90\% of the sample disintegrate, i.e. } N_0 \times \frac{90}{1N} = \frac{9}{10} N_0 = 0.9\, N_0$$

Amount of undecayed substance $N = N_0 - 0.9\, N_0 = 0.1\, N_0$

$$N = N_0\, e^{-\lambda t}$$

$$N = \frac{N_0}{e^{\lambda t}} \quad \text{or} \quad e^{\lambda t} = \frac{N_0}{N} \quad \text{or} \quad \lambda t = \log_e \frac{N_0}{N} = \log_e \frac{N_0}{0.1\, N_0}$$

$$\text{or} \quad \lambda t = \log_e \frac{N_0}{\frac{1}{10} N_0} = \log_e 10$$

$$t = \frac{\log_e 10}{0.495\times10^{-10}} = \frac{2.30\times10^{10}}{0.495} = 4.646\times10^{10} = 4.65\times10^{10}$$

$$t = \frac{1}{\lambda}\log_e\left(\frac{N_0}{N}\right) = \frac{1}{0.495\times10^{-10}}\times\log_e\left(\frac{N_0}{0.1\,N_0}\right) = \frac{1\times10^{10}}{0.495}\times\log_e 10 = \frac{10^{10}}{0.495}\times2.30 = 4.646\times10^{10}$$

**44.** *Calculate the time required for 10% of a sample of thorium to disintegrate. Assume the half life of thorium to be 1.4 × $10^{10}$ years.*

**Solution:**

$$\text{Half life } (T) = 1.4\times10^{10}, \qquad \therefore \text{ Decay constant } (\lambda) = \frac{0.6931}{1.4\times10^{10}} = 0.495\times10^{-10}$$

If $N_0$ be the amount of radioactive substance initially (at $t = 0$), then the amount left after '$t$' time be $N$.

$$N = N_0\, e^{-\lambda t}$$

$$\text{Here 10\% of the sample disintegrate } \textit{i.e. } \frac{10}{100} N_0 = 0.1\, N_0$$

Amount left $N = N_0 - 0.1\, N_0 = 0.9\, N_0$

$$N = N_0\, e^{-\lambda t}$$

$$\text{or} \quad N = \frac{N_0}{e^{\lambda t}} \quad \text{or} \quad e^{\lambda t} = \frac{N_0}{N} \quad \text{or} \quad \lambda t = \log_e \frac{N_0}{0.9\, N_0} = \log_e\left(\frac{1}{0.9}\right)$$

$$\text{or} \quad t = \frac{1}{\lambda}\log_e\left(\frac{1}{0.9}\right) = \frac{1.4\times10^{10}}{0.6931}\times\log_e 1.111 \quad t = \frac{1\times10^{10}}{0.495}\times0.10536$$

$$= 0.2108\times10^{10}$$

$$= 2.108\times10^{9} \text{ years.}$$

**45.** *Calculate the time in which the activity of a sample of thorium reduces to 90% its original value. Assume the half life of thorium to be 1.4 × 10^10^ yrs.*

**Solution.**

Here half life ($T$) = $1.4 \times 10^{10}$ years

The decay constant: $= \dfrac{0.6931}{1.4 \times 10^{10}} = 0.495 \times 10^{-10}$

Initial amount of the sample ($N_0$) = 100

Amount reduced to ($N$) = 90 $\quad \therefore \dfrac{N}{N_0} = \dfrac{90}{100} = 0.9$

or $\dfrac{N_0}{N} = \dfrac{100}{90} = 1.11$

$$t = \frac{2.303}{\lambda} \log \frac{N_0 N}{}$$
$$= \frac{2.303}{0.495 \times 10^{-11}} \log 1.11$$
$$= 4.65 \times 10^{100} \times 0.045$$
$$= 0.2108 \times 10^{10} \times 2$$

We have $N = N_0 \, e^{-\lambda t}$

or $\dfrac{N}{N_0} = \dfrac{1}{e^{\lambda t}}$ or $e^{\lambda t} = \dfrac{N_0}{N}$

or $\lambda t = \log_e \left( \dfrac{N_0}{N} \right)$

or $t = \dfrac{1}{\lambda} \times \log_e \left( \dfrac{N_0}{N} \right) = \dfrac{1}{0.495 \times 10^{-10}} \times \log_e 1.11$

or $t = \dfrac{1 \times 10^{10}}{0.495} \times 0.1043 = \dfrac{0.10436 \times 10^{10}}{0.495} = 0.2108 \times 10^{10}$

$= 2.108 \times 10^9$ year

**46.** *The half-life of Polonium is 140 days. In what time will 15g of polonium be disintegrated out of its initial mass of 16g.*

**Solution:**

Let the initial quantity of polonium is $N_0$ *i.e.* 16*g*.

The quantity of the element disintegrated is 15*g*.

So the quantity of the element left is $N = 16 - 15 = 1g$.

Thus the quantity of the element left after '$n$' half-lives will be $N = N_0 \left( \dfrac{1}{2} \right)^n$ or $\dfrac{N}{N_0} = \left( \dfrac{1}{2} \right)^n$

Here $N = 1$ gram $N_0 = 16$ gram.

or $\dfrac{1}{16} = \left( \dfrac{1}{2} \right)^n$ or $\left( \dfrac{1}{2} \right)^n = \left( \dfrac{1}{4} \right)^2$

or $n = 4$.

The half life of polonium is 140 days. Hence the time of disintegration

= half life × number of half-lives

= 140 × 4 = 560 days.

**47.** *The half life of strontium 90 is 30 years. In what time will 0.04 g strontium 90 become 0.005 g after decay?*

**Solution:**

The initial mass of strontium 90 *i.e.* $N_0 = 0.04$ *g*.

After decay mass remain left i.e. $N = 0.005\ g$

$$\therefore \quad \frac{N}{N_0} = \frac{0.005}{0.04} = \frac{5}{40} = \frac{1}{8}$$

The mass remain left after '*n*' half lives.

will be $$N = N_0\left(\frac{1}{2}\right)^n \quad \text{or} \quad \frac{N}{N_0} = \left(\frac{1}{2}\right)^n$$

or $$\frac{1}{8} = \left(\frac{1}{2}\right)^n \quad \text{or} \quad \left(\frac{1}{2}\right)^n = \left(\frac{1}{2}\right)^3$$

or $$n = 3$$

Hence the required time of disintegration = half life × number of half-lives

$$= 30 \times 3 = 90 \text{ years}$$

**48.** *After certain lapse of time, the fraction of radioactive polonium undecayed is found to be 3.125% of its initial quantity. What is the duration of this time lapse if the half life of polonium is 138 days?*

**Solution:**

The initial quantity of polonium is $N_0 = 100$

Undecayed quantity *i.e.* $N = 3.12\ N_0$

Thus the quantity of the element left after '*n*' half-lives will be $N = N_0\left(\frac{1}{2}\right)^n$

$$\frac{3.125}{100} = \left(\frac{1}{2}\right)^n$$

**49.** *The half life of radium is 1600 years. After how many years 25% of a radium block remains undecayded?*

**Solution:**

Let the initial quantity of radium is $N_0$. the quantity left after *n* half-lives will be

$$N = N_0\left(\frac{1}{2}\right)^n$$

Here $$N = 25\% \text{ of } N_0 = \frac{25}{100}N_0 = \frac{N_0}{4}$$

$$\therefore \quad \frac{N_0}{4} = N_0\left(\frac{1}{2}\right)^n$$

or $$\frac{1}{4} = \left(\frac{1}{2}\right)^n \quad \text{or} \quad \left(\frac{1}{2}\right)^n = \left(\frac{1}{4}\right) = \left(\frac{1}{2}\right)^2$$

$$\therefore \quad n = 2$$

∴ time of disintegration = half life × number of half lives

$$= 1600 \times 2 = 3200 \text{ years.}$$

**50.** *The half life of radioactive substance is 15 years. Calculate the period in which 2.5% of the initial quantity will be left over.* ***(Luck U. 1996)***

**Solution:**

Let $N_0$ be the initial quantity of the radioactive, substance. The quantity ($N$) left after a time '$t$'.

$\therefore \quad \frac{N}{N_0} = \frac{2.5}{100} = \frac{1}{40}$, Half life ($T$) = 15 years.

$$\lambda = \frac{0.6931}{15} = 0.0462 \text{ year}^{-1}$$

$$N = N_0\, e^{-\lambda t}$$

$$\frac{N}{N_0} = e^{-\lambda t}$$

or $$\frac{1}{40} = \frac{1}{e^{\lambda t}}$$

or $$e^{\lambda t} = 40$$

$$\lambda t = \log_e 40$$

$$t = \log_e 40 \times \frac{1}{0.0462} = \frac{3.69}{0.0462} = 79.87 \text{ years.}$$

**51.** *The half life of radon is 3.8 days. After how many days will one twentieth of a radon sample be left over?*

**Solution:**

Half life ($T$) = 3.8 days. $\therefore$ decay constant ($\lambda$) = $\frac{0.693}{3.8}$

$$= 0.18236$$

Let $N_0$ be the initial quantity and the quantity left ($N$) after a time '$t$'.

$\therefore \quad N_0 = 1 \qquad N = \frac{1}{20} N_0 = \frac{N_0}{20}$

We have $N = N_0\, e^{-\lambda t}$ or $\frac{N}{N_0} = e^{-\lambda t} = \frac{1}{e^{\lambda t}}$

or $$\lambda t = \log_e \frac{N_0}{N}$$

$$t = \log_e \frac{N_0}{N} \times \frac{1}{\lambda} = \log_e \frac{N_0}{N_0} \times 20 \times \frac{1}{0.18236}$$

or $$t = \log_e 20 \times \frac{1}{0.18236} = \frac{2.995}{0.18236} = 16.42 \text{ days.}$$

$$t = \frac{1}{\lambda} \log_{10} 20 \times 2.302 = \frac{1}{0.18236} \times 1.30102 \times 2.302 = \frac{2.9949}{.18236} = 16.42$$

**52.** *The half life of radon is 3.8 days. After how many days 10% of radon sample will remain behind?*

**Solution:**

Here half life ($T$) = 3.8 decays constant ($\lambda$) = $\frac{0.693}{3.8}$

$= 0.18236$

Let $N_0$ be the initial quantity and the quantity left ($N$) after a time '$t$'

$$\therefore \quad N_0 = 1 \qquad N = \frac{10}{100} N_0 = 0.1\, N_0$$

We have $N = N_0 . e^{-\lambda t}$ or $\frac{N}{N_0} = e^{-\lambda t} = \frac{1}{e^{\lambda t}}$

or $e^{\lambda t} = \frac{N_0}{N}$ or $\lambda t = \log_e \left(\frac{N_0}{N}\right)$

$$t = \frac{1}{\lambda} . \log_e \left(\frac{N_0}{N}\right) = \frac{1}{\lambda} \log_e \left(\frac{N_0}{0.1\, N_0}\right) = \frac{1}{\lambda} \log_e 10$$

$$\therefore \quad t = \frac{1}{0.18236} \log_e 10 = \frac{2.30}{0.18236} = 12.62 \text{ days.}$$

**53.** *The half life of radon is 3.8 days. How long does it take for 60% of a sample of radon to decay?*

**Solution:**

Let $N_0$ be he initial quantity of radon.

The quantity left ($N$) after a time '$t$'.

When 60% of radon decay, 40% is left behind.

$$\therefore \quad \frac{N}{N_0} = \frac{40}{100} = 0.4\, N_0$$

Here half life $T = 3.8$ days $\quad \therefore \lambda = \frac{0.693}{3.8} = 0.1824$

We have $N = N_0\, e^{-\lambda t}$

So $\frac{N}{N_0} = e^{-\lambda t} \qquad \therefore e^{\lambda t} = \frac{N_0}{N}$

$$\lambda t = \log_e \left(\frac{N_0}{N}\right)$$

$$\therefore \quad t = \frac{1}{\lambda} \log_e \left(\frac{N_0}{N}\right) = \frac{1}{0.1824} \log_e \left(\frac{N_0}{0.4\, N_0}\right) = \frac{1}{0.1824} \times \log_e \frac{10}{4}$$

$$t = \log_e 2.5 \times \frac{1}{0.1824} = \frac{0.91629}{0.1824} = 5.02 \text{ days}$$

**54.** *How much of 5.00 gram of polonium will decay in one year? The half life of polonium is 138 days ($\log_e 5.00 = 1.609$, $e^{-0.223} = 0.800$).*

**Solution:**

The decay constant of polonium is

$$\lambda = \frac{0.6931}{138} = 0.005022 \text{ day}^{-1}$$

Let $N_0$ be the initial quantity of polonium. The quantity left ($N$) after a time '$t$'.

Here $N_0 = 5.00$ gram. $t = 1$ year = 365 days.

We have $N = N_0 e^{-\lambda t}$

$$N = N_0 \times \frac{1}{e^{\lambda t}} = \frac{5.00}{e^{0.005022 \times 365}} = \frac{5.00}{e^{1.834}}$$

$$\log_e N = \log_e \left(\frac{5.00}{e^{1.834}}\right) \quad or \quad \log_e N = \log 5 - 1.833$$

$$= 1.609 - 1.833 = -224$$

$$\therefore \quad N = e^{-224} = 0.8$$

**55.** *The half life of $_{11}Na^{24}$ is 15 hours. How long does it take for 93.75 percent of a sample of this isotope.*

**Solution:**

Here half life of $_{11}Na^{24}$ is 15 hours.

$$T = 15$$

Decay constant $$(\lambda) = \frac{0.6931}{15} \text{ hr}^{-1} = 0.0462 \text{ hr}^{-1}$$

Initial amount of sample ($N_0$) = 100

Let '$t$' be the time in which 93.75% of the sample decays *i.e.* 100 – 93.75 = 6.25% of the sample remains left *i.e.* $N$ will be 6.25

*i.e.* $\frac{6.25}{100} . N_0$ or $N = \frac{6.25}{100} N_0$

We have

$$N = N_0 e^{-\lambda t} \quad \text{or} \quad \frac{N}{N_0} = e^{-\lambda t}$$

$$\text{or} \quad \frac{N}{N_0} = \frac{1}{e^{\lambda t}} \quad \text{or} \quad e^{\lambda t} = \frac{N_0}{N} \quad \text{or} \quad \lambda t = \log_e \left(\frac{N_0}{N}\right)$$

$$\text{or} \quad t = \frac{1}{\lambda} \log_e \left(\frac{N_0}{N}\right) = \frac{1}{0.0462} \times \log_e \frac{N_0 \times 100}{6.25\, N_0} = \frac{1}{0.0462} \times \log_e \frac{100}{6.25}$$

$$\therefore \quad t = \frac{1}{0.0462} \times \log_e 16 = \frac{1}{0.0462} \times 2.77 = 59.956 \text{ hours} = 59.96 \text{ hrs.}$$

$$= 60 \text{ hours.}$$

**56.** *The half life of a radioactive substance is 15 years. Calculate the period in which 2.5% of the initial quantity will be over.*

**Solution:**

Let $N_0$ be the initial quantity of the radioactive element. After 't' time 2.5% of the amount of the initial quantity will be over.

$$N_0 = 100 \quad N = \frac{2.5}{100} N_0 = \frac{1}{40} N_0 \quad \text{or} \quad \frac{N_0}{N} = \frac{N_0}{N_0} \times 40 = 40$$

Here $T = 15$ years $$\therefore \lambda = \frac{0.693}{15} = 0.0462 \text{ years}^{-1}$$

We have $N = N_0 e^{-\lambda t}$ or $\frac{N}{N_0} = \frac{1}{e^{\lambda t}}$ or $e^{\lambda t} = \frac{N_0}{N}$ or $\lambda t = \log_e\left(\frac{N_0}{N}\right)$

$$\therefore \quad t = \frac{1}{\lambda}\log_e\left(\frac{N_0}{N}\right) = \frac{1}{0.0462}\times \log_e 40 = \frac{1}{0.0462}\times 3.6888 = 79.844 \text{ yrs} = 79.84 \text{ yrs}$$

**57.** *A freshly prepared radon source loses half its activity in 3.86 days. Calculate the time in which activity would be reduced to 5%.*

**Solution:**

Half life of radon ($T$) = 3.86 days

$$\therefore \quad \lambda = \frac{0.6931}{3.86} = 0.1795$$

Here initial quantity $N_0 = 100$ and $N$ will be 5% i.e. $\frac{5}{100} N_0$

$$N = \frac{5}{100} N_0 = 0.05\, N_0 \quad \text{or} \quad \frac{N_0}{N} = \frac{N_0}{0.05\, N_0} = \frac{1}{0.05} = 20$$

We have

$$N = N_0 \,.\, e^{\lambda t} \quad \text{or} \quad \frac{N}{N_0} = \frac{1}{e^{\lambda t}}$$

or

$$e^{\lambda t} = \frac{N_0}{N}$$

or

$$\lambda t = \log_e\left(\frac{N_0}{N}\right)$$

$$t = \frac{1}{0.1795}\times \log_e 20 = \frac{1}{0.1795}\times 2.9957 = 16.689 \text{ days}$$

$$= 16.69 \text{ days}$$

**58.** *The half life of a radioactive element is 100 years. In how many years its activity will decay to 0.1 times of its initial value.*

**Solution:**

Half life = $T$ = 100 years

$$\text{decay constant } (\lambda) = \frac{0.693}{T} = \frac{0.693}{100} = \frac{693}{10^5}$$

Let the time be '$t$' years.

$N_0$ = Initial amount, $N$ = Amount after decay.

$$t = \frac{1}{\lambda}\log_e\left(\frac{N_0}{N}\right) = \frac{1}{\lambda}\log_e\left(\frac{N_0}{N_0 \times 0.1}\right) = 0.1\, N_0$$

$$= \frac{1\times 10^5}{693}\times \log_e (10) = 144.3 \times 2.303 = 332.32 \text{ years.}$$

**59.** *The half life of radium is 1590 years. In how many years will one gm of pure element (a) loose one centigram and (b) be reduced to one centigram?*

**Solution:**

The half life of radium is 1590 years.

The decay constant $(\lambda) = \dfrac{0.6931}{1590} = 0.00435911$

Here initial amount $(N_0) = 1$ gram. Let at '$t$' time one gram of radium loses one centigram i.e. .01 gram.

$\therefore$ Radium left behind $(N) = 1 - 0.01 = 0.99$ gram.

We have

$$N = N_0 \,.\, e^{-\lambda t} \quad \text{or} \quad N = \frac{N_0}{e^{\lambda t}} \quad \text{or} \quad e^{\lambda t} = \frac{N_0}{N}$$

or
$$\lambda t = \log_e \left(\frac{N_0}{N}\right)$$

$\therefore$
$$t = \log_e \left(\frac{N_0}{N}\right) \times \frac{1}{\lambda}$$

$$t = \log_e \frac{1}{0.99} \times \frac{1}{0.000435911} = \log_e 1.010 \times \frac{1 \times 10^8}{43591}$$

$$= \frac{0.01005 \times 10^8}{43591} = \frac{1005000}{43591} = 23.055 \text{ days}$$

**60.** *$C^{14}$ has a half life of 5700 years. Calculate the fraction of $C^{14}$ atoms that decay (a) per year (b) per minute.*

**Solution:**

(*a*) We know $T = \dfrac{0.693}{\lambda}$ Here $T = 5700$ years

$$\therefore \quad \lambda = \frac{0.693}{5700} = \frac{693}{57 \times 10^5} = 12.157 \times 10^{-5}$$

$$= 12.16 \times 10^{-5}/\text{year}.$$

$$= 1.216 \times 10^{-4}/\text{year}.$$

$1.216 \times 10^{-4}$ atoms per atoms decay on a time period of one year.

or 1 atom out $\dfrac{1}{1.216 \times 10^{-4}}$ atoms decay/year.

$$\frac{1}{1.216 \times 10^{-4}} = \frac{1000 \times 10^4}{1.216} = \frac{10000 \times 10^3}{1216} = 8.2236 \times 10^3 = 8.224 \times 10^3$$

Thus one out of every $8.224 \times 10^3$ or 8224 radio active atoms decay per year.

(*b*) $\lambda = 1.216 \times 10^{-4}$ per year

$$\text{or} \quad \frac{1.216 \times 10^{-4}}{365 \times 24 \times 60} = \frac{1216 \times 10^{-4}}{5256 \times 10^2 \times 10^3} = \frac{1216 \times 10^{-9}}{5256} = .23135 \times 10^{-9}$$

$$= 0.2314 \times 10^{-9} \text{ per minute}.$$

$$= 2.314 \times 10^{-10} \text{ per minute}.$$

$$\text{or} \quad \frac{1}{2.314 \times 10^{-10}} = \frac{1}{2314} \times 10^9 = 4.3215 \times 10^9$$

Thus one out of every $4.3215 \times 10^9$ radioactive atoms decay per minute.

**61.** *Cesium$^{137}$ has a half life of 33 years. Calculate the fraction of Cesium$^{137}$ that decays (a) per year (b) per minute.*

**Solution:**

(*a*) Calculating decay constant ($\lambda$)

$T = 33$ years. $\quad \therefore \lambda = \frac{0.693}{33} = 21 \times 10^{-3}$ year$^{-1}$

$= 2.1 \times 10^{-2}$ year$^{-1}$

It means $2.1 \times 10^{-2}$ atoms per atoms decays in a time period of one year.

or 1 atom out of $\frac{1}{2.1\times10^{-2}}$ atoms decay per year.

or $\frac{1000}{21} = 47.619 = 47.62$

Thus out of every 47.62 radio active atoms decay per year.

(*b*) $\lambda = 2.1 \times 10^{-2}$ yr$^{-1}$ $= \frac{2.1\times10^{-2}}{365\times24\times60} = \frac{2.6\times10^{-2}}{52.56\times10^{2}} = \frac{2.1\times10^{-4}}{5256}$

$= \frac{21\times10^{-5}}{5250} = 0.00399 \times 10^{-5} = 0.004 \times 10^{-5}$

$= 4 \times 10^{-8}$ minute$^{-1}$

Thus one out of every $0.25 \times 10^{8}$ radioactive atom decay per minute.

The age of recent objects such as those animals or vegetable origin like a piece of wood or animal fossil can be determined by radio-carbon dating method:

(*i*) The determination of the age of a sample of wood (i.e. the time elapsed after the death of the living plant) consists of determining the ratio of the amount of ${}_6C^{14}$ to that of ${}_6C^{12}$ in both pieces of wood i.e. in fresh (living) piece and dead (cut) piece.

(*ii*) Plants absorb carbon dioxide $\left(C^{14}_{02} \text{ and } C^{12}_{02}\right)$ from the atmosphere and prepare cell noise (wood). As long as the plant is alive, the ratio of ${}_6C^{14}$ to ${}_6C^{12}$ atoms in the wood of the plant is same as in the atmosphere but when tree is cut (*i.e.* when plant dies), the ratio of ${}_6C^{14}$ to ${}_6C^{12}$ begins to decrease continuously due to the continuous decrease in the amount of ${}_6C^{14}$ in the plant.

(*iii*) This decrease in the amount of ${}_6C^{14}$ due to its continuous disintegration emitting $\beta$ radiation.

$N_0 = C^{14}/C_{12}$ ratio in the living (*i.e.* fresh) plant.

$N = C^{14}/C_{12}$ ratio in the dead (cut) plant

the age of the wood $t$ is given by

$$t = \frac{T}{0.693} \log \left[\frac{C^{14}/C^{12} \text{ ratio in the living fresh plant}}{C^{14}/C^{12} \text{ ratio in the dead plant}}\right] \quad \text{... (i)}$$

If isotopes of $C_{14}$ in fresh and dead is known

$$t = \frac{T}{0.693} \log \left[\frac{\text{Amount of } C^{14} \text{ of living (fresh) wood}}{\text{Amount of } C^{14} \text{ of dead wood}}\right] \quad \text{... (ii)}$$

**62.** *A carbon specimen found in Bora cave of Andra Pradesh contained* $\left(\frac{1}{8}\right)$ as much as ${}_6C^{14}$ *as an equal amount of carbon in living matter. Calculate the approximate age of the specimen. half life period of* $C_{14}$ *is 5568 years.*

**Solution:**

Half life ($T$) = 5568 years.

Age of the specimen = $t$ years

Initial amount of carbon = $N_0$

Amount of carbon remain after disintegration = $N$

According to the problem $N = \dfrac{N_0}{8}$

Disintegration constant ($\lambda$) = $\dfrac{0.693}{5568}$

$$t = \frac{1}{\lambda}\log_e\left(\frac{N_0}{N}\right) = \frac{1}{\lambda}\log_e\left(\frac{N_0}{\frac{N_0}{8}}\right) = \frac{1}{\frac{0.693}{5568}} \times \log_e 8$$

$$= \frac{5568 \times 10^3}{693} \times \log_e 8 = 8.0346 \times 10^3 \times 2.0794$$

$$= 16.70748 \times 10^3$$

$$= 16707.48 \text{ years}$$

$$= 16707.5 \text{ years}$$

**63.** *The amount of ${}_6C^{14}$ isotope in a piece of wood is found to be one sixth present in a fresh piece of wood in Ajanta cave of Maharashtra. Calculate the age of wood. half life ${}_6C^{14}$ = 5577 years.*

**Solution:**

Half life = $T$ = 5577 years

Age of the wood = $t$ = ? $\qquad \lambda = \dfrac{0.693}{T} = \dfrac{0.693}{5577}$

$N_0$ = Initial amount = 1

$N$ = Amount after disintegration = $\dfrac{N_0}{6}$

$$t = \frac{1}{\lambda}\log_e\frac{N_0}{N} = \frac{1}{\lambda}\log_e\left(\frac{N_0}{\frac{N_0}{6}}\right)$$

$$= \frac{1}{\frac{0.693}{5577}} \times \log_e 6 = \frac{5577}{693} \times 10^3 \times \log_e 6 = 8.0476 \times 10^3 \times 1.79176$$

$$= 8047.6 \times 1.79176$$

$$= 14419.36 \text{ yrs}$$

**64.** *The half life of a radioactive substance is 1672 years. If the initial mass of the substance is 1 g, after how many years will only 1 mg of it to be left behind.*

**Solution:**

Half life $T$ = 1672 years

Initial amount $N_0$ = 1 gram

Amount remained $N$ = 1 mg = $\dfrac{1}{1000}$ gram = 0.001

$$\lambda = \frac{0.693}{1672} \text{ yr}^{-1}$$

$$t = \frac{1}{\lambda}\log_e\left(\frac{N_0}{N}\right) = \frac{1672}{0.693}\times\log_e\left(\frac{1}{0.001}\right) = 2412.698 \times \log_e 1000$$

$t = 2412.6984$

$= 2412.6984 \times 6.907755279 = 16666.3$

$= 16666$ years

**65.** *An old piece of wood was found to have $C^{14}$ activity of 6 disintegration per minute per gram of its carbon content. The $C^{14}$ activity of living wood is 16 disintegration per minute per gram. Estimate the is age of the tree. Half life of $C^{14}$ = 5760 years.* ***(H.P.U. 2001)***

**Solution:**

Half life = $T$ = 5760 yr.

Disintegration constant $\lambda = \frac{0.693}{5760} = 1.2 \times 10^{-4}$ per year.

Initial quantity $N_0$ and after disintegration it will be $N$ (present wood).

$$t = \frac{1}{\lambda}\log\left[\frac{\text{Amount of } C^{14} \text{ in living matter}}{\text{Amount of } C^{14} \text{ in ancient mark}}\right]$$

$$= \frac{1}{1.2\times10^{-4}}\times\log_e\frac{16}{6} = \frac{10^4}{1.2}\times\log_e 2.67 = 8333.3 \times 0.982$$

$= 8183$ years.

**66.** *A piece of ancient wood was found to have $C^{14}/C^{12}$ ratio $\frac{1}{8}$ times that in living plant. Half life of $C^{14}$ is 5570 years. Calculate the age of the ancient wood.*

**Solution:**

Let the age of the wood be '$t$' years

Half life ($T$) = 5570 years.

Disintegration constant $\lambda = \frac{0.6931}{T} = \frac{0.6931}{5570}$

Quantity in ancient wood = $R_0$

Quantity in present wood = $R$

$$\frac{R}{R_0} = \frac{1}{8} \quad \text{or} \quad R_0 = 8R \quad \text{or} \quad R = \frac{1}{8}R_0$$

$$t = \frac{1}{\lambda}\log_e\left(\frac{R_0}{R}\right) = \frac{1}{\lambda}\log_e\left(\frac{R_0\times8}{R_0}\right) = \frac{1}{\lambda}\log_e 8$$

$$= \frac{5570}{0.6931}\times\log_e 8 = 0.803635 \times 2.07944 \times 10^4$$

$= 1.677110 \times 104 = 16711.1$ years

**67.** *The normal activity of carbon, in living matter, is found to about 15 decays per minute for every gram carbon. A specimen from Vimvedkar of M.P. gives an activity of 9 decays per minute per gram of carbon. Estimate the approximate age of this ancient culture. Half life of $C^{14}$ is 5730 years.*

**Solution:**

Let the age of Vimvedkar cave culture lose '$t$' years. Decay constant $(\lambda) = \frac{0.6931}{5730} = 1.2095$

$yr^{-1}$

Normal activity of carbon = 15

Activity of carbon in ancient = 9

$$t = \frac{1}{\lambda}\log_e\left[\frac{\text{Amount of } C^{14} \text{ in living matter}}{\text{Amount of } C^{14} \text{ in dead matter}}\right]$$

$$= \frac{1}{1.2095} \times \log_e \frac{15}{9}$$

$$= \frac{1}{1.2095} \times \log_e 1.67 = 0.82678 \times 0.512823 = 4239 \text{ years.}$$

**68.** *An antique wooden piece weighing 50 grams showed $C^{14}$ activity of 320 disintegration per minute. Estimate how old it is, assuming that living plant show a $C^{14}$ activity of 12 disintegration per minute per gram. Half life of $C^{14}$ is 5730 yrs.*

**Solution:**

Activity of wooden piece ($N$) = $\frac{320}{50}$ = 6.4 disintegration pergram per minute.

Activity of living plant 12

Half life ($T$) = 5730 years

$$\lambda = \frac{0.6931}{5730} = 1.21 \times 10^{-4} \text{ yr}^{-1}$$

Let age be $t$

$$t = \frac{1}{\lambda}\log_e\left[\frac{C^{14} \text{ activity at present}}{C^{14} \text{ activity at ancient}}\right]$$

$$= \frac{1}{1.21\times10^{-4}} \times \log_e\left(\frac{12}{6.4}\right) = \frac{10^4}{1.21} \times \log_e 1.875 = 8264.4 \times 0.6286$$

$$= 5195 \text{ yrs}$$

**69.** *A piece of an ancient wooden boat found in the bank of Krishna river showed an activity of $C^{14}$ of 3.9 disintegration per minute per gram carbon. Estimate the age of the boat if half life of $C^{14}$ is 5,568 years. Assume the activity of first $C^{14}$ is 15.6 disintegration per minute per gram.*

**Solution:**

Let the age of the boat be '$t$' years.

$N$ is the activity of ancient wooden boat *i.e.* 3.9 disintegration per minute per gram.

$N_0$ is the activity of present fresh wood *i.e.* 15.6 disintegration per minute per gram.

$$t = \left[\frac{C^{14} \text{ activity at present wood}}{C^{14} \text{ activity at ancient wood}}\right]$$

$$\lambda = \frac{0.6931}{5568} = 1.24479\times10^{-4} \text{ or } \frac{1}{\lambda} = \frac{1\times10^4}{1.24479} = 8033.48$$

$$t = 8033.48 \times \log_e \frac{15.6}{3.9} = 8033.8 \times \log_e 4 = 8033.48 \times 1.386$$

$$= 11134.4 \text{ years}$$

∴ Age of the boat 11134.4 yrs.

**70.** *In an archeological expedition of Calcutta University, Charcoal from an ancient fire pit of Nalanda (Bihar) was excavated. The sample showed a $C^{14}$ activity of 11.3 counts per gram per minute. The absolute activity of $C^{14}$ is constant for all wood sample and is equal to 15.3 counts per gram per second. Estimate the age of the charcoal sample. Half life of $C^{14}$ = 5568 years.*

**Solution:**

Let the age of the charcoal sample be '$t$' years

$N = C^{14}$ activity of charcoal sample (ancient five pit)

$N_0 = C^{14}$ activity of present wood sample.

$$\lambda = \frac{0.6931}{T} \qquad T = 5566 \text{ years}$$

$$= \frac{0.6931}{5568} = 1.2448 \times 10^{-4} \text{ year}^{-1}$$

$$t = \frac{1}{\lambda} \times \log_e \left[\frac{C^{14} \text{ activity of ancient sample}}{C^{14} \text{ activity of present sample}}\right] = \frac{1}{\lambda} \log_e \frac{15.3}{11.3}$$

$$t = \frac{1}{1.2448 \times 10^{-4}} \times \log_e \frac{15.3}{11.3} = \frac{10^4}{1.2448} \times \log_e 1.3539$$

$$t = 8033.41 \times 0.302999 = 2434.11 \text{ years.}$$

**71.** *Calculate the amount of a radioactive element ($t_{1/2}$ = 140 days) to which 1 gram of the element will reduce in 500 days.*

**Solution:**

$N_0 = 1g, \quad t = 560 \text{ days} \quad T \text{ or } t_{1/2} = 140 \text{ days}$

∴ Number of half-lives in 56 days $(n) = \frac{t}{T}$

$$= \frac{560}{140} = 4$$

∴ Amount of the left after 560 days or 4 half period

$$= \frac{N_0}{2^n} = \frac{1}{2^4} \text{ gram} = \frac{1}{16} \text{ gram}$$

**72.** *The half life of a radio active substance is 5 hrs. What will be its one third life time?*

**Solution:**

Half life time $(T)$ = 5 hrs.

We know $\quad T = \frac{0.693}{\lambda} \qquad$ [$\lambda$ = disintegration constant]

$$5 = \frac{0.693}{\lambda}$$

$$\lambda = \frac{0.693}{5} = 0.13862$$

Let $N_0$ be the initial quantity and $N$ be the quantity left after disintegration.

$$\therefore \quad \frac{N}{N_0} = e^{-\lambda t} \qquad \text{Here } \frac{N}{N_0} = \frac{1}{3}$$

$$\frac{1}{3} = e^{-\lambda t} \qquad 3 = e^{\lambda t}$$

$$\therefore \quad \lambda t = \log_e 3$$

$$t = \frac{\log_e 3}{\lambda} = \frac{1.0986}{0.13862} = 7.925 \text{ hrs}$$

$$= 7.93 \text{ hrs}$$

**73.** *A piece of wood was found to have $C^{14}/C^{12}$ ratio 0.7 times that in living plant. Calculate the period when the plant died. Half life of $C^{14}$ = 5760 years.*

**Solution:**

Half life time ($T$) = 5760 years

Decay constant $\lambda = \frac{0.693}{576}$ year$^{-1}$

Let the age of died time of wood was '$t$'

Initial amount is $N_0 = 1.00$

After disintegration the amount $N = 0.7$

$$\therefore \quad t = \frac{1}{\lambda}\log_e\left(\frac{N_0}{N}\right) = \frac{5760}{0.693} \times \log_e\left(\frac{1}{0.7}\right)$$

$$= 8311.68 \times \log_e 1.43 = 8311.68 \times 0.3577 = 2973.0 \text{ year.}$$

∴ The plant was died 2973 years ago.

**74.** *The $C^{14}$ to $C^{12}$ ratio in a certain piece of wood is 13% that of the atmosphere. Calculate the age of the wood. Given that half life of $C^{14}$ = 5580 years.*

**Solution:**

Here $T$ = 5580 yrs. $\qquad \therefore \lambda = \frac{0.693}{5580} = 1.24 \times 10^{-4}$

Initial amount $N_0 = 100$

After disintegration $N = 13$

Let $t$ be the age of the wood

$$t = \frac{1}{\lambda}\log_e\left(\frac{100}{13}\right) = \frac{1}{1.24 \times 10^{-4}} \times \log_e (7.69) = \frac{1 \times 10^4}{1.24} \times 2.039$$

$$t = 8064.5 \times 2.039 = 16443.5 \text{ years.}$$

CHAPTER 9

# Colorimetry, Photometry, Polarimetry and Spectrophotometry

**Colorimetry, Photometry, Polarimetry and Spectrophotometry**

Light is a form of energy. When it is passed through a coloured medium (solution) some wavelengths are more absorbed than others. But all compounds are not coloured but can be made to absorb light in visible range of wavelength by using certain suitable reagents.

**Spectrophotometry**, **Photoelectric colorimeters** have many similarities of function and uses.

**Photometry refers to the measurement of the light transmitting power of a solution in order to determine the concentration of light absorbing substances present within.** It can also be applied to measure the transmission of energy in the ultraviolet, infrared and visible regions of the radiant energy spectrum. Instruments like **photoelectric colorimeter**, **spectrophotometer** are used to measure transmittance of solutions at various wavelengths of light. Here monochromatic light is passed through a coloured solution of fixed depth and directed towards photosensitive device. It converts radiant energy into electrical energy and current is produced and measured by galvanometer. Absorbance is measured by photometer which not only involves the absorbance of solute in the solution but also all molecules of the liquid through which light passes.

- **Basic Principle:**

(*i*) Molecules or atoms are subjected to changes in energy level by absorption of light energy.

(*ii*) Absorption of light energy generally heats up the absorbing material without any change in colouration.

(*iii*) All chemicals absorbs **electromagnetic** radiation of specific wave length. On the basis of this principle UV and visible absorption spectroscopy are operated.

(*iv*) A beam of light (specific quality) is passed through the experimental material (chemical, solution, solid or gas) and the light transmitted by the material is less than the amount of radiation absorbed by the material.

(*v*) The light absorbed by the material is proportional to the concentration. Therefore a **quantitative** analysis of the material is made.

(*vi*) If a series of known concentrations of chemical (materials) is prepared and the light absorbed by the material at a known wave length of light is measured, a standard relationship of light absorbed with respect to the chemical concentration would be known.

(*vii*) Once standard relationship is known, unknown concentration may be analysed ***quantitatively.***

(*viii*) Since different chemicals absorbed light at different wave lengths a qualitative analysis of unknown chemical may be made through the observation of **absorption spectrum** of that chemical.

- **Colorimetry :**

**The method of estimation of concentration of coloured solute in a solution by comparing its colour intensity with that of a standard solution containing a known concentration of the same solute is known as colorimetry.**

(*i*) In visual colorimetry, natural or aritificial white light is generally used as light source and determinations are usually made with a simple instrument known as *colorimeter*.

(*ii*) Photoelectric colorimeter is a device where a photometer is used to compare the standard solution with unknown solution.

**Characteristics :**

(*i*) Here visible light of the same intensity and an identical wavelength is passed through both solutions.

(*ii*) When white light is passed through a coloured solution, some wavelengths are absorbed more than others.

(*iii*) Many compounds are not coloured but can be converted into coloured compound. Even after conversion, some compounds are not coloured but can be made to absorb light in the visible region by reaction with suitable reagents. These reaction are sensitive and specific.

(*iv*) Small quantities of material (millimole per litre) can be measured.

(*v*) Complete isolation of the compound is not necessary.

(*vi*) It obeys two laws *viz.* **Lambert's law** and **Beer's law**.

- **Photometry :**

**The method of estimation of light absorbing solute in a solution by comparing the intensity of light either absorbed or transmitted by it with that of standard solution containing a known concentration of the same solute is known as photometry.**

(*i*) White light changes to coloured light when certain wave lengths have been absorbed and others reflected i.e. not absorbed.

(*ii*) Thus substances show light absorptive capacity and the actual colour depends on the light incident on the material.

(*iii*) Certain wave lengths are preferentially absorbed but the colour which is seen is the characteristic of the wave length not absorbed.

**Characteristics :**

(*i*) Here specific ultraviolet ray, infrared or visible light wavelength of identical intensity is passed through both the solutions.

(*ii*) It obeys **Lambert's** and **Beer's law**.

**Common features :**

(*i*) Coloriometry and photometry, both these methods depend on the **light absorbing** capacity of the substance to be estimated.

(*ii*) The quantity of light absorbed by the substance depend upon.

(*a*) wavelength of light

(*b*) the arrangement of orbital electron.

(*c*) the number of light absorbing particles in the path of light.

(*iii*) Both these methods obey **Lambert's** and **Beer's law.**

- **Lambert's Law :**

**When monochromatic light passes through a transparent medium, the rate of decrease in intensity with the thickness of the medium is proportional to the intensity or light.**

**Or it may be stated that the intensity of a ray of monochromatic light when passes through an absorbing medium (transparent medium) decreases exponentially as the length (thickness) of absorbing medium increases arithmatically.**

We may express the law by differential equation.

$$\frac{dI}{dl} \propto -1 \qquad ...(i)$$

$I$ = Intensity of the incident light of wave langth
$l$ = Thickness of the medium
$k$ = Proportionality factor

or $$\frac{dI}{dl} = -kI$$

Integrating the equation (*i*) and putting $I = I_0$ when $l = 0$

$$\int_{I_0}^{I} \frac{dI}{dl} = -\int_{0}^{l} kI$$

$$ln\,\frac{I}{I_0} = -kl \quad \text{or} \quad \frac{I}{I_0} = e^{-kl} \quad \text{or} \quad I = I_0 \cdot e^{-kl} \qquad ...(ii)$$

This equation express how the original intensity $I_0$ is reduced to $I$ after passing through a thickness $l$ of the medium.

**Characteristics:**

(*i*) According to this law, the degree of absorption of light by a substance as a function of its thickness.

(*ii*) This law also reveals that each successive layer of identical thickness of a pure light absorbing material absorbs the same proportion of the light enter into it.

(*iii*) On the other land the intensity of transmitted light decreases exponentially with the rise in thickness of the light absorbing material.

(*iv*) The proportion of light absorbed by the material is independent of the intensity of incident light.

- **Absorption coefficient:**

**Absorption coefficient is generally defined as the reciprocal of thickness (*l* cm) required to reduce the light to one tenth $\left(\frac{1}{10}\right)$ its intensity.**

On the basis of Lambert's Law

$$I_t = I_0\, e^{-kl} \qquad ...(i)$$

where $I_0$ is the intensity of incident light falling upon the absorbing medium of thickness $l$. It is the intensity of the transmitted light and $k$ is the constant for the wave length and the absorbing medium used.

By changing the equation (i) above from natural to common logarithms. We obtain

$$I_t = I_0 \cdot 10^{-0.434\,kl}$$
$$= I_0 \cdot 10^{-kl} \qquad ...(ii)$$

where $K = \frac{k}{2.3026}$ and is usually termed as **absorption coefficient.**

- **Beer's Law :**

**When a monochromatic light is passed through a solution, the decrease in the intensity of light with the thickness of the solution is directly proportional not only to the intensity of the incident light but also to the concentration (*c*) of the solutions.**

We may express the law in differential equation

$$\frac{dI}{I} \propto -c$$

or $$\frac{dI}{I} = -k'c \quad ...(i)$$

$I$ = Intensity of the incident light of wave legth $\lambda$

$c$ = Concentration of the solution

$k'$ = Proportionality factor

Integrating the equation (*i*) and putting $I = I_0$. When $c = 0$

$$\int_{I_0}^{I} \frac{dI}{I} = -\int_{0}^{c} k'c$$

$$ln\frac{I}{I_0} = -k'c \quad \text{or} \quad \frac{I}{I_0} = e^{-k'c} \quad \text{or} \quad I = I_0\, e^{-k'c}$$

This equation express how the intensity of monochromatic light falls from $I_0$ to $I$ on passing through a solution of concentration *c*.

**Characteristics:**

(*i*) It establishes the relationship between concentration of substance in a dilute solution and amounts of light absorbed by it.

(*ii*) It also reveals that the intensity of transmitted light falls exponentially with the rise in the number of light absorbing particles in its path.

**Optical Density (OD):**

It is the ratio of incident light by a thickness of transmitted light is called optical density.

- **Factors may cause deviation from Beer's law**

**(*a*) Concentration:**

(*i*) High sample concentrations can lead to chemical reaction which will lead to chemical composition of the solution.

(*ii*) Low sample concentrations. Proteins are known to denature at low concentrations and the denatured product has an absorption spectrum that is different from native protein.

**(*b*) Temperature:**

(*i*) The change temperature of an experimented solution results in change of solubility dissociation or association properties of the solute. This change also reflected in the absorbance.

(*ii*) Thus absorbance measurements must always be done at a constant temperature.

**(*c*) Sample instability:**

(*i*) Some coloured compounds are unstable and undergo changes within a short time.

(*ii*) Therefore absorbance measurement should be done within a very short period.

**(*d*) Flourescence:**

(*i*) Some solute fluorescent (drug).

(*ii*) For such cases deviation may occur.

**(e) Turbidity:** Turbid solution has higher absorbance.

**(f) Instrumentation:** Limitations of Instrument may also result in deviation from **Beer's law.**

- **Beer-Lambert Law :**

(*i*) Both the laws (Beer's and Lambert's) are combined and expressed as **Beer-Lambert law.**

(*ii*) Concentration means the quantity of materials are dispersed in a given solution.

(*iii*) If the **concentration is high** (Beer's law) there are large number of molecules present in the solution may interact and absorb the light energy from monochromatic ray passing through it.

(*iv*) While the **width of the absorbing medium is longer**, larger will be the number of absorbing molecules coming across the path of ray passing through it.

**The fraction of the incident light absorb by a solution at a given wavelength is related to the thickness of the absorbing layer and the concentration of the absorbing substance.**

$I_o$ = Intensity of incident light

$I_t$ = Intensity of transmitted light

$l$ = Thickness of the absorbing medium

$c$ = Concentration of the absorbing, substance in moles/litre.

$k$ = absorption coefficient

$$\text{Log}\,\frac{I_o}{I_t} = kl \qquad \text{... }(i)$$

$$\text{Log}\,\frac{I_o}{I_t} = kc \qquad \text{...}(ii)$$

The combined *Beer-Lambert* Law, is as follows.

$$\text{Log}_{10}\,\frac{I_o}{I_t} = klc$$

If $k$ is kept constant by employing a standard value the formula becomes

$$A = \text{Log}_{10}\,\frac{I_o}{I_t} = lc$$

(*v*) This combined law states that the amount of light absorbed is proportional to the concentration of the absorbing substance and to the thickness of the absorbing material (path length).

(*vi*) The quantity $\frac{I_o}{I_t}$ is known as the absorbance or the **optical density** (O.D).

The quantity of light is transmitted $\frac{I_t}{I_o}$ is known as the **transmittances** (T) *i.e.* T (the amount of light which escapes absorption and is transmitted.)

It gives the ability to transmit light as an exponential function of the concentration of solution.

(*vii*) Absorbance shares a linear relationship with the sample concentration. On the other hand, the relationship between transmittance and sample concentration is a non-linear one.

(viii) Therefore, it is easier to use absorbance as an index of sample concentration.

(*ix*) The absorbancy index is defined as

$$\textbf{Absorbancy index} = \frac{\textbf{A (Absorbancy)}}{\textbf{CL}}$$

$$\left[\begin{array}{l}\text{C = concentration of absorbing material in gms/litre}\\ \text{L = distance in cm travelled by the light in solution}\end{array}\right]$$

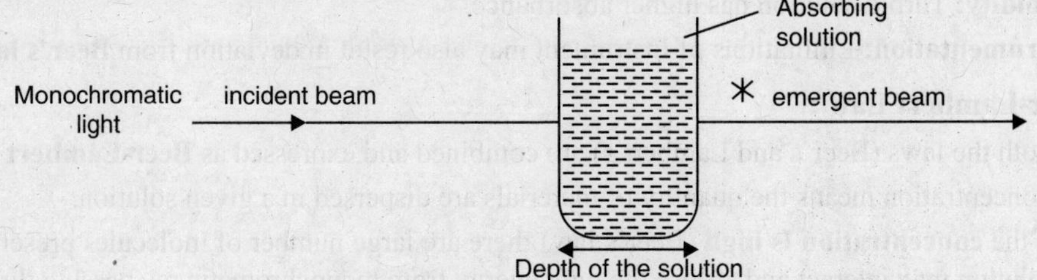


**Fig. 9.1** Lambert-Beer law principle.

**Limitation of Beer-Lambert Law :**

(*i*) Light must be preferably monochromatic and narrowed.

(*ii*) The wavelength of light should be at absorption maximum. It will give greatest sensitivity.

(*iii*) There must be no ionization dissociation or association or solvation of the solute with concentration or time.

(*iv*) The solution showed not be too concentrated to give intense colour.

- **Colorimeter :**

**Principle :**

**It is a simple instrument used for reading of colour, against a blank in the visible range (400-700 nm) by the use of various filters. It provides accurate results of optical density or percentage transmission of a coloured solution.**

(*i*) In colorimetry, the sample is treated with a reagent and it produces a colour with the constituent to be determined.

(*ii*) A standard solution of the substance to be determined is treated in the same way and the colours of the standard and of the unknown are compared in an instrument called "colorimeter".

(*iii*) The intensity of the colour is directly proportional to the concentration of the coloured solution

**(A) Visual Colorimeter instrument :**

**Instrumentation :**

(*i*) Incandescent electric lamp for use of visible light.

(*ii*) Two adjustable cylindrical cups with opaque sides.

(*iii*) Standard and unknown solution (under examination) are taken with the two cups.

(*iv*) A glass rod or plunger is fitted in each cup.

(*v*) Glass plate (splitted into two halves).

(*vi*) Eye piece for visual observation.

**Mechanism :**

(*i*) A visible white light from an incandescent lamp is passed through glass filter.

(*ii*) It then passes through the transparent bases to two cups with opaque sides.

(*iii*) It traverses both the standard and unknown solution and passes through the two prisms (plungers) and converge on the circular glass plate.

(*iv*) Two halves of this glass plate are illuminated by the lights transmitted through the two separate solutions.

(*v*) The intensities of colour in two halves of this glass plate can be compared visually through an eye piece.

**Observation :**

(*i*) By raising or lowering each cup, the depth of the solution, traversed by the light can be adjusted separately.

(*ii*) If plunges the corresponding plunger to respectively greater or lesser depth in the solution until two haves of the glass plate match in colour.

(*iii*) The depth of the solution traversed by the light in each cup at this matching point is recorded from the graduated scale.

(*iv*) If the $S_1$ and $S_2$ are the depth of the unknown and standard solution respectively at the matching point and $C_1$ and $C_2$ are their respective concentration; then the two match solutions have an identical optical density (O.D.) or colour density.

**Calculation :**

According to Lambert-Beer law

$$S_1C_1 = S_2 C_2$$

$$C_1 = \frac{S_2}{S_1} C_2$$

**Inference :**

$$\text{concentration of unknown} = \frac{\text{Reading of standard}}{\text{Reading of unknown}} \times \text{concentration of standard.}$$

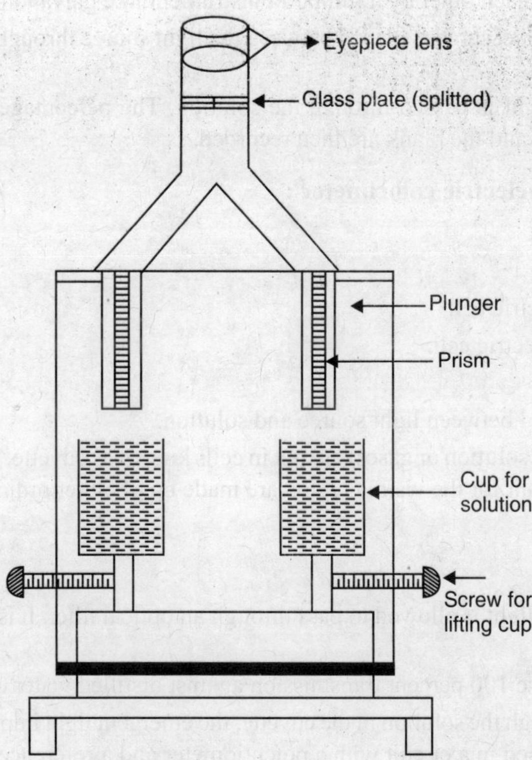


**Fig. 9.2** Diagram of visual colorimeter

**Precautions :**

1. Cups showed not filled to the extent.
2. After using the colorimeter, cups should be rinsed thoroughly with distilled water.

3. The plungers should be rinsed through distilled water and carefully dried.
4. For perfect result mirror, light and eye piece should be properly adjusted.

**Importance :**

(*i*) Small amount of substances are measured.

(*ii*) Traces of highly coloured compounds give highly coloured solutions. Here, on the basis of colour comparison solution substance present in low concentration can be estimated.

**(B) Photoelectric Colorimeter :**

It is used for measurement of colour by electrical devices. It does not cover full colour band.

**Types :**

**(*a*) Single cell type :**

(*i*) Here light is transmitted through solution which is interposed between the light source and Photo cell or Photo emissive cell.

(*ii*) Light is measured directly by the current output of cell and compared with that obtained with blank and standard.

**(*b*) Double cell type :**

(*i*) Here two types of photo cell *viz* working Photoelectric cell and reference Photoelectric cell having same light source are in function.

(*ii*) They are balanced each other by a full point instrument like galvanometer.

(*iii*) The solution to be measured is placed between the light source through coloured filter and one cell

(*iv*) Monochromatic light is passed through the solution. The percentage transmission given by unknown standard and the blank are theu recorded.

**Double cell type Photoelectric colorimeter :**

**Instrumentation :**

(*i*) Light

(*ii*) Working Photoelectric cell

(*iii*) Reference Photoelectric cell.

(*iv*) Galvanometer

(*v*) Optical filter placed between light source and solution.

(*vi*) Cuvette : Sampler (solution or gases) are put in cells known as cuvette. Solutions are dispensed here. Cuvettes meant for the visible region are made up of either ordinary glass or sometimes quartz.

**Mechanism :**

(*i*) A monochromatic light is allowed to pass through an optical filter. It is then transmitted to the solution.

(*ii*) It is adjusted to give 100 percent transmission against distilled water or blank.

(*iii*) After passing through the solution in the cuvette, the emergent light impinges on a photoelectric cell (working) placed in a circuit with a potentiometer and a reference electric cell.

(*iv*) The emergent light excites the photocell on impinging on it. It generates an electric current which is directly proportional to the light intensity falling impinging on it.

(*v*) The potentiometer is adjusted to counter balance this current by an opposite current from the reference cell and to suspend thereby all current flow in the circuit.

(*vi*) The optical density and the percent transmittancy corresponding to this nullifying current are indicated on the calibrated scales of the instrument.

**Observation :**

(*i*) First a blank solution of only reagents and solvent are placed in cuvette. The instrument is adjusted to read either 'O' in the optical density scale or 100 on the percent transmittancy scale or the photometer is set to indicate either O.D (optical density) of the blank as 'O' or its transmittancy ($T_b$) as 100.

(*ii*) The unknown solution (test solution) is put in cuvette and its O.D or percent transmittancy is read and recorded.

(*iii*) Now a standard solution, containing a known concentration of light absorbing substance is put in the cuvette and its O.D or percent transmittancy is also read and recorded.

**Calculation :**

$$\text{concentration of unknown} = \frac{\text{optical density of unknown}}{\text{optical density of standard}} \times \text{concentration of standard.}$$

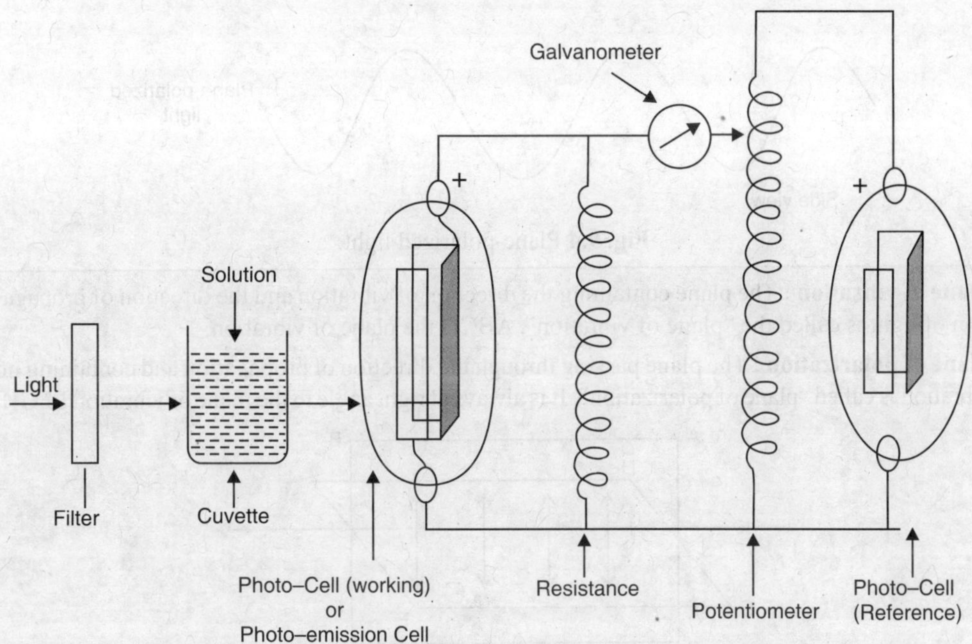


**Fig. 9.3** Schematic representation of a double-cell photometer

**Application of Photoelectric colorimeter :**

1. Biochemical estimation of blood and its constituents.
2. Biochemical estimation different substances present in urine.
3. Biochemical estimation of substances present in C.S.F and other body fluids.

**Monochromatic :** A beam in which all the rays have the same wavelength is known as monochromatic.

**Polychromatic :** A beam of radiation consists of several wavelength, is known as polychromatic.

## Polarimetry :

**The instrument used for measuring the angle of rotation of a plane polarised light is called a "polarimeter."**

The phenomenon in which the light vibrations are confirmed to a single plane is called polarisation of light and the light is called **polarised light.**

(*i*) A beam of ordinary light consists of electro magnetic waves oscillating in many planes. When passed through a polariser (Polaroid lens), only waves oscillating in a single plane pass through.

(*ii*) The emerging beam of light having oscillation in a single plane is said to be **plane polarized.**

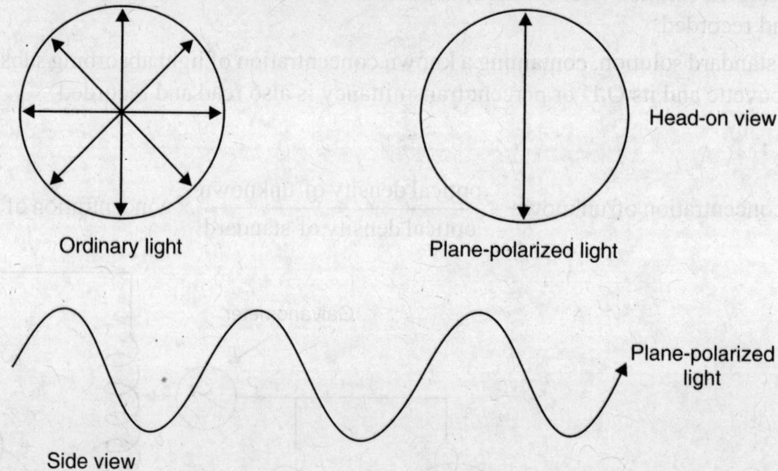


**Fig. 9.4** Plane-polarized light.

**Plane of vibration :** The plane containing the direction of vibration and the direction of propagation of light is called the "plane of vibration". ABCD the plane of vibration.

**Plane of polarization :** The plane passing through the direction of propagation and containing no vibration is called "plane of polarization". It is always at right angle to the plane of vibration EFGH.

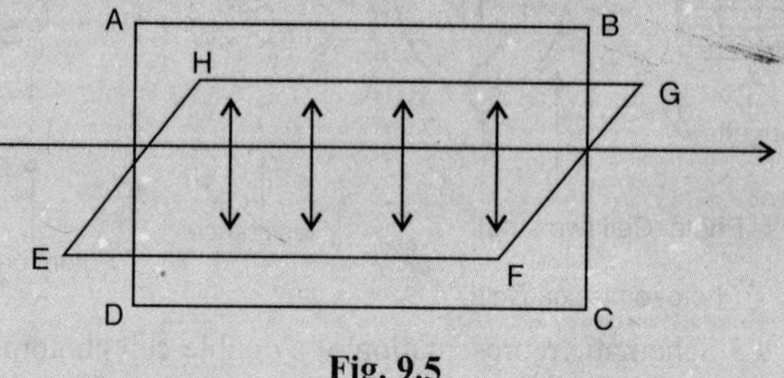


**Fig. 9.5**

- **Optical activity:**

The property of a substance (or complex-compound) to rotate the plane of polarized light through a certain angle either clockwise (+) or dextro-rotatory (right) or anticlockwise (–) or laevo rotatory (left) is called optical activity.

It is a constitute properly and depends upon the arrangement of atom in a molecule.

- **Optically active substance:**

The substances (sugar, quartz etc.) which rotate the plane of polarized light to the right (+) or dextro-rotatory or left (–) or laevo-rotatory are called optically active substance.

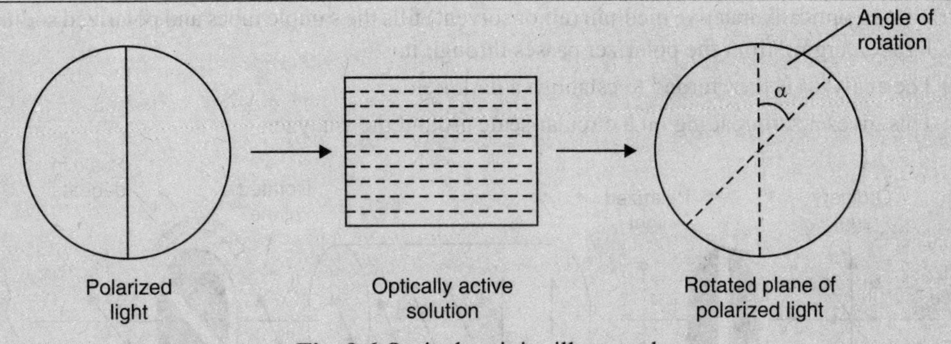


**Fig. 9.6** Optical activity illustrated.

| **Leavorotatory** | **Dextrorotatory** |
|---|---|
| (*i*) A compound which rotates the plane polarized light to anticlockwise (left) to said to be *leavorotatory*. | (*i*) A compound which rotates the plane polarized light to clockwise (right) is said to be *detrorotatory*. |
| (*ii*) By convention it is given minus sign (-). Lactic acid (-) levorotatory | (*ii*) By convention it is given plus sign (+). Lactic acid (+) dextrorotator. |
| (–) – Levorotatory | (+) – Dextrorotatory |

- **Specific rotation:**

(*i*) The rotation of plane polarized light is an intrinsic property of optically active molecules.

(*ii*) When a polarized beam of light is passed through the solution of an optically active substance, its plane is rotated through a certain angle α (angle of rotation).

(*iii*) It depends upon.

(*a*) Nature of optically active substance, (*b*) concentration, (*c*) of the solution (m g/ml) (*d*) length, (*l*) of the sample solution (in dm or 10 cm), (*e*) wave length of light (λ) (*f*) Temperature of the solution (*T*).

*Specific rotation* [α] *can be defined as the observed angle* (α) *of rotation at a concentration* of 1 g/ml (I gram substance is dissolved in one ml solution) and the *path length* (length of the solution) *of 1 dm* (10 cm).

$$[\alpha] = \frac{\alpha}{l \times c} \qquad l = 1, \quad c = 1$$

or $[\alpha]_0^T = \alpha$    *ID* = If *D* line of the sodium light is used.

- **Measurement of optical activity.**

(*i*) Optical activity is measured with the help of an instrument called *Polarimeter*.

(*ii*) This is basically a system of *polarizers* with a sample tube placed between.

(*iii*) First an optically inactive medium (air or solvent) fills the sample tubes and polarized sodium light emerging from the polarizer passes through it.

(*iv*) The analyzer is then turned to establish a dark field.

This gives a zero reading on a circular scale around the analyzer.

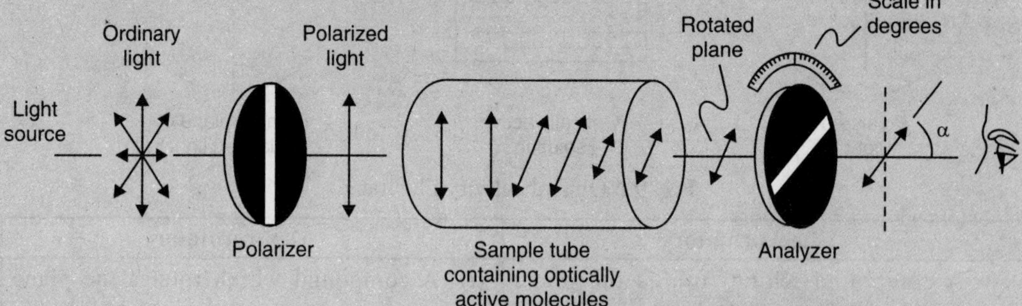


**Fig. 9.7** Schematic diagram of polarimeter.

(*v*) The solution of the given optically active substance is placed in the sample tube.

(*vi*) The plane of polarized light passing through it, is rotated. The analyzer is turned to re-establish the dark field.

(*vii*) The angle of rotation (α) is then noted in degrees on the circular scale. the specific rotation is calculated using the above mentioned equation.

**Optical activity :**

(*i*) The Nicol prism used for polarization is called polarizer.

(*ii*) Two Nicol prisms are placed between the source of monochromatic light (S) and eye (E).

(*iii*) Light passed through the first Nicol Prism, the polarizer (P) and gets polarized.

(*iv*) This plane polarized light passes through the second Nicol Prism, the Analyzer (A) and reaches the eye.

(*v*) No light reaches the eye when the axis of the analyser is perpendicular to the axis of the polarizer.

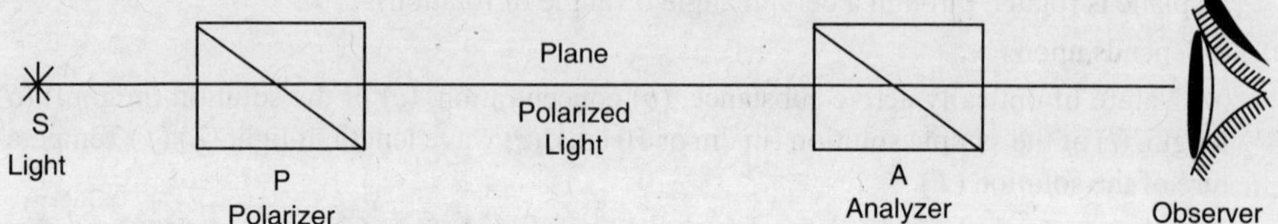


**Fig. 9.8** Plane polarized light passing through Nicol Prism.

**Description of Polarimeter :**

(*i*) It consists of a long hollow barrel carrying a fixed polarizer Nicol Prism and a rotating analyzer Nicol Prism.

(*ii*) A hollow polarimeter tube, encircled by a packet placed inside the barrel between the two prisms.

(*iii*) The tube is filled with a solution for investigation and the jacket around it may be filled with water at a desired temperature.

(*iv*) A condensing lens is placed between light sources and polarizer Prism and a quartz plate is located between the polarizer prism and polarimeter tube.

(*v*) Telescopic eye piece lenses are placed beyond the analyzer prism.

(*vi*) A fixed circular scale graduated in degrees radian is placed at the end of the polarimeter barrel.

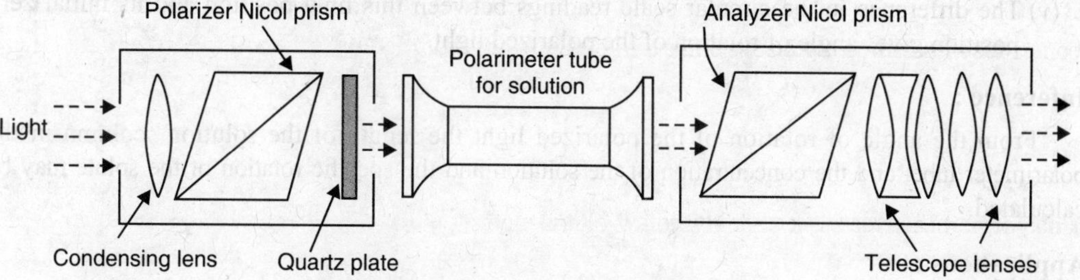


**Fig. 9.9** Sketch diagram of polarimeter

(*vii*) Now the analyzer and the polarizer are arranged for no light reaching the eye. Tube (T) filled with cane sugar solution is interposed between the polarizer and the analyzer. Some light is visible.

(*viii*) The analyzer has to be rotated a bit in order to make the original situation that no light reaching eye.

(*ix*) This is possible only if the plane of polarization is rotated. Here the cane sugar solution has rotated the plane of polarization.

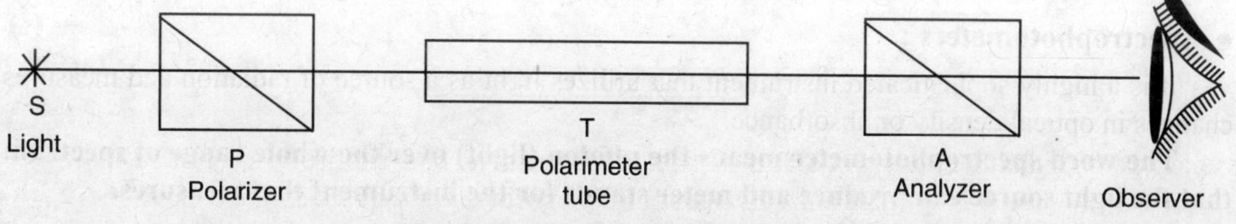


**Fig. 9.10** Polarimeter (schematic)

(*x*) Here the substances like cane sugar can rotate the plane of polarization of light, are known as "optically active" this phenomenon is known as "optical activity".

(*xi*) The substances which rotate the plane of polarization of light towards the right are called "dextrorotatory" where as those rotate towards the left are known as "levorotatory".

**Mechanism :**

(*i*) Monochromatic light from a suitable light sources enters the polarimeter.

(*ii*) It then passes through the polarizer prism and the quartz plate and emerges as plane polarized light with all its waves vibrating in a single plane.

(*iii*) It then passes through the polarimeter tube, analyzer prism and the eye piece lenses to the observer's eye.

**Observation :**

(*i*) Initially the polarimeter tube is empty the analyzer prism is rotated to make both halves of the field of view equally bright. The circular scale reading for their position of the analyzer prism is recorded as zero position.

(*ii*) The polarimeter tube is filled with experimental solution, the jacket around the tube is filled with water at a given temperature and is allowed to stand for sometime to receive that temperature.

(*iii*) Monochromatic light is passed through polarimeter and two halves of the field of view are now unequally illuminated. It is due to the rotation of plane of polarized light by the solution of the tube.

(*iv*) The analyzer prism is rotated until both halves of field of view are equally bright. The circular scale reading for this position of analyzer prism is recorded.

(*v*) The difference in the circular scale readings between this final position and its initial zero position gives angle of rotation of the polarized light.

**Inference :**

From the angle of rotation of the polarized light the length of the solution column in the polarimeter tube, and the concentration of the solution and the specific rotation of the solute may be calculated.

**Applications :**

1. **Optical activity :** (*i*) Most of the biomolecules such carbohydrate, protein, nucleic acids are optically active.
   (*i*) These substances may be either dextrorotatory or levorotatory.
2. **Inversion of sucrose :** Sucrose is dextrorotatory, but boiling with dilute mineral acids it yields an equimolecular mixture of fructose and glucose. This mixture is called "invert sugar" and is 'levorotatory'.
3. **Mutarotation :** The optical rotation of freshly prepared aqueous solution of a reducing sugar (glucose) gradually decreases until the activity stabilizes. This change of optical rotation of reducing sugar is known as mutarotation.

**• Spectrophotometers :**

It is a highly sophisticated instrument that utilizes light as a source of radiation and measures changes in optical density or absorbance.

**The word spectrophotometer means the photon (light) over the whole range of spectrum that the light source can produce and meter stands for the instrument that measures.**

**Instrumentation :**

(*i*) A stable and cheap radiant energy source.

(*ii*) A monochromator to break the polychromatic radiation into component wavelengths or "bands" of wavelengths.

(*iii*) Transparent vessels or cuvettes to hold the sample.

(*iv*) A photosensitive detector.

(*v*) Amplifier and recorder.

**(A) Radiant energy source :**

1. Materials which can be excited to high energy states by a high voltage electric discharge serve as excellent radiant energy.
2. Most commonly used sources of ultraviolet radiations are the hydrogen lamp and deuterium lamp. Xenon lamp may also be used.
3. Tungsten filament lamp is most commonly used for visible radiations.

**(B) Wavelength selectors :**

Wavelength selectors are of two types. Filters and monochromators.

**(*a*) Filters :**

(*i*) It operates by absorbing light in all other regions except for one which they reflect.

(*ii*) Gelation filters are made of a layer of gelation coloured with organic dyes.

(*iii*) Filters resolve polychromatic light into relatively wide band width of about 40 mm.

**(b) Monochromators :**

(i) It resolves polychromatic radiations into its individuals wave lengths and isolator these wavelengths into very narrow bands.

(ii) The essential components of a monochromators are an entrance slit, collimating lens, prism, focussing lens or mirror and exit slit. The entire assembly is mounted in a light tight box.

(iii) Prism disperse polychromatic light from the source into its constituent wavelengths.

(iv) Grattings are often used in monochromators of spectrophotometers operating in ultraviolet, visible and infrared regions.

**(c) Cuvettes :**

(i) Most spectrophotometric studies solutions are used.

(ii) They are dispensed in a cell known as cuvettes.

**(d) Detection Devices :**

(i) Most detectors depends upon photoelectric effect where incident light liberates electrons from a metal or other material surface.

(ii) It is converted into electrical energy in form of current and is proportional to the intensity of incident light.

**Mechanism :**

(i) Light rays are passed through a prism system to produce spectrum from which only a specific wavelength of light is allowed.

(ii) This is done finally by adjustable wavelength selector or filter. Then it is transmitted to a cuvette containing a test solution.

(iii) The selected wavelength is usually the absorption maximum ($\lambda_{max}$) of the solute being estimated in the test solution.

(iv) The emerging from the test solution is focussed by a lens system to a working photo electric cell placed between potentiometric circuit and with a reference electric cell.

**Observation :**

(i) The current generated in the working photoelectric cell by incident light is used in measuring the transmittancy of the searched solute.

(ii) Hence the concentration of the test solution is estimated.

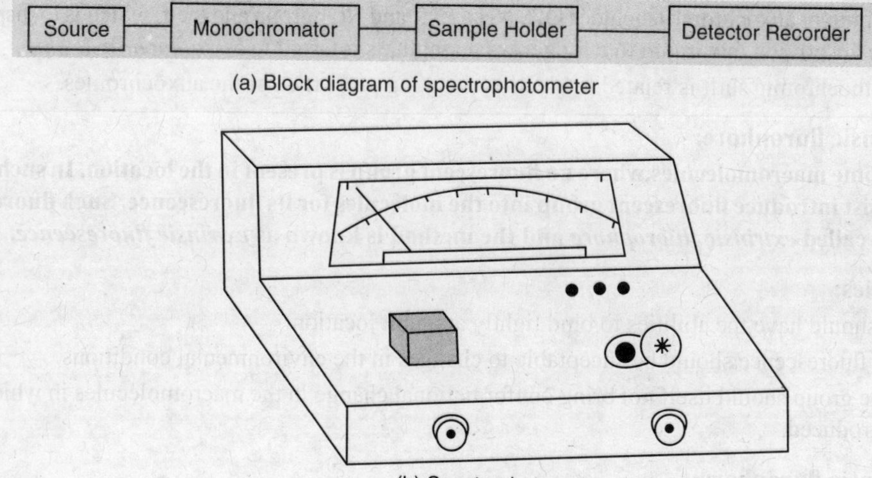


(a) Block diagram of spectrophotometer

(b) Spectrophotometer

**Fig. 9.11**

**Applications :**

1. Many chemicals have characteristic absorption spectra in uv and visible range and they can be identified in a mixture.
   (*a*) Proteins absorb strongly at 280 nm according to their content of amino acid tyrosine and tryptophan.
   (*b*) Nucleic acids and the component bases absorb to maximum in the region of 260 nma.
   (*c*) Haemoglobins absorbs both in visible region and uv region due to their haem group.
2. It can be used in determining the concentration of a compound by measuring the optical density provided absorbancy index is known.
3. The course of reaction can be known by measuring the rate of formation and disappearance of light absorbing material.

- **Absorption and Emission:**

*Absorption* is the phenomenon when a molecule or atom in its ordinary state (ground state) absorbs a quantum of light, then this energy rich molecules or atom is known as excited molecule (atom). It induces the change of electron distribution of the molecule or atom.

The excited molecule or atom undergoes electronic changes as well as physical and chemical properties. This excited molecule with electronic and vibrational changes lead it to emit light and drop to a lower or ground energy level. This phenomenon is known as *emission*.

- **Chromophore:**

(*i*) It is defined as an isolated covalently bonded functional group that shows a characteristic absorption spectrum in the visible and ultraviolet region and imparts colour to the compound.

(*ii*) This group invariably contain double or tripple bonds and include C=C linkage, C≡C bond, the nitro and nitroso groups, the azo group, the carbonyl and thiocarbonyl groups.

(*iii*) If it is conjugated with another of same or different kind, then the absorption is enhanced.

- **Auxochromes**

(*i*) Absorptions of a given molecule may also be enhanced by the presence of groups called auxochromes.

(*ii*) These do not absorb significantly in the ultraviolet region but may have a profound effect on the absorption of the molecule to which they are attached.

(*iii*) Important auxochromes includes OH, $NH_2$, $CH_3$ and $NO_2$ group and their, which is to displace the absorption maximum to a longer wavelength, is referred to as *bathochromic shift*.

(*iv*) Bathochromic shift is related to electron donating properties of the auxochromes.

- **Extrinsic flurophore:**

**In some macromolecules,where no fluorescent group is present in the location. In such cases biochemist introduce fluorescent group into the molecules for its fluorescence. Such fluorescent group is called *extrinsic fluorophore* and the method is known as *extrinsic fluorescence*.**

**Properties:**

(*i*) It should have the abilities to bind tightly at a unit location.

(*ii*) Its fluorescence should be suceptable to changes in the environmental conditions.

(*iii*) The group should itself not bring conformational change in the macromolecules in which it is introduced.

- **Intrinsic flurophore:**

**Intrinsic fluophores are those macromolecules that have their own fluorescent group.**

| | Spectrophotometer | | Photoelectric colorimeter |
|---|---|---|---|
| (*i*) | It uses prism monochromator | (*i*) | Simple selection filter is used. |
| (*ii*) | The wavelength of incident light can be adjusted precisely. | (*ii*) | The wavelength of incident light can not be adjusted. |
| (*iii*) | Greater accuracy is maintained. | (*iii*) | Lasser accuracy is maintained. |
| (*iv*) | Here smaller volumes of liquids and even solid and gasses are used. | (*iv*) | Solid and gases are never used. |

| | Fluorescence | | Phosphorescence |
|---|---|---|---|
| (*i*) | There are certain substances which when exposed to light or certain radiations, absorbs the energy and then immediately start re-emitting the energy. Such substances are called *fluorescent substances* and the phenomenon is known as *fluorescence*. | (*i*) | There are certain substances which continue to glow for sometime even after the external light is cutoff. Such substances are called *phosphores* or *phosphorescent substances* and the phenomenon is called *phosphorescence*. |
| (*ii*) | In this emission wave length are shorter than phosphorescence. | (*ii*) | Here wave length are longer than fluorescence. |
| (*iii*) | It is extremely short lived phenomenon. | (*iii*) | It is a long lived phenomenon. |
| (*iv*) | It is usually seen in liquid medium. | (*iv*) | It is usually seen in rigid medium. |

| Point | Visible light spectrophotometer | Infrared spectrophotometer |
|---|---|---|
| **1. Source** | Tunsten filament lamp is commonly used. | Nrnst Glower and Globar are the most useful resources. |
| **2. Radiation** | Carbon are provide more intense visible radiation. | Globar consist of silicon carbide rod emits radiation. |
| **3. Wave Radiation** | 350 – 2500 nm (4000 - 8000 Å) | 8000 – 1000 Å |

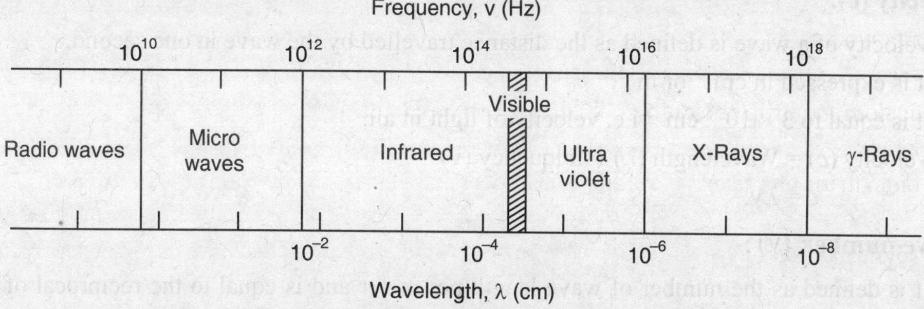


**Fig. 9.12** The electromagnetic spectrum

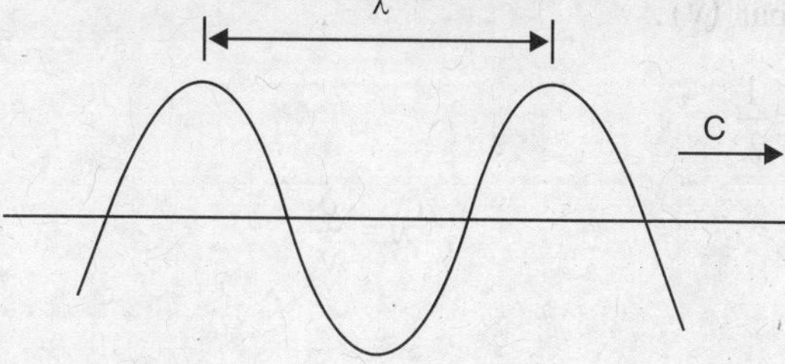


**Fig. 9.13** A single electromagnetic wave

- **Electromagnetic radiation**

(*i*) Electromagnetic radiations consists of electrical and magnetic waves oscillating at right angles to each other.

(*ii*) It behaves as of discrete wave like particles called *Quanta* or *Photons*.

(*iii*) Photon posseses the characteristics of a wave and travel with the velocity of light in the direction of the beam.

(*iv*) The amount of energy corresponding to 1 photon is expressed by Plank's equation.

(*v*) Visible light, X rays, microwaves, radio waves etc. all are electromagnetic radiations. Collectively they make up the ***Electromagentic spectrum*** (Fig. 9.12 and 9.13).

(*vi*) Electromagnetic waves are characterised by frequency (ν), wavelength (λ), wave number $(\bar{\nu})$, velocity (*c*).

- **Wavelength (λ):**

(*i*) It is defined as the distance between two successive crests or two successive troughs of a wave.

(*ii*) It is represented by lambda (λ) and is measured in terms of centimeter (cm) or meter or Angstrom (Å) units. It is also expressed in micrometer (μm) and nanometer (nm).

$$1\ \text{Å} = 10^{-10}\ \text{m} = 10^{-8}\ \text{cm},$$

$$1\ \mu\text{m} = 10^{-6}\ \text{m},\ 1\ \text{nm} = 10^{-9}\ \text{m}$$

- **Frequency (ν):**

(*i*) Frequency is defined as the number of waves or cycles which pass a given point on the wave in one second.

(*ii*) It is also represented by nu (ν) and is measured in terms of cycles (or waves) per second (cps) or Hertz (Hz). A frequency 100 hertz means 100 cycles per second.

(*iii*) A wave of higher frequency has shorter wavelength and vice versa. Thus frequency (ν) of a wave is inversely proportion to its wave length.

$$\nu \propto \frac{1}{\lambda} \quad \text{or} \quad V = \frac{c}{\lambda} \quad c = \text{constant } i.e. \text{ velocity of wave.}$$

- **Velocity (*c*):**

(*i*) Velocity of a wave is defined as the distance travelled by the wave in one second.

(*ii*) It is expressed in $\text{cm}^{-5}$ or $\text{m}^{-5}$.

It is equal to $3 \times 10^{-8}\ \text{cm}^{-5}$ i.e. velocity of light in air.

Velocity (*c*) = Wavelength (λ) × frequency (ν)

$$c = \lambda\nu.$$

- **Wave number $(\bar{\nu})$:**

(*i*) It is defined as the number of wave length per meter and is equal to the reciprocal of wave length (λ).

(*ii*) It is denoted by nu bar $(\bar{\nu})$.

$$\text{Wave number } (\bar{\nu}) = \frac{1}{\lambda}$$

$$c = \lambda\nu \qquad \therefore \lambda = \frac{c}{\nu}$$

$\bar{\nu} = \frac{\nu}{c}$ $\therefore$ Wave number $(\bar{\nu}) = \frac{\text{Frequency } (\nu)}{\text{Velocity } (c)}$

| Wave Parameter | Abbreviation | Unit | Relation |
|---|---|---|---|
| • Wavelength | Lambda = λ | cm / m / Å / nm. [1 cm = $10^{-2}$ m = $10^8$ Å = $10^7$ nm] | |
| • Frequency | Nu = ν | Cycles/Sec. Hz/Sec. | $\bar{\nu} = \frac{1}{\lambda} = \frac{\nu}{c}$ |
| • Velocity | Velocity = $c$ $c = 3 \times 10^{-10}$ cm/sec. $= 3 \times 10^8$ m/sec | cm / sec. or m/sec. | |
| • Wave number | Number = $\bar{\nu}$ | 1 $cm^{-1}$/1 $m^{-1}$. | |

**Light waves are electromagnetic waves:**

(*i*) The vibrations of light are electromagnetic origin. It has two vector component viz a varying electric vector $\vec{E}$ and a varying magnetic vector $\vec{B}$ (same frequency and phase).

(*ii*) $\vec{E}$ and $\vec{B}$ are perpendicular to each other and are in a plane at right angles to the direction of propagation of light.

(*iii*) Experiments have shown that the electric force in a light waves affects a photographic plates and causes **fluorescence** while the magnetic waves plays no role in the effect of light wave.

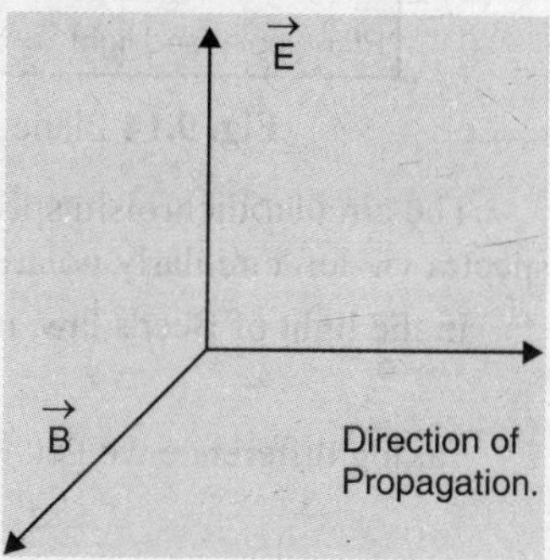


(*iv*) Therefore,the vibrations of electric vector $\vec{E}$ are now selected as the vibration of light.

(*v*) Thus we consider that the planes containing the light vibrations are those in which electric forces are present. Therefore, a plane-polarised beam is one in which the vibration of the electric vector are confined to a single direction.

- **Unpolarised and Polarised light:**

  - In ordinary or unpolarised light vibrations of the electric vector $\vec{E}$ occur symmetrically in all possible directions in a plane perpendicular to the direction of propagation of light.
  - In polarised light, the vibrations of electric vector $\vec{E}$ occur in a plane perpendicular to the direction of propagation of light and are confined to a single direction in the plane (do not occur symmetrically in all possible directions).

- **Dichroism**

(*i*) When a beam of unpolarised light is passed through a crystal like tourmaline, the light splits into two beams, polarised at right angles to each other.

(*ii*) The light with vibrations *perpendicular* to optic axis of the crystal are completely absorbed and the other component with vibrations *parallel* to the optic axis of the crystal are transmitted. Thus emerging light is plane polarised. The selective absorption of the crystal is called *dichroism* and the crystal is called *dichroic*.

- **Circular dichroism (CD):**

**Circular dichroism means the difference in the absorption of right and left handed circularly polarised light by an optically active compound.**

(*i*) An anisotropic (definite orientation of molecule) medium can refract the right ($R$) and left ($L$) components of the plane polarised light to a different extent. So that $n_R \neq n_L$, where $n$ = refractive index.

(*ii*) The refractive index of the right circulatory polarised light ($n_R$) may be different from left circulatory polarised light ($n_L$). This results in retardation of one component more than the other.

(*iii*) Similarly the absorbance $A_R$ for right circularly polarised light may be different from left circularly polarised light.

**The difference in absorbance $A_L - A_R$ is called *circular dichroism* (*CD*).**

$$\Delta A = A_L - A_R \qquad \begin{bmatrix} A_L = \text{absorbance of left circularly polarised light} \\ A_R = \text{absorbance of right circularly polarised light} \end{bmatrix}$$

(*iv*) Circular dichroism may be either positive or negative.

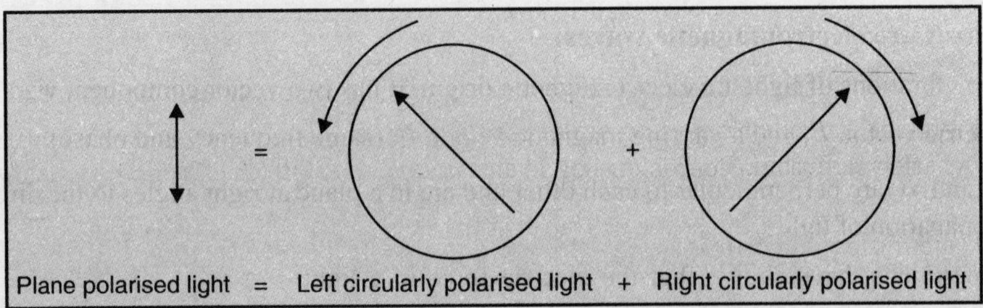


**Fig. 9.14** Plane polarised light can be resolved into two components

The circular dichroism spectra may be thought of as the difference between the two absorption spectra viz left circularly polarized and right circular polarized component.

In the light of Beer's law, it may be expressed as

$$A_L = a_L\, Cb \text{ and } A_R = a_R\, Cb$$

Taking difference we get

$$A_L - A_R = (a_L - a_R)\, Cb$$

$$\therefore\ \Delta a = (a_L - a_R) = \frac{A_L - A_R}{Cb}$$

$\Delta$ = Difference between two extinction coefficients.

(*i*) The usual representation of *CD* spectrum is to demonstrate $\Delta a$ as a function of wavelength.

(*ii*) As circular dichroism depends on absorption of light it will be observed only those spectral region where absorption occurs and will not be observed in the regions where no observation occurs.

- **Optical rotation :**

(*i*) Plane polarised light can be resolved into two beam viz. left circularly polarized light ($n_L$). [Moves counter clockwise] and right circularly polarised light ($n_R$) [Moves clockwise].

(*ii*) The refractive index of the right circularly polarised light ($n_R$) may be different from that for left circularly polarized light ($n_L$).

(*iii*) If the medium preferentially refracts the left circularly polarized light the resultant would be a plane polarised light, rotated some what to the right from its original position and vice versa. This phenomenon is known as *optical rotation*.

(*iv*) Since the refractive index is always dependent on the wavelength, the angle of rotation is also wavelength dependent.

$$\alpha_\lambda = \frac{\pi d}{\lambda}(n_L - n_R) \qquad \left[\begin{array}{l} d = \text{path length} \\ \lambda = \text{wave length} \\ \alpha = \text{angle of rotation} \end{array}\right.$$

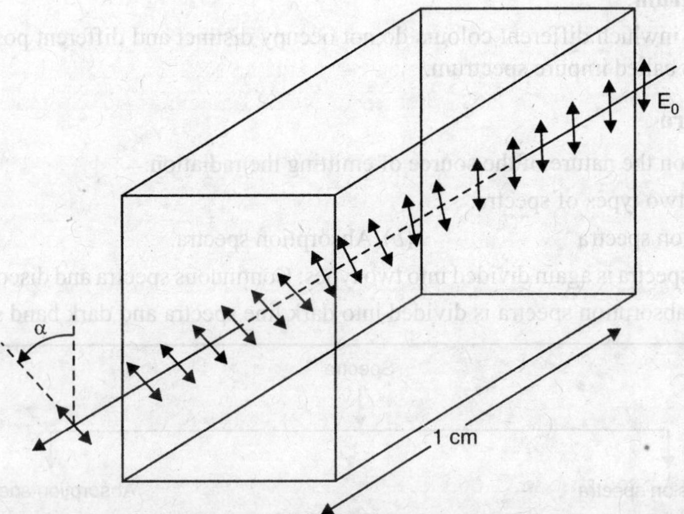


**Fig. 9.15** Optical rotation of plane-polarized light passing through a chiral sample. For the example shown the sign of the rotation angle α is negative.

- **Optical Rotatory Dispersion (ORD)**

Measurement of the change of optical activity with the change in wavelength yields more structural information i.e. called optical rotation dispersion (ORD).

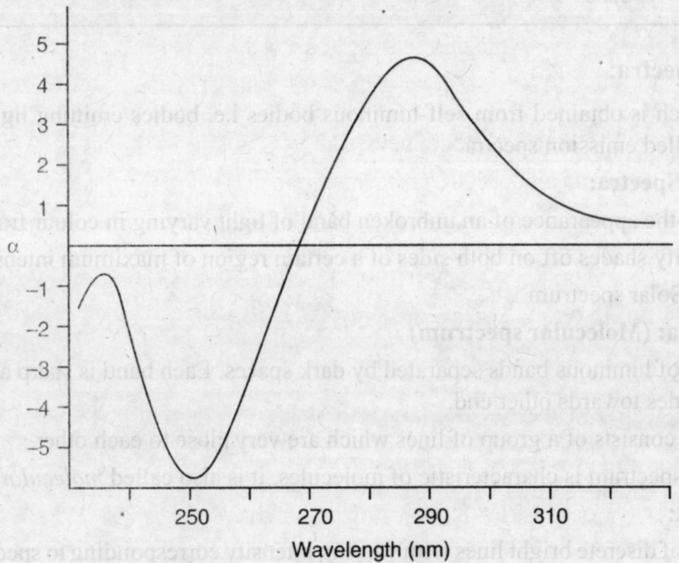


**Fig. 9.16** A typical ORD spectrum

## Spectrum:

A spectrum is an ordered array of waves or particles:

- **Pure spectrum:**

A spectrum inwhich different colours occupy distinct and different position and are distinctly visible is called pure spectrum.

- **Impure spectrum:**

A spectrum inwhich different colours do not occupy distinct and different position and are not visible clearly is called impure spectrum.

## Types of Spectra

Depending on the nature of the source of emitting the radiation.

(*i*) There are two types of spectra.

(*a*) Emission spectra (*b*) Absorption spectra.

(*ii*) Emission spectra is again divided into two types: Continuous spectra and discontinuous spectra.

(*iii*) Similarly absorption spectra is divided into dark line spectra and dark band spectra.

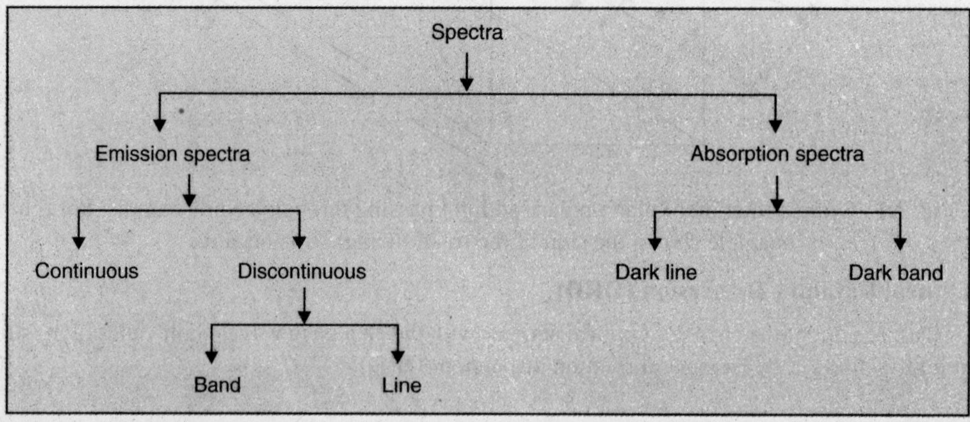


## A. Emission Spectra:

Spectra which is obtained from self-luminous bodies i.e. bodies emitting light themselves on excitation are called emission spectra.

### (*a*) Continuous Spectra:

(*i*) It presents the appearance of an unbroken band of light varying in colour from point to point.

(*ii*) The intensity shades off on both sides of a certain region of maximum intensity.

Example: Solar spectrum

### (*b*) Band Spectra: (Molecular spectrum)

(*i*) It consists of luminous bands separated by dark spaces. Each band is sharp and intense at one end and fades towards other end.

(*ii*) Each band consists of a group of lines which are very close to each other.

(*iii*) Since this spectrum is characteristic of molecules, it is also called *molecular spectrum*.

### (*c*) Line Spectra:

(*i*) It consists of discrete bright lines with varying intensity corresponding to specific wave length.

(*ii*) It is produced by substances in the atomic state. So, it is also known as *atomic spectra*.

Example: Sodium spectra

**B. Absorption Spectra:**

Spectrum of the absorbed light is called absorption spectrum.

(*a*) **Dark line spectra:** Here a continuous range of wavelength is absorbed by the absorber or medium.

(*b*) **Dark band spectrum:** Here a few discrete wave lengths are absorbed by the medium.

**X-ray Spectra:**

(*i*) X-rays emitted from an X-ray tube is analysed under different potential difference applied to the anode, using an X-ray spectrometer.

(*ii*) The target may be made up of molybdenum or tungsten. The intensity of the X-rays were measured.

(*iii*) A graph was drawn by taking the wave length along the X-axis and the intensity of the X-rays along Y-axis.

(*iv*) X-ray spectrum has two parts viz. (*a*) A *continuous spectrum* consisting of radiations of all possible wave lengths with a lower wavelength limit and (*b*) A *line spectrum* or *characteristic spectrum* consisting of definite wave lengths superimposed on the continuous spectrum.

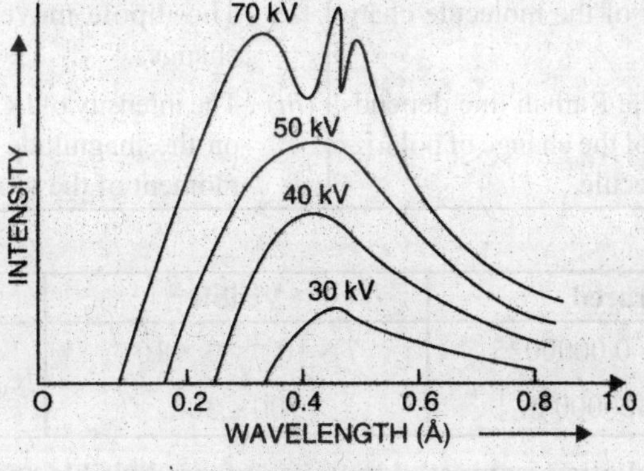


**Fig. 9.17**

- **Raman Spectra**

**While studying scattering of light Prof. C.V. Raman (1928) of Calcutta University found that when a beam of monochromatic light was passed through organic liquid (such as benzene, tolune etc.) the scattered light contained not only the incident light (frequency) but also of other frequency light (higher and lower frequency). This is known as *Raman effect.***

(*i*) When a substance (benzene tolune etc) is irradiated with a monochromatic light of a definite frequency v, the light scattered at right angle to the incident light contained lines, not only of the incident frequency but also of lower and higher frequency.

(*ii*) The lines with lower frequency are called *Stroke's line* and with higher frequency are called *anti stroke's line*. The *stroke's line* are more intense than *anti stroke's line*. Line with the same frequency as the incident light is called *Rayleigh line*.

(*iii*) He also observed that the difference between the frequency of the incident light and that of particular scattered line was constant only *upon the nature of the substance* irradiated and was completely independent of the frequency of incident light.

(*iv*) If $\nu_i$ is the frequency of the incident light and $\nu_s$ that of a particular scattered line, the difference $\Delta\nu = \nu_i - \nu_s$ is called *Raman frequency* or *Raman shift*. Thus the *Raman frequencies* observed for a particular substance are characteristics of that substance.

(*v*) The various observations made by Raman are called Raman effect and spectrum observed is called Raman spectrum.

| | Raman Spectra | | Fluorescent Spectra |
|---|---|---|---|
| 1. | Spectral line have frequencies greater and lesser than the incident frequency. | 1. | The frequencies of the lines in the fluorescent spectrum are always less than the incident frequency. |
| 2. | The frequency shifts of the Raman lines are determined by the scatterer rather than the frequencies themselves. | 2. | Frequencies of the fluorescent lines are determined by the nature of the scatter. |
| 3. | Raman lines are strongly polarised. | 3. | Lines are not polarized. |

| | Raman spectra | | Infra red spectra |
|---|---|---|---|
| (*i*) | It appears due to the scattering. | (*i*) | It arises due to the absorption of radiation by the vibrating molecules. |
| (*ii*) | The polarizability of the molecule changes. | (*ii*) | The dipole movement of the molecules changes. |
| (*iii*) | The intensity of the Raman line depends on the magnitude of the change of polarizability of the molecule. | (*iii*) | The intensity of IR absorption band depends on the magnitude of the change in dipole moment of the molecule. |

| | Infrared | Visible | Ultraviolet |
|---|---|---|---|
| λ m | 0.00003 – 0.0000025 | $7 \times 10^{-7} - 3 \times 10^{-7}$ | $3 \times 10^{-7} - 1 \times 10^{-7}$ |
| λ (nm) | 30000 – 400000 $m^{-1}$ | 700 – 300 | 300 – 100 |

**Electromagnetic radiation, expressed in units, infrared visible and ultraviolet.**

## PROBLEMS

**1.** *Calculate the optical density (absorbance, A) of 0.2 m moles of adenine in a 4 ml volume. The path length is 1 cm. $a_m$ of adenine is 13.3 the measurement was carried out at 260 nm.*

**Solution:**

$A = a_m bc$ — $a_m$ molar extinction coefficient = 13.3

$= 13.3 \times 1 \times 0.05$ — $b$ = Path length = 1 cm

$= 0.665$ — $c$ = Concentration *i.e.* $0.2 = \frac{0.2}{40} = 0.05$

$\therefore$ *OD* of the solution is 0.665.

**2.** *A solution of purified DNA isolated from E. coli gives an absorbance of 0.396 at 260 nm in a 1 cm cubette at pH 4.5. If $E_{1\,cm}^{1\%}$ is 197, calculate the concentration of the solution in gm/ml.*

**Solution:**

Absorbance $(A) = A_s \times b \times c$

or $0.396 = 197 \times 1 \times c$

or $c = \frac{0.396}{197} = 0.0020$ gm/ml

**3.** *A solution having a concentration of 64 μg/ml of a substance of molecular weight 212 has an absorbance of 0.27 at 540 nm in a one cm cuvettle. Find the value of the molar extinction coefficient.*

***(CU (Microbiology): 2011)***

**Solution:**

Concentration $(c) = 64$ μg/ml

$= 64 \times 10^{-3}$ mg/ml

$= 64 \times 10^{-6}$ g/ml

$= \frac{64 \times 10^{-6}}{\text{lit}} \times 10^{3} = 64 \times 10^{-3}$ g/lit.

$A_{540} = As \times c \times b$

$0.27 = As \times 64 \times 10^{-3} \times 1$

$$\left[\begin{array}{l} c = 64 \times 10^{-3} \text{ g/lit} \\ b = 1 \text{ cm} \\ A_{540} = 0.27 \\ As = ? \end{array}\right]$$

or $As = \frac{0.27}{64 \times 10^{-3}}$

$As = \frac{27}{64 \times 10^{2} \times 10^{-3}} = \frac{27}{64 \times 10^{-1}} = \frac{270}{64} = 4.218$

So, $a_m = 4.218 \times 212 = 894.2\ \text{M}^{-1}\ \text{cm}^{-1}$

**4.** *Calculate the transmission, absorbance and absorbance coefficient of a solution which absorbs 90% of a certain wavelength of light beam passed through 1 cm cell containing 0.25 M Solution.*

**Solution:**

It is given $I_{abs} = 0.90\, I_0$ $b = 1$ cm $c = 0.25$ M

$I_t = I_{as} - I_{abs} = I_0 - 0.90\, I_0 = 0.10\, I_0$

$\therefore$ Transmittance $(T) = \frac{I_t}{I_0} = \frac{0.10\, I_0}{I_0} = 0.10$

$$\text{Absorbance } (A) = -\log \frac{I}{I_0} = \log \frac{I_0}{I} = \log \frac{I}{0.1} = 1$$

**Absorbance coefficient**

According to Beer's law

$$\log \frac{I_t}{I_0} = -k'cl$$

or $$\log 0.10 = -k'cl$$

$$-1 = -k \times 0.25 \times 1 = -0.25\, k'$$

or $$k' = \frac{1}{0.25} = 4 \text{ M}^{-1} \text{ cm}^{-1}$$

**5.** *Calculate the value of extinction coefficient of NADH and express its unit. If a 1.37 × 10⁻⁴ of solution exhibit an absorbance of 0.85 at a wave length of 340 nm in a 1 cm cell.*

***(Calcutta Univ. (Microbiology) 2010)***

**Solution:**

Absorbance $(A) = a_m \times l \times c$

$0.85 = a_m \times 1 \times 1.37 \times 10^{-4}$ $\qquad a_m = ?$

$$\therefore a_m = \frac{0.85}{1.37 \times 10^{-4}} = \frac{85 \times 10^2}{10^2 \times 10^{-4}} \qquad A = 0.85$$

$= 85 \times 10^4$ $\qquad l = 1$ cm

$c = 1.37 \times 10^{-4}$

**6.** *20% of incident monochromatic light (Na-D-line) is transmitted through a solution of strength 0.01 (M) contained in a cell of length 1.0 cm. Calculate extinction coefficient.* ***(CU 1989).***

**Solution:**

It is given that 20% of incident light is transmitted

∴ light absorbed is 100 – 20 = 80%

$I_{abs} = 0.80$ $\qquad I_0 = 100 = 1$ $\qquad I_t = 20\% = 0.2$

$l = 1$ cm, $\qquad c = 0.01$ (M)

$$\therefore \text{Transmittance } (T) = \frac{I_t}{I_0} = \frac{0.2}{1} = 0.2$$

According to Beer's law

$$\log \frac{I_t}{I_0} = -k'cl$$

$$\log 0.2 = -k' \times 1 \times 0.01$$

or $$\log 0.2 = -k \times 0.01$$

$$-0.6989 = -k \times 0.01$$

or $$k = \frac{0.6989}{0.01} = 69.89 \text{ lit mol}^{-1} \text{ cm}^{-1}.$$

**7.** *What will be absorbance or O.D. of a solution of concentration 2 × 10⁻⁴ (M) if kept in a cell of path length 5.0 mm (extinction coefficient 5000 lit mol⁻¹ cm⁻¹).* ***(CU 1997)***

**Solution:**

It is given extinction coefficient $(k) = 5000$ lit mol$^{-1}$ cm$^{-1}$.

$c = 2 \times 10^{-4}$ (M) $\quad \lambda = 5.0$ mm = 0.5 cm

$$OD = \log \frac{I_0}{I} = kcl$$

$$OD = 5 \times 10^3 \times 2 \times 10^{-4}$$

$$= 10 \times 10^{-1} = 10 \times \frac{1}{10} = 1$$

**8.** *A certain substance in a cell of length l absorbs 10% of the incident light. What fraction of incident light will be absorbed in a cell which is five times long?*

**Solution:**

**I Case:**

$$I_{abs} = 0.10\, I_0 \qquad I_t = 0.90\, I_0$$

$$\therefore \quad \frac{I_t}{I_0} = 0.90$$

$$\log \frac{I_t}{I_0} = -k'l$$

or $\log 0.90 = -k'l \qquad \therefore -k' = -0.04575$

$$k' = \frac{0.04575}{l} = \frac{0.458}{l}$$

**II Case :**

$$n = 5l$$

$$\log \frac{I_t}{I_0} = -k' \times 5l = \frac{-0.458}{l} \times 5l = -0.2290$$

$$\therefore \quad \log \frac{I_t}{I_0} = -0.229 = -1 + 1 - 0.229 = -1 + 0.771$$

$$\log \frac{I_t}{I_0} = \bar{1}.7710$$

i.e. $\frac{I_t}{I_0} = 0.5902$ or $I_t = 0.5902\, I_0$

$$I_{abs} = I_0 - I_t = I_0 - 0.5902\, I_0$$

$$= 0.4098\, I_0$$

$$= 40.98\%\, I_0$$

**9.** *Monochromatic light is passed through a 1 mm path length cell containing 0.005 moles/dm³ solution. The light intensity is reduced to 16% of its value. Calculate the molar extinction coefficient of the sample. What would be the transmittance if the cell path is 2 mm?* ***[C.U. 1987]***

**Solution:**

Here $\quad c = 0.005$ moles/dm$^3$

$l = 1$ mm = 0.1 cm

Intensity of light reduced 16% of its value.

Absorbance (*OD*) $\quad = \log \frac{I_o}{I_t} = \frac{100}{16}$

$$OD = \log \frac{I_o}{I_t} = \frac{100}{16} = k'cl = k' \times 0.005 \times 0.1$$

$$\log = \frac{I_o}{I_t} = k' \times 0.0005$$

$$\log \frac{100}{16} = k' \times 0.0005$$

$$\text{or } k' = \frac{\log \frac{100}{16}}{0.0005} = \frac{\log 6.25}{0.0005} = \frac{0.79588}{0.0005} = 1591 \text{ lit/mole cm}^{-1}$$

If the path length is 2 mm i.e. 0.2 cm

$$\log \frac{I_o}{I_t} = k \times 0.005 \times 0.2$$

$$\log \frac{I_o}{I_t} = 1591 \times 0.005 \times 0.2 = 1.591$$

$$\log \frac{I_o}{I_t} = 1.591$$

or $$\log \frac{100}{I_t} = 1.591$$

or $$\log 100 - \log I_t = 1.591$$

$$2 - \log I_t = 1.591$$

or $$\log I_t = 2 - 1.591 = 0.409$$

$$I_t = \text{antilog } 0.409 = \mathbf{2.56}$$

**10.** *A 0.01 molar solution of a compound transmits 20% of Na-D line when the absorbing path length is 1.5 cm. What is the molar extinction coefficient of the substance? Solvent is assumed to be completely transparent.* ***[C.U. 1989]***

**Solution:**

Here $l = 1.5$ cm

$c = 0.01$

$I_0 = 100 \quad I_t = 20$

$$OD = \log \frac{I_o}{I_t} = kcl$$

$$= \log \frac{100}{16} = k \times 0.01 \times 1.5$$

or $$k = \frac{\log \frac{100}{20}}{0.01 \times 1.5} = \frac{\log \frac{100}{20}}{0.015} = \frac{\log 5}{0.015} = \frac{0.69897}{0.015} = 46.5\ (\text{M}^{-1})\ \text{cm}^{-1}$$

**11.** *At 460 nm a blue filter transmits 75% of the light and a yellow filter transmits 40% of the light. What is the transmittance at the same wave lengths of the two filter in combination.*

***[V.U. 1996]***

**Solution:**

$I_0 = 100 \quad I_t = 75$

$I_0 = 100 \quad I_t = 40$

OD for blue filter $= \log \frac{I_o}{I_t} = \log \frac{100}{75} = \log 1.333 = 0.1248 = 0.125$

OD for yellow filter $= \log \frac{I_o}{I_t} = \log \frac{100}{40} = \log 2.5 = 0.3979 = 0.398$

As optical density is an additive property, so optical density of combination of blue and yellow filter is (0.125 + 0.398) = 0.523

$$\therefore \quad \log \frac{100}{I_t} = 0.523 \text{ or } \log 100 - \log I_t = 0.523$$

$$\text{or } 2 - 0.523 = \log I_t$$

$$\text{or} \quad \log I_t = 1.477$$

$$I_t = \text{antilog } 1.477 = 29.99\%.$$

**12.** *A 0.003 (M) solution of coloured substance transmits 75% of the incident light of 500 mm, when placed in a cell of length 1 cm. Calculate the molar extinction coefficient, and hence the optical density of a 0.001 (M) solution in the same cell at the same wave length.* ***[C.U. 1974]***

**Solution:**

Here

$$c = 0.003 \text{ (M)}$$

$$l = 1 \text{ cm}$$

$$I_0 = 100 \quad I_t = 75$$

$$\frac{I_t}{I_o} = \frac{75}{100} = 0.75$$

We know that

$$\log \frac{I_t}{I_o} = -k'cl$$

$$\therefore \quad \log 0.75 = -k'cl$$

$$\text{or} \quad -0.1249 = -k' \times 0.003 \times 1$$

$$-0.1249 = -k' \times 0.003$$

$$\text{or } k' = \frac{0.124938}{0.003} = \frac{0.124938}{0.003} = 41.646 \text{ cm}^{-1} (\text{M}^{-1}).$$

$\therefore$ Optical density (OD) $= kcl$

$= 41.64 \times 0.001 \times 1 \text{ cm}$

$= 0.0416$

**13.** *The percentage of transmittance of an aquous solution of a dye at 450 μm and at 25°C is 30% for a $2 \times 10^{-3}$ (M) solution in a 2 cm cell. Calculate the optical density and the molar extinction coefficient. Find the conc. of the same dye in another solution where the percentage transmittancy is 20% in a 1 cm cell at the same temperature and same wave length of light.* ***[C.U. 1980]***

**Solution:**

According to the problem transmittance is 30% *i.e.*

$$\frac{I_t}{I_0} = \frac{30}{100} \text{ or } \frac{I_0}{I_t} = \frac{100}{30}$$

Optical density (OD) $= \log \frac{I_0}{I_t} = \log \frac{100}{30}$

or $\quad$ OD $= \log 10^2 - \log 30 = 2 - 1.47712 = 0.52288$

$= 0.52$

Now

$$\text{OD} = kcl$$

$$\text{K} = \frac{D}{cl}, \text{ Here } c = 2 \times 10^{-3} \text{ (M)}$$

Here, $\quad l = 2$ cm

$D = 0.52288$

$$k = \frac{0.52288}{2 \times 10^{-3} (M) \times 2 \text{ cm}} = \frac{0.52288 \times 10^3}{4} = 0.13072 \times 10^3$$

$$= 130.72 \text{ (cm}^{-1}\text{) (M}^{-1}\text{)}$$

Now $\quad \log \frac{I_t}{I_0} = -kcl$

Transmittancy 20%

$\therefore \quad \frac{I_t}{I_0} = \frac{20}{100}$

$$C = -\frac{\log \frac{I_t}{I_0}}{kl} = \frac{\log 0.2}{130.72 \times 1} = -\frac{-0.6989}{130.72} = \frac{0.6989}{130.72}$$

$= 0.005346542$

$$= \frac{5 \cdot 347}{10^3} = 5.347 \times 10^{-3} \text{ (M)}$$

**14.** *An aquous solution of a compound A of concentration $10^{-3}$ moles/litre absorbs 50% of incident radiation in a cell length 1 cm and another compound B of concentration $2 \times 10^{-3}$ moles/litre absorbs 60% of he incident radiation at a particular wavelength. Calculate the percentage absorbed by a solution containing $10^{-3}$ moles/litre of A and B each in the same cell at the same wave length.*

***[V.U.-1994]***

**Solution:**

According to the problem, for compound *A*, 50% incident light is absorbed.

$$\frac{I_t}{I_0} = \frac{50}{100} \text{ or } \frac{I_0}{I_t} = \frac{100}{50} = 2$$

$$\text{OD} = \frac{\log I_0}{I_2} = \log 2$$

$$\text{OD} = k_A c l \; k_A = ? \qquad c = 10^{-3} \text{ moles/litre}$$

$l = 1$ cm

$$\log 0.5 = k_A \times 10^{-3} \times 1$$

$$k_A = \frac{\log 2}{10^{-3} \times 1} = \frac{\log 2}{10^{-3}} = 0.30102 \times 10^{-3}$$

$$= 301.02$$

$$k_A = 301.02 \text{ lit mole}^{-1} \text{ cm}^{-1}$$

For compound $B$, 60% light is absorbed i.e. 40% transmitted

$$\text{OD} = \log \frac{I_0}{I_t} = \log \frac{100}{40} = \log 0.25$$

$$k_B = \frac{\log 2.5}{2 \times 10^{-3} \times 1 \text{ cm}} = \frac{0.3979}{2 \times 10^{-3}} = \frac{0.3979 \times 10^3}{2} = \frac{397.9}{2}$$

$$= 198.97 \text{ lit mole}^{-1} \text{ cm}^{-1}$$

So, for solution $10^{-3}$ moles litre$^{-1}$ of $A$ and $B$

$$\text{OD} = (k_A + k_M)\, cl$$

$$= (301.02 + 198.97) \times 10^{-3} \text{ mole lit}^{-1} \times 1 \text{ cm}$$

$$= 499.99 \times 10^{-3} = 0.49999$$

$$\therefore \quad \log \frac{I_0}{I_t} = 0.49999$$

$$\therefore \quad \log \frac{100}{I_t} = 0.49999$$

$$\text{or} \quad 2 - \log I_t = 0.4999$$

$$\therefore \quad \log I_t = 2 - 0.4999 = 1.500$$

$$\therefore \quad \log I_t = 1.5001$$

$$\therefore \quad I_t = \text{antilog } 1.5001 = 31.62$$

$\therefore$ 100 – 31.62 = 68.38% light is absorbed.

CHAPTER 10

# Separation Techniques

**Sedimentation :**

**In this process downward movement of solutes or suspended particles in a liquid due to the force of gravity, raising their concentration progressively towards the bottom of the liquid.**

**Characteristics :**

(*i*) Sedimentation is *inversely proportional* to the *absolute temperature* (T).

(*ii*) It is also *inversely proportional* to the viscosity $\eta$ of the liquid medium.

(*iii*) Sedimentation is directly proportional to the acceleration **g** due to gravity .

(*iv*) It is directly proportional to the difference $(\rho_2 - \rho_1)$ between the densities of settling particle and the liquid medium.

(*v*) Sedimentation is also proportional to the effective mass 'm' of the settling particle having volume v where $v = m(\rho_2 - \rho_1)$.

**Explanation**

(*i*) The particle when suspended in a solution are pulled downward by earth's gravitational force. This movement is partially offset by the buoyancy of the particle.

(*ii*) Since the earth's gravitational field is weak, a solution containing macromolecules is usually homogeneous, as a result of the random thermal motion of the molecules.

(*iii*) The rate of sedimentation of the particles increases with the mass of the particle and with the strength of the gravitational field.

(*iv*) When we perform sedimentation experiment with macromolecules, the force of gravitation is insufficient to make the molecules sediment. Under these circumstances the solution containing macromolecules is subjected to centrifugation. It can rotate at high speed.

(*v*) An object moving in a circle at a steady angular velocity will experience a force, 'F' directed outward.

Angular velocity in radian $\omega$. and radius of the rotation in centimeter $r$, the magnitude of force (F) will be $F = \omega^2 r$.

The force on each particle, according to Newton's second law $F = ma$, $m$ = mass of the particle $a$ = linear accelaration

(*vi*) As the particles are spun in centrifuge the radial or centrifugal force ($F_c$) will be

$F_c = m\,\omega^2 r$ $\quad\quad$ $\omega$ = angular velocity

$r$ = radial distance of the particle from the axis of rotation.

(*vii*) As the particles sediment, diffusion effect also take place simultaneously and hence the motion of macromolecules is an ultra centrifuge is the resultant of both effects. Therefore the mass is replaced by the effective mass is $(m - V_p)$ i.e the difference between the weight of the particle and the bouyant force due to displacement of the medium. V is the volume and $p$ is the density of the solvent. $V_p$ is the mass of the solvent displaced by the particle of mass $m$.

Now the radial or centrifugal force ($F_c$)

$$F_c = (m - V_p)\,\omega^2 r$$

(*viii*) This force is opposed by the frictional force ($F_r$) which is proportional to the velocity of the particle

$$F_r = f \times (\text{radial force}) = f\left(\frac{dr}{dt}\right)$$

Where $f$ is the frictional coefficient. When centrifugal force is balanced by the frictional force then $F_c = F_r$ and the particle will sediment with uniform velocity. Since there is no force on them and thus no accelaration.

$$F_c = F_r \text{ or } (m - V_p)\,\omega^2 r = f\left(\frac{dr}{dt}\right) \qquad (\omega = \text{angular velocity})$$

Multiplying both sides by $N_o$ (Avagadro's number)

$$N_o\,(m - V_p)\,\omega^2 r = N_o\, f\left(\frac{dr}{dt}\right)$$

or $$N_o\,(m - V_p) = \frac{N_o\, f\left(\frac{dr}{dt}\right)}{\omega^2 r}$$

(*ix*) The velocity of sedimentation depends on the size and shape of the sedimenting particle and is proportional to the applied force.

$$v = s\,\omega^2 r$$

$s$ = proportionality constant and known as *sedimentation coefficient.*

$$s = \frac{v}{w^2 r}$$

$$s = \frac{\text{velocity}}{\omega^2 r}$$

**Sedimentation coefficient :** Sedimentation velocity of a given species of sedimenting particles per unit field of the sedimenting (centrifugal) force during centrifugal field.

**Sedimentation coefficient (S)**

It is defined as velocity of a particle per unit centrifugal field.

It is measured in Svedberg units ($1S = 10^{-13}$ sec).

It is denoted by the symbol '$S$'

$$S = \frac{v}{\omega^2 r}$$

$v$ = radical velocity in the centrifugal field.

$\omega$ = angular velocity of centrifugation.

$r$ = distance of the particle from the centre of rotation.

- **Characteristic of sedimentation coefficient:**
  - (*i*) It can be considered "Sedimentation Rate" of a particle under centrifugation force.
  - (*ii*) "*S*" is increased for particle of larger mass.
  - (*iii*) "*S*" is increased for particle of larger density.
  - (*iv*) "*S*" is increased for more compact structure of equal particle mass.
  - (*v*) "*S*" is increased with rotational speed.
- **Factors affecting sedimentation coefficient:**
- **Molecular Shape:**
  - (*i*) Large rigid extended molecules will face more friction. While spherical molecule face less friction.
  - (*ii*) Molecules with higher friction have increased value of '*S*' and with lower friction have less value of '*S*'.
- **Molecular Weight:**
  - (*i*) Higher molecular weight molecule sediment faster and diffuses slower than a lower molecular weight molecule of the same density.
  - (*ii*) $M\omega = \dfrac{RTs}{D(1 - \bar{v}\rho s)}$

    Mω = Molecular weight
    $R$ = Gas constant
    $T$ = absolute temperature
    $D$ = Diffusion coefficient
    $\bar{v}$ = Partial specific volume of macromolecule
    ρ = density of the molecule
- **Solute Binding**
  - (*i*) Binding of small solute molecules to large macromolecule affects their '*S*' value.
  - (*ii*) If DNA is suspended in NaCl or CsCl has higher molecular weight and therefore has higher *S* value.

## Centrifugation Technology :

**Principle :**

**The basic principle of centrifugation technology in biology is to be based on sedimentation, separation, purification of mixture of biological as well as particles suspended in liquid medium under the influence of centrifugal field.**

(*i*) These are placed either in tubes or bottles in a rotor in the centrifuge.

(*ii*) Particles different in sizes, shape and density are separated as their sedimentation rate is different.

(*iii*) The centrifugal force $F = m\omega^2 r$ is generated by rotating the rotor of the centrifuge at a high speed.

(*iv*) The gravitational force is increased by increasing the speed of the centrifuge rotor. The particles are sedimentated or settle down faster with the increase in the speed of rotor.

(*v*) As a result of rotation of material a force i.e relation centrifugal force (RCF) acts on the particles.

$F = m\,\omega^2 r$ (centrifugal force)

It is a relative value which is comparable with the value of the gravitational force of the earth.

(*vi*) *F* might be expressed in terms of earth gravitational force if it is divided by mg or $m \times 980$, resultant is referred to as

$$\text{RCF} = \frac{m\,\omega^2 r}{mg} = \frac{\omega^2 r}{g} = \frac{\omega^2 r}{980} \quad \because g = 980$$

(*vii*) In terms of *rpm* (revolution per minute), the operating speed of a centrifuge rotor is expressed. Relationship among radius, rpm on ω are as follows.

$$\omega = \frac{\pi\,(rpm)}{30}$$

substituting for ω, we can write

$$\text{RCF} = \frac{\dfrac{\pi^2\,(rpm)^2}{(30)^2}}{980} \times r = (1.119 \times 10^{-5})\,(rpm)^2\,r$$

The RCF depends upon the *rpm* and the radius of rotation *r*. If *r* is constant, as it may be for a given rotor, variation of *rpm* alone will be cause of for variation in RCF.

(*viii*) *r* would be the distance between the particle in the sample holder and the rotor axis. As centrifugation progresses, the particle will sediment progressively towards the bottom of the sample tube with increasing the distance (radius) from the rotor axis.

(*ix*) Therefore RCF gradually increases. So in practice, radius is calculated from the rotor axis to the middle of the liquid column. This radius is known as the **average radius** ($r_{av}$).

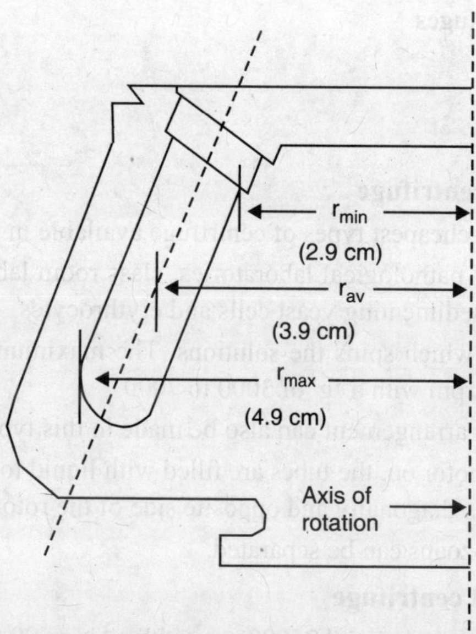


**Fig. 10.1** Cross section of a fixed angle rotor showing how radii at the top and the bottom of the centrifuge differ. The figure shows average radius, $r_{av}$, as calculated from the axis of rotation to the middle of the liquid column.

(*x*) The speed of the the centrifuge rotor might be expressed in terms of *rpm* or RCF. If the material is centrifuge at a speed 1000 *rpm* and maximum distance is 10 cm, between sample and axis of the rotor, the value of *g*

$$\begin{aligned}\text{RCF} &= 1.119 \times 10^{-5}\,(1000)^2\,10 \\ &= 1.119 \times 10^{-5} \times 10^6 \times 10 = 1.119 \times 10^2 \\ &= 111.9 = 112\text{ g.}\end{aligned}$$

(*xi*) So that a particle settle down on the basis of RCF:
But the rate of sedimentation depends upon
(*a*) Centrifugal force (*b*) buoyant force of the medium
(*c*) the frictional resistance of the movement of particle.

- **Instrumentation :**

**Common parts :**

(*i*) Chamber– it encloses the internal parts

(*ii*) Cover with latch.

(*iii*) Centrifuge head with shields or cups.

(*iv*) The shaft and rotor on which the head turns and motor drive assembly. Rotors are three types viz angle, swinging bucket and zonal.

(*v*) It include a power switch, braking devices, speed control, timer and a tachometer (measuring speed rotation).

***Types* :**

On the basis of purpose and suitability of use, the centrifuges can be categorized into four different types:

1. Clinical Bench centrifuge.
2. High speed refrigerated centrifuges.
3. Continuous flow centrifuges.
4. Ultracentrifuges
   (*i*) Preparatory
   (*ii*) Analytical.

**1. Chemical Bench types centrifuge**

(*i*) These are simplest and cheapest types of centrifuge available in the market.

(*ii*) They are used in many pathological laboratories, class room laboratories for routine type of work, particularly for sedimenting yeast cells and erythrocytes.

(*iii*) It consists of a motor which spins the solutions. The maximum speed of this centrifuge is between 4000 to 6000 rpm with a 'g' of 3000 to 7000.

(*iv*) Sometimes the cooling arrangement can also be made in this type of centrifuge.

(*v*) Before switching the motor on, the tubes are filled with liquid to be centrifuged and balanced while keeping the tubes diagonally and opposite side of the rotor.

(*vi*) Here blood cells, protozoans can be separated.

**2. High speed refrigerated centrifuge**

(*i*) It can operate with maximum speed 25000 rpm with a '*g*' of 90,000.

(*ii*) To achieve high speed their load capacity is to maximum 1-5 litres.

(*iii*) They are provided with interchangeable swinging buckets with a fixed angle rotor.

(*iv*) These instruments are routinely used to collect microorganism cell debris, cells, large cellular organelles such as mitochondria lysosomes etc.

**3. Continuous flow centrifuge**

(*i*) It is a kind of high speed centrifuge.

(*ii*) They have longer rotor through which the suspensions of particles flow continually at a rate of about 1-1.5 litre per minute.

(*iii*) Here rotor is not interchangeable.

(*iv*) The suspension medium enters the rotor and the sediment collect on it wall while excess clear fluid is passed through an exit tube.

(*v*) They are largely used to harvest bacteria yeast etc.

**4. Ultracentrifuge**

(*i*) The ultracentrifuge was developed by Svedberg in 1923. This instrument is used to measure **sedimentation coefficient** (*s*).

(*ii*) It is of two types viz **Preparatory & Analytical** Ultracentrifuge.

**(A) Preparatory Ultracentrifuge**

**This is a type of instrument where actual isolation, purification of cells, subcellular organelles polysomes, plasma membrane, chromatin lipoprotein even viruses can be done.**

**Instrumentation :**

(*i*) Preparatory rotor contains cylindrical sample tubes that may have parallel axes or an angular one or perpendicular to rotor's axis of rotation.

(*ii*) It operates on refrigerated condition under vacuum to avoid frictional resistance of the rotor caused by the spinning air.

(*iii*) The whole system is sophisticated with continuously monitoring system of the rotor temperature i.e operates through an infrared temperature sensor.

(*iv*) There is also an overspeed disk system which to check the rotor so that it does not exceed its maximum allowable speed.

(*v*) It operates through some photoelectric devices. For ultrasmooth and quite performance of the rotor, it is attached directly with the motor.

(*vi*) The rotor is friction free as it works on a cushion of air in a non evacuated chamber.

(*vii*) It can attain a maximum speed 80,000 to 1,00,000 and can produce centrifugal field of 600,000 g.

(*viii*) In preparatory ultracentrifugation, the sedimentation occurs through homogeneous medium and so it may be carried out in a solution of inert substances such as sucrose, or CsCl.

(*ix*) The density of such solutions increases from top to bottom of centrifuge tube. These *density gradient* enhance the resolving power of the centrifuge.

(*x*) Preparatory centrifugation methods can be divided into two main technique depending o medium of suspension.

(*a*) separation is carried out in a suspending medium which is homogeneous and is know *differential centrifugation.*

(*b*) separation is carried out on a suspending medium having density gradient and is k as *density gradient centrifugation.*

**(B) Analytical Centrifuge**

**This is an instrument with the help of which sedimentation coefficient of a m determined. It is attached with an optical system through which one can observe t particles and the moving boundaries. The rate at which the boundaries move un conditions helps to find out the sedimentation coefficients. It runs at a speed of abo rpm with about 500,000 g and consists of a specially designed rotator in a special r**

**Instrumentation:**

(*i*) Solid rotor has a hole to hold the sample and is suspended on a wire with sha motor so that the angle of rotation depends on rotor not on shaft.

(*ii*) There are several types of rotor. The simplest one has two cells–one ar sample and other is *counterpoise cell* to *counter balance* the analytical

(*iii*) Analytical cells used with ultraviolet light absorption.

(*iv*) There is an arrangement of special optical system to determine the concentration distributions with in the sample during centrifugation. Schlieren optics or Rayleigh interference optics are used.

(*v*) A thermistor that is temperature measuring device is present at the tip of rotor.

(*vi*) The rotor chamber has an upper condensing and lower collimating lens which are joined with camera lens and emit light on the photographic plate.

(*vii*) But in recent models photographic plate system has been replaced by electronic scanning system.

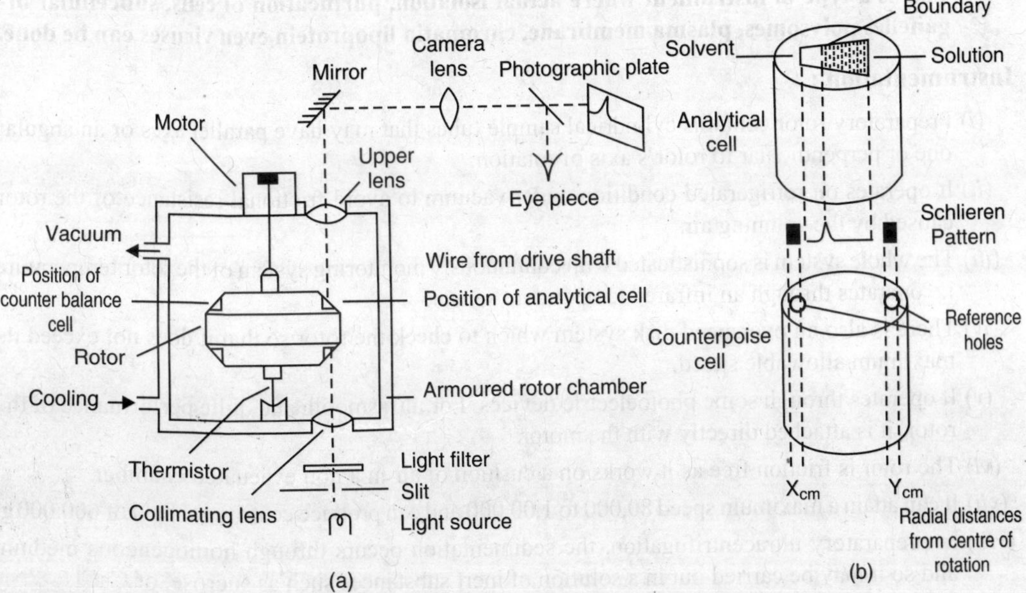


**Fig. 10.2** Schematic representation of Analytical ultracentrifuge (*a*) and (*b*). Details of Analytical and counterpoise cell.

**Mechanism :**

(*i*) In ultraviolet light system a beam of suitable wavelength is passed through the solutions (protein, nucleic acid etc) under analysis contained in the analytical cell.

(*ii*) The intensity of transmitted light is photographically or electronically recorded.

(*iii*) Electronic scanning enables direct visualization and gives a plot of concentration of sample at all levels in analytical cell at any given time. Selecting different wavelengths of light can separately monitor different single components moving in the mixture substance.

(*iv*) In Schlieren optical system, when light pass through uniform concentration does not deviate but pass through the solution of different density zone will be refracted at boundaries between zones.

(*v*) The optical system records the change in the refractory index of the solution and that varies with changes in concentration.

(*vi*) Sediment formation results into the solute free solvent zones. Thus a boundary is formed between the particle free solvent and the sediment material.

*ii*) This state works as a refraction lens and consequently a peak in the final photo sensed image is formed, as an exact record of refractive index gradients and the area left below is proportional to the concentration of the solutes.

(*viii*) Therefore by measuring the refractive index between the reference solvent and the solution, the concentration of solute at any point can be measured.

(*ix*) In Schlieren optical system records refractive index gradient against distance along analytical cell and locates the boundaries in the sedimentation velocity measurements.

In Ray light interference optical system employ one sector containing solvent and the other is solution.

(*x*) But now a days electronic scanning system is operated and directly measure and plot the concentration of the sample at all points in the analytical cell.

- **Chromatography :**

A Russian botanist **Mikhail Tsvet (1906)** first developed chromatographic technique for the separation of different substances present together in a solution or cytosol. The word "Chromatography" originates from two Greek words Khroma (colour) and graphia (writing) and means colour writing.

**It is a process of separation of components of a sample (mixture) depending on their relative movements in two immiscible phases, mobile and stationary according to their affinities for the two phases. A chromatogram is the visual output of chromatography.**

**Features :**

(*i*) The chromatographic separation involves the placing of a sample on to a liquid or solid *stationary phase* and passing a liquid or gaseous *mobile phase* over it or percolates through the interstices of stationary phase, a process known as *eluition.*

(*ii*) Sample components whose distribution ratios between the two phases differ will migrate (be eluted) at different rates and this differential rate of migration will lead to their separation over a period of time and distance.

(*iii*) The sample components which interact more with the mobile phase and least with the stationary phase, migrates fast. The component showing least interaction with mobile phase while inter-acting strongly with the stationary phase migrate slowly.

(*iv*) The quality of chromatographic separation is measured in light of band broadening, or efficiency measured for individual peaks, adjacent peak, the degree of separation & resolution.

(*v*) It can provide qualitative information or the basis of *retenition* parameter & quantitative in formation on the basis of peak heights & peak areas.

(*vi*) It may be regarded as a method of separation where separation of solutes occur between a stationary phase and a mobile phase.

(*vii*) All types of chromatography are based on distribution or partison coefficient which is the expression of the way of the distribution of the compound in two immiscible phases viz mobile & stationary phases.

**Partition coefficient :**

(*i*) It is also known as distribution partition coefficient. It is normally used to describe the way in which a given compound or sample distributes or partitions itself two different extent between two immiscible phases, the mobile and stationary phases.

(*ii*) Immediately after entering into chromatographic system, the solute distributes itself between the stationery and the mobile phases.

(*iii*) If at a given time, during chromatography, the mobile phase is stopped, the compound or sample will be in equilibrium between the stopped mobile phase and the stationary phase. At this stage the concentration of the compound or sample in each of the phases is described by the partition coefficient ($k_d$) *i.e.*

$$K_d = \frac{C_s}{C_m}$$

$C_s$ = Concentration of the compound or sample in the stationary phase.

$C_m$ = Concentration of the compound sample in the mobile phase.

$K_d$ = Distribution/Partition Constant or coefficient.

Now if value of $K_d$ is 1, that will mean that the concentration of a compound or sample in stationary and mobile phase is the same.

**Partition Forces & Interaction :**

(*i*) The distribution of a solute between mobile phase and the stationary phase is a result of the balance of forces between the solute molecules and the molecules of each phase.

(*ii*) The partition or distribution coefficient reflects the relative attraction or repulsion that the molecules of the two phases show for the solute molecules and for themselves.

(*iii*) These attractive or repulsive interactions are accompanied by release or consumption of energy.

(*iv*) The amount of interaction energy provides a measure of the strength of the interaction.

(*v*) In a non polar liquid such as carbon tetrachloride dispersion, interaction is the only force present between two molecules. These non polar molecules do not possesses any permanent dipole moment. The interaction is a resultant of instaneous dipoles formed between the nuclei and electron at zero-point motion of the molecule.

(*vi*) The *Polar Liquid* (solvent) whose molecules have *permanent dipoles* exhibits much more intermolecular interaction as compared to *non polar* molecules.

(*vii*) Those solute molecules which exert either a higher or at least equal attraction with the solvent molecules as compared to the attraction of the solvent molecules for each other will be able to mix with the solvent molecules. Therefore a polar sample component will interact more with the phase which is itself polar & move fast or be retarded more depending on whether polar phase is the mobile phase or stationary phase respectively.

(*viii*) Hydrogen bonding is another type of polar interaction.

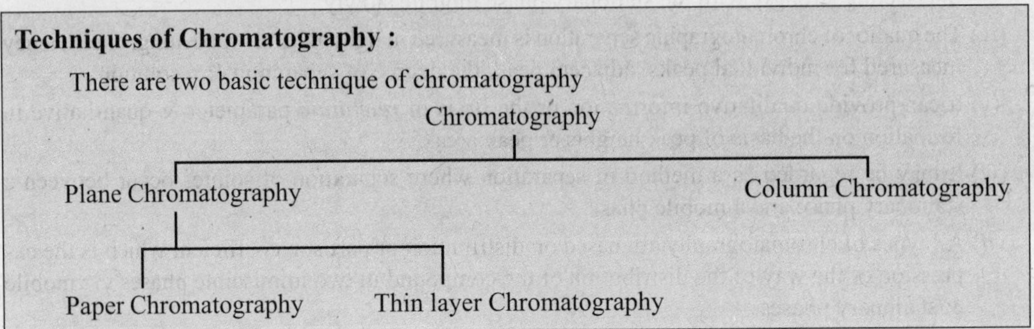


- **Paper Chromatography**

**Izmailov & Schraiber** (1938), the two Russian workers discovered this important technique. This method is very useful for the detection and separation of amino acids.

**Principle :**

**This is a kind of partition chromatography where cellulose fibres in the form of a chromatographic paper act as a support or *stationary phase,* there the samples (substances) are distributed between two liquids *i.e.* one is the stationary liquid (water) which is held in the paper and called the stationary phase, the other is the moving liquid or developing solvent and called the *moving phase.* The components of the mixture to be separated migrate at different rates and appear as spots at different points on the paper.**

**Apparatus :**

(*i*) Air tight cylindrical jar or chamber
(*ii*) Chromatographic paper (Cellulose)
(*iii*) Solvent trough
(*iv*) Solvent (n butanol : acitic acid; water = 4:1:5) which is abbreviated as BAW.

**Method :**

(*i*) A small quantity of sample is applied to a paper as a small spot with the help of platinum loop, capillary tube or micropipett.
(*ii*) The applied spot is about 2cm away from one end of the filter paper.
(*iii*) This is done before dipping the paper into the eluting solvent.
(*iv*) The sample is allowed to dry & then the paper is dipped into the solvent.
(*v*) In both the techniques (ascending & descending) the solvent is placed in the base of the sealed tank or glass jar to allow the chamber to become saturated with the solvent vapour.
The edge of the paper is dipped into a solvent called *developing solvent*.
(*vi*) After equilibration of the chamber the chromatogram may be developed by allowing the solvent to flow along the sheet.

**Detection :**

(*i*) The solvent front is marked and after drying the paper, the position of the components present in the mixture are visualized by a suitable staining reagents (**ninhydrin for amino acids**) called visualising reagents.
(*ii*) Other methods of detection are (*a*) ultraviolet & infrared absorption (*b*) fluorscene & radioacitivity.
(*iii*) The movement of substances retative to the solvent is expressed interms of $R_f$ values *i.e.* migration parameter.
(*iii*) **Migration Parameter :**

(*i*) The positions of migrated spots on the chromatogram are indicated by *viz* $R_f$ $R_n$, $R_m$, & $R_c$.
(*ii*) These parameters are qualitative as well as quantitative.

$R_f$= It is related to the migration of the solute front relative to the solvent front.

$$R_f = \frac{\textbf{Distance moved by the solute from the origin line}}{\textbf{Distance moved by the solvent from the origin line}}$$

**Characters :** (*i*) $R_f$ is the function of the partition coefficient.
(*ii*) It is constant for a given substance where chromatographic system is kept constant.

**Factors regulating $R_f$ value :**

(*i*) The solvent employed
(*ii*) The nature of the mixture
(*iii*) The medium used for separation.
(*iv*) The temperature
(*v*) The size of container/jar/chamber in which the operation is carried out.

**$R_x$ : In some cases the solvent front runs off the end of filter paper, the movement of a substance in such cases is expressed as $R_x$ instead of $R_f$**

$$R_x = \frac{\text{Distance moved by the substance from the origin line.}}{\text{Distance moved by the standard substance x from the origin line.}}$$

**Selection of Solvent System :**

(*i*) The solvents are selected in such a way that the resolution of sample components is satisfactory.

(*ii*) The components of the solvent system should be so chosen that the extent of evaporation of each individual component is more or less similar.

(*iii*) The solvent boiling point should be less than 200°C

(*iv*) The solvent should be stable.

- **Types of Paper Chromatography :**

**(*a*) Descending paper chromatography :**

In the descending paper chromatographic techniques, the development of the paper is done by allowing the solvent to move down the paper.

**Characteristics :**

(*i*) The apparatus consists of a well sealed glass tank of suitable size and shape.

(*ii*) A trough filled with solvent for the mobile phase is placed in the paper portion of the glass tank.

(*iii*) The paper with the sample spotted is inserted with the upper end of the trough containing the mobile phase.

(*iv*) The glass tank or jar has been saturated with solvent vapour.

(*v*) Where the top, end of the paper near which the samples are located is dipped into the trough of the solvent and the rest of the paper is allowed to hang vertically but not in contact with the solvent in the base of the tank.

(*vi*) Development is started by adding the solvent to the trough. Separation of the sample is achieved as the solvent moves downward under gravity.

**Significance :**

(*i*) It is a good method for the mixture containing components with close differences of $R_f$ values.

(*ii*) Here $R_f$ values are not measurable because separation occurs under the influence of gravity.

(*iii*) Therefore distances of separated substances are compared with a suitable standard reference compound such as glucose, if one would like to separate sugar.

$$R_f = \frac{\text{Distance moved by given compound}}{\text{Distance moved by glucose}}$$

(*iv*) Development can be continued indefinitely.

**(*b*) Ascending paper Chromatography :**

In the ascending paper chromatographic technique the development of the paper is done by allowing the solvent to move upward the paper.

**Characteristics :**

(*i*) The apparatus consists of a well sealed glass tank of suitable size and shape.

(*ii*) A trough with solvent for the mobile phase is placed at the bottom of the chamber or tank.

(*iii*) A sheet of filter paper is suspended vertically from a supporting rod inside the sealed glass chamber and the bottom end of the paper i.e. spotted end is placed in contact with trough filled with solvent.

(*iv*) Alternatively the paper may be rolled into a cylinder with its two ends fastened with paper clip or plastic clip.

(*v*) The set up is then closed in an air tight chamber or cylinder which is saturated with solvent vapour.

(*vi*) Then the separation is allowed to go on at a constant room temperature.

**(c) Ascending-Descending Chromatography :**

(*i*) It is the hybrid of the ascending and descending chromatography.

(*ii*) Square the upper part of ascending chromatography can be folded over a glass rod and allow the ascending development.

(*iii*) It then change over into descending development after crossing the glass rod.

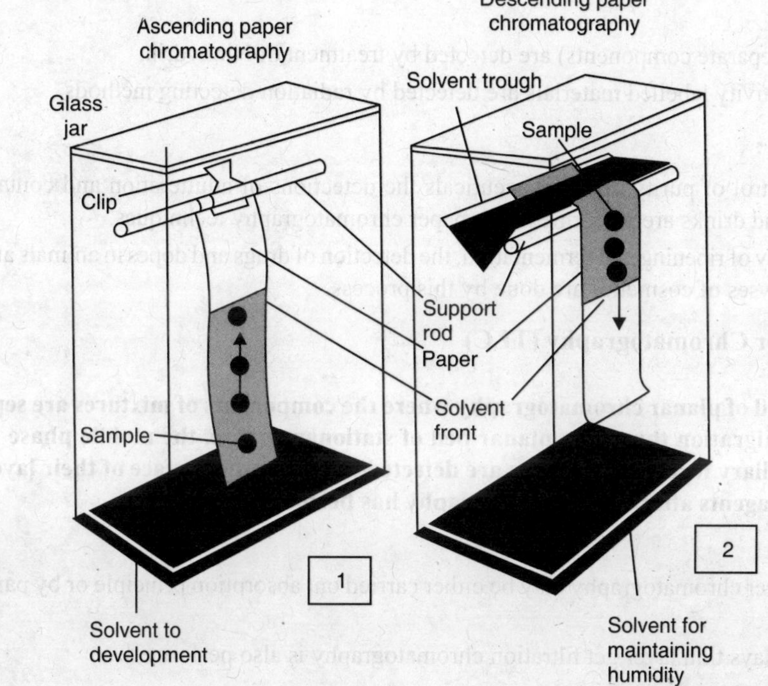


**Fig. 10.3** The instrumentation and working of a paper chromatography set up. 1. Ascending and 2. Descending (arrows show direction of movement of solvent)

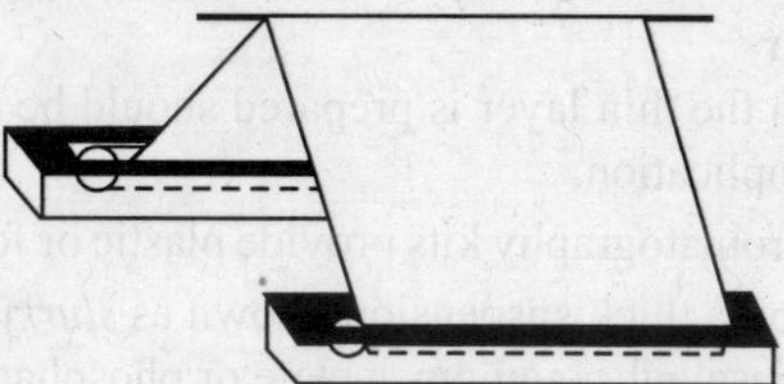

**Fig. 10.4** Ascending-descending chromatography.

**(d) Two dimensional Chromatography :**

In this type of chromatography, a square or rectangular paper is used. The sample is applied at one of the corner and rotated at right angle or 90° and chromatographed again using a second solvent system. This type of chromatography can be carried out with identical solvent system in both the direction or by two solvent system.

**Characteristics :**

(*i*) Filter paper is spotted at one corner of the sheet. Then chromatogram is run parallel to an edge of the paper.

(*ii*) After the first chromatography is completed, the paper is dried.

(*iii*) The filter paper is now rotated at 90°, so that the edge having the series of spots is now at the bottom just above the solvent trough.

(*iv*) Now Chromatogram is run parallel to the second edge *i.e.* right angle to the first axis, in a different solvent.

(*v*) Chromatogram is now having spots of the solute scattered all over the paper.

**Detection :**

(*i*) Spots (separate components) are detected by treatment of Ninhydrin,

(*ii*) Radioactivity labelled materials are detected by radiation detecting methods.

**Applications :**

(*i*) The control of purity of pharmaceuticals the detections of adulteration and contaminants in foods and drinks are performed with paper chromatography techniques.

(*ii*) The study of ripening and fermentation, the detection of drugs and dopes in animals and humans, the analyses of cosmetics are done by this process.

## • Thin Layer Chromatography (TLC)

**Principle :**

**It is a kind of planar chromatography where the components of mixtures are separated by differential migration through a planar bed of stationary phase, the mobile phase flowing by virtue of capillary forces. The solutes are defected in situ on the surface of their layer plate by visualising reagents after the chromatography has been completed.**

**Features :**

(*i*) Thin layer chromatography may be either carried out absorption principle or by partition principle.

(*ii*) Now a days thin layer get filtration chromatography is also performed.

**Method :**

A typical TLC (Thin layer Chromatography) consists of the following steps.

**(*a*) Preparation of thin layer**

(*i*) The glass plate on which the thin layer is prepared should be even and is thoroughly washed and dried before layer application.

(*ii*) Now a days thin layer chromatography kits provide plastic or foil plates instead of glass plates.

(*iii*) This thin layer is made by a thick suspension known as *slurry* which is formed by mixing of water and silica gel or kieselguhr (sodium acetate or phosphate) in a adequate proportion.

(*iv*) This slurry is applied to a plate surface as a uniform thin layer by means of a plate spreader.

(*v*) Sometimes calcium sulphate is added to the supporting adsorbent to adhere the layer to the plate.

(*vi*) For analytical separations the thickness of the layer might be about 0.25mm while for preparative separations the thickness of layer might be about 5mm . If thickness is below 0.2mm the $R_f$ value is affected.

(*vii*) After spreading it is allowed to dry and is then heated in an oven at a standard temperature (105° – 120°C) for activation and then cooled to room temperature.

**(*b*) Development of Chromatogram :**

(*i*) After cooling, the activated plate is spotted with the mixture in a solvent. A minute quantity of mixed sample in solvent is placed at one end of the layer and that end is immersed into the trough containing solvent.

(*ii*) It ascends along the thin layer of adsorbent due to capillary action. It causes migration of the solutes to different spots along the thin layer.

(*iii*) The assembly is kept in a sealed chromatographic chamber saturated with solvent vapour.

**(*c*) Detection :**

After removing the plate and drying it, the separated spots are detected by the following ways:

(*i*) By spraying the plate with chromic sulphuric acids ($H_2SO_4$) or nitric acid ($HNO_3$) and heating to oxidize and charring organic solutes which are revealed as brown or black spot.

(*ii*) Spraying the plate with chromogenic reagent that reacts chemically with solutes to give coloured spots.

(*iii*) Exposing the plate to vaporized iodine (used as universal reagent for compound) which develops dark, brown colouration.

(*iv*) Viewing the plate under UV lamp set an emission wavelength of 254 mm or 370 nm to reveal the solutes as dark spots or bright flurorescent spot on a uniform back ground fluorescent.

(*v*) Scanning the spot with a densitometer, an instrument that measures the intensity of radiation reflected from the surface when irradiated with UV or visible lamp.

(*vi*) Scanning the TLC plate with a Geiger-Muller counter or scintillation counter or autoradiographic method for locating and estimating the radioisotopically labelled solutes in the separated spots.

**Advantages :**

(*i*) The basic technique is very cheap, versatile, transfer more reproducible.

(*ii*) It is often used to determine the complexity of a compound quickly.

(*iii*) It has the ability to run 10-20 or more samples simultaneously for immediate and direct comparison with standard.

(*iv*) All solutes which do not migrate from the origin are detectable.

**Disadvantage :**

Limited reproductibility of $R_f$ values.

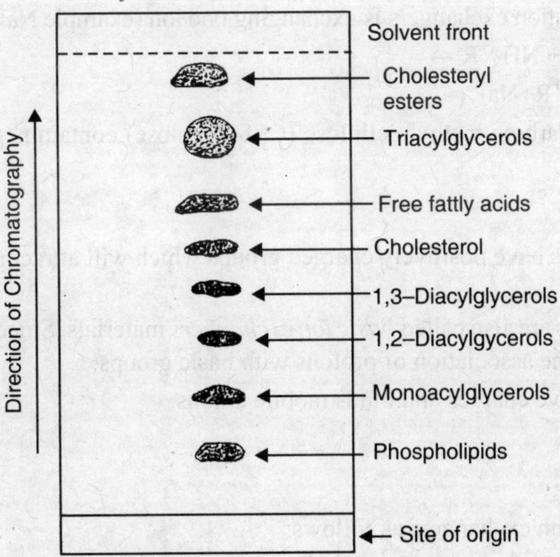


**Fig. 10.5** *Thin layer chromatography of mixture of major lipids*

**Applications :**

(*i*) TLC is applicable to a wide range of organic solutes.

(*ii*) It has been used primarilly in pharcemaceuitical, clinical, forensic areas for qualitative analysis by the comparison of $R_f$ values.

(*iii*) It is useful for checking purity *i.e.* adulteration.

(*iv*) It is used for large scale chromatography.

## • Ion-Exchange Chromatography

**Principle : It may be defined as the reversible exchange of ions in solutions with ions electrostatically bound to inert support medium. The governing factor in ion exchange reaction is the electrostatic force of attraction.**

(*i*) Ion exchange separations are mainly carried out in columns packed with an ion-exchanger.

(*ii*) The ion exchange materials are water insoluble polymers example agarose, dextran, synthetic resins (polysulfonic resins, polycarboxylic resins, polyamine resins) Cellulose (Diethyl aminoethyl cellulose, carboxymethyl cellulose).

(*iii*) The surfaces of ion exchanger have either acidic or basic groups which are fixed. These groups ionize at suitable ph to yield fixed negative or positive chargers on the ion exchangers material.

(*iv*) There are two types of ion exchangers namely cation and anion exchanger.

**A. Cation exchanger :**

(*i*) Cation exchangers posseses negatively charged groups and these will attract positively charged molecules.

(*ii*) These exchangers also called *acidic ion exchanger* materials since their negative charges result from the protolysis of acid groups.

(*iii*) It has fixed negative charges and binds mobile cations.

**(*a*) Synthetic resin :**

(*i*) Polysulfonic resin or polycarboxylic resins.

(*ii*) The action of a cation exchangers is exchanging one ion (example $Na^+$) for another ($NH^+R$) is

Resin-$SO_3^-$ $Na^+$ + $NH_3^+R$ →
Resin-$SO_3^-$ $NH_3^+R$+$Na^+$ ←

(*b*) **Cellulose** - Carboxymethyl cellulose (CM-Cellulose) contain negatively charged groups at neutral pH.

**B. Anion exchanger :**

(*i*) Anion exchangers have positively charged groups which will attract negatively charged molecules.

(*ii*) These exchangers are also called *basic ion exchangers* materials. Since positive charges generally result from the association of protons with basic groups.

(*iii*) It has fixed positive charges and binds mobile anions.

**(*a*) Synthetic resins :**

(*i*) Polyamine resins

(*ii*) The action of anion exchangers as follows:

Resin- $NR_3^+$ + $OH^-$+ $R^1$ $COO^-$ →
Resin- $NR^+_3$ + $R^1COO$ + $OH^-$ ←

(*b*) **Cellulose** - Diethylaminoethyl cellulose (DEAE-Cellulose) contains positively charged groups at pH 7.0.

**Preparation of Exchange Medium :**

(1) Preparation of exchange medium is essential for the satisfactory performance of ion exchange chromatography.

(2) Three steps are essential for the exchanger preparation:

**(*a*) First Step :**

(*i*) Swelling of the medium is known as pre-cycling.

(*ii*) The functional groups are exposed after swelling.

(*iii*) Swelling of anion exchanger is usually carried out by treating it first with an acid (0.5 N HCl) and then with a base (0.5N NaOH). Reverse is the case of cationic exchanger.

(*iv*) Finally if impurities (metal ion) are present, the matrix can be treated with a chelator (Ex. EDTA).

**(*b*) Second Step :**

*Removal of very small particles of the exchanger finers.*

(*i*) It is necessary to remove fines since the presence of large number of such fines which will result in decreased flow rate and unsatisfactory resolution.

(*ii*) Such fines are usually developed during swelling if the exchanger is subjected to vigorous washing and stirring.

(*iii*) To remove the fines, the exchanger is repeatedly suspended in a large volume of water.

(*iv*) Larger polymers are settled down and the slow sedimenting particles (materials) are decanted.

**(*c*) Third Step :**

(*i*) This is made by washing the exchanger with different reagents depending upon the desired counter ion. Example, $NaNO_3$, if $NO_3^-$ is counter ion, formic acid if formate is counter-ion.

(*ii*) Following conversion, excess counter ion are removed by washing the exchanger with large volume of water or dilute buffers.

**Choice of Buffer :**

(*i*) Buffer maintains the pH of the column.

(*ii*) The choice of buffer is made on the basis of type of ion exchange (anionic or cationic) and the nature of compound to be separated.

(*iii*) Anion exchange chromatography should be carried out with cationic buffers. The cation exchange chromatography should be carried out with anionic buffer.

**Some buffers used in ion exchange chromatography**

| ***Buffer*** | ***pH range*** |
|---|---|
| Amonium carbonate | 8-10 |
| Amonium acetate | 4-6 |
| Amonium formate | 3-5 |
| Pyridium acetate | 4-6 |
| Pyridium formate | 3-6 |

**Procedure :**

(*i*) Selection of exchangers depends upon the stability of the component which will be separated.

(*ii*) *Cation exchanger* is used when sample is more stable at a pH below its isoionic point. On the other hand, for a sample exhibits more stability above its isoionic point, an *anionic exchanger* is more useful.

(*iii*) Sample exhibits its stability over a wide range of pH may be separated by using either type of exchanger.

(*iv*) The volume of exchanger used for separation is usually 2-5 folds greater than that needed to bind all of the sample. However excess greater than this are avoided.

(*v*) The pH of the buffer used is usually maintained at about one pH unit more or less than the iso ionic point of the sample component.

(*vi*) In ion exchange chromatography, proteins are separated on the basis of their *over all (net) charge*. If a protein has a net negative charge at pH7, it will bind to a column containing positively charged beads, where as a protein with no charge or a net charge will not bind.

(*vii*) The negatively charged protiens bound to such a column can then be eluted by washing the column with an increasing gradient of a solution of *sodium chloride* ($Na^+$ and $Cl^-$ ions) at appropriate pH.

(*viii*) The chloride ions ($Cl^-$) complete with the protein for positively charged groups on the column. Protein having a low density of negative charge elute first followed by those with a higher density of negative charge.

(*ix*) Columns containing positively charged diethylaminoethyl (DEAE) groups (DEAE Cellulose or sephadex) are used for the separation of negatively charged protein (anionic proteins). This is called *anion exchange chromatography.*

(*x*) Columns containing negatively charged carboxymethyl (CM) groups (CM Cellulose or CM sephadex) are used for the separation of positively charged proteins (cationic proteins). This is called *cation exchange chromatography.*

(*xi*) As an alternative to elution with gradient (increasing concentration) of Na Cl, proteins can be eluted from anion exchange columns by decreasing the pH of the buffer, and from cation exchange column by increasing pH of the buffer, thus altering the ionization state of the amino acid side chain and hence the net charge on the protein.

**Applications :**

(*i*) The ion exchange chromatography is used for amino acid analysis. Infact amino acid "auto analyser" is based on the *ion exchange* principle.

(*ii*) It has been extensively used to determine the base composition of nucleic acids. The mixture of nucleotides as a result of treatment with DNAses and RNAses can be readily separated by ion exchange chromatography.

(*iii*) It is used in all laboratories as fast and effective method of water purification.

(*iv*) Ion exchange chromatography has been used for the separation of many vitamin, biological amines and organic acids and bases.

## • Affinity Chromatography

**Principle : Separation by affinity chromatography is based on the biological properties of macromolecule rather than on a physical property. It is a highly sensitive technique where the separation of sample components depends on their affinities for binding non covalently to specific ligands covalently coupled with surface groups of gel beads of the stationary phase column.**

Binding should be specific and reversible

Macromolecules + Ligand $\longleftrightarrow$ Macromolecules + ligand

**Pre-requisite :**

(*a*) Type of matrix

(*b*) Selection of Ligand

(*c*) Conditions exploited to bind and dissociate the macromolecule.

**(*a*) Supporting matrix :**

(*i*) It should be inert to other molecules to minimize non specific adsorption.

(*ii*) It should be chemically and mechanically stable at varying pH, ionic strength for binding and elution.

(*iii*) It should posseses good flow properties.

(*iv*) It should preferably be highly porus.

(*v*) It should posseses large number of suitable chemical groups for ligand attachment.

(*vi*) The particles which are uniform, spherical and rigid are used.

(*vii*) In practice agarose, cellulose and polyacrylamide gels are used as the insoluble matrix.

**Selection of Ligands :**

(*i*) In this technique ligand is selected after careful consideration. Substrate analogues, enzyme cofactors, receptor antagonist, antigens epitope synthetic antagonists are usually members.

(*ii*) The ligand should interact strongly with the desired macromolecules. It must posseses high specificity and strong affinities for the substance or desired macromolecules.

(*iv*) The ligand should posseses functional groups that can be modified to form covalent linkage with the supporting matrix.

(*iv*) List of Group Specific Ligands :

| *Macromoleculed/cell* | *Ligand* |
|---|---|
| 1. Coagulating factor | Heparin |
| 2. Interferon | Antibody |
| 3. Thrombin | Benzamidine |
| 4. Ribosomal RNA | Lysine |
| 5. Poly (A) messenger RNA | Poly (U) or Poly dT. |
| 6. Chymotrypsin | Tryptophan |
| 7. Glyco proteins | Concanavalin A |

**(c) Ligand attachment**

Covalent coupling of ligand to the supporting matrix involves:

(*i*) Activation of matrix functional groups.

(*ii*) Covalent attachment of the ligand to these activated groups.

(*iii*) The most common method of activation of polysaccharide supports (agarose) involves the treatment with cyanogen bromide (CNBr) at alkaline pH (pH = 11.0)

**Procedure :**

(*i*) Gel beads are swollen before loading into the column.

(*ii*) The buffer which encourages adsorption of the desired molecules on the gel surface is used.

(*iii*) The selected buffer must be supplemented with any cofactors (eg. metal ion) required for ligand-macromolecule interaction.

(*iv*) The buffer should also posseses high ionic strength which minimize non-specific polyelectrolyte adsorption on the charged groups in the ligand.

(*v*) The mobile phase consists of a buffered solvent system in which the sample is dissolved.

(*vi*) The sample is applied at the top of the column and the buffer flow started.

(*vii*) As the solution is passed down the ligand bound gel column, only the specific solute molecules bind to the gel bound ligand molecules to be immobilized on the gel beads and consequently retained in the chromatographic column.

(*viii*) Once macromolecule is bound, the column is eluted with more buffer to remove non specifically bound unwanted molecules.

(*ix*) The purified, bound component may now be eluted (dissociated) by taking recourse to either specific or non specific elution.

(*x*) Non specific elution is carried out by changing either the pH or ionic strength of the buffer. It causes destabilization or dissolution of ligand-macromolecule/solute non covalent bond to make the solute molecules emerge from the column in the effluent eluant.

(*xi*) Specific or affinity elution is carried out by :

(*i*) addition of compounds for which the macromolecule has more affinity.

(*ii*) by addition of compounds for which the ligand has more affinity than it has for the desired macromolecules.

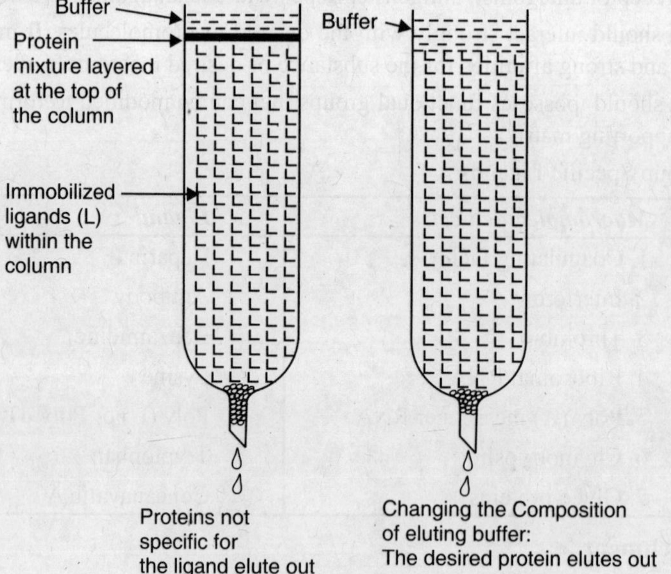


**Fig. 10.6** Principle of affinity chromatography.

**Applications :**

(*i*) A wide range of enzymes and other proteins, immunoglobulins, membrane receptors have been purified by affinity chromatography.

(*ii*) With the aid of this technique messenger RNA is routinely isolated by selective hybridization on poly (U)- Sepharose 4B and viral RNA on the corresponding poly (A).

(*iii*) Immobilised single-standard DNA can be used to isolate complementary RNA and DNA.

(*iv*) It is used for the separation of a mixture of cells into homogeneous population.

(*v*) Metal chelate affinity chromatography is another logical extension of the basic technique. It enables proteins with similar molecular weights and isoelectric points to be separated on the basis of their differential binding to metal ions which have been immobilised by chelation.

- **Gas Chromatography**

(*i*) It is a technique for the separation of volatile components of mixtures by differential migration through a column containing liquid or solid stationary phase. Solutes are transported through the column by a gaseous mobile phase and are detected as they are eluted.

(*ii*) In this case, the substance to be analyzed between mobile phase & a stationary phase.

**Modes (types) of Gas Chromatography :**

**(*a*) Gas-liquid Chromatography (GLC)**

(*i*) It is a form of column chromatography where the *stationary phase* consist of non volatile liquid material coated on inert support.

(*ii*) It is also known as liquid phase.

(*iii*) The mobile phase is an inert carrier gas usualy nitrogen , helium & argon.

**(*b*) Gas- solid Chromatography (GSC) :**

(*i*) Here the liquid phase is absent.

(*ii*) The solid phase which is coated on to the interior of the column is not inert.

(*iii*) It interacts with the sample components carried by the gas by exerting adsorption forces.

**Essential Component of Gas Chromatography :**

**(*a*) Mobile phase :**

(*i*) It is an inert carrier gas, generally nitrogen or argon, helium stored in a gas tank.

(*ii*) It passes through the pressure regulator into the sample injection chamber.

**(b) Sample infection :**

Gaseous , liquid and solid samples are introduced into the flowing mobile phase at the top of the column through sample injection chamber using micro syringe , valve.

**(c) Column and stationary phase :**

(*i*) Two types of columns viz open tubular and packed are commonly used.

(*ii*) The open tubular columns are long narrow, capillary columns or tubes with stationary phase coated on to the inside wall.

(*iii*) Packed columns are stainless steel, copper or glass tube, shorter, larger in diameter & packed with a particulate stationary phase.

(*iv*) Stationary phase are high boiling liquids, waxes.

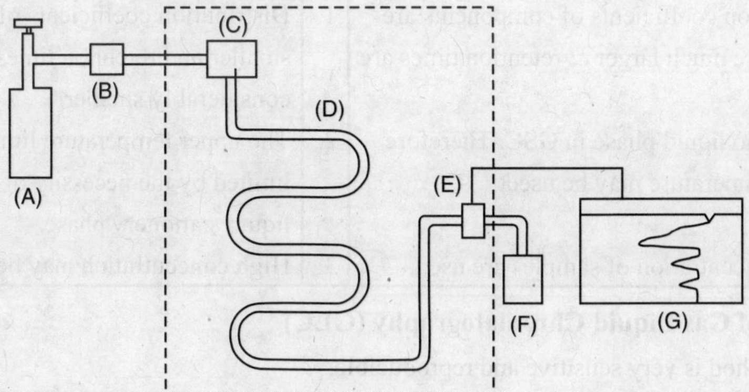


**Fig. 10.7** Schematic of a gas chromatograph. (A) carrier gas tank, (B) pressure regulator, (C) Sample injection chamber, (D) Column, (E) Detector, (F) Fraction collector, and (G) Recorder

**(d) Temperature control :**

(*i*) The column enclosed in a thermostatically- controlled oven that is maintained at a steady temperature.

**(e) Solute detection :**

(*i*) Solutes are detected in the mobile phase.

(*ii*) The detector generates an electrical signal that can be completed & presented in the form of a chromatogram of solute.

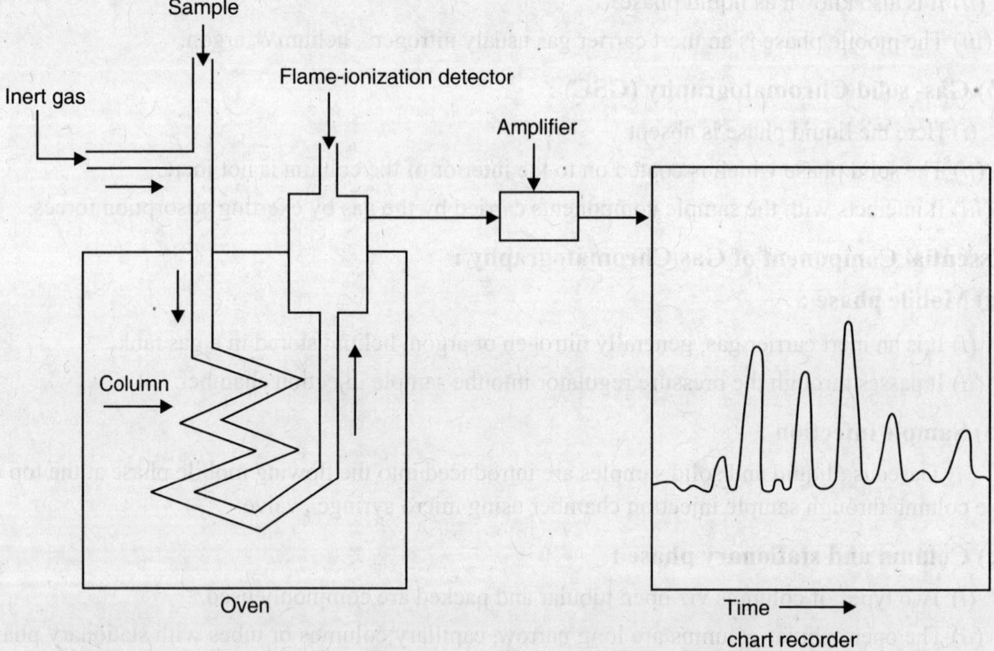


**Fig. 10.8** Diagrammatic representation of Gas-liquid Chromatography

**Difference between GSC and GLC.**

| | **Gas Solid Chromatography (GSC)** | | **Gas Liquid Chromatography (GLC)** |
|---|---|---|---|
| 1. | Distribution coefficients of components are of GSC are much larger & retention times are long. | 1. | Distribution coefficients of compartment smaller and retention times are considerably smaller. |
| 2. | There is no liquid phase in GSC. Therefore higher temperature may be used. | 2. | The upper temperature limit in GLC is limited by the necessity of retaining the liquid stationery phase. |
| 3. | Small concentration of samples are used. | 3. | High concentration may be used. |

**Application of Gas Liquid Chromatography (GLC)**

(*i*) This method is very sensitive and reproducible.

(*ii*) It is widely used in reporting and estimating volatile compounds of fatty acids, steroids and other lipids..

(*iii*) It is used almost as an analytical tool and has the advantage to deal with small amount of sample.

## • Electrophoresis

**Principle :**

**Migration of ions or charged particles or molecules or solutes of a solution towards the opposite electrical poles in an applied electric field at a definite *pH* is called *electrophoresis.***

It may be stated that molecules with different charges or different charge/mass ratio would migrate at different rates in an electric field. The separation of molecules by utilizing these differences in rates of migration is known as electrophoresis.

***Electrophoretic mobilities:***

The rate of electrophoretic mobilities (μ) is expressed in terms of distance per unit time in a unit electric field.

$$\text{Electrophoretic mobility } (\mu) = \frac{\text{Velocity of migration (V)}}{\text{Electric field strength (E)}}$$

(*i*) Consider a spherical molecule of net charge $q$ is placed in an electric field ($E$). The force ($F$) will act upon the particle will depend upon (*i*) the net charge density of the molecule and (*ii*) the strength of the field in which it is placed.

$$F = qE \quad \text{or} \quad F = \frac{\Delta E}{d} q$$

$\frac{\Delta E}{d}$ = Field strength applied

$\Delta E$ = Potential difference between the two electrodes

$d$ = distance between them

(*ii*) Since the particle has been suspended in a solution (fluid), this force ($F$) will accelerate the particle in a solution till a steady state is reached when the frictional force ($f$) is equal and opposite to the applied force.

If $v$ is the steady state velocitiy then $fv = qE$ which is frictional force

$$\text{or} \quad fv = q\frac{\Delta E}{d} \quad \text{or} \quad v = \frac{q\Delta E}{df}$$

(*iii*) According to Stoke's equation, the extent of friction depend upon (*a*) the size and shape of the molecule (*b*) on the viscosity of the medium through which the molecule migrate.

Thus $f = 6\,\pi\, r\, n\, v$

$f$ = Friction exerted on the spherical molecule

$r$ = radius of the molecule

$\eta$ = viscosity of the solution

$v$ = velocity of the migrating molecule.

(*iv*) For a spherical molecule z is the atomic number and e is charge on electron.

we know

$$v = \frac{qE}{f} \quad \text{or} \quad v = \frac{ZeE}{f} \qquad \because q = ze$$

(*v*) The frictional force ($f$) will oppose the accelerating force generated by the electric field. Equating the force of accelaration with stoke's equation we get

$$f = 6\pi\, r\, \eta\, v$$

We know that the force ($F$) act upon the charged particle is $F = qE$ or $F = \frac{\Delta E}{d} q$

These two forces *viz f* and *F* are equal and opposite *i.e* $F = f$

Equating the force with stokes equation we can get $\frac{\Delta E}{d} q = 6\pi r \eta v$

or $v = \frac{\Delta E}{d\, 6\pi r\eta u} q = \frac{\Delta E q}{6\pi r\eta d}$

Thus velocity of the molecule v is proportional to (*a*) the field strength $\left(\frac{\Delta E}{d}\right)$.

(*b*) Charge on the molecule ($q$) but is inversely proportional to

(*i*) Particle size ($r$), viscosities of the solution ($\eta$)

- **Factors affecting electrophoretic mobility:**

**(A) Sample :**

(*i*) Charge/mass ratio of the sample dictates its electrophoretic mobility. The mass consists of not only the size (molecular weight) but also the shape of the molecule.

(*ii*) *Charge* – The higher the charge, greater is the electrophoretic mobility.

(*iii*) *Size* – Larger particles have a smaller electrophoretic mobility.

(*iv*) Shape – Rounded contours elicit lesser frictional and electrostatic retardation compared to sharp contour. Globular protein will migrate faster than the fibrous protein.

**(B) The electric field**

(*i*) The force acting upon an ion of charge $q$ is $\frac{\Delta E}{d} q$.

(*ii*) The rate of migration under unit potential gradient is referred to as mobility of the ion.

(*iii*) An increase in potential gradient increase the rate of migration.

**(C) The Medium**

(*i*) An inert supporting is selected for electrophoresis.

(*ii*) The inert medium can exert adsorption and molecular sieving effects on the particle.

**(D) The Buffer**

**(a) Composition**

(*i*) The commonly used buffers are formate, EDTA, citrate, Tris etc.

(*ii*) The selection of buffer depends upon the type of sample being electrophoresced.

(*iii*) The buffer can affect electrohoretic mobility if it is bound to components of the sample being separated.

**(b) Ionic strength**

(*i*) Increased ionic strength of the buffer means larger share of the current is being carried by the buffer ions results slow migration of the sample.

(*ii*) Decreased ionic strength of the buffer mean a larger share of the current is being carried by the sample ions leads to faster separation.

(*iii*) The chosen ionic strength of the buffer is usually between 0.05 – 0.1M.

**(c) pH**

(*i*) It determines the degree of ionization of organic compounds. It can also affect the rate of migration of these compounds.

(*ii*) Ionization of compound increases alongwith the increases of pH and decreases with the decrease of ionization of compounds.

- **Types of Electrophoresis:**

1. Free electrophoresis or moving electrophoresis.
2. Zone electrophoresis.
3. Disc electrophoresis.
4. Electrofocussing.
5. Electrophoresis convection or ISO electrofocussing.
6. Immuno electrophoresis.

- **Gel Electrophoresis :**

In this type of electrophoresis charged particles may migrate along the gel block or column under the influence of an electric field.

**Types of Gel :**

Various types of gels are available for gel electrophoresis.

**(*a*) Starch gel**

(*i*) It acts as stabilizing medium for zone electrophoresis.

(*ii*) Potato starch is hydrolyzed in acidified acetone at 37°C with taking care in temperature control and the timing of hydrolysis.

(*iii*) The suspension is then neutralized with sodium acetate and washed with large amount of distilled water and dried with acetone.

(*iv*) This hydrolyzed starch when heated and cooled in an appropriate buffer acts as a gel.

(*v*) It has high resolving power.

**(*b*) Agarose gel :**

(*i*) It is cheap, easily available, transparent medium and suitable for photometric scanning.

(*ii*) It is one of the components of agar which is a polysaccharides obtained from sea weeds.

(*iii*) Agarose consists of repeated basic units of agarobiose.

(*iv*) Agar consists of two galactose based polymers agarose and agaropectectin.

(*v*) A solution of agar at 80°c is mixed with an equal volume of 40% polyethylene glycol which precipitates the agarose.

(*vi*) It is then collected, washed with distilled water and dried with acetone.

(*vii*) Dry agarose is first suspended in aqueous buffer and boiled at 40°C to form clear solution. Then it is poured and allowed to come down at room temperature so that to form a rigid gel.

**(c) Polyacrylamide gel :**

(*i*) The most commonly used components to synthesize the matrix (Polyocrylamide gels) are acrylamide monomer. N, $N^1$ methy bisacrylamide, ammonium pursulphate and tetramethylene diamine (TEMFD).

(*ii*) Bis acrylamide is used as a cross linking agent, each molecule of which is a pair of acrylamide molecules linked together by methylene group.

(*iii*) Ammonium persulphate when dissolved in water generates free radicles which can activate acrylamide monomers inducing them to react with other acrylamide monomer forming long chain process.

(*iv*) These chain become cross linked in presence of N N methylene bisacrylamide. TEMED act as catalyst of gel formation.

(*v*) The pore size of the gel is determined by the amount of acrylamide used per unit volume of the reaction medium and the degree of cross linkage.

(*vi*) *Suitable characters* : (*i*) Low adsorption capacity
(*ii*) Lack of electroosmosis (*iii*) Suitabilities for in-situ histochemical and qualitative analysis.

**(*d*) Agarose-acrylamide gel :**

(*i*) It is used to separate high molecular weight nucleic acid (200 Kd or more)
(*ii*) Agarose alone has no molecular seiving action but on mix it provides physical support and acrylamide provides molecular seiving action.
(*iii*) Agarose-acrylamide gel centain 0.5% agarose and 0.5% acrylamide.

**Electrophoretic Mobility of Gel :**

The effect of gel concentration upon a macro molecule's mobility are closely related with pore size i.e molecular sieving action.

$$\text{Log } E = \text{Log } E^1 - KrG$$

E = Electrophoretic mobility, $E^1$= Mobility in sucrose solution. Kr = Retardation coefficient.

G = Total gel concentration.

Retardation coefficient has been described as follows:

$$Kr = C\,(R + r)$$

C = Constant

R = Mean radius of the macromolecule

$r$ = radius of the gel fibers.

**Solubilizers**

(*i*) Several proteins (biologically important) contains more than one polypeptide chain (oligometric protein).
(*ii*) The whole structure is stabilized by hydrogen bonding disulphide bridges or by hydrophobic association.
(*iii*) These proteins migrate as a single band during gel electrophoresis.
(*iv*) A class of substances collectively known as solubilizers which destabilize the protein structure by separating their (protein/subunits).
(*v*) For example core RNA polymerase (2α, β, β′ w) has subunits like α,β and β′ but migrate as a single species in zone electrophoresis. If it is treated with SDS (sodium dodecyl sulphate) a solubilizing agent and electrohoresis led on a gel, the three subunits (αββ′) i.e polypeptide migrate as three different bands.

| Solubilizers | Characteristics |
|---|---|
| 1. Urea | It disrupts hydrogen bonds at concentration from 3-12M. |
| 2. Sodium dodecyl sulphate (SDS) | (*i*) It is an anionic detergent. (*ii*) It disrupts macromolecules whose structure has been stabilized by hydrophobic association. |
| 3. β mercaptoethanol | It disrupts proteins whose polypeptide chains are linked together by disulphide bridges. |

**Procedure :**

(*i*) The electrophoresis apparatus consists of (*i*) a buffer reservoir and (*ii*) D.C. power.
(*ii*) The upper and lower buffer reservoir are connected by the gel.
(*iii*) The platinum electrodes are placed in each reservoir and are connected to terminals extending from top to the unit.

(*iv*) The whole assembly may be covered by a perspex shield.

(*v*) Microgram quantity of sample is dissolved in a small volume of dense sucrose solution or glycerol to prevent its ready mixing with the buffer solution.

(*vi*) The sample solution is placed over the top of the gel column of 'tracking dye' (usually bromophenol blue) is often mixed with the sample.

(*vii*) The extent of migration of the dye gives an index of electrophoretic process. The dye migrates faster than all macromolecules. If the electronphoresis is stopped before or just as the dye comes out of the bottom of the gel which indicates the presence of macromolecules within gel.

(*viii*) The pH is usually fixed out of which provides a net negative charge to most macromolecules. The macromolecular anions of the sample consequently migrate towards the anode which is placed in the lower buffer reservoir.

(*ix*) When the power is switched on, the amount of bubbles generated in the reservoir containing the cathode is more than that of anode. It indicates the flow of current from cathode to anode.

(*x*) Here all the particles have to pass through the gel pore, and consequently larger particles are retarded more than smaller one.

(*xi*) So the rate of migration depends on the charge: mass ratios of the migrating particles.

- **Modes of gel electrophoresis**

**(a) Column electrophoresis**

(*i*) Here the gel is set or polymerized in a column.

(*ii*) Column is then fitted between the upper and lower buffer reservoir.

**(b) Slab gel electrophoresis**

(*i*) Here the gel is set or polymerized into a slab between two glass plates.

(*ii*) The thickness of the slab of the gel can be adjusted by placing spacers of various thickness between the two glass plates.

(*iii*) Once the gel has been cast, a plastic comb is pressed into the gel column and is allowed to remain there till gel is polymerized. Thereafter comb is removed.

(*iv*) This action leaves the wells cut on the gel. Many samples are loaded on the wells simultaneously.

**Detection : Recovery and Estimation**

(*i*) Gels form column are removed by forcing water from a hypodermic syringe around the walls of the column and allows the gels to be extracted under gentle pressure.

(*ii*) Slab gels are removed by introducing a thin metal plate between the two gel plates and coaxing the plates apart.

(*iii*) Before staining the gels may be immersed in a fixative (7% acetic acid) to guard against diffusion of separated components.

(*iv*) *Staining and estimation may be done in the following alternative way:*

(*a*) Staining the proteins in the bands with coomissie blue, silver stain or the fluorescent dye fluorescamine.

(*b*) Elution of the bands with buffer solutions of varying pHs and ionic strength and analysis of separate eluant fractions.

(*c*) Scintillation spectrometry of the bands of radiolabelled components.

(*d*) uv absorbancy of the bands or autoradiography of the gel block for radio active components of its bands.

**Applications of Gel Electrophoresis :**

**(a) Determination of DNA sequences**

(*i*) The two methods of DNA sequencing viz Maxam and Gilbert technique and dideoxy nucleotide technique of sanger are both dependent on high resolution polyacrylamide gel electrophoresis.

(*ii*) They both depend on generating specific sets of radio labelled fragments, each terminating at a particular base.

**(b) Southern and Northern blotting**

(*i*) If the sequence of a portion of desired DNA is known, one can prepare synthetically prepare a complementary oligosucleotide, radiolabelled it and make it to react with the sample which has been separated by agarose gel electrophoresis and transformed to nitrocellulose paper.

(*ii*) DNA-DNA hybridization is known as southern blotting and DNA RNA hybridization is known as northern blotting. Gel electrophoresis is the prime condition for this.

**(c) Restriction mapping of DNA :**

(*i*) Here DNA should be fragmented by restriction endonuclease.

(*ii*) Separation of this fragments are made by gel electrophoresis.

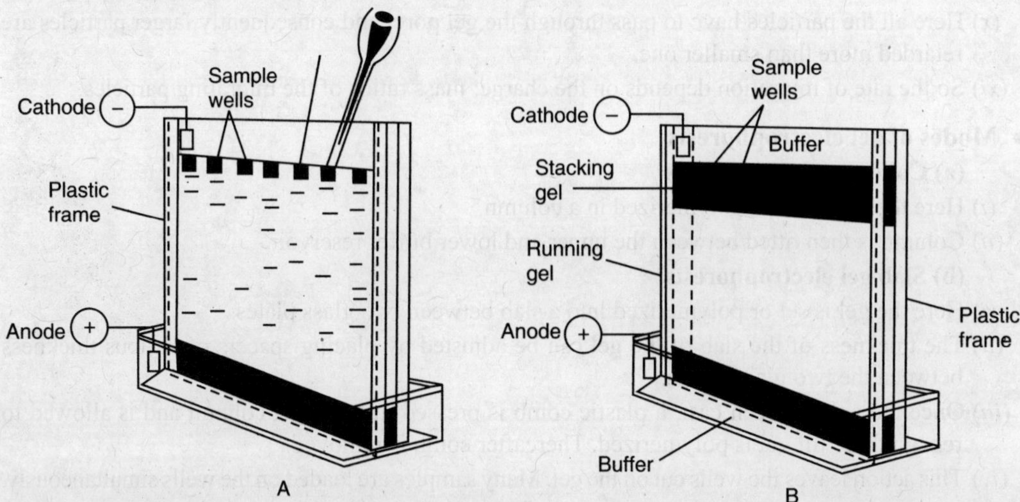


**Fig 10.9** The apparatus set for vertical slab electrophoresis; the sample is dropped into the slots cut (top of slab) and as many samples as the number of slots is in the slab can be simultaneously. B. Disc electrophoresis unit with a stacking gel placed at top of running gel and that makes it different from slab unit.

- **Agarose Gel Electrophoresis:**

**It is a sample method of separating purifying and identifying DNA fragments. Migration of charged molecules *i.e.* ions in an electric field and a definite pH through gel meshwork (agarose) where negatively charged molecules (anion (–)) move towards positive electrode (anode (+)) and positively charged molecules (cataions (+)) move towards negative electrode (cathode (–)) is called gel electrophoresis.**

- **Steps:**

(*i*) The gel of gel electrophoresis consists of agarose, a polysaccharide extracted from seaweed that behave like a gelatin. It forms get with pores of 100 – 300 nm diameter.

(*ii*) Agarose is a powder that dissolve in water or buffer only when heated. After cools, the agarose hardens.

(*iii*) A frame of adhesive tape as surrounding walls is made and placed on a glass or plastic plate. The agarose solution is poured within the frame to form an approximately 3 mm thick layer.

(*iv*) Inserting a plastic comb at one end of the slab before it hardens makes small wells or holes at one end. After the gel solidifies, the comb is removed, leaving small wells or holes at one end.

(*v*) The agarose slab is immersed in a buffer filled tank that has a positive electrode at one end and a negative electrode at the other.

(*vi*) DNA sample is dissolved in loading buffer solution (containing sucrose and glycerol) greater than that of electrophoresis buffer, to increase the density of the solution. So that DNA sample settle to the bottom of well or holes.

(*vii*) DNA samples are loaded into the wells or holes. The bromophenol like dye can be used as monitor. When electric field is applied, the DNA migrates through the gel. Polyacrylamide agarose gel act as seive with small holes between tangled chains of agarose. DNA must migrate through these gaps.

(*viii*) The phosphate backbone of DNA is negatively charged. So that it moves towards the positive electrode. Agarose separate the DNA by size because larger pieces of DNA are slowed down more by agarose. The rate of migration of particular DNA fragment is inversely proportional to the ratio of its molecular weight.

(*ix*) Staining makes the separated DNA visible as bonds in the gel. Most widely used stain is *ethidium bromide*, a fluorescent dye. Therefore, to visualize the DNA, the agarose gel is removed from the tank and immersed into the solution of *ethidium bromide*.

[Ethidium bromide can detect DNA as low as 10 mg/ml]

(*x*) The electrophoretic gel is gently rinsed in as 0.1 μg/ml solution of ethidium brodmide. Now the gel is exposed to uv light of 300 nm, and if (DNA band) fluoresces bright orange.

- **Polyacrylamide Gel Electrophoresis (PAGE)**

(*i*) It is most widely used gel electrophoresis for separation of nucleic acids and proteins. Polyacrylamide gels have smaller pores sizes than agarose gel and therefore it allows precise separation of DNA molecules of 100 – 1000 bp. Therefore it is used in fine scale DNA separation, DNA sequencing etc.

(*ii*) Pure size of polyacrylamide gel can be changed by varying the concentration of acrylamide and bisacrylamide monomers in a fixed volume of gelation solution.

(*iii*) Polyacrylamide gel consists of a chain of acrylamide monomers cross links with bis (N–N′ methylene bisacrylamide unit). Bisacrylamide brings about gelation and cross linking through polymerization. This result the complex network of acrylamide.

- **SDS and PAGE (Sodium Dodecyl Sulphate and Polyacrylamide Gel Electrophoresis):**

(*i*) Unlike DNA, most proteins do not have a net negative charge. Therefore protein samples are treated by boiling with detergent sodium dodecyl sulphate (SDS).

(*ii*) It denatures the sub units and unfolds the polypeptide chains and the entire strands of amino acids is coated with negatively charged SDS.

(*iii*) The amount of SDS and therefore the amount of charge, correlates with length of the protein.

(*iv*) After bring loaded into a sample well of a polyacrylamide gel, the proteins migrate away from the cathode and toward the positive anode.

(*v*) The seiving action of the acrylamide gel allows the smaller protein to move faster than the larger.

(*vi*) Finally the separation of proteins are visualized using either *Coomasic Blue* which binds tightly to all proteins.

- **Pulse Field Gel Electrophoresis**

**It is a electrophoresis technique used for separation of very large fragment of DNA molecule in which pulses of current in the forward direction and shorter pulses in the opposite direction or even side ways direction developed from a hexagonal array of electrodes.**

## 1. Moving Boundary Electrophoresis :

This method of electrophoresis was first developed by A. Tiselius of Sweden in the 1930s.

**Description of the apparatus :**

(*i*) U-tube observation cell.

(*ii*) Two electrodes (anode and cathode)

(*iii*) Thermosat

**Procedure :**

(*i*) The sample or solution under investigation is dissolved in a buffer and poured into the bottom of the U-tube.

(*ii*) Additional quantity of pure buffer solutions are then carefully layered on top to obtain sharp boundaries.

(*iii*) Electrodes are inserted into the side arms to prevent products formed from electrolysis falling into the boundary regions.

(*iv*) The entire apparatus is then immersed in a thermostat.

(*v*) The power is switched on generating an electric field between the electrodes.

(*vi*) Normally in a complex sample containing many macro molecules, some will bear a net negative charge and will move towards the anode, while other bearing net positive charge will move towards cathode. This situation of movement in mutually opposite direction will not be good for satisfactory resolution.

(*vii*) Therefore the pH of the buffer is so selected that all the macromolecules bear a net negative charge. The negative charge macromolecule move towards the anode and they migrate from the macromolecules solution to the pure buffer or into the zone of macromolecule free buffer and form boundary or front.

At a pH above isoelectric point macromolecules (protein) negatively charged & boundaries moves toward anode. At a pH below isoelectric point, the net charge of macromolecule is positive and boundaries move toward cathode. At the isoelectric point, the net charge of molecules is zero & boundaries remain stationary.

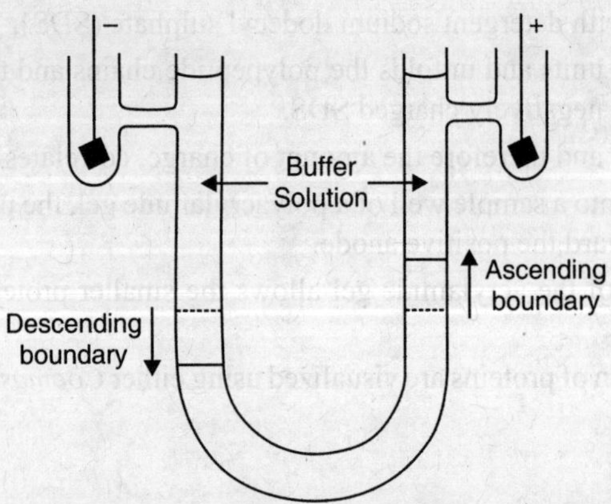


**Fig. 10.10** Moving-boundary electrophoresis apparatus.

(*viii*) As a result, there is sharp change in the refractive index of the solution at this boundary. The refractory index of the macromolecule solution is different from that of pure buffer.

(*ix*) The migration of the boundary is observed by measurement of changes in the refractive index in the electrophoresis cell.

(*x*) The refractive index is measured by direct optical observation usually by "schlieren method."

The electrophoretic patterns that show the direction and relative rate of migration of the major molecules of the sample.

**Comment :**

(*i*) Moving boundary method is very popular for quantitative analysis of macromolecules.

(*ii*) But in recent years it has been superceded by techniques collectively known as *zone electrophoresis.*

### 2. Zone Electrophoresis

**It is known as electrophoresis in stabilized media.**

**It is a kind of separation technique of a sample in an electric media by employing a stabilized media such as paper strip or gel block. It (technique) allows ready separation of components of different mobility into zones.**

**Advantages**

(*i*) Here smaller samples are used

(*ii*) It is a simpler process

(*iii*) It has much more resolving power.

### (A) Paper Electrophoresis

Here charged particles may migrate along a hydrated, porus, inert paper.

**Equipments**

(a) *Filter paper :*

(*i*) It act as a stabilizing medium and very popular for the study of plasma membrane.

(*ii*) Paper of good quality should contain at least 95% cellulose & slight absorption capacity.

(b) *Apparatus :*

(*i*) It consists of power pack and an electrophoretic cell.

(*ii*) The power pack provides a stabilized direct current and controls both voltage and current output.

(*iii*) The electrophoretic cell contains the electrodes buffer reservoirs, a support for paper on a transparent insulating cover.

(*iv*) The electrodes are usually made of platinum.

(*v*) The two buffer reservoirs are usually each partitioned into two interconnected sections, one containing the electrodes and the other in contact with the supporting medium.

(*vi*) Separate compartments are needed so that any change in pH occurs at the electrodes does not affect the buffer in contact with the supporting medium.

(*vii*) The supporting medium is saturated with the buffer prior to the start of electrophoresis process.

(*viii*) Two arrangements of filter strips viz vertical and horizontal are equally viable and are commonly used.

**Process:**

(*i*) The test solution (sample) is placed as a smallspot or thin band or transversely across the central part of a long filter paper strip or a cellulose acetate membrane.

(*ii*) The filter paper is saturated with a buffer solution for maintaining desired pH.

(*iii*) The two ends of the filter paper is dipped into the buffer solution in the anode and cathode chambers.

(*iv*) These chambers are filled with electrodes. When the current is switched on, the charged particles migrates as a separate band along the paper strip towards the opposite electrodes according to the size and charges of the particle.

**Precaution :**

(*i*) Electrophresis process should be held in a cold room. The process usually does not take longer than two hours.

(*ii*) The power is switched off after the electrophoretic run and before the paper is removed.

**Detection and Assay:**

(*i*) After removal, the paper is dried in vacuum oven at 110°c.

(*ii*) If the compounds are not thermostable, the paper is allowed to hang and air dried.

(*iii*) The bands, located on the electrophoregram and the sample components are estimated

(*a*) Spray-painting the bands with suitable reagents and measuring the stain density with scanning densitometer.

(*b*) UV fluorometry of fluorescent components. Example DNA & RNA.

(*c*) *UV absorbancy* of the components like protein peptides and nucleic acids (range-260-280 *nm*).

(*d*) *Scintillation spectrometry* of radioisotopically labelled components.

**Application :**

It can be used in separation oil sugar, dyestuffs, purine and pyrimidiues, amino acids, proteins etc.

**3. Disc (Discontinuous) Gel Electrophoresis :**

This is another modification of conventional zone electrophoresis which allows the sample to enter the gel as a sharp band, thereby helping further resolution.

(*i*) Here the macromolecule (protein) mixture is subjected to an electric field in a retarding gel support which is separated into two parts having different porosity and buffered at different pH.

(*ii*) The macromolecular mixture migrates from the more porous gel into less porous gel accompanied by a change in pH.

(*iii*) As a result each macromolecular species or each protein becomes concentrated as a very thin, sharp band producing much higher resolution than can be acheived in a continuous buffer.

(*iv*) The gel that is preferred for disc electrophoresis is polyacrylamide. There are two different porosity gels viz stacking gel (high porosity) and running or separating (low porosity) are used.

**(a) Stacking gel**

(*i*) The upper, stacking gel is prepared by using about 2-3% acrylamide and is highly porous and devoid of any molecular sieving action.

(*ii*) The buffer used in this gel is usually amino mostly Tris.

(*iii*) The pH is adjusted with hydrochloric acid (HCl) and is about 2 pH units.

**(b) Separating gel**

(*i*) The lower, separating gel is prepared by using about 5-10% acrylamide and pores are numerous and of a smaller diameter imparting molecular sieving property to this gel.

(*ii*) The buffer used in this gel is usually amine such as tris.

(*iii*) The pH is adjusted (8.3) using hydrochloric acid.

(*iv*) In this gel macromolecules are subsequently separated.

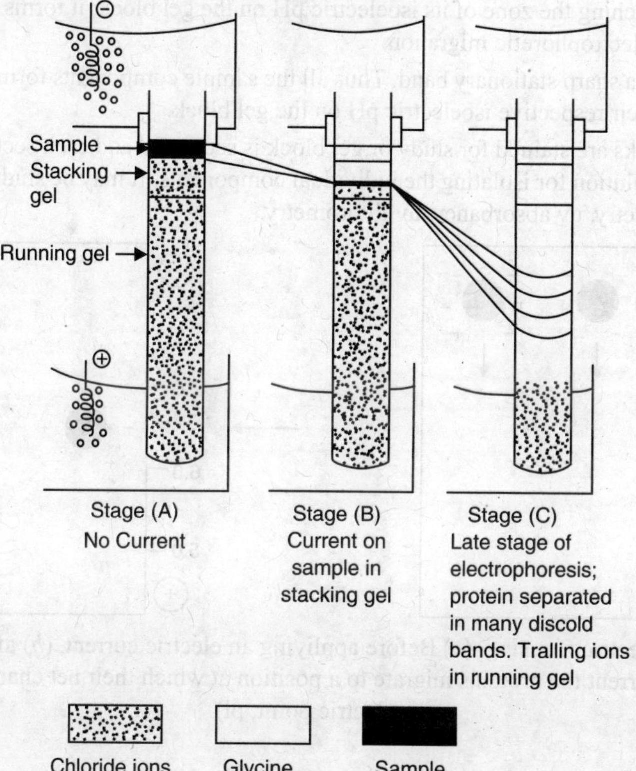


**Fig. 10.11**: Schematic diagram of disc gel electrophoresis showing migration of ions at various stages.

## 4. Electrofocussing or Isoelectric Focussing Electrophoresis :

**It is a kind of electrophresis through a polyacrylamide gel block with a length wise maintained pH gradient alongwith voltage gradient so that each migrant sample component stops migrating and settles into a separate immobile or stationary band on reaching the zone of its isoelectric pH. This technique was discovered by H. Svensson.**

### Pre-requisite :

(*i*) The separation is carried out in a liquid field column or more usually in a horizontal polyacrylamide or agarose gel which has large pores.

(*ii*) The pH gradient is formed by adding to the gel a mixture of low molecular weight polyampholytes which bear both basic amino and acidic carboxy group or both positive and negative charges and consequently arrange themselves in the cathode to anode direction along the gel block in order of their increasing acidic or decreasing basic properties.

### Procedure :

(*i*) Electrophoresis is carried out in a vertical column or a horizontal slab of cross linked polyacrylamide gel.

(*ii*) Stable pH gradient is maintained along the length of the gel block.

(*iii*) Microgram amount of sample is applied on gel block.

(*iv*) The ionized components migrate towards the electrodes with opposite polarity but because of the pH gradient each migrant component progressively looses its charges as well as its rate of migration.

(*v*) Finally on reaching the zone of its isoelectric pH on the gel block, it forms zwitterions which stop further electrophoretic migration.

(*vi*) Now it forms a sharp stationary band. Thus all the sample components form separated band at the sites of their respective isoelectric pH on the gel block.

(*vii*) Then gel blocks are stained for study or gel block is sectioned and each section may be eluted with buffer solution for isolating the individual components. It may be studied using scintillation spectrometry, uv absorbancy, uv fluorometry.

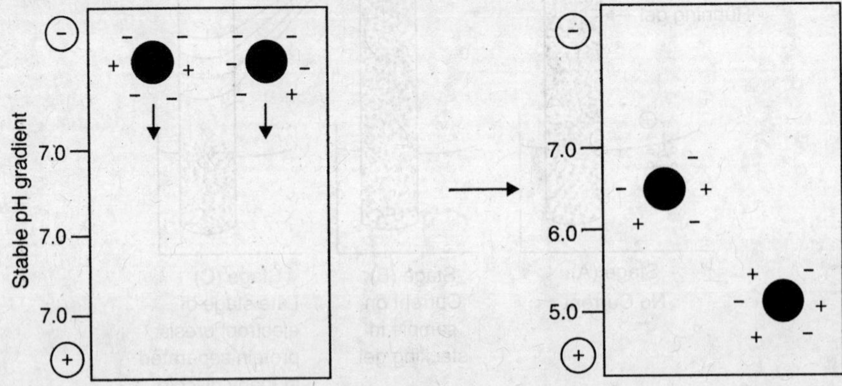


**Fig. 10.12** Isoelectric focusing. (*a*) Before appliying an electric current. (*b*) after applying an electric current the proteins migrate to a position at which their net charge is zero (isoelectric point, pl).

**Application :**

(*i*) It has been widely used for separation and identification of serum proteins.

(*ii*) It is being used by the food and agricultural industries, forensic and human genetics laboratories.

## • IMMUNO ASSAY:

**(*i*) It is a biochemical test that measures the concentration of a substance in a biological liquid (serum, urine) by using the reaction of an antibody to its antigen.**

**(*ii*) It takes advantage of specific binding of an antibody to its antigen.**

### Enzyme Linked Immunosorbant Assay

### (ELISA)

ELISA is a recently developed sensitive method for quantitative measuring of antigen-antibody reaction by using enzyme linked antibody and a colourless substrate that develops a coloured reaction product.

(*i*) This test is done in solid phase

(*ii*) The test can be performed as a direct or indirect assay depending on whether the enzyme is coujugated to primary and secondary antibody.

**Requirements :**

(*i*) A number of enzyme have been employed for ELISA.

(*ii*) Alkaline phosphatase, horse radish per-oxidase (HRP) p-nitro phenylphosphatase.

**Procedure :**

**(a) Indirect ELISA :**

(*i*) It is used to measure antibody, known antigen is coated on the plastic linens of the wells of microfilter plate which is made up of polystyrene latex.

(*ii*) To test for the presence of antibodies against this antigen in the patient, his blood serum is added to the wells.

(*iii*) If the patient's serum contains antibody specific to antigen, the antibody will bind to the adsorbed antigen otherwise not.

(*iv*) After incubation the wells are washed and the enzyme, labelled with anti-human gamma globulin (anti Hgg) is added to the wells.

(*v*) Anti Hgg can react with antigen-antibody complex. The mixture of wells is washed to remove the excess of unbound labelled anti Hgg.

(*vi*) Finally the correct substrate for the enzyme is added which is hydrolysed by the enzyme and developes a colour.

(*vii*) Varying concentrations of antibody in serum shows changes in the intensity of colour.

This method is useful in detection of antibodies to HIV, Salmonella, Yersinia.

**(b) Sandwich ELISA :**

(*i*) Antigen can be determined or quantitated by a sandwich ELISA.

(*ii*) In this technique the antibody (rather than antigen) is immobilized on a microfilter well.

(*iii*) A sample containing antigen is added and allowed to react with the bound antibody.

(*iv*) After the well is washed, a second enzyme-linked antibody (alkaline phosphatase tagged antibody) specific for a different epitope on the antigen is added and allowed to react with the bound antigen.

(*v*) It is again incubated for a few second, the enzyme labelled antibody reacts with the antigen-antibody complex already formed in the wells and results in the development of a 'sandwich'.

(*vi*) The mixture is well washed and any free second antibody is removed by washing, substrate chromogenic (p-nitrophenyl phosphate) is added and the coloured reaction (yellow) product is measured.

(*vii*) The change in colour is measured visually or spectophotometrically. Change in colour shows the presence of desired antigen in the sample.

This technique is useful in detection of toxins of *V.cholerae, E.coli.*

**Radio Immuno Assay (RIA)**

A highly sensitive technique for measuring antigen or antibody that involves competitive binding of radio labelled antigen or antibody.

(*i*) The technique was first developed by two endocrinologists, **S.A. Berson & Rasalyn Yellow** in 1960 to determine levels of insulin and antiinsulin complexes in diabetics. Berson died and yellow was awarded Nobel prize in 1977.

(*ii*) This is highly sensitive technique and can measure even the less concentration (*i.e.* 0.001 ug/ml) of antigen or antibody.

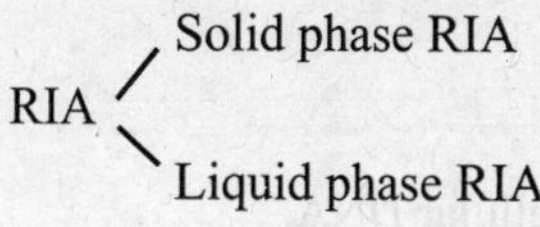

**Solid phase RIA :**

(*i*) In solid phase RIA antibody is linked to an insoluble support (e.g. agarose beads).

(*ii*) The insoluble complex is then mixed with known and unknown antigen.

(*iii*) The separation of antibody-bound labelled antigen from free antigen is done by centrifugation and filtration.

(*iv*) The radioactivity is then measured and percentage of labelled antigen bound to the antibody is calculated.

(*v*) Amount of unknown antigen is determined using reference standard curve.

**Liquid phase RIA :**

(*i*) The liquid phase RIA is based on competitive binding of radio labelled antigen and unlabelled antigen to a high affinity antibody.

(*ii*) The antigen is generally labelled with a gamma-emitting isotope such as $I^{125}$.

(*iii*) The Labelled antigen is mixed with antibody at a concentration that just saturates the antigen-binding sites of the antibody molecule and then increasing amount of unlabelled antigens of unknown concentration are added.

(*iv*) The antibody does not distinguish labelled from unlablled antigen, so two kinds of antigen complete for available binding sites on the antibody.

(*v*) With increasing concentration of unlabelled antigen, more labelled antigen will be displaced from the binding sites on the antibody.

(*vi*) The labelled antigens are made free in solution. The amount of labelled antigen in a solution is measured and the concentration of unlablled antigen can be determined.

**Application of RIA :**

(*i*) This is extremely important in biomedical research technique.

(*ii*) It also helps in clinical practices such as blood banking, allergy diagnosis, endocrinology etc.

## • BIOTECHNOLOGICAL TECHNIQUE :

**Techniques: Molecular Analysis: DNA, RNA and Protein**

(*i*) DNA fragments are obtained from restriction enzyme digestion of genomic DNA.

(*ii*) It is denatured by using alkaline solution and single stranded DNA is obtained.

(*iii*) The single stranded DNA is then separated by electrophoresis in an agarose gel and then blotted into a nirocellulose filter or nylon membrane.

(*iv*) The DNA on the filter paper is then hybridized with unique radio labelled nucleic acid probe and detected by autoradiography. The procedure inwhich DNA is transferred from agarose gel to nitrocellulose filter is termed as **Southern blotting**, after E.M. Southern.

An analogus process in which RNA is transferred from agarose gel to nitrocellulose filter is called *northern blotting.*

Another analogus process in which a protein is transferred from gel to a suitable immobilizing matrix (nitrose cellulose sheet) and then probed with a specific antibody is called *western blotting.*

Southern western screening is a general screening method for identifying a protein which binds to a specific sequence of DNA.

**• Southern blotting:**

This technique is employed for detecting specific genes in cellular DNA.

**Steps:**

1. Extraction of DNA, digestion and loading of agarose gel.
2. Agarose gel electrophoresis and denaturation of DNA.
3. Southern blotting: transfer of DNA fragments from gel to filter.
4. Exposure to probe and hybridization.
5. Autoradiography and autoradiogram.

**1. Extraction of DNA, digestion and loading of agarose gel.**

(*i*) DNA is first extracted from cell and then purified.

(*ii*) The purified DNA is then digested with one or more restriction enzymes in separate reactions.

(*iii*) The restriction enzymes cuts DNA at specific (restricted) sites at several places and produces several fragments (100 to 1000). The average length of the fragments is about 4000 bp.

(*iv*) These digested fragments (double stranded) are then loaded into wells in agarose gel.

**2. Agarose gel electrophoresis and denaturation of DNA**

(*i*) The digested DNA fragments are separated by electrophoresis on an agarose gel.

(*ii*) The DNA fragments of smaller size (0.1 to 20 Kb) are separated by conventional *agarose gel electrophoresis*.

The DNA fragments of larger sizes (40 Kb to several mega base) are separated by *pulse field gel electrophoresis*.

(*iii*) Since *ds* DNA does not undergo hybridization, it has to be denatured to single stranded state by placing the gel in alkaline solution.

**3. Southern blotting: transfer of DNA fragments from gel to filter.**

(*i*) The denatured DNA fragments (ssDNA) are transferred from the agarose gel to a nitrocellulose filter or nylon membrane by capillary action. This blotting procedure is called *southern blotting*.

(*ii*) The gel is laid on the top of a buffer saturated filter paper placed on a solid support of a glass plate, with its two edges immersed in the buffer.

(*iii*) A piece of nitrocellulose filter or nylon membrane of the same size at the agarose gel is laid over the gel.

(*iv*) The nitrocellulose paper is covered with a stack of many papers (paper towels) and a weight placed on the papers. This blotting arrangement is left for a few hours to overnight.

(*v*) The buffer solution is drawn up by filter paper agarose gel, nitrocellulose filter paper and stack of filter papers by capillary action.

(*vi*) The electrophoresed DNA fragments from the agarose gel are carried up and transferred to the nitrocellulose filter or nylon membrane where they bind tightly. It creates a replica of the gel into filter paper. This is *southern blot* in the form of invisible bands on the filter paper.

(*vi*) The nitrocellulose filter paper is removed and wasted thoroughly. It is then baked in vaccum at 80°C for 2-3 hrs for fixing the DNA permanently on to the filter paper.

**4. Exposure to prove and hybridization:**

(*i*) Now the nitrocellulose filter paper (bound nucleic acid) is bathed with a buffered salt solution containing a specific (sequence of interest) radioactive DNA prove for hybridization.

(*ii*) This is kept at 60°C for 16-24 hours.

**5. Autoradiography and autoradiogram**

(*i*) The sheet is removed and washed throughly. As a result only probe molecules that have hybridized to the ss DNA immobilized on the paper remain attached.

(*ii*) The radioactive labelled prove will hybridize only with the DNA molecules that contain a nucleotide sequence complementary to the sequence of the prove.

(*iii*) As a result *radioactive bands* corresponding to the bands containing the gene of interest are formed.

(*iv*) The filter paper is then dried and exposed with an X ray film for autoradiography.

(*v*) After film (autoradiogram) is developed, the dark bands show the position of DNA sequences that have hybridized with the probe.

**Significance :**

(*i*) It is used to map the restriction sites in genomic DNA next to the sequence of the cloned DNA fragments.

(*ii*) Detetion and insertion mutation can be detected.

**Zoo blot:**

(*i*) It is a specialized form of Southern blotting where genomic DNA from variety of different oganisms (animals) hence "Zoo" is examined for sequence similar to human probe sequence.

(*ii*) The more closely related organism, the more likely a related sequence will be found.

(*iii*) The most interesting use for zoo blots is to test for non coding versus coding DNA. Sequence of non coding DNA mutate and change rapidly compared to coding sequences.

- **Northern blotting:**

This technique is a variation of Southern blotting and is employed for detecting RNA instead of DNA.

**Steps:**

1. Collection (extraction) of total cellular RNAs.
2. Agarose gel electrophoresis and denaturation of RNA.
3. Blotting (northern) transfer of RNA fragments from agarose gel to nitrocellulose filter paper.
4. Exposure to cDNA probe and hybridization.
5. Autoradiography and autoradiogram.

**1. Collection of total cellular RNA**

(*i*) Total cellular RNAs are collected or extracted from several tissues (liver, spleen, kidney and muscles) of the species (vertebrate) of interest.

(*ii*) Here the target RNA is usually messenger RNA.

**2. Agarose gel electrophoresis and denaturation of DNA**

(*i*) The RNAs are then subjected to *agarose gel electrophoresis* for separation of different mRNA species according to their size.

(*ii*) Denaturants like formaldehyde, glycol, are used to prevent single RNAs from pairing and hybridizing to themselves.

**3. Blotting (northern): transfer of RNA fragments from gel to nitrocellulose filter paper.**

(*i*) The mRNA is then transferred to nylon membrane or nitrocellulose filter paper.

(*ii*) RNA may bind tightly but not covalently to nitrocellulose or nylon membrane. This blotting procedure is termed as northern blotting.

**4. Exposure to cDNA probe and hybridization**

(*i*) After northern blotting, the RNA on the filter paper is incubated with radio labellede this blotting procedure is termed as **northern blotting** DNA probe.

(*ii*) The probe will hybridize to complementary mRNA molecules and then it is subjected to autoradiography i.e. exposed to film or chromogenic substance.

**5. Autoradiography and autoradiogram**

(*i*) As a result, the radio active RNA bands are formed.

(*ii*) If the attached of mRNA contains sequence similar to the probe film will turn black in that area. After autoradiography, the dark spot (autoradiogram) will show the specific RNA sequence which have hybridized.

**Significance :**

(*i*) It is extremely helpful in studies of gene expression.

(*ii*) It enables the presence or absence of genes in a DNA fragments.

• In large genomes, using mRNA is more efficient because all the introns are removed.

**• Western blotting (Immuno blotting):**

**(*i*) It is a variation of Southern blotting where protein is transferred from an electrophoretic gel to a nitrocellulose membrane by means of electric force.**

**(*ii*) Proteins may be proved with a specific antibody to identify the protein of interest.**

**• Steps:**

1. Extraction of protein.
2. Separation of protein through SDS page
3. Western blotting.
4. Exposure to prove and hybridization.
5. Autoradiography and autoradiogram.

**1. Extraction of protein:**

(*i*) Proteins are extracted from cell. The mixture of proteins in cell extract is dissolved in a solution containing a negatively charged detergent **sodium dodecyl sulphate** (SDS).

(*ii*) SDS denatures the proteins and give it an overall negative charge.

**2. Separation of protein:**

(*i*) The proteins are separated according to size by SDS polyacrylamide gel electrophoresis (SDS-PAGE).

(*ii*) They migrate to the positive electrode (anode), the rate of migration being determined by the size of the molecules.

**3. Western blotting:**

(*i*) The proteins are transferred from the SDS-polyacrylamide gel to the nitrocellulose filter paper. This process is called *Western blotting*.

(*ii*) It is performed by applying electric current for two hours to move the protein from the gel to the surface of the nitrocellulose filter.

(*iii*) As a result proteins bind to the nitrocellulose filter.

**4. Exposure to probe and hybridization:**

(*i*) The nitrocellulose filter is then incubated with radioactively labelled antibodies raised against the protein of interest.

(*ii*) Labelling of the antibodies is done with $I^{125}$.

(*iii*) The antibodies bind or hybridize to the specific protein to the filter.

**5. Autoradiography and autoradiogram:**

(*i*) The antibodies bound to the filter can be detected by *autoradiography*.

(*ii*) Thus the protein against which the antibody was raised can be seen and identified through *autoradiogram*.

(*iii*) The bound antibody can also be detected by staining with **coomassie blue** or by the ELISA technique.

**Significance:**

It is used to detect individual protein.

- **Dot blot:**

**Hybridization technique where multiple DNA samples are attached to a filter paper in small spot. The blot can then be probed with a gene of interest to find matching sequence.**

**Steps:**

(*i*) It begins by spotting several DNA or RNA samples on to nylon membrane or nitrocellulose membrane. Often different concentration of samples are dotted side by side.

(*ii*) Denaturation of the target DNA sequences is done either previously exposing to heat or exposure to a filter containing an alkali.

(*iii*) The membrane along with target DNA is incubated with a radiolabelled probe. This result *hybridization* (target DNA-probe).

(*iv*) The probe solution is decanted and the membrane is washed to remove excess probe.

(*v*) Now it is dried and exposed to an autoradiographic film to obtain an autoradiograph.

(*vi*) Sample contain DNA or RNA complementary to the probe will leave a black spot on the film i.e. seen in autoradioagraph.

**Significance:**

1. Dot blots are a quick and easy way to determine if the target sample has a related sequence.
2. Here multiple samples can be processed in a smaller amount of space.

**In Situ Hybridization (ISH)**

**It is a version of hybridization analysis for determining the original site (in situ) of specific nucleic acid (DNA/RNA) sequences in chromosome by hybridizing labelled DNA or RNA to denatured DNA in chromosome preparation and visualizing the labelled probe by autoradiography.**

**Procedure:**

(*i*) DNA in the chromosome must be made single stranded i.e. denatured by breaking the hydrogen bonds between the base pairs.

(*ii*) Here a sample is dried on a microscopic slide and then is treated with formaldehyde.

(*iii*) A radioactive labelled probe (DNA or RNA) is hybridized with the denatured chromosomal DNA.

(*iv*) The position at which the probe hybridize to the chromosomal DNA is visualized through autoradiography.

It provides the information about the map location of gene.

**Fluorescent In Situ Hybridization**

**A technique of insitu hybridization in which a probe labelled with fluorescent dye (quinacrine and quinacrine mustard) binds to specific sequence of nucleic acid (DNA/RNA) of chromosome and makes itself visible by its fluorescence.**

**Procedure:**

(*i*) First drop of cells (samples) arrested in the metaphase stage of the cell cycle is dried on a microscopic slide. It is then treated with formaldehyde, so that DNA in the chromosome become denatured by breaking the hydrogen bonds between the base pairs but do not lose their characteristic metaphase morphologies or cells must be heated to make the DNA single stranded.

(*ii*) Samples where RNA is the target do not need to be heated.

(*iii*) A probe labelled with fluorescent tag is added to denatured chromosome on the slide and incubate the probe for hybridization with the complementary sequences of the DNA region of the chromosome.

(*iv*) After words wash away unhybridized probe.

(*v*) Now place the slide under special microscope that focuses ultraviolet (uv) light on the chromosomes. The uv light causes the bound probe (hybridize probe) to fluoresce in the visible range of the spectrum.

(*vi*) By detecting the fluorescent signal, the position at which the probe hybridize to the complementary sequence of DNA is visualized.

**• Chromosome Painting:**

**It is a variant of fluorescent in situ hybridization (FISH) where colourful images are created by treating chromosomes spreads with fluorescently labelled DNA fragments that have been isolated and characterised in the laboratory.**

**Process:**

(*i*) Chromosome painting is done at the stage of cell cycle when chromosomes are specially compact and easy to visualize.

(*ii*) Sometimes selective painting is done in the interphase.

(*iii*) The probes which are used for chromosome painting are specific for sites scattered along the length of each chromosome.
The DNA fragment is chemically labelled with fluorescent dye in the laboratory that act as probe.

(*iv*) The labelled probe is then applied to the chromosome that have been spread on a glass slide.

(*v*) Under right conditions, the probe (DNA fragment) will bind to chromosomal DNA that is complementary to it in sequence.

(*vi*) As a result of binding, the chromosomes are labelled with fluorescent dye that is present in the DNA fragment.

(*vii*) After the probe has bound, the chromosome spreads are irradiated with light of an appropriate wavelength.

(*viii*) The resulting bans or dots of colour reveal where the complementary DNA sequence the target of the probe is located on the chromosome.

**Significance:**

1. Useful in Karyotyping the chromosome.
2. Used in the detection of genetic disorder.

**• DNA Fingerprinting or Typing Profiling:**

**The nucleotide form the code, therefore the coded genetic information can be profiled to produce most authentic identity card of any organism. The molecular technology that facilitates the identification of an individual based on the unique characteristic of their DNA, observed by the banding pattern of DNA fragments obtained by restriction enzyme digestion and followed by southern blot hybridization using minisatellite probe is known as DNA finger printing or DNA typing.**

- **Principle:**

DNA finger printing is based on *sequence polymorphism* that occurs in the human genome and genome of other organisms. The sequence polymorphisms are slight sequence differences, usually single base pair changes that occur from individuals to individuals once in every few hundred base pairs on average. Each difference from the concensus human genome sequence is generally present in only a fraction of the total population but every individual has some of them. These polymorphic locus is called minisatellite or VNTR (Variable Number Tandem Repeat) and forms a haplotype which shows Mendelian inheritance among their offspring. This locus is made up of a variable number of identical sequences joint together in tandem.

One family of minisatellites in human genome share a common **'core' sequence**. The core is **GC** rich sequence of 10 – 15 bp. These repeats are written as $(C-A)_n\ (G-T)_n$ occur in 100000 blocks in every genome and uniformly distributed throughout the genome.

**Steps:**

(*i*) The DNA fingerprints can be prepared from small amount of blood, semen, hair bulbs or other cells.

The DNA is extracted from blood stain, semen stain or hair roots of disputed children or suspected persons and the same from parents or close relatives.

(*ii*) Sometimes DNA sample is isolated from preserved tissues of individuals long after their death.

(*iii*) It is then amplified by **PCR (Polymorase Chain Reaction)** and analyzed with carefully chosen **DNA probes.**

(*iv*) Now the isolated and amplified DNA is cut with restriction enzymes and subjected to agarose gel electrophoresis for separation of the fragment according to their size. These *ds* DNA are denatured by soaking the gel in alkali to make it ssDNA.

(*v*) DNA from forensic samples are amplified by PCR and are used as probes. This probe is more radioactive ($P^{32}$).

(*vi*) The DNA fragments are transferred to nitrocellulose membrane or paper by the southern blot technique. The DNA bands appearing on the membrane is then immersed in a solution containing a radio active labelled DNA probe for hybridization.

(*vii*) It is then washed in water to remove unhybridized DNA and passed through X-ray film.

(*viii*) The hybridized complementary DNA sequences develop images (print). Identical prints that contain specific DNA sequences appearing on the X ray films are identified. Thus identity is confirmed.

> In CCMB (Hyberbad) BKM-DNA probe is used for hybridization of DNA. BKM = banded krait (Poisonous snake–*Bungarus tasciatius*) minor satellite.

**Significance:**

(*i*) It helps in separation of proteins and amino acids.

(*ii*) Identification of criminal.

- **DNA Foot Printing:**

It means finding the target DNA sequence or binding site of DNA binding protein.

**It is a technique to determine the length of the DNA sequence that binds to a particular protein, based on the idea that phosphodiester bonds in the region covered by the protein are protected from degradation by *DNase*. The resulting (remaining or protected) DNA fragments are electrophorised, the protein binding site shows up a *gap* or *foot* print in the pattern where the protein protected the DNA from degradation.**

**Steps:**

(*i*) A pure double stranded DNA is labelled at one end with $P^{32}$ is isolated.

(*ii*) Bind a protein to DNA.

(*iii*) Next digest the DNA-protein complex under mild conditions with DNase

(*iv*) Remove the protein from the DNA molecule, the DNA molecule is then denatured to separate into two strands.

(*v*) DNase cut the DNA at regular intervals except where the proteins bound and protected the DNA.

(*vi*) Resultant fragments from the labelled strand are separated through gel electrophoresis and is then defected by autoradiography.

(*vii*) The pattern of bands from DNA cut in the presence of a DNA binding protein is then compared with that from DNA cut in its absence.

The protein covers the nucleotides at its binding site and protects DNA from attack by DNase.

(*viii*) The labelled DNA fragments that shows no cleavage will show an area which is missing in gel electrophoresis. It leaves a 'gap' or "footprints" on the DNA.

**Significance:**

(*i*) Length of the DNA sequence can be observed.

(*ii*) It is also used to locate the binding sites of protein on RNA.

**DMS Foot Printing:**

In this method, DNA methylating agent DMS (Dimethyl sulphate) instead of DNase is used to attack the DNA protein complex. The DNA is then broken at the methylated sites. Unmethylated or hypermethylated sites showup on electrophoresis of the labelled DNA fragments and demonstrate where protein bound to the DNA.

**Polymerase Chain Reaction (PCR):**

**It is a rapid versatile invitro method for amplifying single molecule of DNA, involving repeated cycle of denaturation, hybridization to oligonucleotide primers and synthesis of polynuclestide by the action of taq polymerase at high temperature.**

**Requirements:**

(*i*) DNA template.

(*ii*) DNA polymerase: The used head stable DNA polymerase is *Taq polmerase* (obtained from *Thermus aquaticus*).

(*iii*) Primers: Two oligonucleotides.

(*iv*) Deoxy nucleotide triphosphate dATP, dTTP, dCTP and dGTP.

(*v*) Thermal cycler: Reaction is done in a thermal cycler and the process is automated by computer device.

| **Thermostable DNA Polymerase** | |
|---|---|
| **Polymerase** | **Source** |
| • Pfu polymerase | *Pyrococcus furiosus* |
| • Tli polymorase | *Thermococcus litoralis* |
| • Tfl polymerase | *Thermus flavus* |
| • Tma polymerase | *Themologa maritima* |

- **Steps:**

(*a*) Isolation and purification of DNA segment to be amplified.

(*b*) Cyclical amplification of DNA.

(*i*) Denaturation.

(*ii*) Annealing.

(*iii*) Chain extension.

(*c*) Screening of amplified DNA.

In PCR, the reaction takesplace in an appendroff tube, where the primer strand is added from outside in a form of deoxyoligonucleotide and DNA polymerase enzyme is added to help in polymerization.

**(*a*) Isolation and purification of DNA segment:**

**(*b*) Cyclical amplification of DNA:**

**(*i*) Denaturation:**

(*i*) The target DNA (isolated DNA) containing the segment to be cloned is heated to 92 – 95°C (average 95°C for 30 seconds) to denature its complementary strands.

(*ii*) This process is called **melting** of target DNA.

(*iii*) After denaturation, each strand act as template for DNA synthesis.

- **Annealing:**

(*i*) The denatured DNA is annealed to an excess of the synthetic oligonucleotide primers by incubating them together at 50-60°C temperature for 30 seconds. Here primer anneals (hybridise) each template.

(*ii*) The ideal annealing temperature depends on the base composition of the primer. The annealing temperature (in 0°C) can be calculated using the formula: $T = 2\ (AT) + 4\ (G + C)$.

- **Chain extension (Polymerization)**

(*i*) Nucleoside triphosphates (dATP, dTTP, dCTP and dGTP) and themostable *taq polymerase* are added to the reaction mixture.

(*ii*) DNA polymerase is used to replicate the DNA segment between the sites complementary to the oligonucleotide primers.

(*iii*) The primer provides the free 3′ OH required for covalent extension and the denatured genomic DNA provide the required template function.

(*iv*) Polymerization is usually carried out at 70-72°C for 1.5 minutes. The product of the first cycle of replication are then denatured, annealed to oligonucleotide primers and replicated again with DNA polymerase.

(*v*) This process is repeated many times (25-30 cycles) until the desired level of amplification is achieved.

## Inverse PCR:

It is the method for using PCR to amplify unknown sequences by circularizing the template molecule.

- **Step:**

(*i*) A restriction enzyme is chosen that does not cut within the stretch of known DNA.

(*ii*) The length of the recognition sequence should be six or more base pairs inorder to generate reasonably long DNA segments for amplification by PCR.

(*iii*) The target DNA is then cut with this restriction enzyme to yield a piece of DNA that has compatible sticky ends, one upstream of the known sequence and one downstream.

(*iv*) The two ends are ligated to form a circle. The PCR primers are designed to recognize the end regions of the known sequence.

(*v*) Each primer binds to a different strand of the circular DNA and they both point "outwards" into the unknown DNA. PCR then amplifies the unknown DNA to give linear molecules with short stretches of known DNA at ends and the restriction enzyme site in the middle.

Inverse PCR sequence DNA near a known sequence by finding a restriction enzyme recognition sequence a way in the unknown region. Cutting out this template, and amplifying the entire piece with Taq Polymerase.

- **Reverse Transcriptase PCR (RT-PCR):**

**A two step method for detecting and quantitating a particular RNA in an RNA mixture by first converting the RNAs to DNA by reverse transcriptase and then performing polymerase chain reaction using primers specific for the RNA of interest.**

- **Steps:**

(*i*) In the first step DNA is synthesized from RNA using **reverse primer** (oligo (dT) for example mRNA) and enzyme **reverse transcriptase.** Then uses a forward primer to make ssDNA to dsDNA.

(*ii*) In the second step, the specific cDNA is amplified by PCR (Tth DNA polymerase is used).

**Properties of RT-PCR Enzyme:**

Tth DNA polymerase is isolated from *Thermus Thermophilus* bacterium.

- It can reversely transcribe DNA from RNA.
- It can multiply DNA into DNAS by its polmerase activity.
- It can withstand a temperature of 98°C like Taq polymerase.

- **RT-PCR (Real Time PCR)**

**It is a technique of quantifying the amplification of specific region of a DNA molecule.**

- **Steps:**

(*i*) In the first step **reverse primer** and **reverse transcriptase** is used. Then uses forward primer to make it double stranded.

(*ii*) In the next step DNA amplification is done in the presence of highly sensitive fluorescent dye ***SYBR green*** which binds to double stranded DNA emits 1000 fold greater fluorescence.

(*iii*) As the DNA is amplified by PCR, the staining is measured in real time (hence the name of the technique) by means of special thermal cycler that uses laser detection of the fluorescence.

(*iv*) The amount of experimental mRNA is quantitifed and is measured by comparing the rate of production of stained amplified DNA with the control of known amount of starting mRNA.

## DNA Isolation and estimation of fragment size by Gel Electrophoresis

Probably the most common method of DNA extraction is phenol extraction. The fundamental aim of phenol extraction is the deprotenization of an aqueous solution containing the desired nucleic acids. In simple terms the phenol reagent is mixed with the sample under conditions which favor the dissociation of proteins from nucleic acids. Centrifugation of the mixtures yields two phases: a lower organic protein phase carrying the protein and some lipids, while the aqueous layer containing the nucleic acids remains on the top. The denatured proteins and lipids form an insoluble layer at the *interphase.* The whole theme behind is that, following extraction with phenol and centrifugation, the protein component of a sample is denatured or rendered insoluble, while the nucleic acid remains in the aqueous phase.

**Caution!** *A large interphase between the phenol and the aqueous layers indicates incomplete extraction of the nucleic acid. Additional rounds of extraction of the aqueous layer and the interphase may be required. Quality of phenol is very important.*

Many commercial suppliers provide liquefied phenol, which can be used directly, without further purification, it is clear and colourless. If however, the commercially supplied phenol is either crystalline or has a yellow or pink coloration, it is necessary to purify by distillation. The coloration is indicative of the phenolic oxidation product (quinones. diacids and others) which form during storage. These contaminants must be removed because they cause cleavage of the phosphodiester bonds and cross linking of the DNA strands.

Caution! *Distillation must he done with experienced hand because phenol causes serious burns.*

## Saturation of phenol and Antioxidant of phenol

The distilled phenol should be saturated by mixing 1M Tris buffer pH 8.0 (@ 3ml of Tris buffer/10 ml of phenol) under constant stirring on magnetic stirrer. Allow to stand, two layers separate, pour off the top layer. After this add 0.1M Tris buffer pH 8.0 stir, allow to stand, layers separate, pour off the top layer. Now add a few drops of 0.1% of 8-hydroxyquinoline (w/v) to retard oxidation. Add again 0.1M Tris buffer pH 8.0 and store the saturated phenol under it at 4°C.

### *Chloroform:*

Used in conjunction with phenol, chloroform improves the efficiency of nucleic acid extraction. This is principally due to its ability to denature protein, thereby, assisting the dissociation of nucleic acid from protein. The high density of chloroform also enhances the separation of the phases, facilitating the removal of aqueous phase with little cross contamination with organic material. Removal of lipid is also another benefit of using chloroform.

In most situations in which retrieval of DNA from an *in-vitro* reaction is required, a single extraction with phenol (Tris saturated), chloroform-isoamyl alcohol (25:24: Iv/v/v) equal in volume lo the volume of tlie aqueous sample is sufficient.

### *Isoamyl alcohol:*

Usually isoamyl alcohol is added to the chloroform to prevent frothing.

### *Ethanol precipitation:*

The DNA in the aqueous phase can subsequently, be recovered by ethanol precipitation. Traces of phenol and chloroform must be removed from the DNA before it is used in enzymatic reaction or applied filters. The most common method for removing these solvents is precipitation of nucleic acids with chilled ethanol. in the presence of high salts. In brief term, a precipitate is formed by leaving a mixture of the sample, salt and ethanol at a low temp. (- 20°C or lower). The precipitated salt of nucleic acid is then sedimented by centrifugation, the supernatant is then removed and the nucleic acid pellet resuspended in a buffer appropriate to it's subsequent use.

## The General Procedure To Be Followed In The Laboratory For DNA Isolation

Animal and human tissues are convenient source of genomic DNA e.g. Ig of liver tissue contains about $10^9$ cells. To obtain an equal number of cells from blood we require 1 litre of blood.

To minimize degradation of nucleic acid the tissues should be kept in ice or quick *frozen* in liquid nitrogen and processed as soon as possible.

### *Step-I*

The tissues should be finely pieced with the help of sharp scalpel and then homogenized adding minimum amount of homogenizing buffer also called ***Nuclei extraction buffer***.

Composition of homogenizing buffer:

| 1) 0.35M Sucrose (MW 342.30) | 11.98 g | Rupture cell membrane |
|---|---|---|

2) 0.05M Tris (MW 121.11) 0.60 g Buffering reagent
3) 0.025M KC1 (MW 74.56) 0.186 g Maintains ionic strength
4) 0.05M Mg acetate (MW 214.46) 0.07 g Maintains ionic strength

Dissolve in 90 ml of deionized water, adjust pH to 7.4 with 1.0 M HC1 (if required). Final volume make up to 100 ml.

***Step-II***

Filter the materials through the cheese cloth and wait for 30 min.

***Step-III***

Spin the homogenate for 8 min at 8000 rpm.

Sucrose in the homogenizing buffer forms a concentration gradient while the salt maintain a suitable pH, when the cell membrane is broken and the organelles are precipitated.

***Step-IV***

Take the pellet and suspend it to equal volume of DNA extraction buffer for 15 min; vortex.

Composition of DNA extraction buffer.

1) 0.15 M NaCl (MW 58.44) 0.876 g Maintains ionic strength
2) 0.1 M EDTA (MW 372.24) 3.724 g Chelating agent
3) 0.05 M Tris (MW 121.1) 0.600 g Buffer

Dissolve in 90 ml of deionised water. Adjust pH to 8.2 with 1M NaOH. Final Volume make up to 100 ml.

**Note:** The NaCl offers a suitable pH while EDTA acts as a checlating agent. $Mg^{2+}$ can mediate aggregation of nucleic acid to each other and to protein, EDTA (ethylene diamine tetra acetic acid) the $Mg^{2+}$ chelator, in the DNA extraction buffer prevents it. If necessary dissociation of protein can further be improved by incubation of the cell extracts with proteinase K in the presence of SDS. This solution also provides the salts necessary to precipitate DNA in a salt form, during subsequent steps.

***Step-V***

To this solution add 1/10 vol. of 20% SDS and mix for 10 min by rolling at room temperature.

SDS is a detergent that causes rupturing of the nuclear membrane and helps in denaturation.

***Step-VI***

Add Phenol + (Chloroform + Isoamyl alcohol) mixture equal volume.

**Note:** Phenol chloroform- isoamyl alcohol should be in ratio of 25:24:1. Phenol chloroform togather denatures the DNA bound proteins when they are precipitated. Isoamyl alcohol prevents frothing due to the presence of SDS in solution.

***Step-VII***

Centrifuge the solution at 8000 rpm for 10 min at room temperature.

***Step-VIII***

Pipette out the aqueous phase and repeat the chloroform isoamyl alcohol step.

Note: Chloroform brings down the traces of phenol from the DNA and forms the bottom layer.

***Step-VI* to *Step-VIII* can be repeated.**

***Step-IX***

Add 1/10 volume of 5M Sodium acetate and mix for 10 min by rolling.

**Note:** Sodium acetate increases salt concentration. Replaces water molecules attached to nucleic acids to precipitate DNA.

***Step-X***

Add 2.5 volume of chilled ethanol DNA is precipitated. Keep at - 20°C for 20 mm. to complete precipitation.

**Note:** The precipitate can be recovered by centrifugation, the pellet is washed in 70% alcohol (this removes any solute trapped in the precipitate) these contaminants make the DNA difficult to dissolve and can inhibit enzymatic reactions. Dry the pellet in a lyophilize or vacuum dessicator and resuspend the DNA in TE (pH 8.0) buffer (10M Tris-HCI at pH 8.0: 0.5mM EDTA).

**1 kb ladder:**

1kb ladder of US patent no. 4,403,036 is suitable for sizing linear double stranded DNA fragments from 500 bp to 12 kb. Prepared from a plasmid containing repeats of a 1,018 bp DNA fragment, the ladder consists of 12 fragments ranging from 1,018 bp to 12,216 bp. In addition to these 12 bands, the ladder contains vector DNA contents that range from 75 to 1,636 bp.

The double stranded DNA can be visualized on 0.5% to 1.5% agarose gels after cthidium bromide staining and viewing against ultraviolet light. This ladder can be radio-labeled using T4 polynucleotide kinase, T4 DNA polymerase, DNA poll or the large fragment of DNA polymerase l(Klenow fragment).

**Agarose Gel Elcctrophorcsis:**

The standard method to separate, identify and purify DNA fragments is to electrophoresis through agarose gels. The location of DNA in gel can be determined directly. Bands of DNA are stained with low-concentration of the fluorescent, intercalating dye cthydium bromide; as little as 1 ng of DNA can then be detected by direct examination of the gel in ultraviolet light (Sharp *et al,* 1973).

The electrophorelic migration rale of DNA through agarose gels is dependent upon parameters; They are

1. Molecular size of the DNA.
2. The agarose concentration.
3. The conformation of the DNA and
4. The applied current.

A DNA fragment of a given size migrates at different rates through gels containing different concentrations of agarose. There is a linear relationship between the log of the elcctrophoretic mobility of DNA ($\mu$) and gel concentration ($\tau$). which is described by the equation:

$$\log \mu = \log \mu_0 - k_r \tau,$$

where $\mu_0$ is the free elcctrophorctic mobility and $k_r$ is the retardation coefficient, a constant that is related to the properties of the gel and the size of the migrating molecule. Thus by using gels of different concentrations, it is possible to resolve a wide *size* range of DNA fragments.

**Agarose percentage in gels for the respective separation of the size of different linear DNA Molecules**

| Agarose percent in the gel | For the size of DNA molecules in klis |
|---|---|
| 0.3 | 60 - 5 |
| 0.6 | 20- 1 |
| 0.7 | 10-0.8 |
| 0.9 | 7-0.5 |
| 1.2 | 6 - 0.4 |
| 1.5 | 4-0.2 |
| 2.0 | 3-0.1 |

(genomic DNA - usually checked at 0.8% and PCR product at 1.5%)

**Gel clcctrophoresis tanks and buffers:**

At present, almost all agarose gel electrophoresis is carried out with horizontal slab gels. In almost all the system the gel is cast on a removable cast plate, which is a glass plate. The plate is installed on a platform so that the gel remains submerged just below the surface of the electrophoresis buffer.

**Tank Buffer: (after Maniatis *et al.*).**

For historical reasons, Tris-acetate is the most commonly used buffer, its buffering capacity is rather low. In comparison, Tris-borate has significantly higher buffering capacity and gives good resolution of DNA fragments. Tris-phosphate is sometimes preferred when one needs to separate DNA fragments from gel. We have used Tris-borate EDTA (TBE). The composition is given below :

*Tris-borate (TBE) 5x Concentration / litre*

| | |
|---|---|
| 0.089M Tris-borate | 54 g Tris base |
| 0.089M Boric acid | 27. 5g Boric acid |
| 0.002M EDTA | 20 ml 0.5M EDTA (pH 8.0) |

## DNA Size Standard

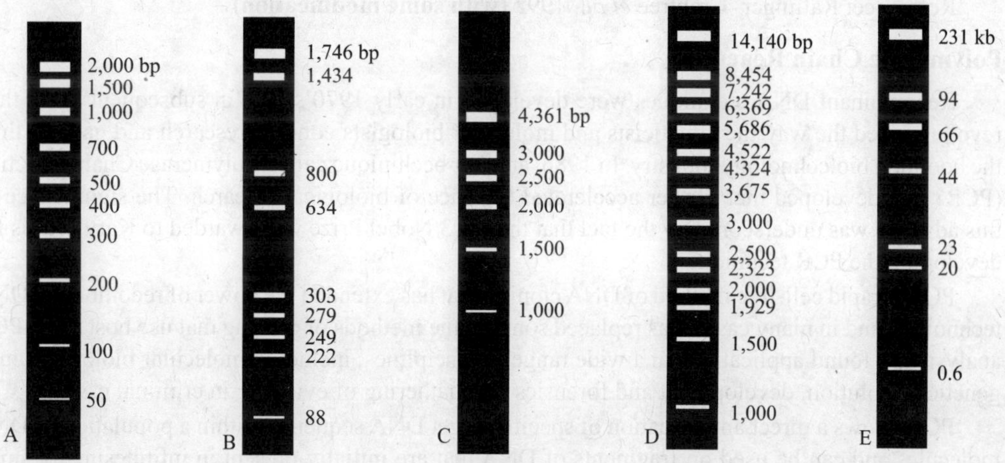


**Fig. 10.13.**

**Electrophoretic profiles of the DNA size standards.** A. AmpliSize 50-2,000 bp ladder on 4% AmpliSize agarose, B. pBR322 digest on 1.8% Molecular Biology Certified agarose, **C.** 1-4.2 kb graded series on 1% Molecular Biology Certified agarose, D. supplemented λ *BstE* II digest on 1% Molecular Biology Certified agarose, and E. λ *-Hind* III digest on 0.75%. Molecular Biology Certified agarose.

**Polymerase Chain Reaction (PCR) high light it.**

**Preparation of PCR mixture: For one (0.5 ml PCR tube (480 Perkin Elmer Machine)**

| *No.* | *Reagents* | *Stock Concentration* | *Working Concentration* | *You have to add* |
|---|---|---|---|---|
| 1 | Buffer | 10X | IX | 2.5uJ - |
| 2 | dNTPs | 1OmM | 0.2mM | 4µl (1×4) |
| 3 | $MgCl_2$ | 25mM | 2mM | 0.5µl |
| 4 | BSA | Img/ml | 0.04mg/ml | lµl |
| 5 | Primer | | 200ng | 1.5µl |
| 6 | Taq | 3u/µl | lµ/µl | 0.5µl |
| 7 | Water | | | 13/14µl |

Tlic amount of water depends upon the amount of DNA to be added for PCR. If you add 2µl of DNA then the amount of water will be 13µl or if you add 1 µl of DNA then the amount of water will be 14µl. In general the total reaction mixture will be of 25µl.

After preparing the mixture in the tubes the content should be covered with mineral oil to prevent evaporation.

**Temperature Profile**

| Temperature (°C) | Time (min.) | No. of cycle/s |
|---|---|---|
| 94 | 2 | 1 |
| 94 | 1 | |
| 40 | 2 | 40 |
| 72 | 2 | |
| 72 | 10 | 1 |
| 4 | No limit (to preserve) | |

Reference: Rallingcr- Crabtree *et al.,* 1992 **(with some modification)**

## Polymerase Chain Reaction

Recombinant DNA techniques were developed in early 1970's. and in subsequent years they revolutionized the way the geneticists and molecular biologists conduct research and gave birth to the booming biotechnology industry. In 1986, another bechiniquc called Polymerase Chain Reaction (PCR) was developed that further accelerated the pace of biological research. The significance of this advance was underscored by the fact that the 1993 Nobel Prize was awarded to Kary Mullis for developing the PCR technique.

PCR is rapid cell-free method of DNA cloning that has extended the power of recombinant DNA technology and in many cases, has replaced some of the methods of cloning that use host cells. PCR analysis has found applications in a wide range of disciplines, including molecular biology, human genetics, evolution, development and forensics (the gathering of evidence in criminal cases).

PCR allows a direct amplification of specific target DNA sequence within a population of DNA molecules and can be used on fragments of DNA that are initially present in infinitesimally small quantities. In order to amplify a sequence by PCR some information about the nuclcotide sequence is required. This information is used to synthesize two oligonucleotide primer that arc added to a denatured DNA (single-stranded) sample. The primers hybridize to complementary sequences that flank the single-stranded sequence to be amplified. A heat-stable of DNA polymerase called *Tag* polymerase is used to synthesize a second strand of the target DNA.

There are three basic steps in the PCR reaction, and the amount of amplified DNA produced is limited theoretically by the number of times these steps are repeated:

1. In the first step, the DNA to be amplified is denatured into single strands. This DNA does not have to be purified and can come from any number of sources, including genomic DNA. forensic samples such as dried blood or semen dried samples stored as part of medical reports, single hairs, mummified remains of the fossils. The double stranded DNA is denatured by heating at 90-95°C. until it dissociates into a single strand (usually about 5 minutes).
2. The temperature of the reaction is lowered to somewhere between 50 and 70°C and primers are allowed to bind to the denatured DNA. These primers are synthetic oligonucleotides (10-30 nucleotides long) that bind specifically to sequences flanking the segment to be amplified. These primers are designed to serve as the starting points for the synthesis of new strands complementary to the target DNA.
3. A heat stable form of DNA poiymerase *(Taq* polymerase) derived from a bacterium that lives in hot springs *(Thermus aquaticus),* is added to the reaction mixture and DNA synthesis is carried out at temperature between 70-75°C. The polymerase adding nucleotide in the 5'- to- 3' direction, making a double stranded copy of the target DNA.

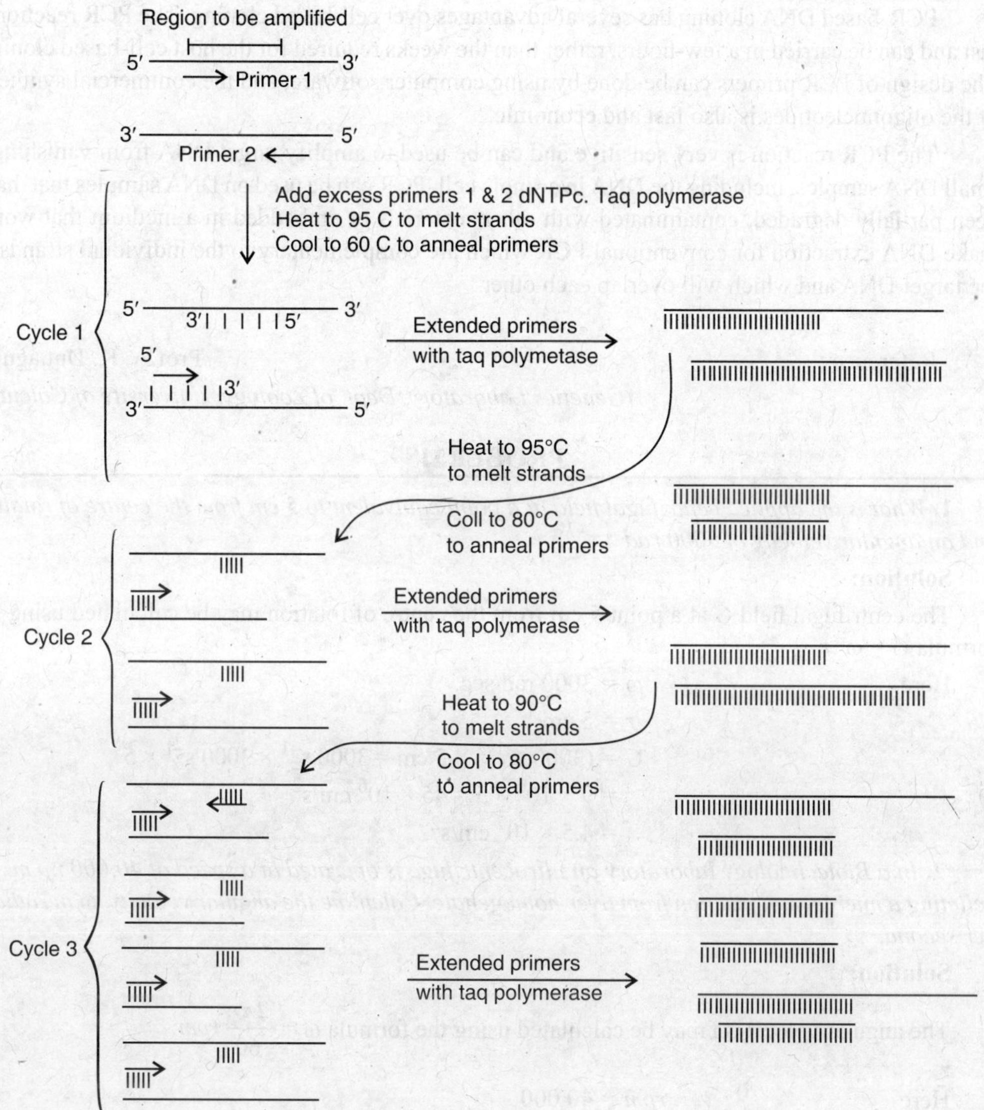


**Fig. 10.14**

## The 'Cycle':

Each set of three steps - *denaturation* of the double-stranded product, *annealing* of primers and *extension* by polymerase - is referred to as a cycle.

## PCR is a chain reaction:

This is because the number of new strands is doubled in each cycle, and the new strands serve as templates in the next cycle. Each cycle, which takes 4 to 5 minutes, can be repeated by carrying out each of the steps again.

25-30 cycles results in an over 1,000,000-fold increase in the amount of DNA. The process has been automated through the development of machines called *thermocyclers* that can be programmed to carry out a predetermined number of cycles, yielding a large number of target DNA, which can be used in other procedures such as cloning, sequencing, clinical diagnosis and genetic screening.

PCR-based DNA cloning has several advantages over cell-based cloning. The PCR reaction is fast and can be carried in a few-hours, rather than the weeks required for the host cell-based cloning. The design of PCR primers can be done by using computer software, and the commercial synthesis of the oligonucleotides is also fast and economic.

The PCR reaction is very sensitive and can be used to amplify target DNA from vanishingly small DNA samples, including the DNA in a single cell. PCR can be used on DNA samples that' have been partially degraded, contaminated with other materials or embedded in a medium that would make DNA extraction for conventional PCR which are complementary to the individual strands of the target DNA and which will overlap each other.

Reference : Prof.A. K. Duttagupta
*(Genetics Laboratory, Dept. of Zoology, University of Calcutta)*

## PROBLEMS

**1.** *What is the applied centrifugal field at a point equivalent to 5 cm from the centre of rotation and an angular velocity of 3000 rad$^{-1}$ s$^{-1}$?*

**Solution:**

The centrifugal field $G$ at a point 5 cm from the centre of rotation may be calculated using the formula $G = \omega^2 r$.

Here $\omega = 3000$ rad/sec.

$r = 5$ cm

$$\therefore \quad G = (3000\ s^{-1})^2 \times 5\ cm = 3000\ s^{-1} \times 3000\ s^{-1} \times 5$$

$$= 9 \times 10^6 \times 5 = 45 \times 10^6\ cm/s^2$$

$$= 4.5 \times 10^7\ cm/s^2.$$

**2.** *In a Biotechnology laboratory an ultracentrifuge is operated at a speed of 40,000 r.p.m. for pelleting a microsomal fraction from liver homogenate. Calculate the angular velocity, ω in radians per second.*

**Solution:**

The angular velocity ω may be calculated using the formula $\omega = \frac{2\pi r}{60}\ rpm$

Here $rpm = 40{,}000$

$\pi = 3.1416$

$$\therefore \quad \omega = \frac{2 \times 3.1416 \times 40 \times 10^3}{60} = \frac{2 \times 31416 \times 4 \times 10^4}{10^4 \times 60}$$

$$= \frac{251328}{6 \times 10} = \frac{41888}{10}$$

$$= 4.1888 \text{ rad/sec.}$$

**3.** *In the laboratory of PG department of Serampore College an ultracentrifuge is operating at 58000 RPM.*

(*a*) Calculate ω in radians per second.

(*b*) Calculate centrifugal force at a point 6.2 cm from the center of rotation. How many "g"'s is this equivalent to?

**Solution:**

$$(a)\ \omega = \frac{RPM}{60}\ 2\pi = \frac{58000 \times 2 \times 3.14}{60} = 966.67 \times 2 \times 3.14 = 6070.68$$

$= 6070.68$ radians/sec.

$(b)\ a = \omega^2 r = (6070.68)^2 \times 6.2 = 6070.68 \times 6070.68 \times 6.2$

$= 36853155.66 \times 6.2$

$= 228489565.1 = 2.284 \times 10^8$ cm/sec$^2$

(c) The earth gravitational field $= g = 980$ cm/sec$^2$

$$a = \frac{2.284 \times 10^8}{980} = 233061.2 \times g$$

**4.** *In the biochemistry laboratory of calcutta university, it is found that a protein has sedimentation coefficient value of 3.12 × $10^{-13}$ sec. in water. Its diffusion coefficient in water is found to be 8.2 × $10^{-7}$ cm$^2$/sec. Both the above values have been corrected for 20°C in water. The partial specific value of the protein is 0.735 and the density of water at 20° is 0.9982. Determine the molecular weight of the protein corrected R = 8.314 × $10^7$ erg/mole/degree.*

**Solution:**

$$M\omega = \frac{RTs}{D\,(1 - \bar{v}\rho)},$$

Here $R = 8.314 \times 10^7$ erg per mole per degree

$D = 8.2 \times 10^{-7}$ $T = 273 + 20 = 293$

$\rho = 0.9982$

$\bar{v} = 0.732$

$$M\omega = \frac{8.314 \times 10^7 \times 293 \times 3.12 \times 10^{-13}}{8.2 \times 10^{-7}\,(1 - 0.732 \times 0.9982)}$$

$$= \frac{8314 \times 293 \times 312 \times 10^{-13} \times 10^{-5} \times 10^6 \times 10^7}{82\,(1 - 0.7306824)} = \frac{760032624 \times 10^{-5}}{82 \times 0.2693176} = \frac{760032624 \times 10^{-5}}{22.0840432}$$

$$= \frac{760032624 \times 10^{-5} \times 10^5}{2208404} = 3441547$$

**5.** *In a pathological laboratory of calcutta the following data are obtained from human serum at 20°C temperature.*

*The gas consant (R) = 8.314 × $10^7$ ergs/mole/degree*

*Diffusion coefficient (D) = 6.1 × $10^{-7}$ cm$^2$/sec.*

*Sedimentation coefficient (s) = 4.6 × $10^{-13}$*

*Density of water at 20°C (ρ) = 0.998*

*Specific volume ($\bar{v}$) = 0.74 at 20°C*

*Calculate Mω of albumin.*

**Solution:**

$$M\omega = \frac{RTs}{D\,(1 - \bar{v}\rho s)}$$

$R = 8.314 \times 10^7$ erg/mole/deg.

$T = 273 + 20 = 293$

$D = 6.1 \times 10^{-7}$

$P = 0.998$

$s = 4.6 \times 10^{-13}$

$\bar{v} = 0.74$

$$M\omega = \frac{8.314\times10^{7}\times293\times4.6\times10^{-13}}{6.1\times10^{-7}\ (1-(0.74)\ (0.998)}$$

$$= \frac{8314\times293\times46\times10^{7}\times10^{-13}\times10^{-4}\times10^{8}}{61\ (1-0.73852)}$$

$$= \frac{112056092\times10^{15}\times10^{-17}}{61\times0.26148} = \frac{112056092\times10^{-12}}{15.95028} = \frac{112056092\times10^{-2}\times10^{5}}{1595028}$$

$$= 70.25336985\times10^{3} = 70{,}253$$

**6.** *The following results are obtained from a research laboratory of Pune. The experiment was done on protein at 20°C. Sedimentation coefficient was 4.6 × $10^{-13}$ sec, the diffusion coefficient was 5.96 × $10^{-7}$ cm$^2$/sec. the partial specific volume was 0.736 and density of water at 20°C was 0.9982. Determine the molecular weight.*

**Solution:**

$R = 8.314\times10^{7}$
$T = 273 + 20 = 295$
$D = 5.96\times10^{-7}$
$S = 4.6\times10^{-13}$
$\rho = 0.9982$
$\bar{v} = 0.736$

$$M\omega = \frac{RTs}{D\ (1-\bar{v}\rho)}$$

$$= \frac{8.314\times10^{7}\times295\times4.6\times10^{-13}}{5.96\times10^{-7}\ (1-(0.736)\ (0.9982)}$$

$$= \frac{8314\times295\times46\times10^{-6}\times10^{-4}}{569\times10^{-9}\ (1-0.7346752)} = \frac{112820980\times10^{-10}\times10^{9}}{596\ (0.2653248)}$$

$$= \frac{112820980\times10^{-1}}{596\times0.2653298} = \frac{112820980}{158.1335808\times10} = \frac{112820980\times10^{6}}{1581.335808}$$

$$= 71345$$

**7.** *In the PCR process, if we assume that each cycle takes 5 minutes, how many folds amplification would be accomplished in 1 hour.*

**Solution:**

Each cycle takes 5 minutes and doubles the DNA.

In one hour there would be $\frac{60}{5} = 12$ cycles

So the DNA would be amplified $2^{12} = 4096$ fold

**8.** *In PCR process, how many copies of DNA molecules will be produced after 5th and 16th cycle of amplification?*

**Solution:**

After one cycle of replication, one ds DNA will yield 2 ds DNA *i.e.* $2^n = 2^1 = 2$ ds DNA

After 5th cycle of replication it will yield

$$2^5 = 2\times2\times2\times2\times2 = 32$$

After 16 cycle it will yield $2^{16} = 1024$.

CHAPTER 11

# Thermodynamics

The term **thermodynamics** is concerned with the flow of heat and it deals with the relationship between heat and work. It is now applied to study the energy changes associated with all physical and chemical processes and also the mutual transformation of different kinds of energy.

**Some Basic Terms :**

- **System :**

A quantity of matter or an object chosen for observation isolated from the rest of the universe, with a bounding surface is called a system. A system has definite amount of one or more substances.

- **Surroundings :**

All other parts of the universe (rest of the universe) outside the boundary of the system are called surroundings.

**Types of Systems :**

- **Open System :**

A system which can exchange both matter and energy across the boundary is called an open system.

**Example :** ice in an open beaker.

- **Close System :**

A system that can exchange energy with the surroundings but not the matter is called close system.

**Example :** ice in a closed beaker.

- **Isolated or Insulated System :**

A system that can exchange neither matter nor energy in the surrounding medium is called isolated or insulated system.

**Example :** ice in a thermo flask.

- **Homogeneous System :**

  - A system is said to be homogeneous when it has same chemical composition uniform throughout.

  **Example :** mixture of gases.

- **Heterogeneous System:**

(*i*) If a system is not uniform throughout, it is said to be heterogeneous.

(*ii*) A heterogeneous system consists of two or more different phases which are homogeneous in themselves and are separated from one another by definite bounding surfaces.

**Example :** ice in contact with water.

- **Macroscopic System :**

A system which consists of a large number of atoms, ions or molecules is called macroscopic system.

- **Macroscopic Properties :**

The measurable properties associated with macroscopic system are called macroscopic properties.

**For example :** Pressure, volume, temperature, mass, density.

- **Extensive Property :**

A property which depends upon the amount of substance(s) present in the system is called extensive property.

*e.g.* mass, volume, energy, work, etc.

- **Intensive Property :**

A property which is independent of the amount of substance(s) present in the system is called intensive property.

*e.g.* temperature, pressure, density, etc.

**Thermodynamic Equilibrium :**

A system is said to be thermodynamic equilibrium when any of its macroscopic properties such as volume, temperature, pressure, etc. does not undergo any change with time.

**(*a*) Thermal equilibrium :**

If there is no flow of heat from one portion of the system to another, it is said to be in thermal equilibrium. This can happen if the temperature of the system remains constant.

**(*b*) Chemical equilibrium :**

If the chemical composition of various phases of the system does not change with time, the system is said to be in chemical equilibrium.

**(*c*) Mechanical equilibrium :**

If no mechanical work is done by one part of the system on another part of the system, the system is said to be in mechanical equilibrium.

**Thermodynamic Process :**

The operation by which a thermodynamic system changes from one state to another is called a thermodynamic process.

**(*a*) Isothermal process:**

A process in which the temperature of the system remains constant throughout, the process is called isothermal process.

In **Exothermic process,** the evolved heat is given out by the system instantaneously and thus the temperature of the system does not rise at any stage of the process.

In **Endothermic process,** the required amount of heat is absorbed instantaneously by the system from the surroundings and thus the temperature of the system does not fall at any stage of the process.

Here the change of temperature $(dT) = 0$

**(*b*) Adiabatic process :**

A process during which no heat enters or leaves the system during any step of the process is known as adiabatic process.

Such process is often carried out in a closed insulated container such as thermos bottle.

Here the change in heat $(dq) = 0$

**(*c*) Isobaric process :**

A process during which pressure of the system remains constant throughout the reaction is known as isobaric process.

Expansion of gas in open system is an example of isobaric process.

Here $dP = 0$

**(*d*) Isochoric process :**

A process during which volume of the system remains constant throughout the reaction is known as isochoric process.

*e.g.* Change taking place in closed system.

For an isochoric process $dV = 0$

**(*e*) Cyclic process :**

A process during which system comes to its initial state through a number of different processes is called cyclic or cycle process.

For cyclic process $dE = 0$, $dq = dw$.

**(*f*) Reversible process:**

(*i*) It is a process which is carried out infinitesimally slowly so that the properties of the system virtually remain unchanged (equilibrium) at any stage of the change or at all times.

(*ii*) A reversible process is always in a state of equilibrium at each of the small stages.

(*iii*) If the infinitesimally small driving force is reversed, the process will be carried out in reverse direction.

*Example :* Applying small amount of heat, water vaporises and on cooling same amount of heat, vapour condenses to liquid water.

**(*g*) Irreversible process:**

(*i*) If a process occurs suddenly or spontaneously such that equilibrium is disturbed during the transformation, it is called irreversible process.

(*ii*) All natural processes are irreversible.

**Example:** Water flows from higher altitude to lower altitude. Flow of hot from heat end to cold end.

| | Reversible Process | | Irreversible Process |
|---|---|---|---|
| (*i*) | It is in equilibrium state at all stages of the operation. | (*i*) | It is in equilibrium only at the initial and final stages of the operation. |
| (*ii*) | All changes can be reversed; if the process is reversed. | (*ii*) | All changes do not return to initial state. |
| (*iii*) | It is extremely slow process. | (*iii*) | It has moderate speed. |
| (*iv*) | It requires many steps. | (*iv*) | It requires finite step. |
| (*v*) | Work done in reversible way is greater. | (*v*) | Work done is less in irreversible way. |

| Isothermal Process | Adiabatic Process |
|---|---|
| (*i*) There is no change in temperature of the system. | (*i*) No change in heat between system and surroundings. |
| (*ii*) Increase or decrease of internal energy does not occur in this system. | (*ii*) Increase or decrease of internal energy occurs in this system. |
| (*iii*) This process is often carried out by placing the system in a thermostat (a constant temperature bath). | (*iii*) Such process is often carried out in a closed insulated container such as thermos bottle. |
| (*iv*) In isothermal process change in temperature $(dT) = 0$ | (*iv*) In adiabatic process the change in heat $(dq) = 0$ |

- **Concept of Work, Energy and Heat :**

**Work (*W*):**

It is a fundamental physical property of thermodynamics and is defined as the product of force (*F*) applied on a body and the displacement (*d*) of the body along the direction of force.

$$W = F \times d$$

**Sign convention:**

(*i*) If work is done on the system by the surrounding and the energy of the system is increased, then *w* is positive (+).

(*ii*) On the other hand, if work is done by the system on surrounding and the energy of the system is decreased, then *w* is negative(–). (New UPAC convention)

- **Units of Work**

(*i*) In C.G.S. it is measured in ergs.

(*ii*) In S.I. units the work is measured in Jules.

**Heat (*H*):**

Heat is a mode of energy exchanged between the systems and surroundings as a result of difference of temperature between them.

It is usually represented by letter *q*.

**Sign convention:**

(*i*) When heat is absorbed by the system from the surroundings, it is given a positive (+) sign.

(*ii*) When heat is given out of the system to the surroundings, it is given a negative (–) sign.

- **Units of Heat**

(*i*) It is erg in C.G.S. units.

(*ii*) It is Jules in S.I. units.

- **Energy (*E*):**

It may be defined as the capacity to do work. Every system stores some amount of energy.

**Internal Energy:**

(*i*) A Thermodynamic system by virtue of its existence must possess a store energy. Therefore the total energy stored in a substance by virtue of its chemical nature is called its internal energy. It is usually denoted by ***E***.

(*ii*) Internal energy is a state property and its absolute value cannot be determined. However, change in internal energy can be determined experimentally by bomb calorimeter.

(*iii*) Internal energy of a system depends upon

(*a*) the quantity of the substance (*b*) its chemical nature

(*c*) temperature, pressure, volume.

(*iv*) If $E_1$ is the initial state of internal energy and $E_2$ the final state, then change of internal energy *i.e.* $\Delta E = E_2 - E_1 \quad (E_2 > E_1)$.

**Sign conventions:**

(*i*) If $E_1 > E_2$ the extra energy possessed by the system in the initial state would be given out and $\Delta E$ will be negative.

$\Delta E$ is negative if energy is evolved.

(*ii*) If $E_2 > E_1$ the extra energy will be absorbed by the process and $\Delta E$ will be positive.

$\Delta E$ is positive if energy is absorbed.

- **Laws of Thermodynamics :**

**First Law :**

**It states that energy can neither be created nor be destroyed, although it can be converted from one form to another.**

(*i*) The total energy of an isolated system always remains constant, although it may undergo transformation from one form to another.

(*ii*) Total energy of a system and surroundings remains constant.

(*iii*) Whenever certain quantity of some form of energy disappears, an exactly equal amount of some other form of energy must appear.

(*iv*) It is impossible to construct a perpetual motion machine that can produce work without spending energy on it.

According to Einstein (1905) expression "*energy can be produced by the destruction of mass.*" The energy produced ($E$) by the destruction of mass ($m$) is given by the equation

$$E = mc^2 \quad (c = \text{velocity of light}).$$

The total mass and energy of an isolated system remains constant. (modified first law).

**Explanation of First Law of Thermodynamics :**

(*i*) When mechanical work is done heat is produced and that the work done is proportional to the heat produced.

$W$ = Mechanical work done.

$H$ = Amount of heat produced.

$W \propto H$ or $\frac{W}{H}$ = constant.

(*ii*) When heat is applied to a body, a part of it is used to increase the internal energy *i.e.* the temperature of a body and a part is used to do some external work.

Heat supplied = Increase in internal energy + External work done

$$Q = du + dw$$

**Mathematical Statement :**

(*i*) Let us consider a system in a state $A$ and in a state $B$. Let the energy of this system in these states be $E_A$ and $E_B$ respectively.

(*ii*) Suppose that when the system undergoes change from state $A$ (*Initial State*) to state $B$ (*Final State*), it absorbs heat ($q$) from the surroundings and also performs some work ($w$).

(*iii*) The absorption of heat by the system tends to raise the energy of the system while the work done by the system tends to lower the energy of the system, since the performance of work requires expenditure of energy.

(*iv*) Thus the change of internal energy ($\Delta E$) of the system accompanying this process will be equal to $E_B - E_A$ or $q - w$ *i.e.*

$$\Delta E = E_B - E_A = q - w$$

or $\Delta U = q - w$ $\Delta U$ = Internal energy

When small changes are involved it can be written as $dE = dq - dw$

or $dU = dq - dw$. (in the differential form).

(*v*) If heat '*q*' is supplied to a system and suppose work '*w*' is done on the system, then formula for the change of internal energy of the system is as follows:

$$\Delta E = q + w \quad \text{or} \quad \Delta U = q + w.$$

- **Enthalpy or Heat Content :**

Enthalpy is defined as the heat content (*H*) of a system given by sum of the internal energy (*U*) and the pressure volume (*pV*) work.

$$H = E + PV \quad \text{or} \quad H = U + PV$$

(*i*) The pressure (*P*) and volume (*V*) of a system are thermodynamic variables and their product *PV* is expressed as energy. The sum of the two energy terms associated with the systems namely, internal energy '*U*' and the *PV* energy, is universally represented by *H* and is called enthalpy or heat content of the system.

(*ii*) $H = U + PV$

In the differential form $dH = dU + PdV + VdP$

- **Change in enthalpy of a reaction :**

(*i*) Suppose a chemical system with internal energy and volume $V_1$ undergoes reaction at constant pressure to form a system of the products having internal energy and volume $V_2$. If enthalpies are $H_1$ and $H_2$ respectively, then

$$H_1 = U_1 + PV_1$$

$$H_2 = U_2 + PV_2$$

$$H_{2\text{Products}} - H_{1\text{Reactants}} = U_2 + PV_2 - (U_1 + PV_1)$$

or

$$H_2 - H_1 = U_2 + PV_2 - U_1 - PV_1$$

$$= U_2 - U_1 + P(V_2 - V_1)$$

$$\Delta H = \Delta U + P\Delta V$$

(*ii*) **$\Delta H$ at constant pressure and $\Delta U$ at constant volume**

$$\Delta U = q - P\Delta V \text{ (from first law of thermodynamics)}$$

If the volume is constant, $\Delta V = 0$

$$\Delta H = \Delta U + P\Delta V$$

Substituting the value of $\Delta U$

$$\Delta H = q - P\Delta V + P\Delta V = q$$

$$\Delta H = q_p \text{ at constant pressure.}$$

**Units and Sign Conventions of ΔH**

In C.G.S. system = K. Calories

In S.I. = K Joules.

For endothermic reactions **$\Delta H$ is positive** and for exothermic reaction **$\Delta H$ is negative.**

**Limitations of the First Law :**

(*i*) It does not predict whether a certain process can occur spontaneously or not and if so in which direction.

(*ii*) It does not tell whether a gas can diffuse from a lower pressure to higher pressure or not.

(*iii*) It does not explain why the water does not flow from lower level to higher level.

(*iv*) First law states that energy from one form is converted into another form of an equivalent amount of energy. However, it has been observed that various forms of energy are completely transformed into other, heat energy cannot be completely converted into an equivalent amount of work without producing some changes elsewhere.

(*v*) It does not tell in which state it will occur.

**Application of I^st Law:**

**(*a*) Change at constant volume**

(*i*) For an infinitesimally small change $dE = dq - dw$. If the system be a gaseous one and is under a pressure $P$ and the increase in volume be $dv$, then the work done = $P \,.\, dv$. So the equation will be $dE = dq = P \,.\, dv$ or $dq = dE = P \,.dv$.

(*ii*) At constant volume $dv = 0$ $\quad \therefore\ dq = dE$

(*iii*) So for a gaseous system, the heat absorbed at a constant volume is equal to the increase in internal energy of the system.

**(*b*) For a cyclic process**

For a cyclic process, there will be no change in internal energy of the system. *i.e.* $\Delta E = 0$

$\therefore\ \Delta E = q - w$ or $0 = q - w$ $\quad \therefore\ w = q$

*i.e.* work done for a cyclic system is equal to the amount of heat absorbed.

**(c) For adiabatic process**

No heat exchanged *i.e.* $q = 0$

$\therefore\ \Delta E = q - w$ or $w = -\Delta E$

*i.e.* for an adiabatic system the work done is equal to the decrease in internal energy of the system.

**• Zeroth Law of Thermodynamics:**

When two bodies have equality of temperature with a third body, they in turn have equality of temperature with each other.

**Physical Significance of Enthalpy**

(*i*) Every system or substance has some definite energy stored in it, called the internal energy.

(*ii*) The energy stored within the system or substance that is available for conversion into heat is called heat content or *enthalpy* of the system or substance.

(*iii*) Absolute value of the heat content or enthalpy cannot be measured. In thermodynamic process, only with the changes in enthalpy ($\Delta H$) can easily be measured.

**• Heat capacity:**

The heat capacity of a system is defined as the amount of heat required to raise the temperature of the system through 1°C.

$$C = \frac{q}{T_2 - T_1} = \frac{q}{\Delta T} \qquad \begin{bmatrix} q = \text{amount of heat supplied to the system} \\ T_2 - T_1 = \text{Temperature rises from } T_2 - T_1 \end{bmatrix}$$

- **Heat capacity at constant volume (*Cv*):**

(*i*) It may be defined as the rate of change of internal energy with temperature at constant volume.

(*ii*) $Cv = \frac{dE}{dT}$

- **Heat capacity at constant pressure (*Cp*):**

It may be defined as the rate of change of enthalpy with temperature at constant pressure.

$$Cp = \frac{dH}{dT}$$

**Second Law:**

**All spontaneous processes (natural occurring) are thermodynamically irreversible and are accompanied by heat that increases entropy of the universe continuously.**

(*i*) All spontaneous processes like flow of water down the hill, flow of heat from hot end to cold end, diffusion of gases from high pressure to low pressure etc are thermodynamically irreversible.

(*ii*) The complete conversion of heat into work is impossible without leaving some effects elsewhere.

(*iii*) It is impossible to construct a machine, functioning in cycles which can convert heat completely into the equivalent amount of work producing changes elsewhere.

(*iv*) Heat cannot be transferred to a hotter body from a colder one without the help of external agency (machine).

(*v*) Entropy of universe continuously increases by every spontaneous process.

- **Spontaneous Process:**

It may be defined as a process which proceeds of its own accord without the help of external agency. It is also known as natural or irreversible process.

- **Characteristics :**

(*i*) All spontaneous processes are unidirectional.

(*ii*) It may or may not be instantaneous, some of this process may be slow or rapid.

(*iii*) All spontaneous processes have a natural urge to occur.

(*iv*) Once a system attains equilibrium, no further spontaneous change will take place.

(*v*) Entropy of a system will increase by spontaneous process.

**Example:**

(*i*) When common salt is dropped in water, it dissolves spontaneously.

(*ii*) A piece of zinc reacts with copper sulphate spontaneously.

$$Zn + CuSO_4 \longrightarrow ZnSO_4 + Cu$$

**Factors Determining the Spontaneous Process:**

1. Enthalpy factor
2. Randomness factor
3. Driving force

- **Plank's definition:** It is impossible to construct a machine working in cycle which can convert heat completely into an equivalent amount of work without producing any additional changes elsewhere.

- **Kelvin's definition:** It is impossible to construct a heat engine operating in cycles which can perform work at the expense of heat obtained from a thermal reservoir.
- **Clausius's definition:** It is impossible for a self-acting machine, unaided by any external agency, to transform heat from a colder body to a hotter body.

- **Explanation of the Second Law of Thermodynamics :**

(*i*) According to this law, a process can take place spontaneously if the total entropy of the system and its surroundings increases during the process.

$$(\Delta S_{\text{System}} + \Delta S_{\text{Surroundings}}) > 0$$

(*ii*) The entropy of the universe tends to rise progressively to a maximum, because each spontaneous process in it increases the entropy of the universe.

$$\Delta S_{\text{Universe}} = (E\Delta S_{\text{System}} + E\Delta S_{\text{Surroundings}}) > 0$$

(*iii*) Under constant temperature and pressure, the relationship between the free energy change ($\Delta G$) of a reacting system and the change of entropy ($\Delta S$) is given by

$$\Delta G = \Delta H - T\Delta S$$

$\Delta G$ = Change of free energy.
$\Delta H$ = Change in enthalpy of a system.
$\Delta S$ = Change of entropy.
$T$ = Absolute temperature.

(*iv*) A physical or chemical process proceeds with a decline in free energy until they reach an equilibrium where the free energy of a system is at minimum. In case of biochemical reactions, $\Delta H$ is approximately equal to $\Delta E$, the total change in internal energy of a reaction. The above equation may be expressed as

$$\Delta G = \Delta E - T\Delta S$$

(*a*) If $\Delta G$ is negative, then the system will function spontaneously with decrease in free energy (*exogenic*)

(*b*) If $\Delta G$ is positive, then the system will function only if energy is supplied to it (*endogenic*)

(*c*) If $\Delta G$ is zero, the system is at *equilibrium* and no net change takes place.

(*v*) Thus a decline in $\Delta G$ is accompanied by an increase in $T\Delta S$. These are equal if there is no heat transfer between the system and surroundings.

If a reaction proceeds with a decline in free energy we call it spontaneous process.

Clausius combined the first and second laws as the energy of this universe remains constant while the entropy of the universe tends towards maximum.

- **In Ecosystem :**

$\Delta G$ = Represents the metabolizable energy or the material assimilated.
$H$ = The energy actually available for growth and maintenance.
$\Delta S$ = The energy loss through entropy (*respiratory energy*)

(*i*) The ecosystem can be considered as *thermodynamic unit* with negative $\Delta G$. It functions as long as it receives energy.

(*ii*) The energy received from solar radiations is channelled through the food chain. The food chain has different trophic levels and energy is transferred from one trophic level to the other.

(*iii*) Each organism in the food chain obtains an amount of energy at a particular level and transfers an amount to the next in accordance with the laws governing energy transfer, such that energy gain and loss are balanced, the starting point being transduction of light energy into chemical potential energy by plants.

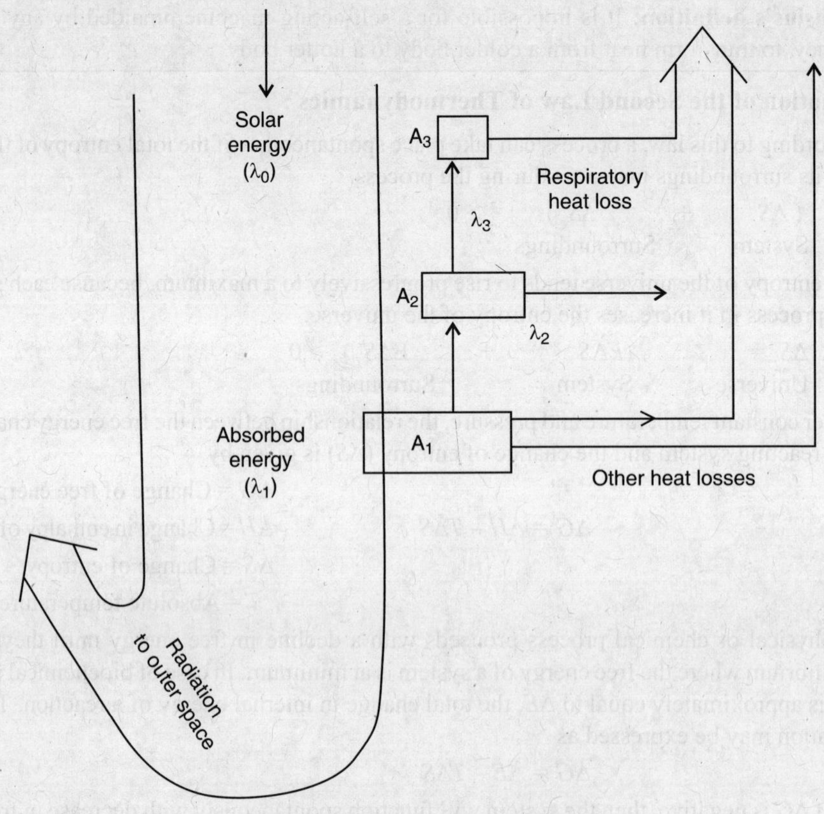


**Fig 11.1** Trophic dynamic concept

This flow diagram shows that at each transfer, heat energy (random form) dissipates. Hence the energy transfer is not 100% efficient and there is degradation of energy from a non-random to random form (Second law of thermodynamics).

- **Internal Energy:**

(*i*) It is an energy associated with a system by virtue of its molecular constitution and the motion of the molecules.

(*ii*) The energy gained by the molecular constitution is known as *internal potential energy* and *internal kinetic energy* is developed due to the motion of the molecules.

$$\text{Total Internal Energy} = \text{Internal Potential Energy} + \text{Internal Kinetic Energy}$$

(*iii*) Internal energy is symbolized by $E$ or $U$.

**Entropy :**

**Definition : It is defined as a property of a system which measures the extent of disorder or randomness of the system and becomes maximum in a system as it approaches true equilibrium.**

(*i*) This term is introduced by R.J.E Clausius and denoted by symbols.

(*ii*) During a spontaneous change the entropy of an isolated system increases.

(*iii*) The change of entropy of a system depends only on the initial and final states of the system and does not depend on the path. So the change of entropy can be expressed as

$$\Delta S = S_{\text{final}} - S_{\text{initial}}$$

(*iv*) **The change of entropy of a system is defined as the integral of all terms involving heat exchanged (*q*) divided by the absolute temperature (*T*) during each infinitesimally small change of the process being carried reversibly.**

The heat changed $dq$ and the temperature $T$ are thermodynamic quantities. The system has a thermodynamic function whose change is measured by $\frac{dq}{T}$, is independent of the path transformation of the system. This function has been called entropy by Clausius and is denoted by the symbol ***S***. Its change $dS$ is measured by the ratio of the reversible heat change and the temperature at which the heat change occurs *i.e.*

$$dS = \frac{dq_r}{T} \qquad [r = \text{ indicating reversibility}]$$

**Characteristics of Entropy Function :**

(*i*) Entropy of a system is a measure of its **disorderliness** or **randomness**. With the increase in the disorderliness of a system, the value of its entropy also increases and hence the value of entropy change ($dS$) becomes positive ($dS > 0$).

(*ii*) The entropy of a system is a definite property of the system just like its **internal energy** and **enthalpy**.

(*iii*) The change in the value of entropy of a system in going from one state to the other is independent of the path through which the change takes place.

(*iv*) The value of entropy depends on the independent variables which are used to define the state of the system.

(*v*) Entropy of the universe is increasing.

(*vi*) The value of entropy change ($dS$) for a cycle change of the state (where the system comes back to its initial state) is always zero (0).

(*vii*) As long as the spontaneous process in a system continues, its entropy goes on increasing and when the process reaches the equilibrium state, the entropy of the process becomes maximum.

- $\Delta S$ should be positive (*i.e.* $\Delta S > 0$) in a spontaneous and irreversible process in an isolated system.
- For any reversible process change of entropy is zero. *i.e.* $\Delta S = 0$
- For any reversible process total change of entropy.

  $\Delta S_{\text{total}} = \Delta S_{\text{system}} + \Delta S_{\text{surounding}} = 0$
- In a spontaneous process total change of entropy.

  $\Delta S_{\text{total}} = \Delta S_{\text{system}} + \Delta S_{\text{surrounding}} > 0$
- In any spontaneous process,the entropy of universe increases.

  $\Delta S_{\text{univ}} > 0.$

**Significance :**

(*i*) The spontaneous process is accompanied by an increase in the 'disorder' or "randomness" of the molecule. Therefore entropy may be regarded as "*a measure of disorder or randomness of the molecular arrangements in a system*".

(*ii*) A spontaneous process always proceeds from a less probable state to a more probable state by the interplay of natural forces. Therefore, entropy is a function of the probability of the thermodynamic state.

(*iii*) *Melting of solid:* In solid state, there is no disorder in molecules or atoms or ions but when it changes to liquid randomness is increased.

(*iv*) The vaporization of liquid is accompanied by a net increase of entropy.

**Standard Entropy of a Substance ($S^{\circ}$):**

Entropy of 1 mol of a substance in its pure state at 1 atm and 25°C is called the standard entropy of that substance and it is denoted by $S^{\circ}$.

- **Standard Entropy Change ($\Delta S^{\circ}$) of a Reaction**

(*i*) When each of the reactants and products in a chemical reaction is in its standard state (1 atm. pressure and 25°C), the entropy change is called standard entropy change and is denoted by $\Delta S^{\circ}$.

(*ii*) $\Delta S^{\circ} = \sum S^{\circ}_{\text{Products}} - \sum S^{\circ}_{\text{Reactants}}$

**Residual Entropy:** The entropy of a system as it approaches absolute zero is called residual entropy.

**Units of Entropy :**

Since $\Delta S$ or $dS = \dfrac{q}{T} = \dfrac{dq}{T} = \dfrac{\text{Heat change}}{\text{Absolute temperature}}$

(*a*) In C.G.S. $dS = \dfrac{dq}{T}$ = calories × degrees$^{-1}$ = cal. deg$^{-1}$

(*b*) In S. I. units $dS = \dfrac{dq}{T} = \dfrac{\text{Joules}}{\text{degrees in Kelvin}}$ = Joules. Kelvin$^{-1}$ = J.K$^{-1}$

**Entropy in Biological System:**

(*i*) Biological organisms (plants and animals) are able to maintain their biology by taking in free energy of nutrients (or sunlight) from environment and they return equal amount of energy (in the form of heat) in less useful form to their environment.

(*ii*) It implies that living organisms constantly produce entropy by dissipating less useful energy in the environment.

**Example:**

Oxidation of glucose in respiration:

$$C_6H_{12}O_6 + 6O_2 \longrightarrow 6CO_2 + 6H_2O + \text{Energy} \begin{cases} \nearrow \text{Heat} \\ \searrow \text{ATP} \end{cases}$$

(*i*) The energy is conserved in the form of ATP which is used in various endogenic biochemical reactions. On the other hand, heat is produced as less useful energy. It results in increase of *entropy*.

(*ii*) Furthermore, number of molecules increases in the reactions *i.e.* one glucose molecule on oxidization with 6 molecules of oxygen yields 6 molecules of $CO_2$ and 6 molecules of water *i.e.* (6 + 6) = 12 molecules. Thus total 7 (1 + 6) reactants molecules produce 12 (6 + 6) molecules product *i.e.* increase in the number of molecules in this process results in molecular disorder *i.e.* increase of *entropy*.

(*iii*) Bioenergetic equation is $\Delta G = \Delta H = T\Delta S$

$\Delta S$ can be calculated if the values of $\Delta G$, $\Delta H$ and $\Delta T$ are known.

At constant temperature (298 K) and pressure (1 atm) various values of oxidation of glucose are given below.

$\Delta G = -686000$ cal/mol.

$\Delta H = -673{,}000$ cal/mol.

T = 298 K

$$\Delta S = \frac{\Delta H - \Delta G}{T} = \frac{-673000 - (-686000)}{298\,K} = \frac{13000}{298} = 43.62 \text{ cal/mol.}$$

Entropy increases by 43.62 cal/mol.

- **Entropy of the Universe is increasing:**

**(Delhi Univ. 1995, 1998)**

(*i*) All processes are thermodynamically irreversible and these are accompanied by increase of entropy.

(*ii*) The main ideas of the first and second laws of thermodynamics may be summed up as "the energy of the universe is constant whereas entropy of the universe is continuously increasing and tends to a maximum value."

(*iii*) Hence it may be concluded that entropy of the universe is increasing.

**Third Law:**

**Every system has a finite positive entropy, but at the absolute zero temperature, the entropy may become zero and does so in the case of pure crystalline substance.**

(*i*) It may be stated as 'the entropy of all perfectly crystalline solids may be taken as zero at absolute zero temperature.'

(*ii*) At absolute zero there is no thermal motion of a substance if in a crystal the atoms or ions are in a highly regular arrangement, so that there is no disorder, then such crystalline substance has zero entropy.

(*iii*) In can be stated that the entropy of all pure crystalline substances may be zero at absolute zero.

(*iv*) It may be stated that during a physical or chemical transformation the change of entropy tends to zero as the temperature tends to absolute zero. This is also known as **Nernst heat theorem**.

- **Mathematical statement:**

Entropy of a substance at solid state *i.e.*

$$\Delta S = S - S_0 = \int_0^T \frac{C_p}{T}\,dT$$

Since $S_0$ is zero at 0 K for perfectly crystalline substances, then

$$S = \int_0^T \frac{C_p}{T}\,dT$$

$S$ = Entropies at $T$ K
$S_0$ = Entropies at 0 K
$C_p$ = Heat capacity

**Gibbs Free Energy Function:**

The spontaneity of any reaction or process cannot be predicted by enthalpy (*H*) or entropy (*S*) of a system.

(*i*) Internal energy (*E*), enthalpy (*H*) and entropy (*S*) are the linear combination of *thermodynamic function*.

(*ii*) Helmholtz's free energy and Gibbs free energy functions are thermodynamic potential.

(*iii*) Free energy may be regarded as a measure of potential energy of the substances but its quantity cannot be measured directly.

**Definition: It is defined as the maximum amount of energy available to a system that can be converted into useful work.**

(*i*) It is denoted by the symbol '*G*' and is given by relation

$$G = H - TS \qquad \begin{bmatrix} G = \text{Free energy} \\ H = \text{Enthalpy of the system} \\ T = \text{Absolute temperature} \\ S = \text{Entropy of the system} \end{bmatrix}$$

(*ii*) Gibbs free energy determines the ability of a system for doing useful work.

(*iii*) A part of the total work done by a system may be used to perform the pressure, volume work of the system.

**Mathematical representation of free energy:**

(*i*) Since *H*, *T* and *S* are the function of the state only, therefore '*G*' is also function of the state of the system only. Hence for isothermal process occurring at temperature *T*, we can write:

$$G_1 = H_1 - TS_1 \qquad ...(i) \text{ for initial state.}$$
$$G_2 = H_2 - TS_2 \qquad ...(ii) \text{ for final state.}$$

(*ii*) **J.W. Gibbs** introduced the free energy function by combining the first and second laws of thermodynamics.

From equations (*i*) and (*ii*)

$$G_2 - G_1 = (H_2 - TS_2) - (H_1 - TS_1) = H_2 - TS_2 - H_1 + TS_1$$
$$= (H_2 - H_1) - T(S_2 - S_1)$$

$$\Delta G = \Delta H - T\Delta S \qquad ...(iii) \qquad \begin{bmatrix} \Delta G = \text{Change in Gibbs's free energy of the system.} \\ \Delta H = \text{Enthalpy change of the system.} \\ \Delta S = \text{Entropy change of the system.} \end{bmatrix}$$

(*iii*) Equation (*iii*) provides quantitative interrelationship between the changes in energy, heat and entropy in chemical reactions taking place at constant temperature (*T*) and pressure (*P*) prevailing in biological system.

(*iv*) The term $T\Delta S$ is that fraction of $\Delta H$ cannot be converted into useful work. These theoretically available useful work cannot be utilized.

(*v*) The enthalpy change, $\Delta H$ is given by the following equation:

$$\Delta H = \Delta E + P\Delta V. \qquad \begin{bmatrix} \Delta E = \text{The change in internal energy of reaction.} \\ \Delta V = \text{The change in the volume of reaction.} \end{bmatrix}$$

As the volume changes, $\Delta V$ has been found to be small (all biochemical reactions) hence $\Delta H$ has been nearly equal to the change of internal energy $\Delta E$.

$$\therefore \quad \Delta G = \Delta E - T\Delta S \qquad [\because \Delta H = \Delta E]$$

(*vi*) The change in free energy reaction *i.e.* $\Delta G$ will depend both on the change in internal energy and on the change of entropy of the system.

- **Gibbs-Helmholtz equation:**

At constant temperature

$$\Delta G = \Delta H - T(S_2 - S_1) = \Delta H - T\Delta S$$

This equation is known as Gibbs Helmholtz equation.

- **Prediction of ΔG in the light of Gibbs's-Helmholtz equation**

If $\Delta G$ is negative or $\Delta G < 0$ reaction would be spontaneous. If $\Delta G$ is positive or $G > 0$ reaction would be non-spontaneous.

(*i*) The process is continuous if $\Delta G$ is negative.

(*ii*) The process is in equilibrium if $\Delta G = 0$. At equilibrium the concentration of each species of both reactants and products will remain constant. There is no net reaction on either direction.

(*iii*) If $\Delta G$ is positive, the process is non-spontaneous. So the process cannot take place in the forward direction. It may take place in the reverse direction.

(*iv*) If $\Delta H$ is negative and $\Delta S$ is positive then $\Delta G$ is negative. So the reaction is spontaneous at any temperature.

(*v*) If $\Delta H$ is positive and $\Delta S$ is negative, the $\Delta G$ is negative, the $\Delta G$ is positive at any positive at any temperature, then the reaction is non-spontaneous at any temperature.

(*vi*) If $\Delta H$ is negative and $\Delta S$ is negative then the reaction is spontaneous only at low temperature *i.e.* $\Delta G$ is negative. But $\Delta G$ is positive at high temperature in this case. So in that case reaction is non-spontaneous.

(*vii*) If $\Delta H$ is positive and $\Delta S$ is positive then $\Delta G$ is negative at high temperature. So the reaction is spontaneous at high temperature. But in this case $\Delta G$ is positive at low temperature. So in that case the reaction is non-spontaneous.

- **Free Energy and Chemical Equilibrium:**

A reaction is said to be in equilibrium if $\Delta G = 0$. In such case, the reaction can proceed in both directions.

$$\Delta G = \Delta H - T\Delta S \qquad \text{(Gibbs's-Helmholtz reaction)}$$

At equilibrium, $\Delta G = 0$, $T\Delta S = \Delta H$

or $$T_{eq} = \frac{\Delta H}{\Delta S}$$

(*i*) When $\Delta H$ is –ve and $\Delta S$ is +ve, $T_{eq}$ will be negative.

$$T_{eq} = \frac{-\text{ve}}{+\text{ve}} \quad i.e.\ T_{eq} = -\text{ve}\ i.e. \text{ the reaction is never at equilibrium.}$$

(*ii*) When $\Delta H$ is +ve and $\Delta S$ is –ve then $T_{eq}$ will be negative.

$$T_{eq} = \frac{+\text{ve}}{-\text{ve}} = -\text{ve}\ i.e. \text{ reaction is also never at equilibrium.}$$

(*iii*) When $\Delta H$ and $\Delta S$ both are +ve. Then $T_{eq}$ will be positive.

$$T_{eq} = \frac{-\text{ve}}{+\text{ve}} = +\text{ve}$$

Now if the reaction is carried out at a temperature higher than $T_{eq}$. ($T > T_{eq}$) then $\Delta G$ will be negative (–ve) and the reaction would be spontaneous.

On the other hand, the reaction will be non-spontaneous if the reaction is carried out at a temperature lower than $T_{eq}$ ($T < T_{eq}$), the reaction will be spontaneous in backward direction.

(*iv*) When $\Delta H$ and $\Delta S$ both are –ve, then $T_{eq}$ will be positive (+ve). If the reaction is carried out at a temperature greater than $T_{eq}$ $(T > T_{eq})$, $\Delta G$ will be +ve and reaction will be non-spontaneous. On the other hand, $\Delta G$ will be –ve and the reaction will be spontaneous, if the reaction is carried out at a temperature lower than $T_{eq}$ $(T < T_{eq})$.

- **Physical Significance of Gibbs's free Energy:**

At constant temperature ($T$)

$$\Delta S = \frac{\Delta Q_{rev}}{T}$$

or $$\Delta Q_{rev} = T\Delta S$$

At constant pressure ($P$)

$$\Delta H = \Delta E + P\Delta V \quad ...(i)$$

We know $\Delta G = \Delta H - T\Delta S$

Substituting the value of $T\Delta S$ and $\Delta H$ from equation (*i*)

$$\Delta G = \Delta E + P\Delta V - \Delta Q_{rev}$$

$$\Delta G = \Delta E - \Delta Q_{rev} + P\Delta V \quad ...(ii)$$

Now according to the first law of thermodynamics,

$$\Delta Q_{rev} = \Delta E + w_{max.}$$

$$\Delta E - \Delta Q_{rev} = -w_{max.}$$

Substituting the value in equation (*ii*), we get

$$\Delta G = -w_{max.} + P\Delta V$$

$$-\Delta G = w_{max.} - P\Delta V$$

(*i*) But $P\Delta V$ is the work of expansion done by the system corresponding to the increase in volume $\Delta V$.

(*ii*) Hence $(w_{max.} - P\Delta V)$ stands for the maximum work other than the work of expansion. This is called the *maximum* useful work *available from the process*.

(*iii*) **Hence we conclude that for a process occurring at constant temperature and constant pressure the change in Gibbs's free energy is equal to the maximum useful work obtainable from the process *i.e.* the total work minus the pressure volume work.**

- **Free energy:**

According to Gibbs-Helmholtz. $G = H - TS$.

Since $H = U + PV$ [Enthalpy = Some of the internal energy and the pressure volume work]

Substituting the value of $H$ in the equation of $G = H - TS$

$$G = U + PV - TS$$

This is the mathematical statement of free energy.

- **Free Energy Change : ($\Delta G$)**

We know $G = U + PV - TS$

$$\therefore \quad \Delta G = \Delta U + (PV) - \Delta(TS).$$

If the process takes place at constant pressure, the above equation will be

$$\Delta G = \Delta U + P\Delta V - T\Delta S$$

As we know that

$$\Delta H = \Delta U + P\Delta V$$

After substituting the value of $\Delta U + P\Delta V$ the equation becomes

$$\Delta G = \Delta H - T\Delta S$$

This equation is known as Gibbs's-Helmholtz equation.

**Net Work:**

The quantity $P\Delta V$ is a work done by the gas on expansion against external pressure $P$. Thus $\Delta G$ gives the maximum work obtained from a system other than that due to change in volume at constant temperature and pressure. The work other than that due to change of volume is called net work.

Mathematically net work $(-\Delta G) = W_{rev} - P\Delta V$

$$-\Delta G = W_{rev} - P\Delta V.$$

- **Standard free energy formation of a compound ($G°$)**

It is the free energy formation of a compound at 25°C at 1 atm. pressure.

- **Standard free energy change ($\Delta G°$)**

The standard free energy change of a chemical reaction may be defined as the free energy change when the reactants and products are in their standard state (298 K, 1 atm. pressure).

**Table 11.1 Free Energy Reaction: Effect of Sign of $\Delta H$ and $\Delta S$**

| | $\Delta H$ | $\Delta S$ | | $\Delta G = \Delta H - \Delta S$ |
|---|---|---|---|---|
| 1. | –ve | +ve | –ve | The reaction is spontaneous at all temperature. |
| 2. | +ve | –ve | +ve | The reaction is non-spontaneous at all temperature. |
| 3. | –ve | –ve | –ve (at low temp.) | The reaction is spontaneous at low temperature. |
| 4. | –ve | –ve | +ve (at high temp.) | The reaction is non-spontaneous at high temperature. |
| 5. | +ve | +ve | +ve (at low temp.) | The reaction is non-spontaneous at low temperature. |
| 6. | +ve | +ve | –ve (at high temp.) | The reaction is spontaneous at high temperature. |

- **Differences between Gibbs's free energy and Helmholtz free energy.**

(*i*) Gibbs's free energy is represented by the symbol $G$ and Helmholtz free energy is represented by the symbol $A$.

$$A = E - TS$$

$$G = H - TS$$

(*ii*) $-\Delta A$ gives the maximum work obtainable from a system whereas $-\Delta G$ gives the maximum useful work obtainable from the system.

**The feasibility of a reaction with the help of free energy change ($\Delta G$)**

1. For the spontaneous irreversible reaction $\Delta G < 0$.
2. For the reversible spontaneous reactions $\Delta G = 0$.
3. For the non-spontaneous reaction $\Delta G > 0$; in other words $\Delta G$ is positive.

- **Carnot cycle:**

It is a process by which maximum conversion of heat into work occurs in a cyclic process in which all the necessary intermediate steps are carried out reversibly. Such reversible cycle is called *Carnot cycle*. The engine working on the basis of this cycle is called Carnot engine or heat engine.

- **Efficiency of Carnot cycle or Heat engine: ($\Sigma$)**

(*i*) Efficiency of heat engine is defined as the fraction of heat absorbed by the engine which it can convert into work.

(*ii*) It is represented by Σ *i.e.* $\Sigma = \frac{q_2 - q_1}{q_2} = \frac{T_2 - T_1}{T_2}$

$q_2$ = Heat absorbed by the engine at higher temperature $T_2$

$q_1$ = Heat evolved by the engine at lower temperature $T_1$.

(*iii*) Since $\frac{T_2 - T_1}{T_1}$ is always less than unity, the efficiency of heat engine is always less than unity.

- **Joule-Thomson effect:**

Joule-Thomson studied the relation of fall of pressure with lowering temperature (accompanied by cooling) during adiabatic expansion by porns plug experiment. This is phenomenon of change of temperature produced when gas expands adiabatically from higher pressure to lower pressure is known as *Joule-Thomson effect*.

- **Ideal gas shows neither heating nor cooling in Joule-Thomson experiment.**

(*i*) The intermolecular forces of attraction (*i.e.* Van der Waal's force) in an ideal gas are negligible.

(*ii*) Hence no energy is used up in overcoming these forces of attraction when the gas expands adiabatically.

Thus the internal energy of the gas does not fall and therefore the temperature also does not fall.

- **Joule-Thomson Coefficient (μ):**

Joule-Thomson coefficient (μ) may be defined as the temperature change in degrees produced by a drop of one atmospheric pressure when the gas expands under conditions of constant enthalpy.

$$\mu = \left(\frac{dT}{dP}\right)_H$$

(*i*) For cooling, μ will be positive (because $dT$ and $dP$ both will be negative)

(*ii*) For heating, μ will be negative (because $dT$ is positive while $dP$ is negative).

(*iii*) If $\mu = 0$, the gas gets neither heated up nor cooled on adiabatic expansion (because $\mu = 0$ only if $dT = 0$ for any value of $dP$).

- **Inversion temperature:**

(*i*) Every gas has a definite temperature (at a particular pressure) at which μ (*J*-Thomson coefficient) = 0. Below this temperature μ is positive and above this temperature μ is negative.

(*ii*) The temperature (at a particular pressure) at which $\mu = 0$ *i.e.* the gas neither cooled down nor heated up on adiabatic expansion and below which μ is positive on adiabatic expansion and above which μ is negative is called *inversion temperature*.

- **Important formulae based on the first law of Thermodynamics:**
  - Isothermal process : $q = w\ (\Delta E = 0)$
  - Adiabatic process : $\Delta E = -w\ (q = 0)$
  - Cyclic process : $q = w\ (\Delta E = 0)$
  - Isochoric process : $q = \Delta E\ (\Delta V = 0, \therefore P\Delta V = w = 0)$
  - Isobaric process : $q_p = \Delta H\ (\Delta P = 0)$

- **When ΔH = ΔU?**

For a reaction at constant pressure involving ideal gases if $\Delta n$ (difference of total number of moles of products and reactants) is zero, then

$$\Delta H = \Delta U$$
$$\Delta H = \Delta U + RT_{\Delta n}$$

if $$\Delta n = 0, \Delta H = \Delta U$$

**Answer with reason? (Statement is correct or incorrect)**

(*a*) *Every isolated system is closed.*

**Ans:** (*i*) For closed system, mass is not exchanged with surroundings.

(*ii*) For isolated system, neither mass nor heat is exchanged with surroundings. So the statement is correct.

(*b*) *Every closed system is isolated.*

**Ans:** (*i*) *The statement is incorrect.*

(*ii*) Closed system exchanges heat with surroundings but isolated system neither exchanges heat nor mass with the surroundings.

- **Bond Energy**

**How does bond energy help to determine the heat of a reaction?**

The bond energy of a particular bond is defined as the average amount of energy released when one mole of bonds are formed from isolated, gaseous atoms or the amount of energy required when one mole of bonds are broken so as to get the separated gaseous atoms.

(*a*) (*i*) For diatomic molecules (like $H_2$, HCl, etc.), the bond energy is equal to the *dissociation energy* of the molecules.

(*ii*) But for a polymorphic molecule like $CH_4$ the bond dissociation energies of the four C–H bonds are different. Hence an average value is taken.

(*b*) (*i*) Heat reaction can be calculated from bond energy data.

$\Delta H_{reactant}$ = Σ bond energies of reactants – Σ Bond energy of products.

## PROBLEMS

**1.** *A system is provided 100 Joules of heat and work done on the system is 20 Joules. What is the change in the internal energy of the system?*

**Solution:**

$\Delta Q$ = 100 Joules $\Delta W$ = 20 Joules

$$\Delta U = \Delta Q - \Delta W$$
$$= 100 - 20$$
$$= 80 \text{ Joules}$$

**2.** *A gas is contained in a vessel fitted with a movable piston. The container is placed on a hot stove. A total of 100 Cal of heat is given to the gas and the gas does 40 J of work in the expansion resulting from heating. Calculate the increase in internal energy in the process.*

**Solution:**

Heat given to the gas is $\Delta Q$ = 100 Cal = 418 J [1 Cal = 4.186 Joule]

Work done by the gas is $\Delta W$ = 40 J

$$\Delta U = \Delta Q - \Delta W$$
$$= 418 \text{ J} - 40 \text{ J} = 378 \text{ J}$$

**3.** *Calculate standard entropy change in the following reaction.*

$$2FeS(s) + 3O_2(g) \longrightarrow 2FeO(s) + 2SO_2(g)$$

*Given that* $S^\circ_{FeS} = 67.4\ J\ K^{-1},\ S^\circ_{O_2} = 205.0\ J\ K^{-1}$

$S^\circ_{FeO} = 54\ J\ K^{-1}$ *and* $S^\circ_{SO_2} = 248.5\ J\ K^{-1}$

**Solution:**

We know
$$\Delta S^\circ = \sum S^\circ_{\text{Products}} - \sum S^\circ_{\text{Reactants}}$$
$$= [2S^\circ_{FeO} + 2S^\circ_{SO_2}] - [2S^\circ_{FeS} + 3S^\circ_{O_2}]$$
$$\Delta S^\circ = [2 \times 54 + 2 \times 248.5] - [2 \times 67.4 + 3 \times 205.0]$$
$$= [108 + 497] - [134.8 + 615]$$
$$= 605 - 749.8 = 144.8 \text{ J K}^{-1}$$

**4.** *Calculate standard entropy change in the following reaction.*

$$Ag_2O(s) \longrightarrow 2Ag(s) + \frac{1}{2}O_2(g)$$

*Given that* $S^\circ_{Ag_2O} = 121.75\ J\ K^{-1},\ S^\circ_{Ag(s)} = 42.76\ J\ K^{-1}\ S^\circ_{O_2(G)} = 205.01\ J\ K^{-1}$

**Solution:**

$$\Delta S^\circ = \sum S^\circ_{\text{Products}} \sum S^\circ_{\text{Reactants}}$$
$$= \left[2S^\circ_{Ag} + \frac{1}{2}S^\circ_{O_2}\right] - 2S^\circ_{Ag_2}$$
$$= \left[2 \times 42.76 + \frac{1}{2} \times 205.01\right] - 121.75$$
$$= (85.52 + 102.505) - 121.75$$
$$= 188.025 - 121.75$$
$$= 66.275$$

**5.** *Enthalpies of formation of $C_2H_5OH(l)$, $CO_2(g)$ and $H_2O(l)$ are –277.0, –393.5 and –285.8 kJ mole$^{-1}$ respectively. Calculate enthalpy change for reaction.*

$$C_2H_5OH(l) + 3O_2(g) \longrightarrow 2CO_2(g) + 3H_2O(l)$$

**Solution :**

ΔH for the reaction $C_2H_5OH(l) + 3O_2(g) \longrightarrow 2CO_2(g) + 3H_2O(l)$ is

$$\Delta H = \Delta H^\circ_{f\,(\text{Products})} - \Delta H^\circ_{f\,(\text{reactants})}$$
$$= [2\Delta H^\circ_f(CO_2) + 3\Delta H^\circ_f(H_2O)] - [\Delta H^\circ_f(C_2H_5OH) + 3\Delta H^\circ_f(O_2)]$$
$$= [2 \times (-393.5) + 3 \times (-285.8)] - [-277.0 + 3 \times 0]$$
$$= [-787 - 857.4] - [-277]$$
$$= -1644.4 + 277 = -1367.4 \text{ kJ}$$

**6.** *In an experiment in the Biochemistry laboratory, the following reaction has been taken place:*

$$CO_2(g) + H_2(g)\ CO(g) + H_2O(g).$$

*Given that $\Delta H^\circ_f$ for $CO_2(g)$, $CO(g)$ and $H_2O(g)$ are –393.5, –111.3 and –241.8 kJ/mole respectively. Calculate $\Delta H^\circ$ for the above reaction.*

**Solution :**

ΔH° for the reaction $CO_2(g) + H_2(g) \longrightarrow CO(g) + H_2O(g)$ is

$$\Delta H^\circ = [H^\circ_{f\,(\text{products})} - H^\circ_{f\,(\text{reactants})}]$$
$$= [-111.3 + (-241.8)] - [-393.5 + 0]$$
$$= (-111.3 - 241.8) + 393.5$$
$$= -353.1 + 393.5$$
$$= 40.4 \text{ kJ}$$
$$\Delta H^\circ = +40.4 \text{ kJ.}$$

[Enthalpy of every element in standard state is answered as zero.]

**7.** *If a man takes diet equivalent to 10000 kJ per day and consumes energy in all forms to a total of 12500 kJ per day, what is the change in internal energy per day? If the energy lost was stored as sucrose (1632 kJ per 100 g), how many days should it take to lose 1 kg (Ignore water loss)?*

**Solution :**

Energy taken by a man = 10000 kJ

Energy consumed in doing work = 12500 kJ

Change in internal energy per day 12500 – 10000 = 2500 kJ

The energy is lost by the man as he expends more energy than he takes.

Now 100g of sucrose = 1632 kJ loss in energy

1000g sucrose = 16320 kJ

∴ Number of days required to lose 1000 g of weight

or $\quad 16320 \text{ kJ of energy} = \dfrac{16320}{2500} = 6.52 \text{ days}$

$= 6.5 \text{ days}$

**8.** *Gargi needs about 10000 kJ energy per day. How much carbohydrates (in mass) will she have to consume; assuming that all her energy needs are met only by carbohydrates, in the form of glcuose? The enthalpy of combustion of glucose is 2900 kJ mol$^{-1}$.*

**Solution :**

Molecular mass of glucose,

$$C_6H_{12}O_6 = 6 \times 12 + 1 \times 12 + 16 \times 6$$
$$= 72 + 12 + 96 = 180$$

Combustion of glucose can be represented as

$$C_6H_{12}O_6(s) + 6O_2(g) \longrightarrow 6CO_2(g) - 6H_2O(g) + 2900 \text{ kJ}$$

2900 kJ energy is produced by 180g of glucose.

$$10{,}000 \text{ kJ energy is produced by glucose} = \frac{180}{2900} \times 10{,}000 \text{ cm} = 620.78.$$

**9.** *In the Dept. of Microbiology of R.K. Mission Vidyamandira, bio-mass gas is formed by bacterial fermentation of animal refuse in a biomass plant. It mainly contains methane and its heat of combustion is – 809 kJ mol$^{-1}$ according to the following equation.*

$$CH_4 + 2O_2 \longrightarrow CO_2 + 2HO_2;\ \Delta H = -809 \text{ kJ}$$

*How much biomass would have to be produced per day for a small village of Arambagh of 50 families, if it is assumed that each family requires 20,000 kJ of energy per day? The methane content in biomass is 80% by mass.*

**Solution :**

Energy consumption of 50 families per day

$$50 \times 20{,}000 \text{ kJ}$$
$$= 10 \times 10^5 \text{ kJ} = 1 \times 10^6 \text{ kJ}$$

809 kJ of energy is obtained by burning methane = 16 g. ($CH_4 = 12 + 4 = 16$)
$1 \times 10^6$ kJ of energy will be obtained by burning

$$= \frac{16}{809} \times 10^6 \text{ g} = 0.019777 \times 10^6$$
$$= 1.98 \times 10^4 \text{ g}$$
$$= \frac{1.98 \times 10^4}{10^3} = 19.8 \text{ kg}$$

Since methane content in biomass is 80% by mass, hence the mass of biomass needed

$$= \frac{100}{80} \times 19.8 = 24.75 \text{ kg}$$

**10.** *The heat evolved in the combustion of glucose shown in the following equation.*

$$C_6H_{12}O_6 + 6O_2(g) \longrightarrow 6CO_2(g) + 6H_2O(g);\ \Delta H = -2840 \text{ kJ}$$

*What is the energy requirement for production of 0.36 g of glucose by the reverse reaction.*

**Solution :**

The equation provided

$$C_6H_{12}O_6 + 6O_2(g) \longrightarrow 6CO_2(g) + 6H_2O(g);\ \Delta H = -2840 \text{ kJ}$$

Writing the equation in reverse order

$$6CO_2(g) + 6H_2O(g) \longrightarrow C_6H_{12}O_6(s) + 6O_2(g);\ \Delta H = 2840 \text{ kJ}$$

For production of 1 mole of $C_6H_{12}O_6 = \begin{Bmatrix} 12 \times 6 + 12 \times 1 + 6 \times 16 \\ 72 + 12 + 96 = 180 \end{Bmatrix}$

(180 g) heat required (absorbed 2840 kJ)
For 180g 2840 kJ heat absorbed

For 1g $\frac{2840}{180}$ kJ heat absorbed

For 0.36 g the heat absorbed is $= \frac{2840}{180} \times 0.36 = 5.68$ kJ

**11.** *Calculate the amount of heat supplied to Carnot's cycle working between 368 K and 288 K if the maximum work obtained is 895 Joules.*

**Solution :**

Here $T_2 = 368$ K, $W = 895$ Joules
$T_1 = 288$ K

The formula for efficiency of engine

$$\eta = \frac{W}{q_2} = \frac{T_2 - T_1}{T_2}$$
$$= \frac{895}{q_2} = \frac{368 - 288}{368}$$

or $$\frac{895}{q_2} = \frac{80}{368} \text{ or } q_2 = \frac{895 \times 368}{80} = 895 \times 4.6$$
$$= 4117 \text{ Joules}$$

CHAPTER 12

# Nuclear Magnetic Resonance (NMR)

Nuclear Magnetic Resonance (NMR), an important experimental technique currently used in Biophysics, is based on absorption spectroscopy. In Physics, NMR is used to study the nature and interaction of the atomic nucleus. On the other hand, in Chemistry it is used to determine the chemical structure of small molecules. In Biophysics and Biochemistry NMR is used to determine the structure of large molecules (proteins) and their dynamic behaviour.

**Magnetic poles:** Poles of a magnet are definite regions near the two ends of a magnet with maximum power of attraction.

When a magnet is suspended freely, a particular pole of the magnet always points towards the north and other towards the south. The former is called the *north seeking* or simply the north pole and the latter the *south seeking* or simply the south pole.

- **Pole strength:** (*i*). It measures the attractive power of a pole of a magnet. (*ii*) It is denoted by *m* or *p* or Qm. S.I unit of pole strength is Am.
- **Magnetic axis:** The line obtained by joining the poles of a magnet (NS) is called the axis of the magnet.
- **Neutral region:** At the centre of the magnet the magnetism is almost zero and is considered as neutral region.
- **Magnetic length:** The distance between the two poles of a magnet is called the magnetic length of the magnet.

**Permanent magnet and temporary magnet :**

- A magnet which can retain its magnetism for a long time is called a permanent magnet.
  **Example:** steel or tungsten steel.
- A magnet which cannot retain its magnetism for a long time is called a temporary magnet.
  **Example:** a rod of soft iron.
- **Unit pole:** If two identical poles 1 cm apart in air repel each other with a force of 1 dyne, each is said to have unit strength and is called unit pole.
- **Magnetic field:** The space surrounding a magnet in which magnetic force is exerted is called a magnetic field.

**Intensity of magnetic field**

The intensity of a magnetic field at a point is defined as the force experienced by a unit north pole placed at that point.

Let pole strength $m$ intensity at a point distance $r$, $F = \frac{m}{\mu r^2}$ *i.e* intensity at a distance $r$ is $\frac{m}{\mu r^2}$.

In the C.G.S. system the unit of intensity is one dyne per unit pole and is called *Oersted* (Oe).

- **Magnetic moment**

**The moment of a magnet (magnetic moment) is equal to the moment of the mechanical couple required to keep the magnet at right angles to a field of unit intensity,**

Magnetic moment (M) = Pole strength (m) × Effective length (2l)

$$M = 2ml.$$

***Magnetic dipole:*** A combination of two isolated, equal and opposite poles separated by a small distance constitutes a magnetic dipole. If the separation be $d$ and the strength of each pole $m$, then the moment of the dipole = m.d. The direction of this moment is from south pole to north pole.

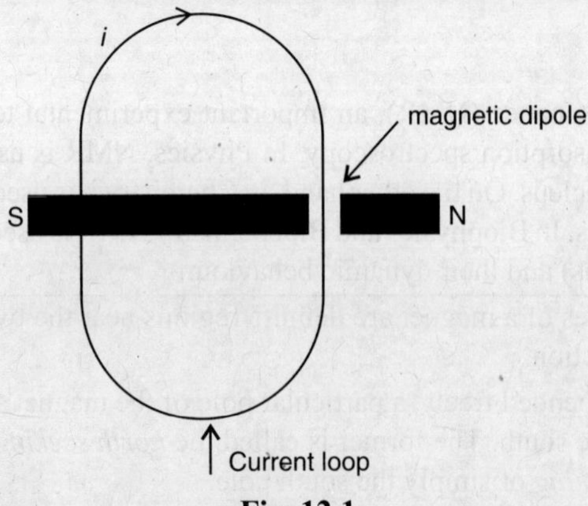


**Fig. 12.1**

**According to the modern theory of magnetism, the magnetism arises due to the orbital motion of the electrons around the nucleus and the spinning motion of the electrons.**

The magnetic moment of the electron due to its orbital motion is called orbital magnetic momentum.

(*i*) Any particle moving in a circular path has an angular momentum.

(*ii*) Let us consider an electron of mass $m$, charge $e$ moving in a circular path of radius $r$ around the nucleus. The speed of the electron is $v$. The angular momentum of the electron $L$ = m. $v$. r. It is a vector.

(*iii*) Since the electron is negatively charged, its motion is equivalent to a current $i = \frac{e}{T}$, where $T$ is the time taken for one rotation, with a speed $v$.

$$\text{i.e.} \quad v = \frac{2\pi r}{T}, T = \frac{2\pi r}{v} \quad \text{so} \quad i = \frac{ev}{2\pi r}$$

(*iv*) The magnetic moment of this current loop is $M$

$$M = iA = \frac{ev}{2\pi r} \times \pi r^2 = \frac{evr}{2}$$

(*v*) The magnetic moment $M$ is a vector, acts along a line passing through the centre of the loop, and directed vertically downwards (Fig. 12.1).

(*vi*) The angular momentum of the electron is $L = mvr$.

The orbital magnetic moment is $M = \frac{evr}{2}$.

They are oriented together in Fig. 12.1

$$L = mvr$$

or $$vr = \frac{L}{m}, \quad M = \frac{evr}{2}$$

$$\therefore \quad M = \left(\frac{e}{2m}\right)L \qquad \left[\because vr = \frac{L}{m}\right]$$

This equation shows that the magnetic momentum $M$ of the electron is proportional to its angular momentum $L$.

(*vii*) According to quantum mechanics, the angular momentum $L$ of an electron cannot take all possible values.

Its values are restricted or quantised.

$$L = mvr = n\frac{h}{2\pi} \text{ where n = 1, 2, 3 ...}$$

$$M = n\left(\frac{eh}{4\pi m}\right) = \mu_B \quad \text{where, } \mu_B = \frac{eh}{4\pi m}$$

The least value of $n = 1$ and the corresponding magnetic moment is called *Bohr Magneton* $\mu_B$. Thus *Bohr magneton* is equal to the orbital magnetic moment of an electron circulating in an orbit with the smallest allowed value of orbital angular momentum. S.I. unit of $\mu_B$ is $Am^2$ or $JT^{-1}$.

The equation $M = n\mu_B$ means that the atomic magnetic moment is quantised.

**Spin magnetic moment of an electron:**

(*i*) Spin is a property possessed by only certain atomic nuclei (Hydrogen nucleus H or the phosphorus isotope $^{31}P$).

(*ii*) Spin is the rotation of a body about an axis that passes through its centre. Similarly the electrons rotate about their own axes called spin in addition to their orbital motion.

(*iii*) The electron has an angular momentum due to its spin. According to quantum theory the magnitude of the spin angular momentum is $S = \frac{1}{2}\frac{h}{2\pi} = \frac{h}{4\pi}$.

(*iv*) The electron can be considered as charged sphere spinning about its own axis while its orbit around the nucleus. So the electron has an intrinsic magnetic moment due to its spin. Spin magnetic moment of the electron $= M_s = \frac{1}{2} \times \frac{eh}{4\pi m} = \left(\frac{e}{m}\right)S$.

(*v*) The ratio of the orbital magnetic moment ($M$) to the orbital angular momentum ($L$) of the electron $= \frac{M}{L} = \frac{e}{2m}$

$$\left[\begin{array}{l} \because L = mvr \\ M = \frac{evr}{2} \cdot \frac{M}{L} = \frac{evr}{2mvr.} \end{array}\right]$$

The ratio $\frac{e}{2m}$ is called *gyromagnetic ratio* of the electron.

(*vi*) The ratio of the spin magnetic moment ($M_s$) to the spin angular momentum (S) $= \frac{Ms}{s} = \frac{e}{m}$.

- **Energy Levels Inside an Atom**

The space or chamber of an electron, round the nucleus having *n*, *l* and *m* of definite energy (quantum) level is called **orbital.**

(*i*) Each energy level inside an atom can be classified into **SHELLS**.

(*ii*) Each shell contains a number of **SUB SHELLS**.

(*iii*) Each subshell in its turns contains a number of discrete energy levels called **ORBITALS** i.e. a **hall of residence of electrons.**

- **SHELLS (Principal Quantum Number):** *n*

(*i*) It determines the successive major energy levels of electrons in an atom.

(*ii*) It is designed by a quantum number "*n*"

| Shell: | K – L – M – N | |
|---|---|---|
| Number of electrons | 2 8 18 32 | [$2n^2$] |

- **SUBSHELL (Angular Momentum Quantum Number]:** $l$

(*i*) It determines the magnitude `*L*' of the orbital angular momentum of the electron during its motion around the nucleus.

$$L = \sqrt{l\,(l + I)}\,\frac{h}{2p}$$

- **ORBITALS (Magnetic Quantum Number):** $m_l$ **or** $m$

(*i*) It represents the space orientation of the electrons in *s*, *p*, *d*, *f* sub shells under the influence of magnetic field.

**Spin Quantum Number (s):**

(*i*) An electron, while moving in an orbit round the nucleus, also rotates or spins about its own axis either in clockwise direction or an anticlockwise direction.

(*ii*) It has two values namely + 1/2 (clock) and – 1/2 (anticlock)

An electron spining in a clockwise direction represented by an arrow pointing upwards (↑) while that spining in a anticlockwise direction representing an arrow downwards (↓).

***Pauli Exclusion Principle*: "*No two electrons* in *an atom can exist in the same quantum state.*"**

*According to Pauli's Principle*:

(*i*) When there are two electrons in the same orbital they have opposite spins or antiparallel spins $+\frac{1}{2}$ and $-\frac{1}{2}$, which oppose and cancel each other.

(*ii*) Since the magnetic field produced by one electron cancelled by that produced by the other electron.

(*iii*) Thus an electron pair in an orbital is represented as ↑↓. Here the two electrons which have opposite spins are said to be paired up electrons.

(*iv*) The state of opposite spins (↑↓) gives lower energy to the system) and has greater stability. The state of parallel spin (↑↑) gives higher energy to the system and has less stability.

**Magnetic moment of nucleus :**

1. The nucleus contains protons and neutrons. In addition to charge and mass, a proton posseses angular momentum or spin. The spining charges generates a magnetic field.

2. The strength of magnetic field is expressed as *magnetic moment* μ. It is oriented in earth's magnetic field. Nuclear magnetic moment is very small and is usually neglected.

3. The protons and neutrons of the atom have spin property. When protons and neutrons of an atom are present in a pairs in the nucleus, there will be no spin.

But if there is any unpaired protons, that will impart a *magnetic moment* which can interact with an applied magnetic field i.e. the nuclei absorb energy and may lie either in low energy state (nuclear spin parallel with the field) or in a higher energy state (anti parallel to the field).

4. This interaction of unpaired proton with the magnetic field is the basic principle of NMR.

- **Concept of Resonance:**

According to *resonance concept* if two (or more) alternate valence bond structures can be written for a molecule, the actual structure of the molecule is said to be a *resonance* or *mesomeric hybrid* of all these alternate structures.

> Resonance is the description of the electronic structure of a molecule or an ion by means of several schemes of pairing of electrons, with the features of each scheme contributing in the description.

**Conditions:**

(*i*) In all the canonical forms, the constituent atoms must have the same relative position though the electron distribution may be different.

(*ii*) The different canonical forms should be virtually the same or nearly the same.

(*iii*) They must have the same number of unpaired electrons. The ozone can be represented by two canonical forms.

$$:\!O-O=O\!: \longleftrightarrow :\!O=O-O\!:$$

(Double headed arrow (↔) is placed between the two contributing form)

(*iv*) These two canonical forms are equivalent. It gives additional degree of stabilization to ozone but also makes the two O–O bonds in ozone equivalent.

(*v*) Each bond having the characteristics between those of a single bond and a double bond.

- **Features:**

1. The actual molecule is more stable than what is expected from any of its resonating structure.

2. The bond distance will be shorter than the normal value.

3. The bond order changes *i.e.* the type of C–C bond may change from that of a pure single bond to one having partial double bond.

- **Resonance Energy:**

(*i*) It has been found that in every resonance molecule the observed (experimental) heat formation of the molecule is greater than the calculated heat formation i.e. the actual molecule is more stable than the resonating structure considered for calculating the heat formation.

(*ii*) The difference between observed and calculated heat formation is called *resonance energy* and is represented by $\Delta E$.

(*iii*) Thus $\Delta E$ is the difference between the bond energy of the actual structure ($E_{actual}$) and that of the most stable resonating structure.

**Calculation of Resonance Energy**

$$\ddot{\underset{..}{O}}=C=\ddot{\underset{..}{O}} \longleftrightarrow :\overset{+}{O}\equiv C-\overset{-}{\ddot{\underset{..}{O}}} \longleftrightarrow :\overset{-}{\ddot{\underset{..}{O}}}-C\equiv\overset{+}{O}:$$

(I) (Symmetrical structure) covalent bond; (II) (III) (Unisymmetrical structures) ionic bond

Among the three structures (*I*) has maximum bond energy and stable

$\Delta E = E_{actual} - E_{calculated}$

$E_{actual} = 1602.8$ and $E_{calculated} = 2 \times 732.3 = 1464.6$

$\Delta E = 1602.8 - 14464.8$

$= 1328.2$ kJ/mole

## • Basic Principles of NMR : (Nucleo Magnetic Resonance)

**The interaction of electromagnetic radiation with matter is essentially a quantum phenomenon. It is dependent both upon the properties of radiation and the appropriate structural parts of the material involved.**

(*i*) An atom is an aggregation of electrons, protons and neutrons.

(*ii*) In order to explain the magnetic properties of certain nuclei it is necessary to assume that the nuclear charge is spinning around an axis. While spinning on their axes, the atomic particles generate magnetic fields showing specific *magnetic moments.* The spinning nucleus can be thought of as a small bar magnet.

(*iii*) Protons and neutrons are found in the nucleus and electrons are found in the orbits around the nucleus.

(*iv*) It is believed that protons and neutrons present in the nuclei of atoms also spin about their axes.Their spin quantum number is equal to ½. If the spin of all the particles are paired, there will be no net spin and the nuclear spin quantum number I will be zero. This type of nucleus is said to have *zero spin.*

(*v*) Spin quantum number for the nucleus can have values 0, ½, 1, 1½, 2 etc. depending upon the number of neutrons and protons having parallel and antiparallel spin. It is represented by I.

(*vi*) When I is ½ , there is one net unpaired spin and this unpaired spin imparts a *nuclear magnetic moment* μ to the nucleus. The distribution of positive charge in a nucleus of this type is spherical. The properties for I = ½ are represented symbolically as a spinning sphere.

(*vii*) When $1 \geq 1$, the nucleus has a spin associated with it and the nuclear charge distribution is non-spherical.

(*viii*) Magnetic nuclei assume discrete orientations with corresponding energy levels under the influence of external magnetic fields. The value of I determines the number of quantized energy levels available. This is given by the series

$$I, I-1, I-2, \ldots -1$$

The number of possible orientation is given by (2I + 1) so that the nucleus with spin $\frac{1}{2}$ *i.e.* only for $^{13}C$, $^{15}N$, $^{19}F$, $^{91}P$.

(*ix*) These levels are described as (*a*) aligned (*parallel*) with the applied field (*lower energy*) (*b*) opposed (*antiparallel*) to the field (*higher energy* ). These two levels will have the energy of –μHo and + μHo and for low and high energy respectively, where Ho is the intensity of the applied magnetic field and μ is the nuclear magnetic moment. The difference in energy $\Delta E$ is then equal to 2 μHo

$$\Delta E = E_2 - E_1 = \mu Ho - (-\mu Ho) = \mu Ho + \mu Ho = 2\mu Ho. \text{ [See Fig. : 12.2]}$$

(*x*) Transition from low energy state to high energy state can occur when nuclei absorb sufficient or appropriate quantum energy. If a nuclei (proton) is precessing in the aligned orientation (low energy state) it can absorb energy and pass into opposed orientation and vice versa by losing energy.

(*xi*) If we irradiate (electromagnetic) the precessing nuclei with a beam of suitable radio frequency, the low energy state may absorb the energy and elevate to higher energy state.

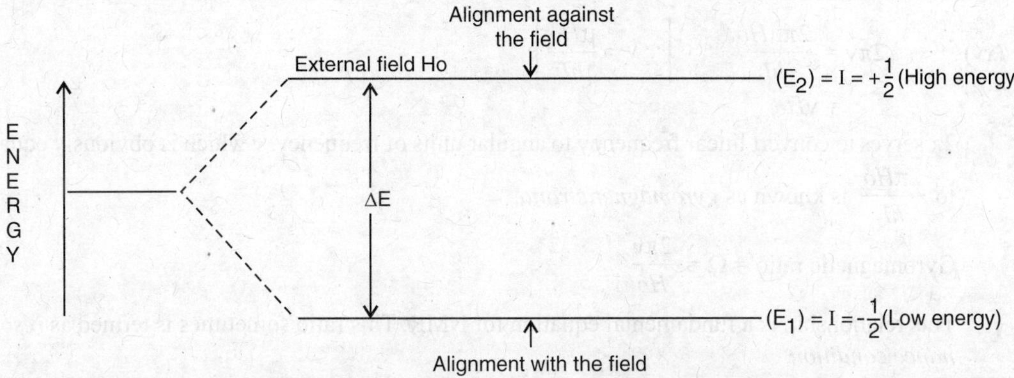


**Fig. 12.2** Energy states for nuclei in a magnetic field.

(*xii*) As the energised nuclei cannot remain in the excited (higher) state for a long time, they tend to lose energy and fall back to ground (lower) state. The energy absorption process starts again by the nuclei and falls back to the ground state again. This process is repeated several times.

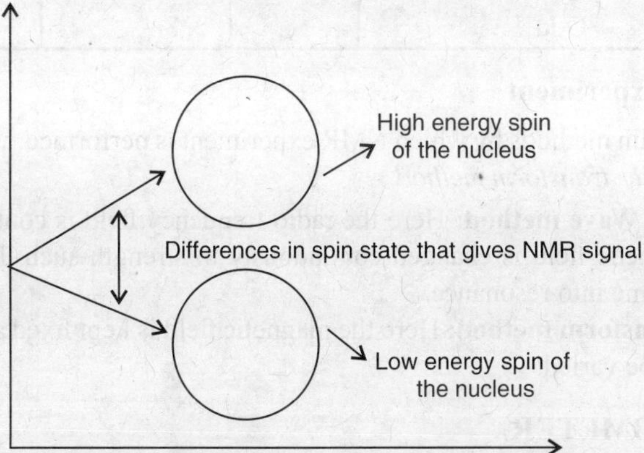


**Fig. 12.3** Magnetic Field Principles of NMR

(*xiii*) **The precessing nuclei (protons) will absorb energy from the radio frequency source. If the precessing frequency is the same as the frequency of the radio frequency beam, the nuclei and the radio frequency beam are said to be in *resonance*. Hence it is known as "*Nuclear_Magnetic Resonance" (NMR).***

(*xiv*) We know from the above $\Delta E = 2\mu Ho$ $\left(\text{where } I = \frac{1}{2}\right)$. The energy of radio wave is given by the Einstein-Planck equation as

$$E = h\nu$$

In general, for a nuclei with spin quantum number I, the difference in energy is given by

$$\Delta E = \frac{\mu\, Ho}{I} \quad (\textit{i.e.}\text{ energy transition})$$

When resonance takes place, this energy is absorbed to bring about the nuclear transition. therefore at resonance, *Energy of radiation = Transition energy.*

$$h\nu = \frac{\mu Ho}{I}$$

or

$$\nu = \frac{\mu Ho}{h.I} = \frac{\mu}{hI} Ho$$

(*xv*) $$2\pi\nu = \frac{2\pi\mu Ho}{hI} \qquad \left[\because \nu = \frac{\mu Ho}{hI}\right]$$
$$= \nu Ho$$

$2\pi$ serves to convert linear frequency to angular units of frequency. $\nu$ which is obviously equal to $\frac{2\pi Ho}{hI}$ is known as *gyromagnetic ratio.*

$$\text{Gyromagnetic ratio} = \Omega = \frac{2\pi v}{Ho}$$

This relationship is a fundamental equation for NMR. This ratio sometimes is termed as *resonance condition.*

| Mass number | Atomic number | Spin quantum number | Atomic nucleus |
|---|---|---|---|
| Odd | Odd or even | $\frac{1}{2}, \frac{3}{2}, \frac{5}{2}$ | $^{1}H$, $^{11}B$, $^{13}C$, $^{19}F$ |
| Even | Even | 0 | $^{12}C$, $^{16}O$, $^{32}S$ |
| Even | Odd | 1, 2, 3 | $^{2}H$, $^{10}B$, $^{14}N$. |

**Methods of NMR experiment**

There are two main methods by which NMR experiment is performed, viz, the *continuous wave method* and *the Fourier transform method.*

(A) **Continuous Wave method:** Here the radio frequency field is continuously applied to the sample and the magnetic field is changed continuously in strength such that all the nuclei in the sample are able to come into resonance.

(B) **Fourier Transform method:** Here the magnetic field is kept fixed and the oscillating radio frequency field may be varied.

## • NMR SPECTROMETER

This instrument consists of the following components :

**(*a*) Electromagnets:**

(*i*) Heavy and powerful electromagnets ($10^3 - 10^4$ kg) are used in the NMR instrument.

(*ii*) Permanent magnets have also been used for NMR imaging.

(*iii*) Their field strength varies from 500 to 3000 gauses.

(*iv*) Super conducting magnets are used instead of resistive coil magnets.

(*v*) Super conducting magnets are useful in cases of certain materials (zero resistance at low temperature). At zero resistance a current will flow continuously without input of power.

**(*b*) Radio Frequency (RF):**

(*i*) A *radio frequency transmitter* is used to generate the monochromatic generation.

(*ii*) A sample of volume at least 0.3 ml is sealed in a glass tube and inserted into a probe containing the radio frequency (RF) transmitter and receiver coils which are fitted orthogonally or at right angle to the main magnetic field.

(*iii*) Various types of radio frequency coils are used.

(*iv*) In some NMR spectrometers, the field is kept constant and the frequency of RF radiation is varied.

**(*c*) Sample:**

1. A sample of volume 0.3 ml is taken in a thin walled cylinder glass tube.

2. This glass tube is placed between the poles of the magnet.
3. There is a single coil wrapped around the sample probe act as radio receiver for detection of absorption signal.

**(*d*) Sweep generator:**

(*i*) A pair of 'sweep' coils are mounted on either side of the sample.

(*ii*) It enables the field to 'sweep' through the resonance while the RF frequency is held constant.

## Mechanism:

(*i*) A radio frequency source of 60 mHz is applied on the sample.

(*ii*) The sample is dissolved to a very high concentration in a solvent which lacks protons such as $D_2O$.

(*iii*) When radio frequency is applied on the sample it causes small oscillating electromagnetic field on the sample and the energy difference between nuclear spin level matches with the radio frequency.

(*iv*) This EM field induces a transition between energy levels when transition occurs in sample.

(*v*) The oscillation in the field induces a voltage oscillation in the radio frequency receiver coil which is amplified and detected.

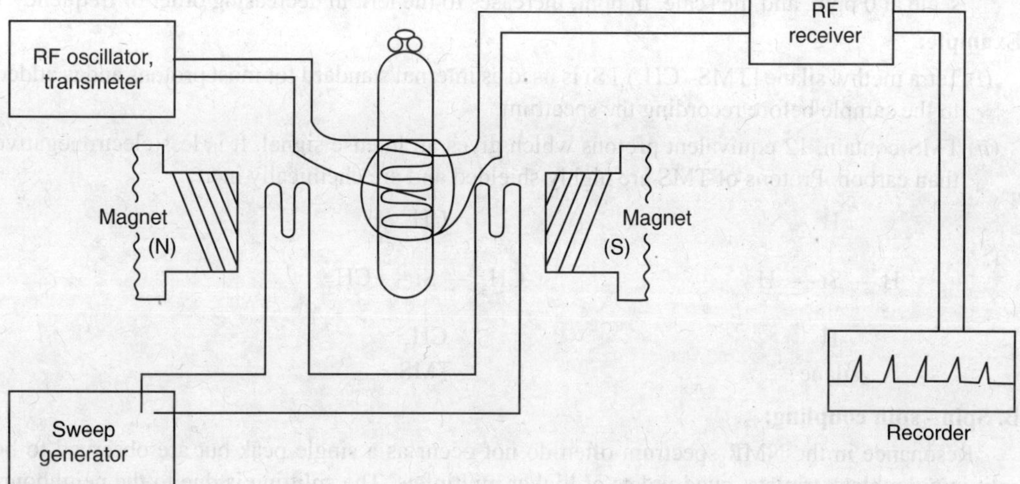


**Fig. 12.4** A line diagram of the instrument of NMR spectrophotometer along with its components

## NMR Parameter:

### A. Chemical Shift:

The nuclei of an element (proton) gives rise to distinct spectral lines in different chemical environments, because the nuclei in such environment resonate at different frequency, the difference in resonance frequency are expressed as *chemical shift.*

It may be expressed as the shifts in position of NMR absorptions as compared to the standard reference are known as *chemical shift.*

(*i*) It arises due to the shielding of nuclei (proton) from the external magnetic field by electrons.

(*ii*) It is the basis of NMR spectroscopy.

(*iii*) The effect of external applied magnetic field on each of the protons in a molecule will not be the same. An additional small field is created just around the nucleus due to the electron currents surrounding each nucleus.

(*iv*) The total effective field at each nucleus will thus depend on the neighbouring atoms and chemical groups. Since the magnetic field is different at each chemically distinct nuclei, the frequency at which the resonance occurs will also be different at each nucleus.

## Measurements of Chemical Shift:

(*i*) It is measured with respect to some arbitrarily chosen reference point in the spectrum. Usually an inert compound with well known resonance position is used as standard.

(*ii*) The standard may be used externally or internally. Chemical shifts are expressed in the dimensionless units of parts per million (ppm) and are normalised to be independent of the applied magnetic field. The magnitude of the chemical shift is expressed as δ value.

$$\delta = \frac{V_s - V_R}{V_R} \times 10^6 \left[\begin{array}{l} V_s = \text{Resonance frequency of the sample} \\ V_R = \text{Resonance frequency of the reference} \end{array}\right] \text{in ppm}$$

(*iii*) Most chemical shifts have δ values between 0 and 10.

(*iv*) Another scale, known as τ/tau, is used for chemical shift. It is related to δ scale by the following relationship.

$$\tau = 10 - \delta$$

(*v*) By convention, the resonance frequency of the standard is placed at the right-most end of the scale at 0 ppm, and the scale, in ppm, increases to the left, in decreasing order of frequency.

**Example:**

(*i*) Tetra methyl silane [TMS-$(CH_3)_4$] Si is used as internal standard for most protons and is added to the sample before recording the spectrum.

(*ii*) TMS contain, 12 equivalent protons which gives an intense signal. It is less electronegative than carbon. Protons of TMS are highly shielded and are chemically inert.

```
      H                      CH3
      |                      |
  H — Si — H         CH3 — Si — CH3
      |                      |
      H                      CH3
    silane                   TMS
```

**B. Spin - spin coupling:**

Resonance in the NMR spectrum often do not occur as a single peak but are observed to be split into doublets, triplets, quadruplets or higher multiplets. The splitting is due to the neighbouring spins that are transmitted via the electrons of the covalent bonds between the atoms. *Spin-spin* coupling is thus a '*through bond*' coupling and occurs only due to nuclei within the molecule or it is the phenomenon which involves the interaction between like and different spins of nearby nuclei. It is known as *spin - spin coupling:*

(*i*) The strength of the coupling is denoted by 'J'. It is measured by hertz (Hz) and is independent of the applied magnetic field strength.

(*ii*) Proton-proton couplings are usually transmitted only through two or three bonds although weak couplings are often transmitted further.

(*iii*) In certain rigid structures of favourable geometry coupling through four bonds may be reasonably large.

(*iv*) The extent of direct spin-spin interaction or coupling is large for hydrogen nuclei. When liquid or gaseous samples are considered, however, this direct spin coupling mechanism is found not to apply because of random molecular motion within the sample. Nonetheless, the n.m.r spectra of liquids do show the phenomenon of coupling but effect is very much smaller.

(*v*) Spin-spin coupling can be first order or second order. In the case of first order splitting the difference in frequency between the resonance peaks involved in the coupling is much greater than the coupling constant 'J'. In the case of second order this is not so. The second order spectra will be much more difficult to interpret than first order spectra.

(*vi*) Coupling constants are affected by the following factors:

(*a*) Dihedral angle between the nuclei;

(*b*) The electronegativity of the substituents on the moeity;

(*c*) Valence angles; and

(*d*) Bond lengths.

***$^{13}C - {}^{2}H$ coupling:***

(*i*) The carbon atom of $CDCl_3$ is attached to a deuterium nucleus. Deuterium ($^2H$) has a spin quantum number I = 1 and can therefore exist in three spin states in a magnetic field. There are three possible energy levels for a deuterium atom placed in a magnetic field.

(*ii*) The carbon atom therefore experiences three slightly different magnetic fields depending upon the state of deuterium nucleus to which it is attached.

(*iii*) Since the difference in energy between the three states is very small, there is essentially equal probability that a carbon atom will be bonded to a deuterium in any one of the three states. The result is that the carbon nucleus comes into resonance at three frequencies with equal probability,

(*iv*) The $CDCl_3$ signal is three equally spaced weak lines at δ77. The carbon is said to be coupled to the deuterium, and the separation of the lines in Hz is called coupling constant J. Because there is only one bond between the carbon and deuterium, it is further qualified as $^1J_{CD}$.

(*v*) Carbon - deuterium coupling is much less important than carbon - hydrogen coupling, but it is given as an example because it is visible in a spectrum.

**Relaxation:**

**The mechanism by which a nucleus returns from higher to lower energy state and excess spin energy is shared either with the surroundings or with the other nuclei is referred to as *relaxation process.* The time taken for a fraction $\frac{I}{e}$ = 0.37 of the excess energy to be dissipated is called *relaxation time.***

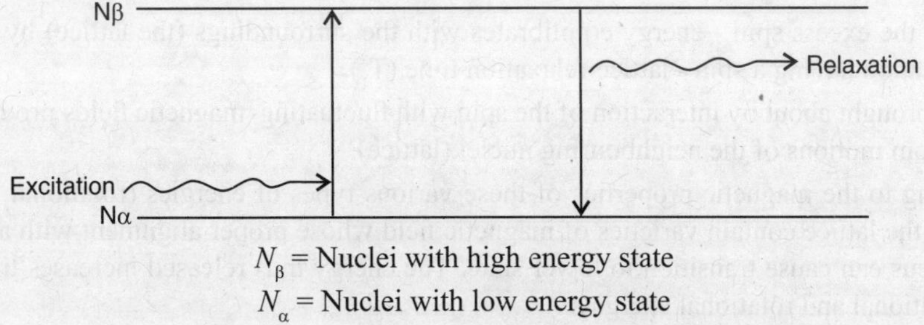


$N_\beta$ = Nuclei with high energy state

$N_\alpha$ = Nuclei with low energy state

**Fig. 12.5** Showing Relaxation state of nuclei ($N_\beta \rightarrow N_\alpha$)

(*i*) Upon irradiation of a particular nucleus, the rate of absorption of energy is initially greater than the rate of emission because of the slight excess of nuclei in the lower energy state.

(*ii*) However, the absorption signal rapidly attains some finite value. Only if relaxation back to the lower energy state can occur at least as rapidly as absorption, will the intensity of nuclear absorption at a given frequency remain constant. Otherwise in time, the radio frequency wave field would equalize the population of the energy levels and the spin system would become saturated.

(*iii*) Two types of relaxation spins are operative: **spin - spin relaxation** and **spin - lattice relaxation.**

**(*a*) Spin - spin relaxation:**

(*i*) This is also known as *transverse relaxation.* It is characterised by time constant. $T_2$ known as spin - spin or transverse relaxation time.

(*ii*) It involves the mutual exchange or sharing of excess spin energy between the two proximal precessing nuclei.

(*iii*) Here magnetization component in the X and Y direction immediately after the application of radio frequency (rf) pulse, the net magnetization is given by a maximum value.

(*iv*) When the rf field is off from precission frequency, the spin tend to fall out of phase through dipole - dipole interaction and the net magnetization in the X or Y direction falls towards zero.

(*v*) A nucleus in the upper energy state can transfer its energy to a neighbouring nucleus by a mutual exchange of spin. Recording the dispersion mode signal as a function of time provides a curve from which $T_2$ can be calculated,

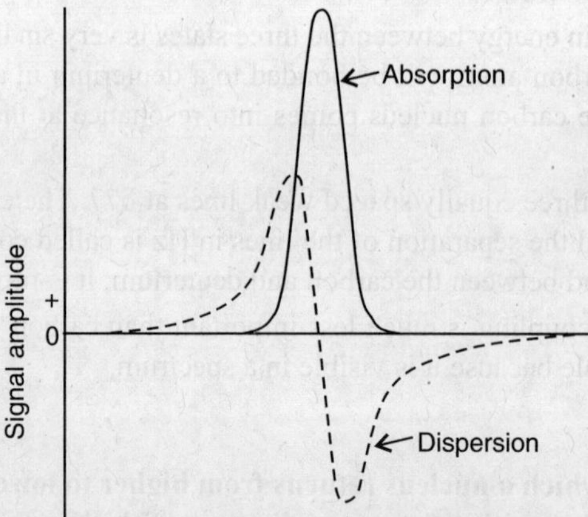


**Fig. 12.6** Line shapes of the two observable NMR

**(*b*) Spin - lattice relaxation:**

*It is also called longitudinal relaxation.*

(*i*) Here the excess spin - energy equilibrates with the surroundings (the lattice) by spin-lattice relaxation having a spin - lattice relaxation time ($T_1$).

(*ii*) It is brought about by interaction of the spin with fluctuating magnetic fields produced by the random motions of the neighbouring nuclei (lattice).

(*iii*) Owing to the magnetic properties of these various types of energies (rotational, vibrational etc), the lattice contain varieties of magnetic field whose proper alignment with a precessing nucleus can cause transition to lower state. The energy thus released increases translational, vibrational and rotational energies.

## Applications of NMR:

**(*a*) Structural diagnosis:**

(*i*) The NMR spectroscopy is mainly applied to study the structure of small organic molecules and small globular proteins.

(*ii*) Structural information about small proteins (various cytochromes, hen egg white lysozyme and calcium binding proteins) have been obtained by using NMR spectroscopy.

(*iii*) It is used in identification of structural isomars such as

$CH_3 - CH_2 - CH_2\ OH$ — 4 signals — n-propanol

$CH_3 - CH(OH) - CH_3$ — 3 signals — iso propanols.

**(*b*) Quantitative studies :**

(*i*) NMR has been used to determine the concentration of metabolites.

(*ii*) Concentration of phosphocreatine in muscles are measured by NMR technique.

**(*c*) Biological structures and functions:**

(*i*) NMR has been used to study the conformation of the lipid head groups of the biological membranes and their interaction with integral proteins of the membrane.

(*ii*) It has also been used to study alanine and lactate transport in human erythrocyte.

**(*d*) Studies of intact organs:**

(*i*) 31p has been applied to study of intact biological organs viz heart, kidney etc.

(*ii*) With the help of 31p resonances small molecules such as ATP (adenosine triphosphate), ADP (adenosine-diphosphate), creatinine phosphate, etc. have been studied.

**(*e*) Study of dynamic characteristics of protein structure:**

NMR has been used to study the dynamic characteristics of protein like histone, prothrombine, cytochrome $B_5$ and plasminogen.

(*i*) Nuclei that contain odd number of protons and/or neutrons show NMR spectra, since they have half integral spin.

(*ii*) Nuclei that have integral spin (in particular zero spin) are those which contain even number of protons or neutrons. These nuclei do not show NMR.

(*iii*) For example $^{12}C$ ($6_p$ 6*n*), $^{12}O$ (8*p*, 8*n*), do not show NMR

(*iv*) $^{1}H$, $^{2}H$, $^{13}C$, $^{14}N$, $^{15}N$, $^{19}P$, $^{31}P$ show NMR

- **Tetramethylsilane (TMS) is used as reference spectroscopy:**

(*i*) It contains 12 protons and all are chemically equivalent.

(*ii*) It gives a single well defined peak in the NMR spectrum.

(*iii*) The protons are highly shielded by their electrons from external magnetic field so that it shows NMR at a very high external magnetic field strength as compared to other protons.

(*iv*) It is chemically inert.

(*v*) Its boiling point is low (27°C) so that it evaporates off quickly and the sample compound can be easily recovered.

$Si(CH_3)_4$ : $CH_3$ above, $CH_3 - Si - CH_3$, $CH_3$ below

Tetramethylsilane

**Shielding and deshielding of protons:**

(*i*) The electrons involved in covalent bonding are paired and usually do not have a magnetic field. An applied magnetic field can creates additional modes of circulation of these electrons. Thus it generates a small localized magnetic field.

(*ii*) Different protons have different electronic environment i.e. the electrons present in the bond close to the proton and the electrons present in the neighbouring bonds show different effects. These electrons induce their own magnetic field which may reinforce the applied magnetic field or oppose it.

(*iii*) Now for NMR to occur the proton has to flip its spin from lower energy level to higher energy level *i.e.* it has to change its spin from ↑ to ↓.

(*iv*) If the induced magnetic field reinforces the applied magnetic field, a smaller external magnetic field will be required to flip the spin of the proton to make the nuclei to resonate. Such a proton is said to be *deshielded* and the absorption is said to be *downfield.*

(*v*) On the contary, if the induced magnetic field opposes the external magnetic field, a stronger magnetic field will be required to make the nuclei to resonate. *i.e.* NMR. Therefore, the proton has to flip its spin from ↑ to ↓. The proton is then said to be *shielded* and the absorption is said to be *upfield.*

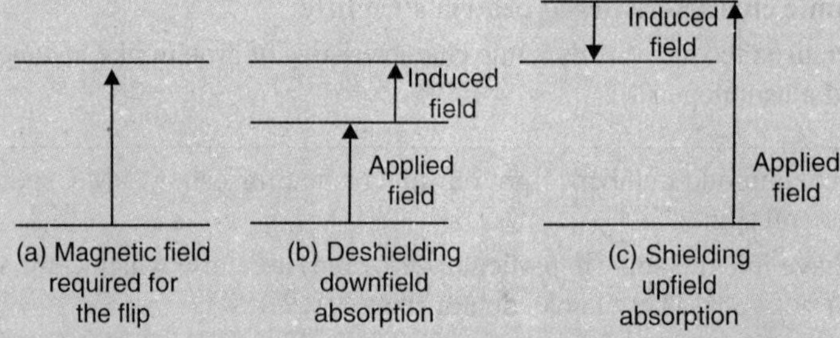


- **Molecular Energy level:**

Internal energy of a molecule is of three types: (*a*) Rotational energy (*b*) Vibrational energy and (*c*) Electronic energy.

**(*a*) Rotational energy:**

(*i*) It involves the rotation of molecules about the centre of gravity or parts of the molecule.

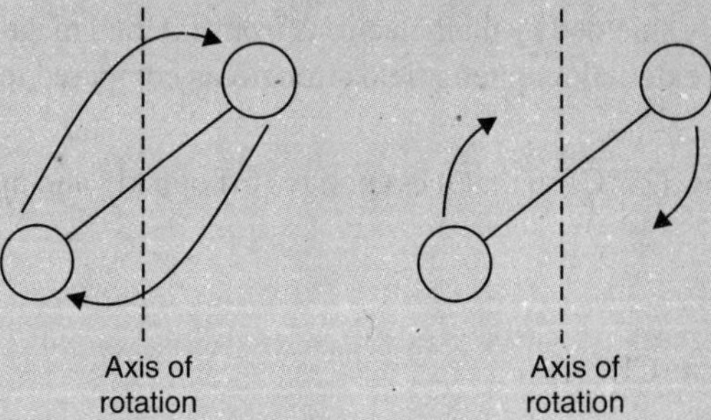


**Fig. 12.7** Molecular rotations in a linear molecule

**(*b*) Vibrational energy:**

(*i*) It is associated with stretching, contracting and bending of covalent bonds in molecules.

(*ii*) The bond behaves as spirals made of wire.

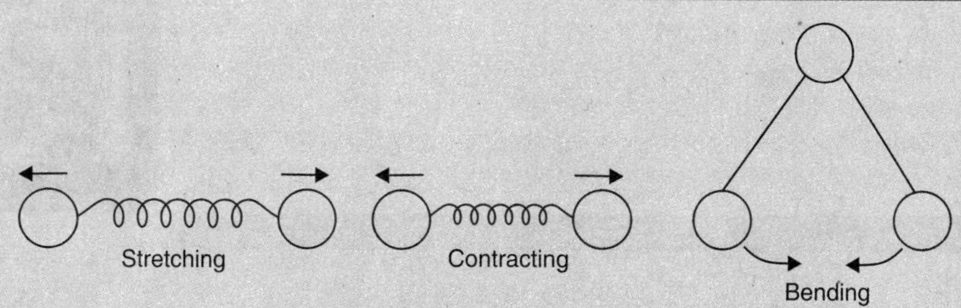


**Fig. 12.8** Vibrations within molecules

**(*c*) Electronic energy:**

It involves in the distribution of electrons by splitting of π bonds by absorbing energy or the promotion of energy into higher levels.

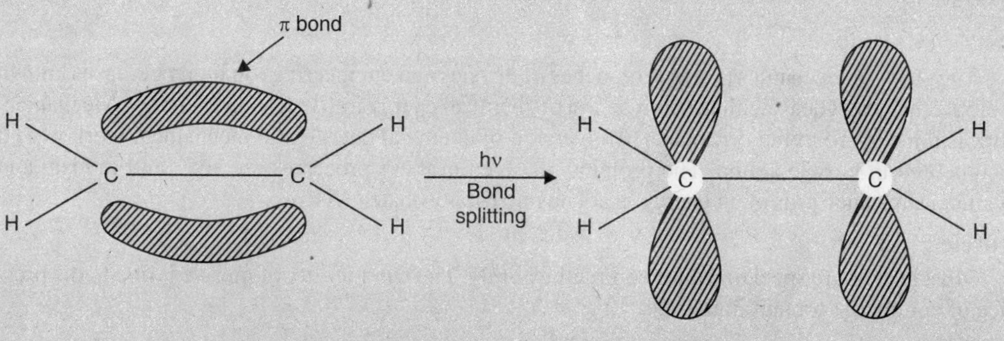


**Fig. 12.9** Splitting of pi bond by absorption of energy.

CHAPTER 13

# Biomechanics

The study of mechanics (various branches of physics and engineering) on living beings is known as biomechanics. It deals with the conversion of chemical energy into mechanical energy. Mechanical work is done by all living organisms. Conversion of chemical energy into mechanical work is well documented in muscle action. The proteins involved in these processes are actin and myosin and chemical substance providing energy is adenosine triphosphate (ATP).

**Muscle :**

Muscle cells are specialized for contractile ability. The contractility of muscle is due to the presence of contractile protein in muscle.

**Types :**

(*a*) **Smooth muscle :**

(*i*) This type of muscle has no striation and so it is known as *non-striated* or *smooth muscle.*

(*ii*) These muscles are not contracted at will, so this type of muscle is also known as *involuntary muscle*.

(*iii*) They are widely distributed in the visceral organ, so usually referred to as *visceral muscles.*

(*b*) **Striated muscle :**

(*i*) They display longitudinal and cross striations of alternating dark and light bands. Thus they are named *striated muscles.*

(*ii*) These are attached to the bones of the body and so this muscle is also regarded as *skeletal muscle.*

(*iii*) They are contracted at the 'will' of the animal and hence this muscle is also known as *voluntary muscle.*

(*c*) **Cardiac muscle :**

(*i*) These are involuntary because the activities of these muscles are not under the control of will.

(*ii*) Structurally they are *striated.*

- **Phasic (Twitch) muscle :**

(*i*) They are locomotory muscles with origin and insertion on the endoskeleton structures.

(*ii*) They are often arranged in antagonistic pairs.

(*iii*) They are relatively rapid in their contraction.

- **Tonic muscle:**

(*i*) They are normally arranged in soft organs or hollow structures such as gastrointestinal tract or urino genital tract.

(*ii*) One part of the muscle often inserts on another part of the muscle.

(*iii*) They contract slowly.

- **Properties of muscle :**

1. **Excitability :** It receives and responds to stimuli.

2. **Conductibility :** Stimulated by a stimulus of adequate strength and is conducted to all its other parts within no time.

3. **Contractibility :** Ability to shorten and thicken when sufficient stimulus is received.

4. **Elasticity :** It is the ability of a muscle to return to its original shape after contraction or extension.

5. **Extensibility :** It can be stretched passively when relaxed.

6. **Tonicity :** The state of mild contraction is known as tonicity.

- **Function of muscle :**

1. Motion, including locomotion.
2. Maintenance of posture.
3. Heat generation.

- **Structure of Striated muscle :**

(*i*) Each muscle cell is elongated and possess fibre-like property. So a muscle is known as **muscle fibre**.

(*ii*) A striated muscle fibre is cylindrical, multinucleated and is covered by a thin transparent membrane known as *sarcolemma.*

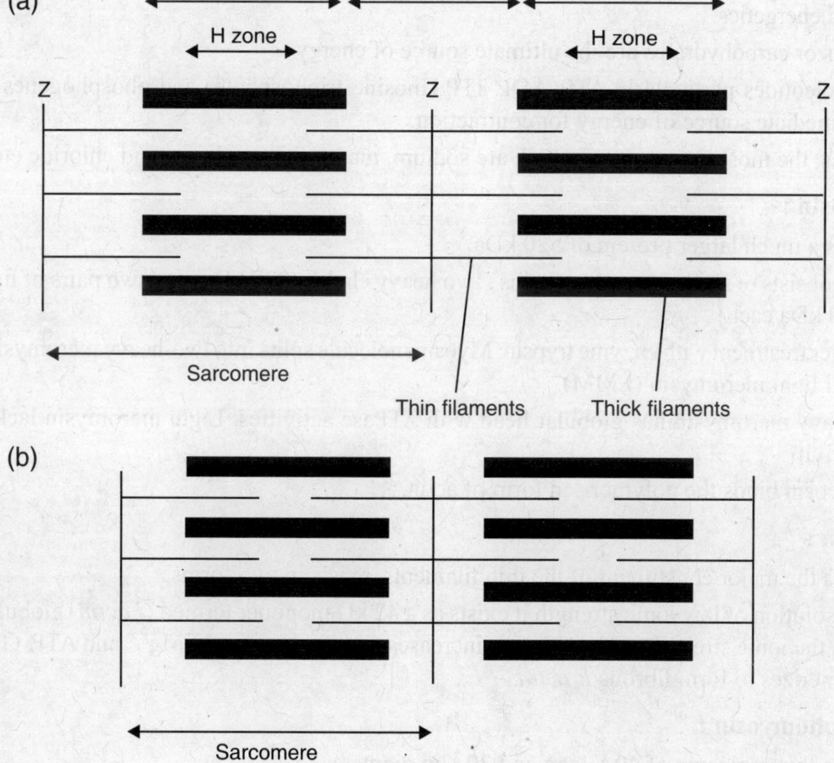


**Fig.13.1** Schematic diagram of striated muscle, under phase contrast microscope. (*a*) Relaxed (*b*) Contracted

(*iii*) Each cell contains within its sarcoplasm (*cytosol*) many parallel myofibrils, many mitochondria (sarcosomes), endoplasmic reticulum, glycogen, ATP, phosphocreatine and glycolytic-enzymes.

(*iv*) Longitudinal section of myofibrils under electron microscope shows repeating functional unit called *sarcomere.* A sarcomere has a banded appearance due to regions with different optical properties. A dark *A band* and a light *I band* alternate regularly along the length of the myofibrils.

(*v*) The middle of the *A zone* is less dense known as *H zone* and a dark *M line* is found in the middle of the H zone. The *Z* lines bisect the *I* bands.

(*vi*) There are two kinds of filaments in the sarcomere. The *thick filaments* make up the *H* zone of the *A* band and only *thin filaments* are found in the *I* band. These filaments are protein filaments of diameter 150 Å and 70 Å respectively.

(*vii*) Thick filament consists mainly of the protein *myosin* while the thin filament consists of *actin, tropomyosin* and the *troponin.*

(*viii*) Each thick filament is surrounded by six thin filaments in the region of overlap, and each thin filament has three neighbouring thick filaments. The two different kinds of filaments interact with each other by *cross bridges.*

- **Molecular Components of Muscle :**

(*i*) Proteinaceous components viz. myosin, actin, troponin and tropomyosin are the building blocks of myofilaments.

(*ii*) Contractile proteins of which myosin, actin and tropomyosins are predominant.

(*iii*) Enzymes and other protein such as troponin that are associated with contraction, relaxation and energetics.

(*iv*) Fats or carbohydrates are the ultimate source of energy.

(*v*) Nucleotides particularly ATP, ADP, ITP (Inosine triphosphate) and phosphogenes are more immediate source of energy for contraction.

(*vi*) Ions, the most important of which are sodium, magnesium, calcium and chloride etc.

**(A) Myosin :**

(*i*) It is a much larger protein of 520 kDa.

(*ii*) It consists of six polypeptide chains : two heavy chains (220 kDa) and two pairs of light chains (20 kDa each)

(*iii*) After treatment with enzyme trypsin, Myosin molecule splits into two-heavy meromysin (HMM) and light meromysin (LMM).

(*iv*) Heavy meromysin has globular head with ATPase activities. Light meromysin lacks ATPase activity.

(*v*) Myosin binds the polymerised form of actin.

**(B) Actin :**

(*i*) It is the major constituent of the thin filaments, exists in two forms.

(*ii*) In solution of low ionic strength it exists as a 42 kD mononer termed *G actin* (globular shape). As the ionic strength of the solution increases and in presence of $Mg^{2+}$ and ATP, G actin polymerizes to form fibrous *F actin.*

**(C) Trophomyosin :**

(*i*) It is thin molecule of 40Å long and 20Å in diameter.

(*ii*) There are two known forms viz. *tropomyosin* A and *tropomyosin B.*

(*iii*) It is of 70 kDa.

**(d) Troponin :**

It is complex of three subunits *viz, 'Tn'* C (18kDa), Tn I (24kDa) and Tn T (37kDa).

- **Energy for contraction of muscle**

(*i*) It is obvious that the ATP is the main source of energy for contraction. It ultimately comes from the oxidative metabolism of glucose **(glycolysis)**. Anerobically, one molecule of glucose yields two molecules of ATP, whereas in aerobic oxidation the yield is 38 ATP molecule.

(*ii*) Striated muscles contain high levels of phosphocreatine which breaks down to release energy.

$$\text{Phosphocreatine} + \text{ADP} \rightleftharpoons \text{Creatine} + \text{ATP}$$

Vertebrate skeletal muscles contain about 5 times more phosphocreatine than ATP.

- **Contraction of muscles :**

The filaments interact with each other through cross bridges. The process by which shortening of the contractile element in muscle is brought about by sliding of the thin filaments over the thick filaments. The width of the A band is constant, whereas the Z lines move closer together when the muscle contracts and further apart when it is stretched. During muscle contraction I zone shrinks.

The interaction between **F actin** and myosin forms a complex known as **actomyosin** which is highly viscous. In presence of ATP and $Mg^{++}$, **actomyosin** dissociates.

$$\text{Actomyosin} \xrightarrow[\text{Mg}^{++}]{\text{ATP}} \text{Actin} + \text{Myosin} + \text{ADP} + i\text{P}$$

(*i*) According to sliding model, the length of the filaments does not change but the length of the **sarcomere** decreases as the filaments slide past each other during contraction. Thus, though the length of the thick and thin filaments does not change, the size of H zone and I band decreases. The cross bridge changes angle during the sliding motion of the filament.

(*ii*) The sliding during muscle contraction is produced by breaking and reforming of the cross linkages between actin and myosin. The heads of the myosin molecule link to actin at a 90° angle, produce movement of myosin on actin by swiveting and then disconnect and reconnect at the next linking site, repeating the process in serial fashion.

(*iii*) ATP is the main source of muscle contraction. In muscle, hydrolysis of ATP is catalyzed by ATP found in the heads of the myosin molecule where they are in contact with actin.

(*iv*) The actomyosin complex remains in two forms *viz. activated* and *inactivated form.* The binding of myosin-ATP complex to thin filaments in the presence of $Ca^{2+}$ ions is the activated form.

(*v*) The onset of muscle contraction is mediated by $Ca^{2+}$ ions. The nerve excitation triggers the release of $Ca^{2+}$ ions bind to Tn C component of troponin and causes conformational changes which are transmitted to tropomyosin and actin.

(*vi*) Consequently tropomyosin moves towards the helical groove of the thin filament which enables $S_1$ heads of myosin molecules to interact with actin unit of thin filament. Contraction takes place with concomitant hydrolysis of ATP and again $S_1$ head is blocked by tropomyosin.

(*vii*) Thus the binding of $Ca^{2+}$ to troponin causes the activated myosin ATP binding site on actin to be exposed.This is otherwise blocked by troponin and tropomyosin.

$$Ca^{2+} \rightarrow \text{Troponin} \rightarrow \text{Tropomyosin} \rightarrow \text{Actin} \rightarrow \text{Myosin}$$

- **Biomechanics in vertebrates :**

Physical forces are permanent part of an animal's environment. The study of how physical forces affect and are incorporated into animal designs is termed as *biomechanics.*

- Basic concepts dealing with biomechanics are *length, time* and *mass*.
- *Length* is a concept of distance, *time* is a concept of flow of events and *mass* is a concept of inertia.
- *Velocity* is the rate of change of an object in an object position and *acceleration* in turn is the rate of change of velocity.

• *Force* describes the effects of one body acting on another through their respective and accelerations.

• *Density* is mass divided by volume.

• *Work.* Whenever a force acting on a body displaces it work is said to be done. (w = F.S. s distance in the direction of force).

• *Power* is defined as the time rate of doing work. Unit is watt.

• *Friction.* When a body slides or rolls over another body or tries to do so then a force opposing the motion acts between those surfaces of the body which are in contact. This force is called *frictional force.*

• *Vectors.* The physical quantities having magnitude as well as direction are called vectors. They are added and substracted according to special laws, such as *law of parallelogram* of forces, *law of triangle* of forces, *law of polygon* of forces.

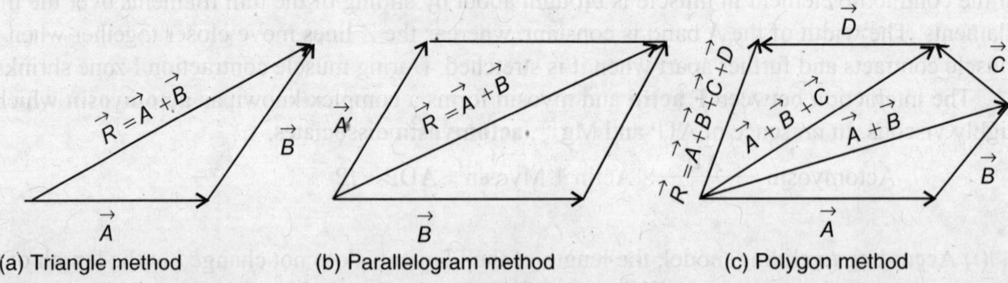


**Fig. 13.2**

**Basic force of laws :**

**1. First law of inertia :**

Because of its inertia, every body continues in a state of rest or in a uniform path of motion until a new force acts on it to set it in motion or change its direction.

***Inertia :* It is the tendency of a body to resist a change in its state of motion. If the body is at rest, it will resist being moved and if it is in motion, it will resist being diverted or stopped.**

**2. Second law of motion :**

**The change in an object's motion is proportional to the force acting on it or the rate of change of momentum is directly proportional to the applied field.**

$\overrightarrow{F} = \overrightarrow{ma}$ $F$ = force $m$ = *mass* $a$ = *acceleration*

**3. Third law :**

**Between two objects in contact, there is for each action an opposite and equal reaction.**

**4. Centre of mass :**

(*i*) It is the point in the body at which total mass of the body is supposed to be concentrated.

(*ii*) As a moving animal changes the configuration of its parts, the position of its centre of mass changes from one instant to the next.

• **Snake movement by lateral undulation :**

(*i*) The body is thrown into serpentine loops *i.e.* right and left. The animal locates with its coils several projections such as pebbles or plant stems. The body then presses sideways against these objects in a direction that is obliquely backward in relation to the direction the snake is to move.

(*ii*) **Biomechanical analysis :**

(*a*) The thrust of the snake against the object ($F_t$) is opposed by an equal and opposite force ($F_o$) *i.e.* reaction of object on snake.

(*b*) This force ($F_o$) has two components *viz.* lateral components ($F_e$) which is reduced by the friction and the forward component ($F_t$) in the direction of movement of snake.

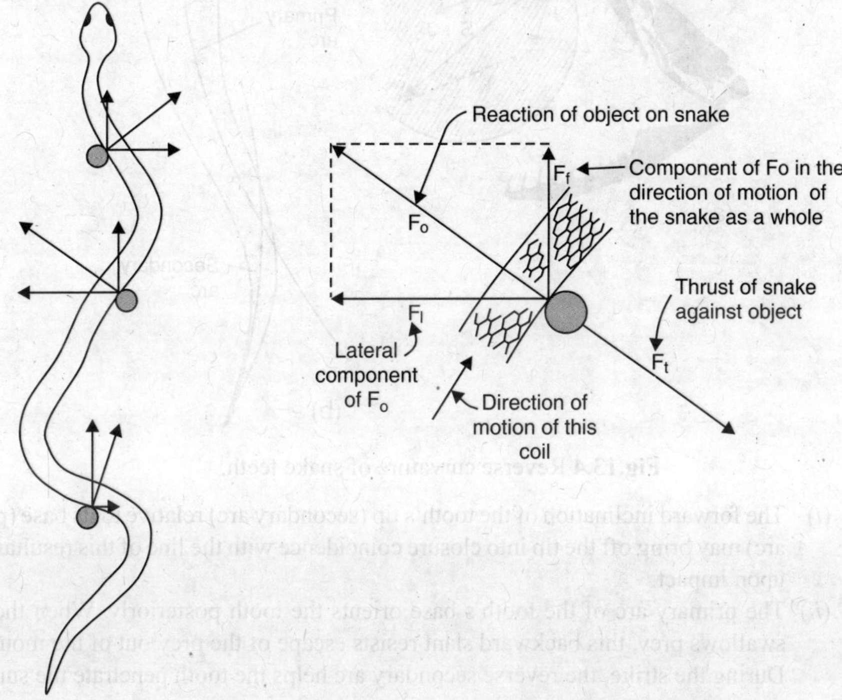


**Fig. 13.3** Diagram of Lateral Undulation of a Snake.
(The value of $F_t$ is somewhat reduced by sliding friction.)

1. **Cranial Kinesis :** Movement between the upper jaw and brain case about joints between them. Cranial kinesis provides a way to change the size and configuration of mouth rapidly. It helps in engulfing the prey. It also allows the tooth bearing bones to move quickly into strategic positions during rapid feeding.
2. The venomons viper erects the maxillary bone bearing the fang and swings it from a folded position along its upper lip to the front of the mouth, where it can easily deliver venom into prey.
3. (*a*) When snake launches its head at and closes its jaws on prey, two forces are transmitted through the tip of the anterior teeth. These component forces are represented here by vectors (Fig. 13.4).

    (*i*) One vector represents the force arising from the forward momentum of the skull (*S*).

    (*ii*) The other represents the force of jaw closure (*J*). The resultant force on impact is $S + J$.

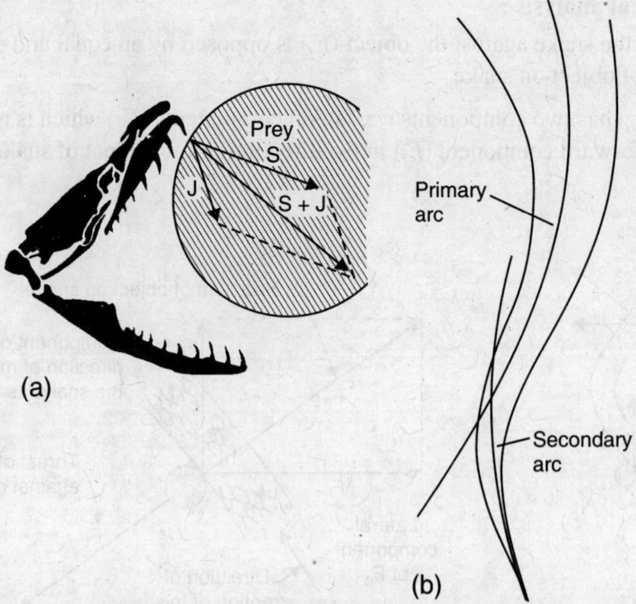


**Fig.13.4** Reverse curvature of snake teeth.

(*b*) (*i*) The forward inclination of the tooth's tip (secondary arc) relative to its base (primary arc) may bring off the tip into closure coincidence with the line of this resultant force upon impact.

(*ii*) The primary arc of the tooth's base orients the tooth posteriorly. When the snake swallows prey, this backward slant resists escape of the prey out of the mouth. During the strike, the reverse secondary arc helps the tooth penetrate the surface of the prey.

**Biomechanics**

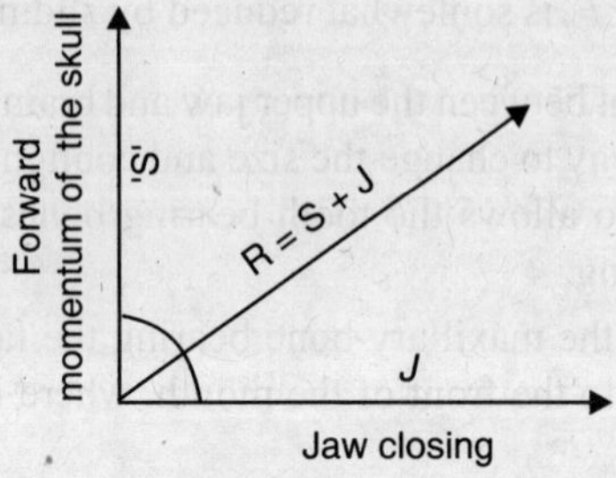


Resultant force transmitted through the tip of the anterior teeth ($T$)

$$R = \sqrt{a^2 + b^2 + 2ab \cos \theta}$$

Here $R = T$

So, $T = \sqrt{S^2 + J^2 + 2SJ \cos \theta}$

When the snake tries to close the mouth, the angle tends to zero (0). At that time the movement of the skull and jaw closing force tends to be same direction .

$$T = \sqrt{S^2 + J^2 + 2.SJ \cos \theta}$$

$$= \sqrt{S^2 + J^2 + 2.S.J.I.}$$

$$= \sqrt{S^2 + J^2 + 2SJ}$$
$$= \sqrt{(S + J)^2}$$
$$\therefore \quad T = S + J$$

- **Life in fluids :**

Water and air are fluids. Certainly air is thinner and less viscous than water but it is a fluid none the less. The physical phenomena that act on fishes in water generally apply to birds in air. Air and water differ in viscosity.

When a body moves through a fluid, the fluid exerts a resisting force in the opposite direction to the body's motion. This resisting force is known as *drag,* may arise from various physical phenomena, but forces caused by *friction drag* and by *pressure drag* are usually the most important. As animals move through a fluid, the fluid exerts a *resisting force* (*drag*) on the surface of the animal where they make contact. This force creates *friction drag.* It depends on viscosity of the fluid, area of the surface, surface texture, and the relative speed of fluid and surface.

Together friction and pressure drag contribute to profile drag which is related to the profile or shape of an object presents to the moving fluid.

In general *four physical characteristics* affect how the fluid and the body dynamically interact. These are (*i*) density or mass per unit volume of the fluid (*ii*) size and shape of the body as it meets the fluid (*iii*) velocity of the fluid and (*iv*) viscosity of the fluid refers to its resistance to flow.

These four characteristics are brought together in a ratio known as the *Reynold's number.*

$$Re = \frac{\rho LU}{\mu}$$

$\rho$ = is the density of the fluid, $\mu$ = measure of viscosity, $L$ = body's shape and size, $U$ = velocity through the fluid.

The Reynold's number tells us how properties of an animal affect fluid flow around it. In general, *at low Reynold's numbers, skin friction* is of great importance, *at high Reynold's number, pressure drag* might predominate.

**(A) Swimming of fish :**

(*i*) Fish swimming raises very difficult problems in hydrodynamics. When a solid body moves through a fluid the pattern of flow around it depends on its shape.

(*ii*) The resistance that a medium (here water) offers to the motion of an object is called *drag.* It may be *frictional* or *pressure drag.*

(*iii*) The frictional drag is the product of half the density of the fluid medium, times the area of the object, times a value called *drag coefficient.*

(*iv*) The drag coefficient must be calculated in each instance and varies with shape and surface texture of the object and with value called *Reynold's number.*

(*v*) For bodies moving in water, the Reynold's number is $10^6$ (*length in m*) × (*velocity m* $S^{-1}$). A large (1m) fish swimming fast, at 10m $S^{-1}$ has a Reynold's number of $10^7$. A small (5 cm) fish swimming slowly, at 5cm$S^{-1}$, has a Reynold's number of 2500.

(*vi*) We know that when a body moves through a fluid, the fluid exerts a force on it in the direction opposite to the motion *i.e.* called *drag.* Consider a body moving at a velocity U through fluid of density $\rho$. Let its frontal area A, *i.e.* the view of the body, seen along the direction of motion.

$$\text{Drag} = \frac{1}{2}\rho U^2 A C_D \qquad [C_D = \text{Drag coefficient}].$$

$C_D$ is constant for bodies of the same, shape attitude and Reynold's number.

**Propulsion :**

(*i*) Fish propel themselves through undulatory or oscillatory mechanism.

(*ii*) They move forward by thrusting a propulsor (*i.e. fin, paddle or body segment*). The forward motion in fish is initiated at tail region.

(*iii*) The tail fin constantly thrusts obliquely against the water with a force ($F_t$). The inertia of the water causes it to push with equal force in the opposite direction ($F_w$). This force has two components *viz.* forward component ($F_f$) and a lateral component ($F_l$).

(*iv*) Because of the streamlined shape of the fish , the water offers little resistance. The lateral force vectors ($F_l$) add nothing toward progression but the foward force vectors ($F_f$) drive the fish.

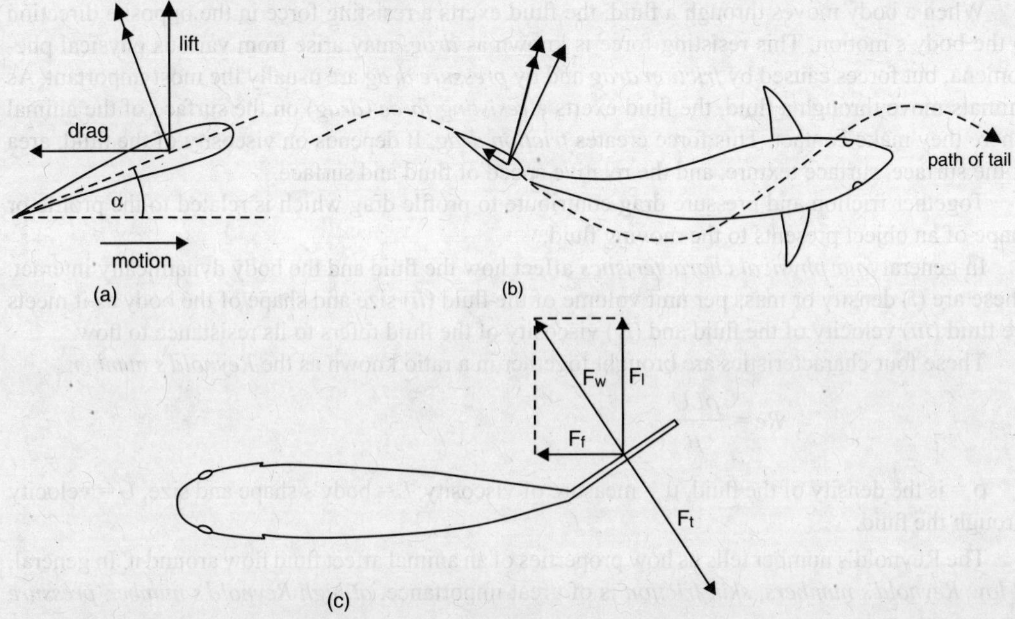


**Fig. 13.5** (*a*) A diagrammatic section through a hydrofoil, showing the components of hydrodynamic force which acts on it.
(*b*) A diagram showing how fish use their tails as hydrofoils for propulsion.
(*c*) A diagram showing some principal forces which act on a fish tail for initiation of forward motion.

**Aerodynamic Basis of Flight :**

During horizontal flapping flight, four forces act on a bird at equilibrium.

The upward *lift* (*L*) is opposed by the weight (*mg*) tending to pull the bird down. *Drag* (*D*) acts in the direction opposite to the direction of travel and wings generate *thrust (T)*, a forward force component. The angle at which the wing meets the airstream is its *angle of attack.*

(*i*) *Increasing the angle of attack* increases lift, but only up to a point. As the angle of attack increases, drag increases as well.

(*ii*) An aerofoil is an object that, when placed in a moving stream of air, produces a useful reaction.

(*iii*) All the forces acting on the wing that are derived from its motion can be divided into a component called *Drag* (*D*) which is opposite to the direction of *flight* and a component called *lift* (*L*) which is right angle to *D*.

(*iv*) Total drag is the overall force that resists movement of an animal through a fluid (air), *Parasitic drag* is an animal's resistance to passage through a fluid (air). Several types of resistance contribute to parasitic drag.

*Profile Drag* is the portion of this resistance caused by the shape of the animal moving through the fluid (air). *Friction drag* is caused by shear, stress at the boundary layer. *Induced drag* is associated with the production of lift. The vector difference between lift and its effective vertical component represents the induced drag. Therefore induced drag is the vector component of the lift force acting opposite to the direction of travel.

(*v*) If an airfoil or wing is convex on its upper surface it is called *camber.* The line joining front and back edges of a *cambered* airfoil (*wing*) is called the *chord.* Narrowness of wing is expressed as aspect ratio. **A** which is the span of the wings (*i.e.* tip of one wing to tip of other) divided by the average width of the *chord.* **A** (aspect ratio) = $\frac{Span}{Chord}$

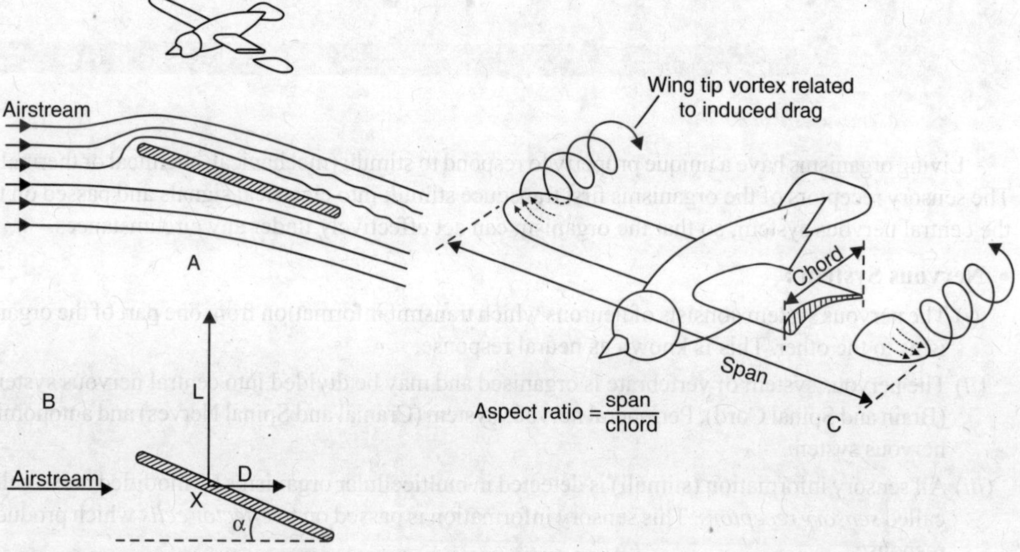


**Fig 13.6** Some factors related to lift (A& B) Nature of the aspect ratio and wing tip vertex (*C*).

(*vi*) The drag on an aerofoil of plan area '*S*' moving at a velocity '*U*' relative to a fluid density *p* and generating lift *L* is expressed in the following equation.

$$\frac{1}{2}\rho U^2 SC_{DO} + (2KL^2 / \pi R_\rho U^2 S)$$

$C_{DO}$ = is the zero-lift drag coefficient, *R* = aspect ratio and *K* is a factor which depends on how the wings taper.

$\frac{1}{2}\rho U^2 SC_{DO}$ = profile drag ($D_P$); $2KL^2 / \pi R_\rho U^2 S$ = induced drag ($D_i$). If an complete aircraft is considered rather than just an aerofoil, drag on the fuselage has also to be considered. It is known as parasite drag. If the frontal area of fuselage is A and its drag coefficient $C_{DF}$. the parasite drag $\frac{1}{2}\rho U^2 AC_{DF}$.

Thus total drag = Parasite Drag + Profile Drag + Induced Drag

$$= \frac{1}{2}\rho U^2 (AC_{DF} + SC_{DO}) + (2KL^2 / \pi R_\rho U^2 S)$$

(*viii*) The total drag like lift increases with the angle of attack.

CHAPTER 14

# Neurobiophysics

Living organisms have a unique property to respond to stimuli (mechanical, chemical or thermal). The sensory receptors of the organisms first transduce stimuli into electrical signals and passed on to the central nervous system, so that the organism can act effectively under any circumstances.

- **Nervous System :**

(*i*) The nervous system consists of neurons which transmit information from one part of the organisms to the other. This is known as neural response.

(*ii*) The nervous system of vertebrate is organised and may be divided into central nervous system (Brain and Spinal Cord). Peripheral nervous system (Cranial and Spinal Nerves) and autonomic nervous system.

(*iii*) All sensory information (stimuli) is detected in multicellular organisms by modified nerve cells called *sensory receptors*. This sensory information is passed on to *effector cells* which produce response.

(*iv*) Interposed between the receptors and effectors are the conductile cells of the nervous system the neurons.

(*v*) **Sensory (affarent) neurons** carry impulses from the visceral organs to the central nervous system.

**Motor (efferent) neurons** carry impulses from the central nervous system to peripheral nervous system.

**Inter neurone** is placed between the sensory and motor neurons in the central nervous system.

**Neurone :** A nerve cell with all its processes (*axon and dendron*) that forms the structural and functional unit of nervous system is known as neurone.

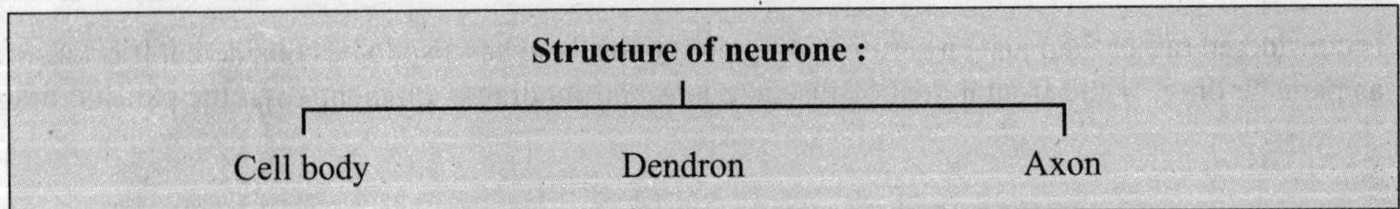


(*a*) **Cell body :** It is known as soma or neurocyton or perikaryon. It contains a well defined nucleus and nucleolus surrounded by a granular cytoplasm. It may be star shaped, oval, round or flask shaped.

(*i*) **Neurolemma :** It covers the cell body and made of lypoprotein.

(*ii*) **Cytoplasm :** It is known as neuroplasm. It consists of neurofibrils, Nissil granules (**ribo-nucleo protein**) ribosome, golgibodies, mitochondria, centrosome (*inactive*).

(*iii*) **Nucleus :** There is a single, oval or round nucleus in each cell body.

(*b*) **Dendron :** Short fibers with branching processes that arises from the cell body is known as dendron.

(*i*) It contains Nissil granules.

(*ii*) Thin terminal branches are known as dendrites.

(*iii*) Neurofibrils are present.

(*c*) **Axon :** A long and elongated generally branchless cytoplasmic process that comes out from the cell body is known as axon.

(*i*) It arises from the conical elevation of the cell body and this point is known as **axon hillock**.

(*ii*) Axon and axon hillock are devoid of Nissil granules.

(*iii*) It is generally long and may have collateral branches.

(*iv*) It's cytoplasm is known as axoplasm which is covered by **axolemma**.

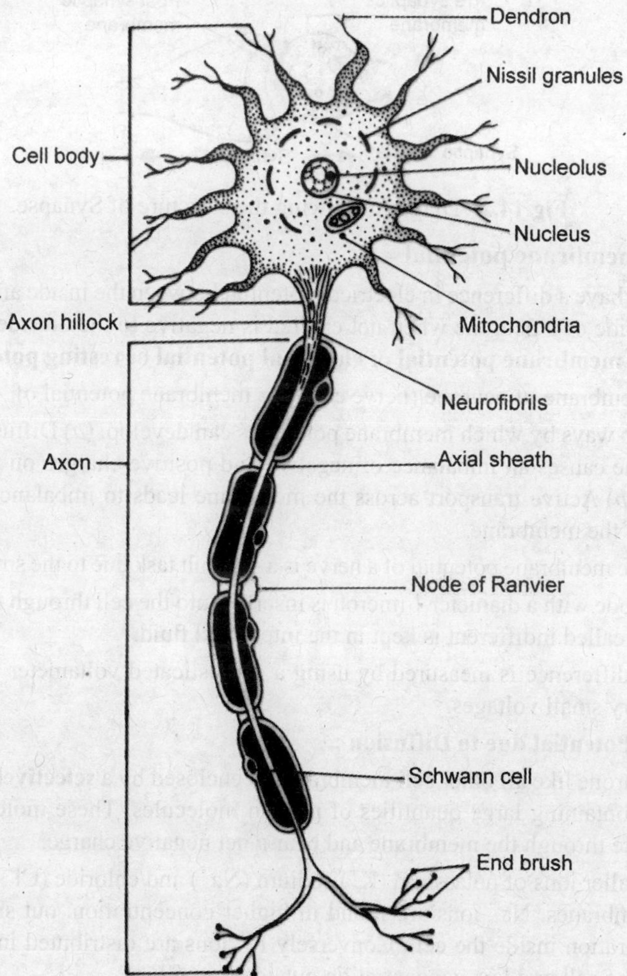


**Fig. 14.1.** Structure of a typical neurone.

(*v*) The axon is generally covered with a myelin sheath (**medulated**). Does not form a continuous envelope but leaves a naked depression at regular intervals known as **nodes of Ranvier**.

(*vi*) The terminal portion of axon undergoes numerous branches which are known as **end brush** or **axon terminals**.

**Synapse :**

In the nervous system the message is transmitted from one point to another. The point at which two neurones (**axon-dendron, axon somatic cell, axon-axon**) associated with each other is called synapse.

(*i*) The terminal swelling part of an axon is known as **synaptic knob**. These synaptic knobs are covered by a **pre-synaptic** membrane.

(*ii*) The membrane of soma and dendrite is known as **post synaptic** membrane.

(*iii*) Pre synaptic knobs are separated from dendrite and soma by a minute space (200Å) known as **Synaptic Cleft**.

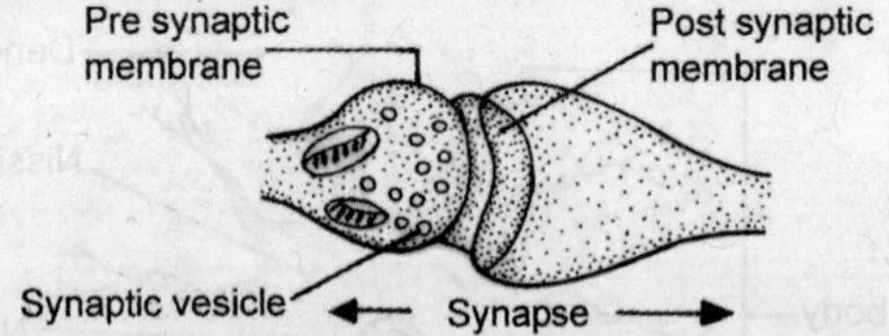


**Fig.14.2.** Diagram showing the structure of Synapse.

- **Biophysics of membrane potential**

All living cells have a difference in electrical potential between the inside and outside of the cell membrane. The inside of a neurone when not excited is negative to the outside. This difference in charge is known as **membrane potential** or **electrical potential** or **resting potential**.

The plasma membrane of neurone (nerve cell) has membrane potential of —70mV across it.

(*i*) There are two ways by which membrane potentials can develop. (*a*) Diffusion of ions through the membrane causes an imbalance of negative and positive charges on the two sides of the membrane. (*b*) Active transport across the membrane leads to imbalance of the charges on either side of the membrane.

(*ii*) Measuring the membrane potential of a nerve is a difficult task due to the small size of the fibres.

A micro electrode with a diameter 1 micron is inserted into the cell through the membrane while the other electrode called indifferent is kept in the interstitial fluid.

The potential difference is measured by using a sophisticated voltameter which is capable of measuring even very small voltages.

(*a*) **Membrane Potential due to Diffusion :**

(*i*) The neurone like all other cell membranes is enclosed by a **selectively permeable** membrane containing large quantities of protein molecules. These molecules are unable to penetrate through the membrane and bear a net negative charge.

(*ii*) The smaller ions of potassium ($K^+$) sodium ($Na^+$) and chloride ($Cl^-$) can diffuse across the membranes. $Na^+$ ions are found in higher concentration, out side the cell and low concentration inside the cell. Conversely $K^+$ ions are distributed in high concentration inside the cell and low concentration outside the cell.

(*iii*) German physiologist Bernstein (1902) suggested that the nerve cell membrane is selectively permeable to $K^+$ ions and relatively impermeable to $Na^+$ ions and not all permeable to large protein molecules.

(*iv*) Potassium ($K^+$) ions which are smaller can flow out to make the outside more and more positively charged i.e., electropositivity and thereby building up negative potential i.e., electronegativity inside. Thus a potential difference is developed and this prevents further outflow of $K^+$ ions. The potential across the membrane is now known as the *Nernst potential for potassium ions.*

(*v*) Sodium ($Na^+$) ions concentration is more in outside. Sodium ions move inward and a membrane potential with reverse polarity is generated when the membrane potential rises high enough to prevent further diffusion of $Na^+$ ions, it is known as the *Nernst potential for sodium ions*, *The magnitude of the potential is given by

$$\mathbf{EMF(in\ mV) = -\ 61 \times \log\left[\frac{Concentration\ of\ positive\ ions\ (inside)}{Concentration\ of\ negative\ ions\ (outside)}\right]}$$

When the concentration of positive ions inside is ten times that of negative ions outside, then log(10) is 1, and the EMF = –61 mV. The equation is Nernst equation. For potassium, Nernst potential ≅ –94 millivolts and for sodium it is ≅ + 61 millivolts.

Under resting state the membrane potential coverages to –90 millivolts which is near the Nernst potential for potassium. This is because at resting stage the membrane is more permeable to $K^+$ and only slightly permeable to $Na^+$ ions.

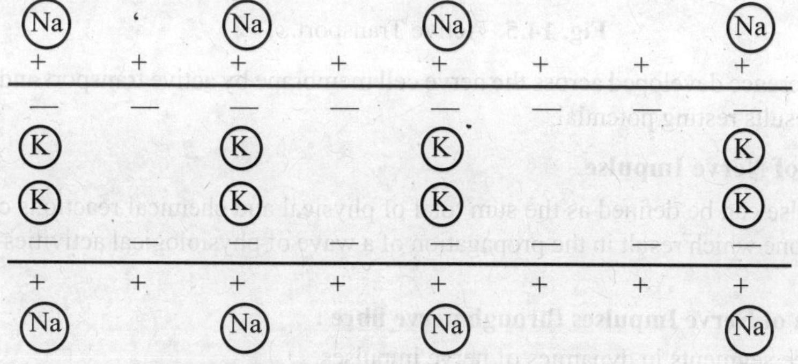


**Fig. 14.3.** Distribution of ions during resting potential

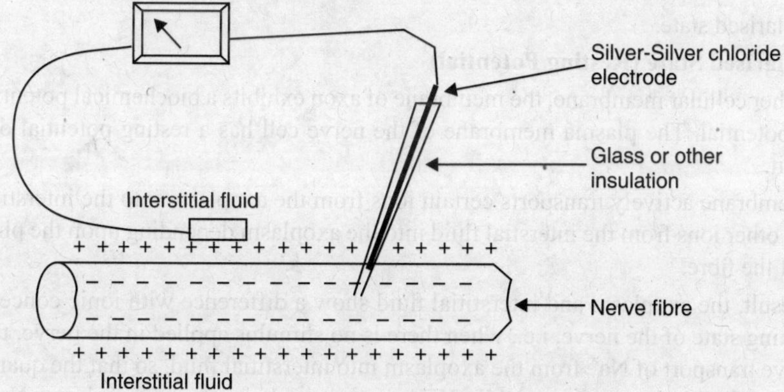


**Fig. 14.4** Measuring membrane potential

(*b*) **Active Transport**

(*i*) Like cell membrane it can actively pump sodium and potassium across its borders. Thus $Na^+$ is moved from inside the cell to the outside and $K^+$ is transported from inside to the outside of the membrane.

(*ii*) They are moved against their concentration gradients, so that $Na^+$ accumulates in the extra-cellular fluids and $K^+$ accumulates in the intracellular fluids. This process requires metabolic energy and is called active transport to distinguish it from diffusion which is passive transport.

(*iii*) Specific carrier molecules are involved in the active transport system. Let the carrier molecule ($x$) is attached to the $Na^+$ in the intracellular fluids. This $x\,Na^+$ would pass from inside to outside of the membrane and release $Na^+$ to the extracellular fluid.

(*iv*) The carrier ($x$) would be converted into another carrier ($y$) by enzymes. It attaches on to the $K^+$ in the extracellular fluids and passes across the membrane as $y$K. It then releases $K^+$ inside the cell. It is again converted to $x$ enzymatically.

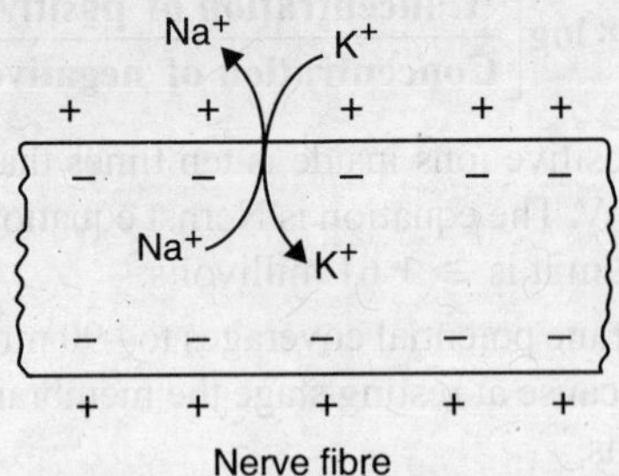


**Fig. 14.5.** Active Transport.

The ionic difference developed across the nerve cell membrane by active transport and the Donnan equilibrium results resting potential.

## • Transmission of Nerve Impulse.

A nerve impulse can be defined as the sum total of physical and chemical reactions created by stimulus in a neurone which result in the propagation of a wave of physiological activities along the nerve fibre.

### A. Conduction of Nerve Impulses through nerve fibre :

There are three segments in dynamics of nerve impulses.

(*a*) A polarised state (*b*) Depolarised state and

(*c*) Repolarised state.

**(*a*) A Polarised State (Resting Potential)**

(*i*) Like other cellular membrane, the membrane of axon exhibits a biochemical potential or membrane potential. The plasma membrane of the nerve cell has a resting potential of —90 mV across it.

(*ii*) The membrane actively transports certain ions from the axoplasm into the interstial fluid and certain other ions from the interstial fluid into the axoplasm depending upon the physiological state of the fibre.

(*iii*) As a result, the axoplasm and interstitial fluid show a difference with ionic concentration. In the resting state of the nerve, i.e., when there is no stimulus applied in the nerve, there occurs an active transport of $Na^+$ from the axoplasm into interstitial fluid, so that the quantity of $Na^+$ ions is negligible in axoplasm while $K^+$ ions are much greater in exoplasm. Normally, the concentration of Na+ ions in the interstitial fluid are much greater than that inside the axon. This is due to the active transport $Na^+$ ions from **axoplasm** to interstitial fluid. This transport of $Na^+$ ions, is known as **Sodium Pump.**

(*iv*) On the other hand, $K^+$ ions are distributed in high concentration in axoplasm and low concentration in interstitial fluid. It flows from inside to outside to make outside more positive charged i.e., **electropositivity**.

(*v*) Due to sodium pump the interior of the axon becomes strongly **electronegative** *i.e.*, $Na^+$ ion concentration low inside.

This is the **polarised state** or **resting potential** or membrane potential of nerve at resting state.

(*b*) **Depolarised State :**

(*i*) It is the most familiar activity of the stimulated neuron is the **action potential**. At this stage (**activated axon**) the sodium gates open, sodium ions ($Na^+$) flows (diffuse) into the axoplasm through the membrane starting at point.

(*ii*) The membrane potential rises in the positive direction from the resting level zero and then becomes 20 to 35 mV. This reversal of polarity is called **overshoot-potential.**

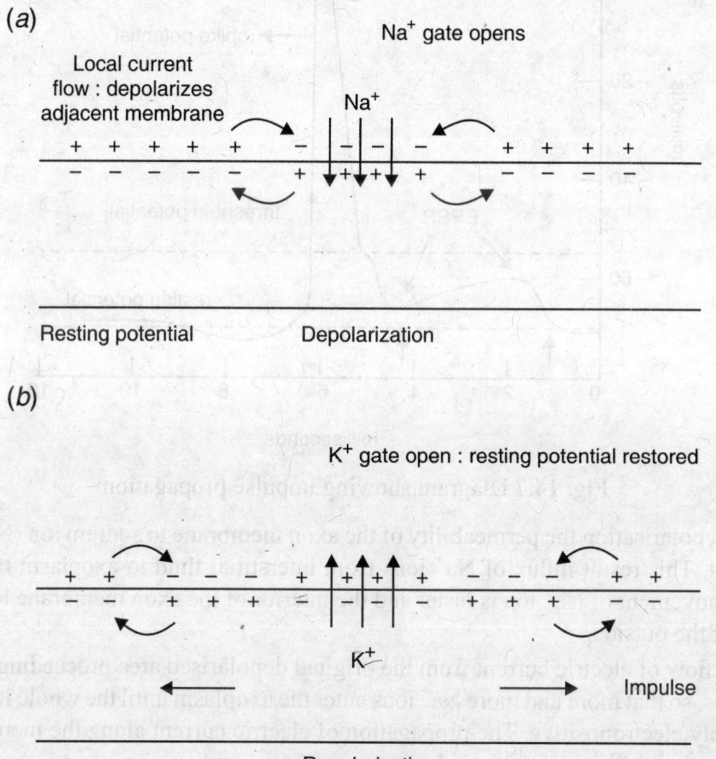


**Fig. 14.6 :** Action potential in a mammalian nerve (in an intact animal the properties of the synapse mean that the inpulse only travels in one direction). (*a*) Activation ; (*b*) recovery.

(*iii*) Suddenly the potential then reverts to the resting level i.e., about —70mV. In certain cases it may fall even beyond that. All these changes together called **action potential**. The rise and fall of the action potential is called **spike potential**. The short phase of the action potential before reaching the resting potential is called **negative after potential** and the curve which falls below the resting level is called **positive after potential.**

(*iv*) Depolarization only occurs when the stimulus attains a threshold strength, then the action potential is generated in **all or none law fashion.**

(*v*) The influx of $Na^+$ ion does not proceed indefinitely and is halted near the peak of action potential. As a result of this $Na^+$ ion permeability falls to its resting level and $K^+$ ion also move outward.

(*vi*) Soon after attaining the peak of the spike, $Na^+$ ion permeability stops and $K^+$ ion permeability rises above the resting state. Consequently efflux of $K^+$ ion takes place and restore the membrane potential.

(*vii*) At the end of the spike, original level of potential is restored but the $Na^+$ ion permeability is in an active state. This is the **absolute refractory period** and the nerve is insensitive to further stimulation.

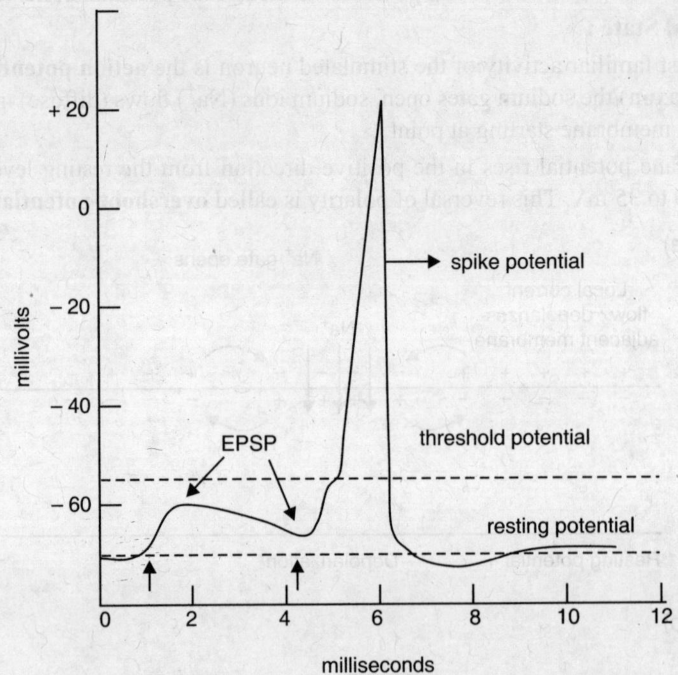


**Fig. 14.7** Diagram showing impulse propagation

(*viii*) During depolarisation the permeability of the axon membrane to sodium ion ($Na^+$) is suddenly increased. This result influx of $Na^+$ ions from interstitial fluid to axoplasm takes place. The inward movement of $Na^+$ ion is faster and the interior of the axon membrane become positive to that of the outside.

(*ix*) Now the flow of electric current from the original depolarised area proceeding further in both directions, so that more and more $Na^+$ ions enter the axoplasm until the whole interior becomes completely electropositive. The propagation of electric current along the membrane is called depolarisation wave or nerve impulse.

(*c*) **State of repolarisation :**

(*i*) Soon after depolarisation wave has spread completely over the entire length of the fibre, no more sodium ions ($Na^+$) can enter the axoplasm.

(*ii*) But a large quantity of potassium ions diffuse through the membrane and $Na^+$ ions begin to diffuse out resulting in electronegativity inside the membrane and electropositivity outside the membrane. This process is known as **repolarisation**.

(*iii*) It starts exactly on the same point in the nerve fibre at which depolarisation has started and it extends in both directions. In other words, it starts once again and the whole process of depolarisation and repolarisation is completed within a fraction of second.

**B. Conduction of Nerve Impulse through Synapse :**

Synapse is the functional junction of the two neurone where one end is occupied by axon terminus and the other end is occupied by dendron apex of the next neurone.

Ultrastructurally synapse shows three components like **presynaptic knob** (axon end), **synaptic cleft** and **post synaptic knob** (dendron apex). Synaptic cleft contains several parallely placed synaptic filaments.

**Mechanism of impulse propagation :**

(*i*) From nerve cell perikaryon, neurohumor vesicles are formed where impulse transmitting components remain loaded. These vesicles comming at the axon end *i.e.*, at the presynaptic knob, diffuses and the component is released in the synaptic cleft.

(*ii*) The transmitting component is chemically **acetylcholine** which is as soon as reaches the post synaptic knob, increases the permeability of post synaptic knob. Then $K^+$ ion exist out and $Na^+$ ion comes in. **Depolarization** appears at the post synaptic knob and the action potential is initiated. After a while post synaptic knob turns to **repolarization**.

(*iii*) **Acetylcholine,** a neurotransmitter, i.e., the action potential causing chemical which is short lived and very soon it is hydrolysed by the enzyme **choline estarase** produced from presynaptic knob. This enables the post synaptic knob to reach in repolarized state to receive further impulses.

(*iv*) The neurotransmitter break down products are recycled after diffusing back into the presynaptic axon knobs.

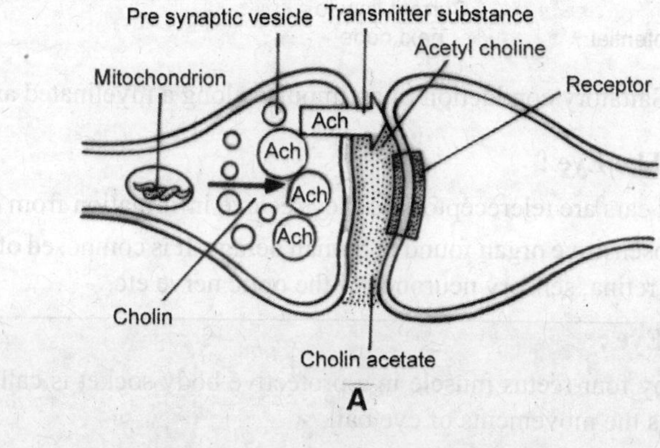


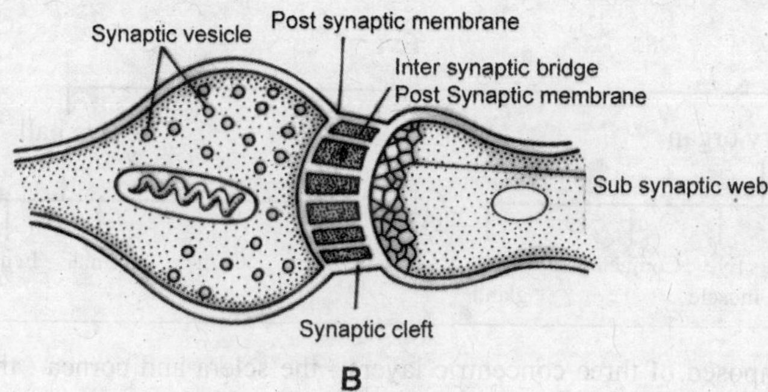


**Fig. 14.8.** Impulse propagation through synapse

- **Language and Speed :**

(*i*) The frequency of impulses is known as *language of nervous system*. The frequency is proportional to the strength of the stimulus above threshold level.

(*ii*) The **velocity** of impulse is proportional to the cross sectional area of axon.

(*iii*) In non myelinated nerve fibres the velocity of conduction is much slower. In most animals (vertebrates) myelination increases conduction velocity (diameter 10μ can have conduction velocity 120 m per second).

(*iv*) The myelin acts as an effective insulator and action potentials are generated only at the nodes of Ranvier where the axon plasma membrane is exposed.

(*v*) Ions cannot pass through the myelin sheath and nodes of Ranvier permit ions to pass through it more easily. Therefore, depolarization at the node leads to flow of current between it and the next node which depolarizes in turn. This jumping or keeping of depolarization from node to node is known as **saltatory** propagation that facilitates great rapidity of conduction.

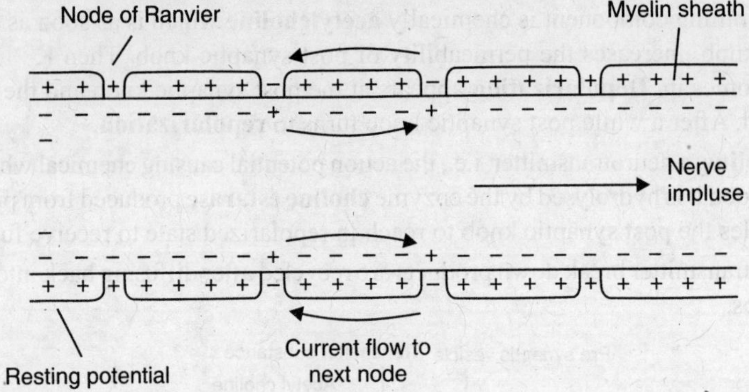


**Fig. 14.9.** Saltatory conduction of an impulse along a myelinated axon

**Sensory Mechanism : The Eye :**

Organs like eyes and ears are telereceptors as they receive information from distant object.

Eye is the main photosensitive organ found in human beings. It is composed of many sense cells, the rods and cenes of the retina, sensory neurones of the optic nerve etc.

**Structure of Human Eye :**

The eyes are held by four rectus muscle in a protective body socket is called orbit and two oblique muscles controls the movements of eye ball.

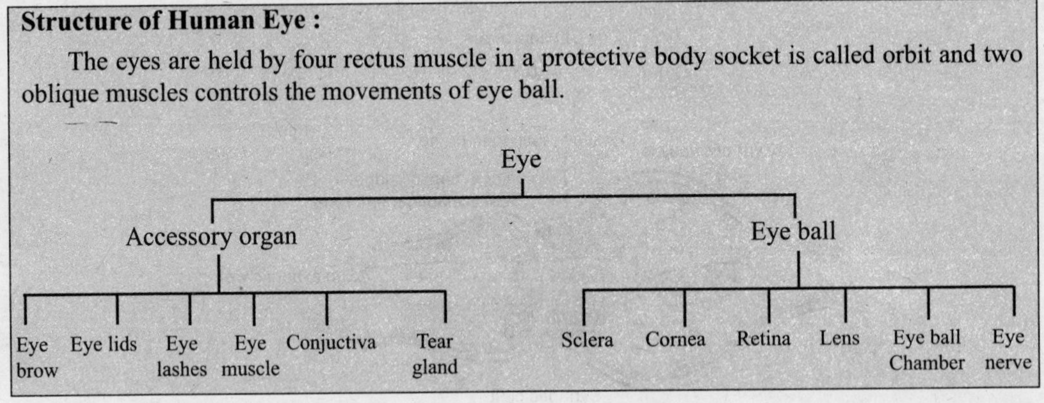


The eye is composed of three concentric layers : the sclera and cornea ; the choroid, ciliary body, lens and iris and the retina and is supported by the hydrostatic pressure of the aqueous and vitreous humours.

- **Sclera :** External covering of the eye, formed of tough connective tissue. It protects and maintains shape of the eye ball.
- **Cornea :** Transparent front part of the sclera. The curved surface of the cornea helps in focussing light.
- **Conjunctiva :** Thin transparent layer of cells protecting the cornea and continuous with the epithelium of eyelids.
- **Eyelid :** It protects the cornea from mechanical and chemical damage of the retina.
- **Choroid :** Rich in blood vessel and pigment supply to the blood vessel.
- **Ciliary muscles :** Circular sheet of smooth muscle fibres forming circular and radial muscle.

It alters the shape of the lens during accommodation.

- **Lens :** Transparent, elastic and biconvex structure.
- **Retina :** Light sensitive portion of the eye containing rod (sensitive to low light) and cone (sensitive to bright light) cells.
- **Aqueous humour :** Clear solution of salt secreted by the ciliary body & presence in between cornea and irris.
- **Vitreous humour :** Transparent, jelly like solution reside behind the lens & retina.
- **Iris :** Circular muscular diaphragm containing the pigment giving eye its colour.
- **Pupil :** Opening in iris all light enters through this.

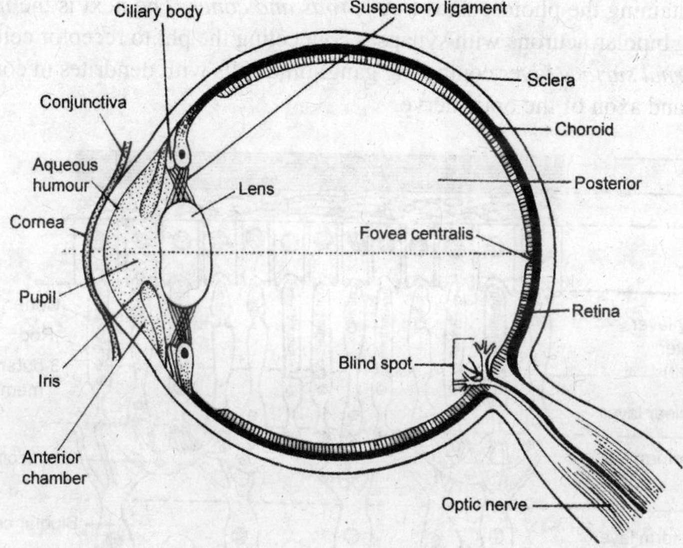


**Fig. 14.10.** The structure of mammalian eye.

- **Photoreceptors :**

Retina is the light sensitive layer of the eye. It converts light energy into nerve impulses which are transmitted to the brain through the optic nerves.

(*i*) The vertebrate retina consists of two kinds of photoreceptors called **rods and cones.** Rods are responsible for black and white vision when illumination is dim, while cones are responsible for high visual activity and colour vision. Humans have 100 million rods and 3 million cones in each retina. Most cones are located in the central region of retina known as *fovea centralis* where the eye forms its sharpest image. Rods are completely absent from fovea.

**Differences between rods and cones.**

| Rods | Cones |
|---|---|
| (*i*) They are at the periphery of the retina. | (*i*) They are tightly packed together at the fovea. i.e.centre of the retina. |
| (*ii*) More numerous than cones i.e., $120 \times 10^6$ | (*ii*) Less numerous ; $6 \times 10^6$. |
| (*iii*) Very sensitive to low level of illumination. | (*iii*) Very sensitive to bright light. |
| (*iv*) Contain only one visual pigment. | (*iv*) Contain three visual pigments. |

| | |
|---|---|
| (*v*) Principally used for night vision. | (*v*) They are used principally in day light. |
| (*vi*) It possesses is rapid regeneration power. | (*vi*) It possesses slow regeneration power. |

(*ii*) In rods, photopigment is called **rhodopsin**. It consists of the protein opsin bound to a molecule of cis retinal which is produced from vitamin A.

(*iii*) The photopigments of cones is called **photopsins** and is structurally very similar to rhodopsin cone posseses three types of pigments. **Retinal** is a prosthetic group for each pigment. Absorption maximum of three kinds of cone photopsin are 420 nm (blue absorbing) 530 nm (green absorbing) and 560 nm (red absorbing).

(*iv*) The retina is composed of three layers of cells. First there is *photoreceptor layer* (*outer most layer*) containing the photosensitive cells *rods and cones*, The next is the *intermediate layer* containing bipolar neurons with synapses connecting the photo receptor cells. The third layer is the *internal surface* layer containing ganglionic cells with dendrites in contact with bipolar neurones and axon of the optic nerve.

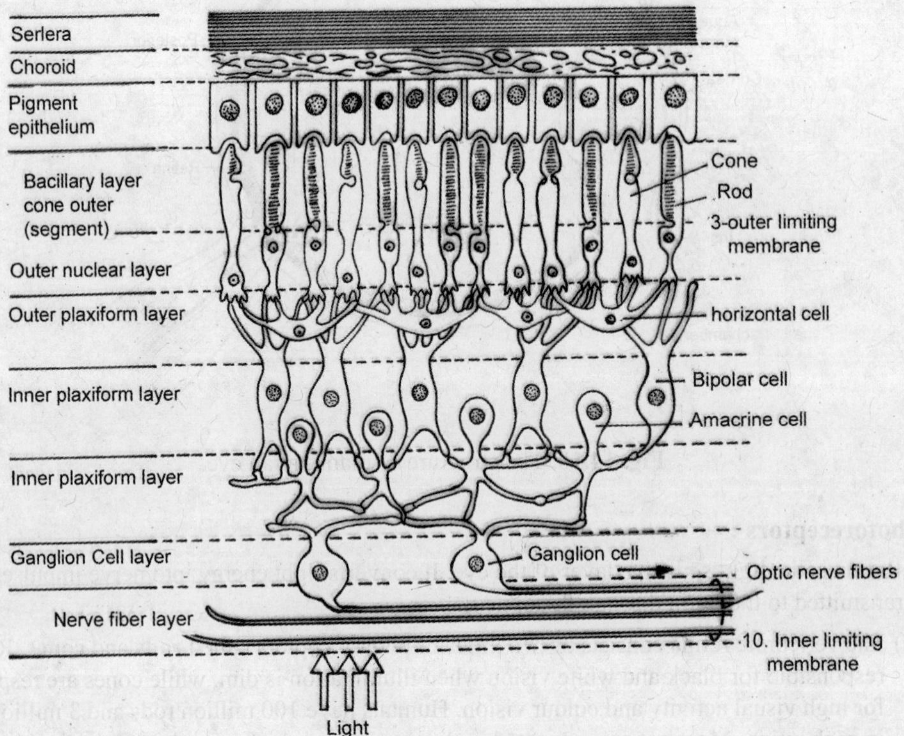


**Fig. 14.11** Detailed structure of retina.

- **Sensory Transduction in Photoreceptors**

The retina act as a photoneural transducer i.e., it converts light energy into a neural response which is in the form of spike potential and travel along the optic nerve.

(*i*) Rod and cone contains many $Na^+$ channels in the plasma membrane of its outer segment. In the dark, many of these channels are open. As a result $Na^+$ ions continuously diffuses into the outer segment. The flow of $Na^+$ ion that occurs in the absence of light is called **dark current**. It causes depolarization of the membrane of photoreceptor.

In the light, the $Na^+$ ion channel in the outer segment rapidly close, reducing the **dark current**. It results in *hyperpolarization.*

(*ii*) Activated rhodopsin activates many molecules of a protein called *transducin*. Transducin activates an enzyme that breaks down cyclic GMP (*c*GMP). GMP is the guanine containing cousin of AMP. *c*GMP is required to keep the $Na^+$ channel open. The channel will close if the *c* GMP is converted to GMP.

(*iii*) When a photopigment absorbs light, II-cis retinene changes to trans retinene and dissociates from opsin. This is known as *belching reaction.*

(*iv*) The recovery process involves the reduction of retinal to 'all trans' vitamin A. Diffusion of vitamin A into the epithelium and reconversion of vitamin A by *light* a enzmatically to the cis. II form. The II cis. vitamin A is then oxidized to "II. cis retinal" and linked to protein opsin. Thus Detachment and reattachment of the chromophore is crucial part in photochemistry of vision.

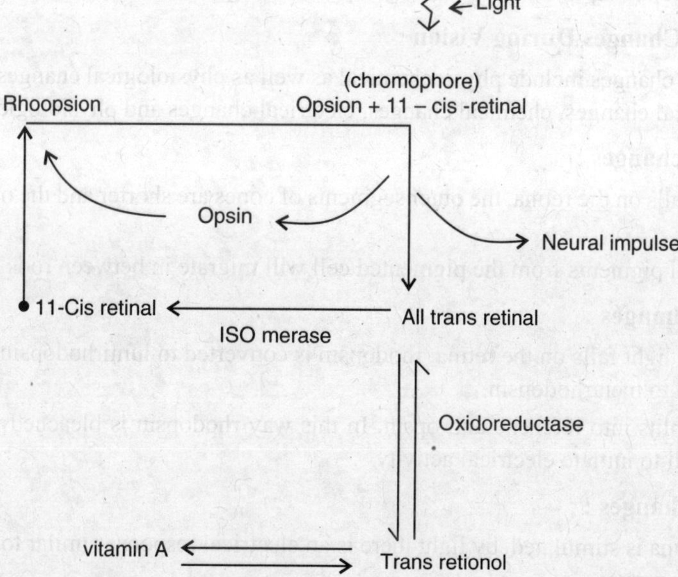


**Fig. 14.12.** Visual Cycle

- **Image Formation :**

(*i*) During vision, light emitted from the objects fall on the 1 cms and then are focused clearly on retina. So that Photoreceptive rods and cone cells are stimulated.

(*ii*) A dioptric apparatus is formed by the cornea, aqueous humour, tens and victreous humour which are all transparent. The dioptic apparatus is concerned with the focussing of image on the retina.

(*iii*) The most important role in vision is contributed by the corner and lens. The corner help in placing the image on the retina which the lens is responsible in making adjustment for sharp focussing.

- **Biophysical Aspect of Vision :**

(*i*) The eye is a self regulatory system which is operated by focussing the image on the retina and by regulating the amount of light falls on the retina.

(*ii*) About 50% of light passing through the cormeal surface is absorbed by the occular media (aqueous humour, lens, vitreous humour)

(*iii*) The aqueous and vitreous humour are somehow less transparent than corner. All rays above 2700 mμ are absorbed by aqueous humour and all the rays above 1600 mμ are absorbed by vitreous humour.

(*iv*) The lens absorbs all rays below 30 mμ or above 2500 mμ. The lens converts harmful shorter wavelength to larger one.

(*v*) The eye is most sensitive at wavelength of about 507 mμ. The minimal energy [(2.1 to 5.7) $\times 10^{-10}$ ergs] is required for that wavelength.

(*vi*) A large proportion of light pass through the transparent retina before it is absorbed by the photopigment rhodopsin. Not all the lights is absorbed in rods by rhodopsin. Most of the rays (80%) is absorbed by retina and chroid coat.

(*vii*) For the evokation of sensation of light 2 – 14 quanta of light is required. Stimulated rod-gives its signal to the bipolar cell when single quanta of light has been absorbed.

- **Biophysical Changes During Vision :**

Biophysical changes include physiochemical as well as physiological changes. It occurs in four steps viz structural changes, chemical changes, electrical changes and physiological changes.

**(*a*) Structural changes :**

(*i*) As light falls on the retina, the outer segments of cones are shorter and the outer limbs of rods swell.

(*ii*) The visual pigments from the pigmented cell will migrate in between rods and cones.

**(*b*) Chemical changes :**

(*i*) When the light falls on the retina, rhodopsin is converted to lumirhodopsin. It is then further converted to metarhodopsin.

(*ii*) Then it splits into retenene and opsin. In this way rhodopsin is bleached which excites the retinal cell to initiate electrical activity.

**(*c*) Electrical changes :**

(*i*) When retina is stimulated by light there is an electrical response similar to the current of action in the nerve.

(ii) Incase of any one colour, a geometric rise of intensity couses an arithmatic increase in the current with coloured rays of apparent equal intensity, the yellow rays are said to give a large current.

**(*d*) Physiological changes :**

(*i*) The central part of the retina is called **fovea centralis**. There remains only cone cells which are responsible for bright light.

(*ii*) The peripheral part of retina is called **macula lutea**. There remains both rod and cone cells. There remains both rod and cone cell. They are responsible both dim and bright light.

- **Neural aspect of Vision :**

(*i*) The retina is a photosensitive layer and act as photo neural transducer (*converts light energy into a neural response*). The neural response travels through optic nerve and reaches to central nervous system (*CNS*). From the central nervous system, the neural response reaches to the specific areas of cerebral cortex. Thus the information is analysed both in retina and in CNS to generate shape, colour & brightness of the object.

(*ii*) The responses from either eye for an object are protected on to the occipital lobe of the cerebral cortex. There are three layers of cells viz., retinal cells, intermediate cells and a third layer of cells which are important on this function.

(*iii*) These layers are interconnected and the information flows from retinal cells to the brain through these layers of cells and optic nerve.

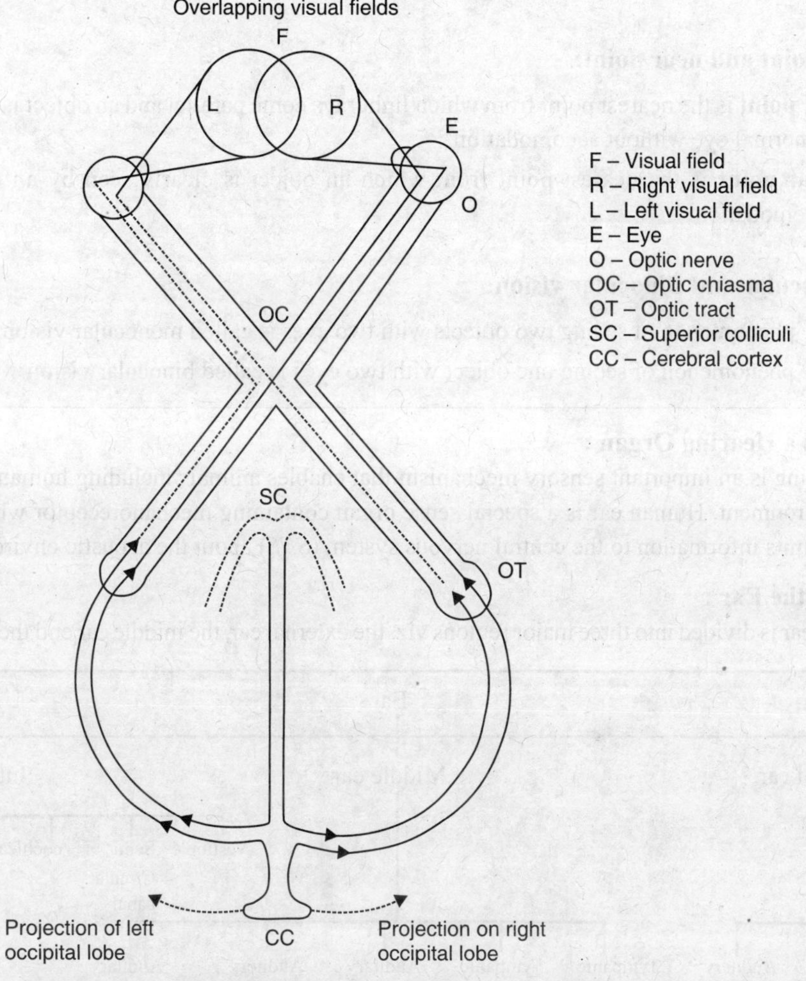


**Fig. 14.13** Pathways of vision in the brain.

(*iv*) The information coming out of the optic nerve divides into two bundles. Visual cortex in the brain is responsible for analysis of these impulses and subsequent vision. The few fibres go to the mid brain. Intensits of light, clearity of vision etc. are processed in the mid brain and a feedback is sent to the appropriate regions of the eye.

(*v*) The left side of the brain receives information from the right side of the visual field and the right side of the brain receives signals from the left side of the visual field.

They run through the optic chiasma and the information from both the eyes are put together to acheive binocular vision. Every point in the retina seems to have a corresponding point in the visual cortex. The point that are close together on to the surface of the retina are also close together in the visual cortex.

- **Accomodation of eye:**

(i) It means adjustment of the optical apparatus for near vision.

(ii) This is accomplished by increasing the curvature and refracting power of the lens.

- **Far point and near point:**
  - **Far point** is the nearest point from which light rays come parallel and an object is clearly seen by normal eye without accomodation.
  - **Near point** is the nearest point from which an object is clearly seen by an eye without accomodation.

- **Monocular and Binocular vision:**

The phenomenon of seeing two objects with two eyes is called monocular vision.

The phenomenon of seeing one object with two eyes is called binocular vision.

- **Ear as a Hearing Organ :**

Hearing is an important sensory mechanism that enables animals including human to perceive their environment. Human ear is a special sense organ containing mechanoreceptor which extracts and transmits information to the central nervous system (*SNS*) about the acoustic environment.

**Parts of the Ear :**

The ear is divided into three major regions viz. the external ear, the middle ear end the internal ear.

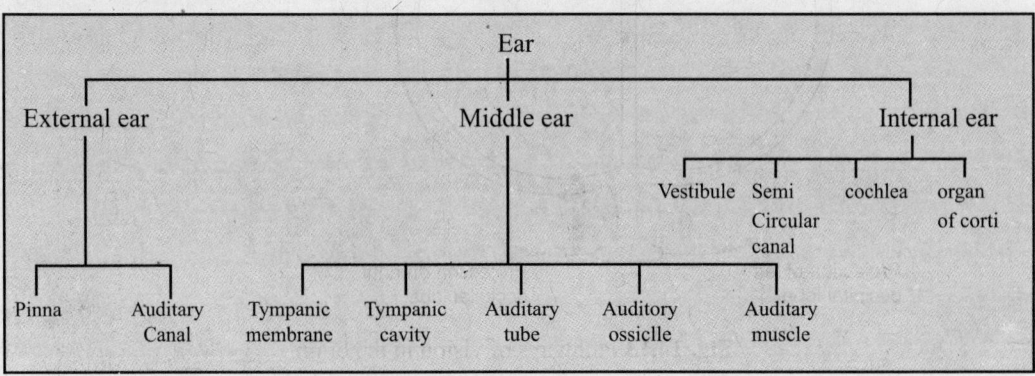


**Fig. 14.14** Schematic diagram of basic structure of the mammalian ear.

**A. External Ear :**

(*i*) It consists of **pinna** and **auditory canal**.

(*ii*) Pinna is made up of fibrocartilage and covered by skin.

(*iii*) It receives sound wave and transmit it to the external auditory meatus or canal.

(*iv*) Auditory canal is a 'S' shaped canal that ends at the ear drum or tympanic membrane.

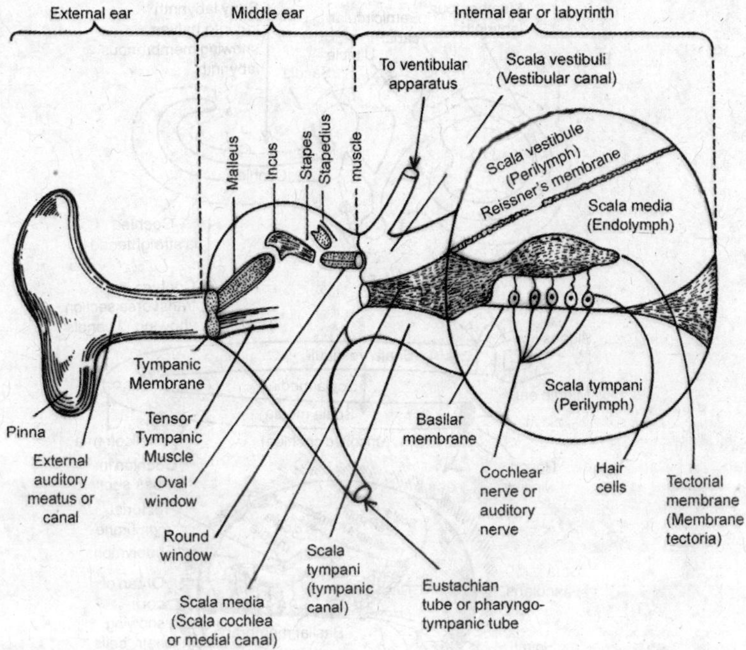


**B. Middle Ear :**

(*i*) Tympanic membrane, Tympanic cavity, Auditory (eustachian) tube and auditory ossicles.

(*ii*) **Tympanic membrane** forms the outer wall of the middle ear. It has three layers (*outer, middle and inner*) sound waves strike over it.

(*iii*) **Tympanic Cavity** is a air filled space in the temporal bone separated from external auditory canal by tympanic membrane and form internal oval and round windows.

(*iv*) **Auditory Ossicles** has three bones or ossicles malleus (hammer), incus (anvil) and stapes (stirrup). The base of the stapes remain linked with *oval window* which separates middle ear from internal ear. Below oval window, *round window* is marked.

(*v*) **Ossicular muscles** are *tensor tympanum* and *stapedius muscle.*

**C. Internal Ear**

(*i*) It is also called *labyrinth*. The labyrinth, comprise the cochlea semi-circular canal and vestibule.

(*ii*) **Vestibule** constitutes the oval centre portion of the labyrinth. It consists of two sacs. viz., ***utricle*** and *saccule* which are filled by *endolymph* and surrounded by *perilymph.*

(*iii*) **Semi-circular Canals** are three small ducts that are lying at right angle to each other and connected with vestibule. They remain suspended in *perilymph*. Each duct has an expanded end. The *ampulla* which contains a receptor *crista ampukaris*.

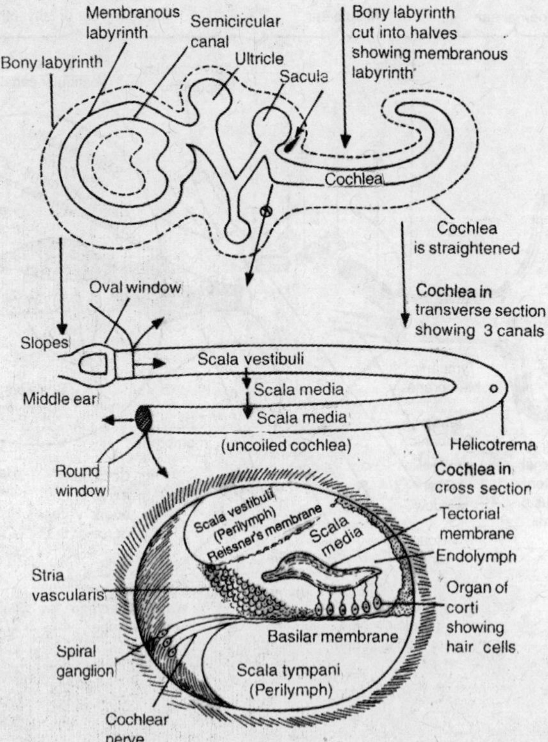


**Fig. 14.15:** *Schematic representation of internal ear with different sections.*

(*iv*) **Cochlea**, the spiral structure containing perilymph filled *scala vestibuli* and *scala tymphari* endolymph filled *scala media* (cochlearduct). Scala media is separated from scala vestibuli by *vestibular membrane* (*Reissner's membrane*) and from scala *tymphani* and by *bastibular membrane. Scala vestibuli* and *Scala tymphani* communicate each other by an aperture *helicotrema.* The basilar membrane contains the sound receptor *organ of corti.*

(*v*) **Organ of Corti** (organ of hearing) is a complex structure consisting of *supporting cells* and hair cells. Hair cells are arranged in rows and connected basally with the fibres of VIII cranial nerves. The hair cells have bristles like sensory hairs *stereocilia.*

## Bio Physical Aspects of Hearing :

Hearing is an important sensory mechanism. The process of hearing can be considered as the conversion of mechanical energy in the form of pressure. Variations in the air to electrical energy in the form of nerve impulse and are transmitted by auditory nerve to the brain for processing. The ear is a hearing apparatus. The entire range of sound perception in humans is between 20 to 20,000 Hz.

### A. Perception and Transmission.

(*i*) Perception of sound wave is made by coordinate activity of external and middle ear. Sound waves that reach the ear are directed by pinna into the external auditory canal.

(*ii*) When the sound waves strike the tympanic membrane, the alternate *compression* and *decompression* of the air causes the tympanic membrane to vibrate. The central area of tympanic membrane (ear drum) is attached with malleus which also begins to vibrate. The vibration is then picked up by the incus which transmit, the vibration to the stapes.

(*iii*) As the stapes moves back and forth, it pushes the *oval window.* The oval window separates the air filled middle ear from the liquid filled channels of the inner ear. Ear ossicles amplify the vibrations and transmit it to the inner ear.

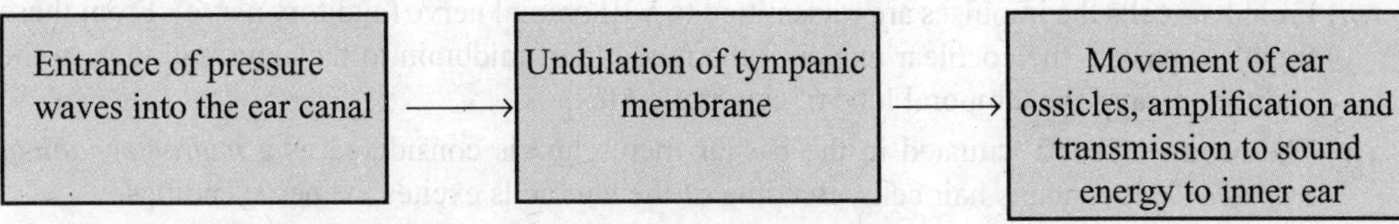


(*iv*) The amplified vibrations are transmitted across the middle ear lead to the movement of the foot plate of stapes. If then strikes over the oval window.

**B. Transmission of Travelling Waves :**

When the foot plate of the stapes is pushed into the oval window, the pressure of the cochlear fluid increases and as the stapes is withdrawn it decreases. The movement of foot plate of stapes produces a series of *"travelling waves"* in the perilymph of the scala vestibulli.

(*i*) The height of the waves increases to a maximum as the wave travels in the cochlea and drops within a very short time.

(*ii*) The sound vibration are transmitted to the perilymph of scala vestibuli by the movement of foot plate of stapes over oval window. It extends from scala vestibuli to scala tympani via *helicotrema* and these strike over round window. The vibrations move across the Reissner's membrane to endolymph of scala media leading to the movement of *basilar membrane.*

(*iii*) Due to the elasticity of the basilar membrane, it bends towards the round window and initiate the pressure of wave. Low frequency vibrations are transmitted from upper canal to lower canal throughout length of the basilar membrane. High frequency vibrations are restricted to the close of the oval window.

(*iv*) **Generation of Action Potential :**

(*i*) Normally endolymph of scala media is rich in $K^+$ ion and poor in $Na^+$ ion, while perilymph is rich in $Na^+$ ion and poor in $K^+$ ion. Thus polarization persists having inner positive (+) charge and outer negative (—) charges of sacla media. The potential difference of this site appears as 80 mV.

(*ii*) When the bascilar membrane vibrates, the hair cells of the organ of corti move against the tectorial membrane. The movement of hairs develop receptor potential result rapid influx of $K^+$ ion. The *depolarization* spread throughout the cell and causes opening of Calcium ion ($Ca^{++}$) Channels. That results that an influx of $Ca^{++}$. This leads to release of neuro transmitter from hair cell membrane which excites the sensory nerve.

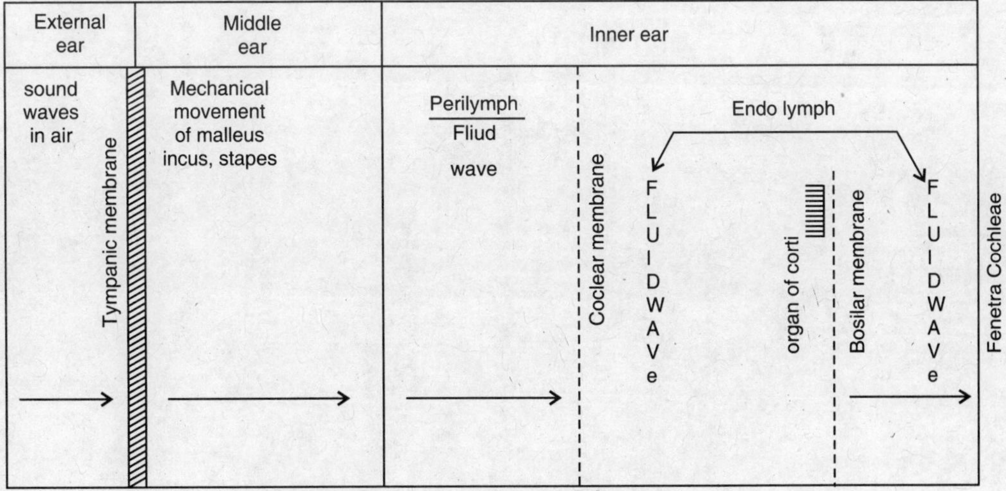


**Fig. 14.16** Transmission of Sound through the ear.

(*iii*) From hair cells the impulses are transmitted to VIII cranial nerve (auditory nerve). From these impulses pass to the cochlear nuclei of the medulla of midbrain to thalamus and then to the auditory area of the temporal lobe of cerebral cortex.

(*iv*) The **organ of corti**, situated in the basilar membrane is considered as a *neuromechanical transducer*. It contains hair cells. Bending of the hair cells excites the nerve endings.

- **Middle ear ossicles:**

  Malleus, incus and stapes.

- **Vestibular apparatus:**

  (*i*) Vestibular apparatus is the non auditory part of the internal ear.

  (*ii*) It consists of three semicircular canals and two sac like structures called utricle and saccule.

  (*iii*) It contains propioreceptors which is concerned with kinesthetic sensation and maintains posture and equilibrium.

- **Fenestra ovalis and Fenestra rotunda:**

| Name | Character | Location |
|---|---|---|
| **Fenestra ovalis** | oval window : | Junction of middle ear and scala vestibuli of cochlea. |
| **Fenestra rotunda** | round window : | Junction of middle ear and scala tymp. |

CHAPTER 15

# Nanotechnology

***Nanotechnology*** is an umbrella term that covers many research dealings with objects that are measured in nanometer. It is the science of understanding the structure and behaviour of materials at the atomic or molecular level. The technique used for manipulating atoms and molecules to fabricate materials, devices and systems is known as nanotechnology.

The main practical objectives of **nanobiotechnology** are using biological components to achieve nanoscale tasks.

High power scanning microscopes and special types of probes help the scientists to have clean observation of individual atoms and molecules. Thus it has become possible to manipulate and move atoms and molecules to form new structures and thus design new materials that are built from atomic level constituents. This is term "bottom up" approach and the study of the materials is "***nanotechnology***".

**The prefix nano means a billionth ($1 \times 10^{-9}$).**

Many internal component of biological cells are in the nanoscale range. The symbol used is 'n', Nanobiotechnology links biotechnology and genetic engineering.

- **Nanotechnology:**

It is the study of manipulating matter on an atomic and molecular scale.

(*i*) It deals with various structures of matter having dimensions of the order of a billionth of a matter.

(*ii*) Nanotechnology refers to the projected ability to construct items from the bottom up using technique and tools being developed today to make complete, high performance products.

(*iii*) The main fundamental principle of nanotechnology is high position control.

- **Nanoparticles**

(*i*) They are generally considered to be a number of atoms or molecules bonded together in a cluster with a radius from 5 nm to 100 nm (i.e. less than < 100 nm).

(*ii*) The metal oxide of nanoparticles of different sizes in stained-glass windows produce different beautiful colours because a particle scatters only that wave length which compares with its sizes.

(*iii*) Such small groups of atoms (clusters) go by different names viz nanoparticles, nanocrystals, quantum dots and quantum boxes.

- **Nano materials:**

The size of the atom clusters (particles) that constitute the material decides the characteristics of nano materials.

(*i*) Made up of small crystalline grain.

(*ii*) They have a relatively larger surface area and make the materials more chemically reactive.

(*iii*) Quantum effect can begin to dominate the behaviour of the matter at the nanoscale affecting the optical, magnetic behaviour of the matter.

- **Nano rods:**

(*i*) These are cylindrical shaped nanoparticles.

(*ii*) The diameter must be in nanoscale.

- **Nano shells:**

These are hollow nano sized particles that can hold different materials.

- **Nano tubes:**

(*i*) Cylindrical shaped, made of pure carbon with diameter 1 to 50 nanometers.

(*ii*) They have novel properties that make them potentially useful in a wide variety of application.

- **Nanocarpets:**

(*i*) This structure is formed by stacking a large number of nanotubes together with their cylindrical axes aligned vertically.

(*ii*) The carpet has antibacterial qualities and the ability to change colour.

- **Nanowires:**

(*i*) These are wires of dimensions in the order of a nanometer ($10^{-9}$ meters) range.

(*ii*) They can be metalic, semiconducting and insulating.

(*iii*) These can be designed of repeating organic units such as DNA.

- **Structure of Nanoparticles**

(*i*) They are usually spherical but rods, plates and other shapes are sometime used.

(*ii*) They may be solid or hollow and are composed of a variety of materials often in several discrete layers with separate functions.

(*iii*) Typically there is a central functional layer, a protective layer and an outer layer allowing interaction with biological world.

(*iv*) The central functional layer usually display some optical or magnetic behaviour.

(*v*) The protective layer shields the functional layer from chemical damage by air, water or cell components and conversely shields the cell from any toxic properties of the chemicals composing the functional layer.

(*vi*) The outer layer(s) allow nanoparticles to be "biocompatible". This generally involves two aspects, water solubility and specific recognition.

(*vii*) Nanoparticles are often made water soluble by adding hydrophilic outer layer for biological use.

(*viii*) In addition chemical groups must be present on the exterior to allow specific attachement to other molecules or structures.

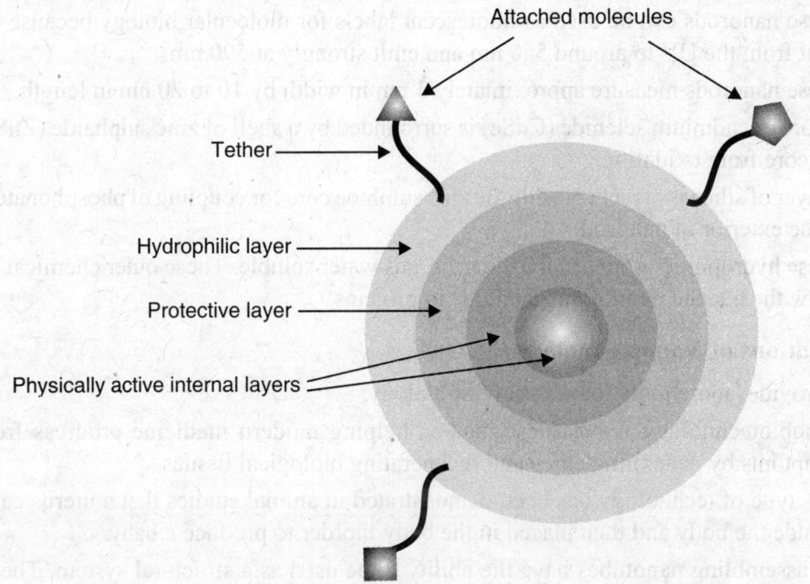


**Fig. 15.1** Typical layered structure of Nanoparticles: Several layers surround the physically active core. Chemical groups are often added to the exterior to allow attachement of biological molecules.

**Attachement of molecules:**

- Fluorescent labeling and optical loding
- Detection of pathogenic organisms
- Delivery of pharmaceuticals or genes.
- Purification and manipulation of biological components
- Tumor destruction by chemicals
- Contrast enhancement in magnetic resonance imaging (MRI).

- **Nanoparticles for labeling:**

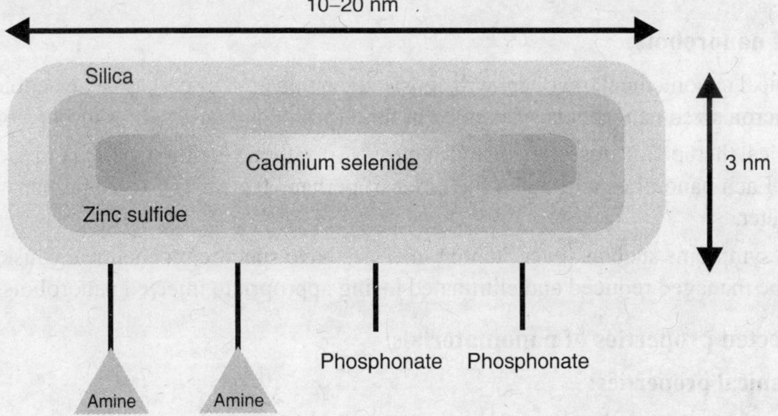


**Fig. 15.2** CdSe Nanorods: Luminescent CdSe nanorods are encased in protective layers of zinc sulfide and of silica. Hydrophilic chemical groups on the outside allow proteins or other biological molecules to be attached.

(*i*) These nanorods can be used as fluorescent labels for molecular biology because they absorb light from the UV to around 550 nm and emit strongly at 590 nm.

(*ii*) These nanorods measure approximately 3 nm in width by 10 to 20 nm in length.

(*iii*) A core of cadmium selenide (CdSe) is surrounded by a shell of zinc sulphaide (ZnS) to protect the core from oxidation.

(*iv*) A layer of silica is present outside the zinc sulphide core for coupling of phosphonates or amines to the exterior of nanorod.

(*v*) These hydrophilic groups make the nanorods water soluble. These outer chemical groups also allow the attachement of the nanorods to proteins.

- **Applications of Nanotechnology**

(*i*) It provides more tools for the study of biology.

(*ii*) Nanobiotechnology is best described as helping modern medicine progress from treating symptoms by generating cures and regenerating biological tissues.

(*iii*) This type of technology has been demonstrated in animal studies that a uterus can be grown outside the body and then placed in the body inorder to produce a baby.

(*iv*) Set assembling nanotubes have the ability to be used as a structural system. They would be composed together with rhodopsins; which would facilitate the optical computing process.

(*v*) DNA can be used as structural proteomic system–a logical component for molecular computing.

(*vi*) The $P^{53}$ protein, a product of nanobiology, can literally shut down the metabolism of living cells. This protein is considered to be a prime candidate as a cure for certain cancer.

- **Carbon nanotube:**

(*i*) It is a cylindrical rolled up sheet of grapheme, which is a single layer of graphite atoms arranged in hexagonal pattern.

(*ii*) Their hexagonal structure gives them great tensile strength and elastic properties.

(*iii*) Carbon nanotubes are strong and when bent, they are very resilient i.e. they spring back to their original shape.

(*iv*) They also transfer heat very efficiently.

- **Role of nanorobots:**

(*i*) A typical nanomedical treatment will consist of an injection of perhaps a few cubic centimeter of micron sized nanorobots suspended in fluid (probably water/saline suspension).

(*ii*) A typical therapeutic dose may include upto 1 – 10 trillion (1 trillion = $10^{12}$) individual nanorobots. Each nanorobot will be on the order of perhaps 0.5 micron up to perhaps 3 micron in diameter.

(*iii*) Most symptoms such as fever, itching, measles have specific biochemical causes which can also be managed reduced and eliminated losing appropriate injected nanorobots.

**Some selected properties of nanomaterials:**

**(*a*) Mechanical properties:**

(*i*) They are made up of small crystalline grains.

(*ii*) If these grains can be made very small or even nanoscale in size, the interface area within the material greatly increases which enhances its strength.

- **Chemical properties:**
  - (*i*) As the nanoparticle size decreases, the interparticle spacing decreases in metals and surface area increases. As a result material properties changes. The electronic properties also changes.
  - (*ii*) Therefore, nanoparticles can be arranged into layers on surfaces, providing large surface area and hence enhanced activity.
- **Magnetic properties:**
  - (*i*) A normal ferromagnet contains domains each containing several thousand atomic spins. within the domains the spins are aligned in a straight direction while different domains point in different direction.
  - (*ii*) However nanostructured magnetic materials exhibit totally new class of magnetic properties. As the particle size decreases, the formation of domain wall becomes energetically unfavourable and the particles are called single domain particles with critical diameter.
  - (*iii*) In this situation the magnitisation changes do not take place through domain wall motion but require the coherent rotation of spins resulting in large coercivities.
- **Distinguish between atomic clusters and nanoparticles**

| Atomic cluster | Nanoparticle |
|---|---|
| (*i*) The general practice used by the scientist for investigating the materials is to break them into fundamental building blocks and grouping them (atomic cluster) and study them at each stage. | (*i*) They may be defined as those materials whose structural element, crystallities or molecules have dimensions in the 1 to 100 nm range. |
| (*ii*) Attempt is made to explain the structure and properties of bulk materials. | (*ii*) Clusters of atoms consisting of typically hundred to thousand on nanometer scale are called nanoclusters (nanoparticles) others with much greater dimension called cluster. |

**Bottom up method:**

(*i*) The procedure is to collect; consolidate and fashion individual atoms and molecules into structure.

(*ii*) This is carried out by a sequence of chemical reactions controlled by catalysts.

**Top down method:**

(*i*) It starts with a large-scale object or pattern and gradually reduces its dimensions.

(*ii*) This is accomplished by a technique called *lithography* which shines radiation through a template on to a surface coated with a radiation-sensitive resist.

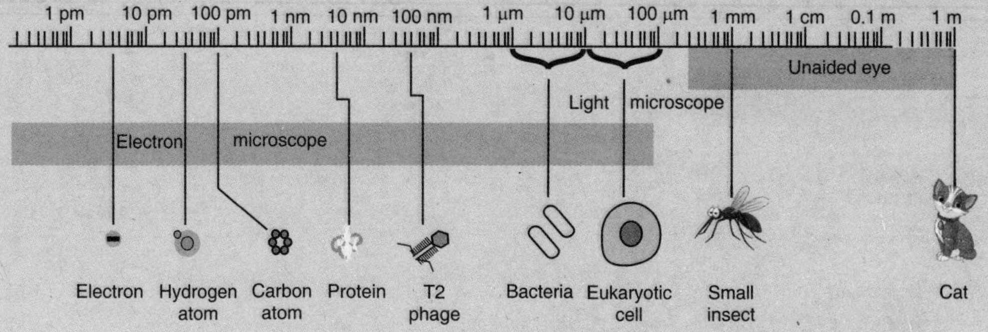


**Fig. 15.3** Objective size (range 1 m to 1 Pictometer)

**Table 15.1: Prefixes and sizes**

| Length Unit | Meters | Examples |
|---|---|---|
| 5.9 terameters | | mean distance from sun to Pluto |
| Terameter | $10^{12}$ | |
| 150 gigameters | | distance to the sun |
| Gigameter | $10^{9}$ | |
| 380 megameters | | distance to the moon |
| 6.3 megameters | | radius of the earth |
| 3.2 megameters | | length of Great Wall of China |
| Megameter | $10^{6}$ | |
| Kilometer | $10^{3}$ | |
| 30 meters | | blue whale |
| Meter | 1 | large dog |
| Millimeter | $10^{-3}$ | small insect |
| Micrometer | $10^{-6}$ | bacterial cell |
| 500 nanometers | | wavelength of visible light |
| 100 nanometers | | size of typical virus |
| 3.4 nanometers | | one turn of DNA double helix |
| Nanometer | $10^{-9}$ | molecules |
| 350 picometers | | molecular diameter of water |
| 260 picometers | | atomic spacing in solid copper |
| 77 picometers | | atomic radius of carbon (= resolution limit of atomic force microscope as of 2004) |
| 32 picometers | | atomic radius of hydrogen |
| Angstrom | | = 100 picometers = $10^{-10}$ meter |
| 2.4 picometers | | wavelength of electron |
| Picometer | $10^{-12}$ | |
| Femtometer | $10^{-15}$ | radius of atomic nucleus |
| Attometer | $10^{-18}$ | radius of proton |
| Zeptometer | $10^{-21}$ | |
| Yoctometer | $10^{-24}$ | radius of neutrino |

# Appendix 1 : Some Useful Tables

**Table 1. Seven base unit**

| *Physical Quantity* | *Name of Unit* | *Official S.I. symbol* |
|---|---|---|
| Length | metre | m |
| Mass | kilogram | kg |
| Time | second | s |
| Electric current | ampere | A |
| Temperature | kelvin | K |
| Luminous intensity | candela | cd |
| Amount of substance | mole | mol |

**Table 2. Some commonly used derived units**

| *Physical Quantity* | *Name of Unit* | *Symbol* |
|---|---|---|
| Area | Square metre | $m^2$ |
| Volume | Cubic metre | $m^3$ |
| Density | kg per cubic metre | $kg\ m^{-3}$ |
| Velocity | Metre per second | $ms^{-1}$ |
| Acceleration | Metre per sec per sec | $ms^{-2}$ |
| Force | Newton | $N = kg\ ms^{-2}$* |
| Power | Watt | $W = Js^{-1}$ |
| Pressure | Pascals (Newtons per sq. metre) | $Pa = Nm^2 = kg\ m^{-1}\ s^{-2}$ |
| Work or Energy (Heat etc) | Joule | $J = Nm = kg\ m^2\ s^{-2}$** |
| Magnetic flux density | Tesla | $T = kg\ s^{-2}\ A^{-1}$ |
| Frequency | Hertz | $Hz = s^{-1}$ |
| Electric charge | Coulomb | $C = As$ |

(*Contd.*)

* Newton is defined as the force that gives a mass of 1 kg an acceleration of 1 m $sec^{-2}$ so that $f = ma$ = (1 kg) (1 m $sec^{-2}$) = 1 kg m $sec^{-2}$ = 1 N.

** Joule is the work done when a displacement of 1 metre takes place by a force of 1 newton so that $w = fd$ = (1 N) (1 m) = Nm = (1 kg m $sec^{-2}$) (1 m) = 1 kg $m^2$ $sec^{-2}$ 1 J

| *Physical Quantity* | *Name of Unit* | *Symbol* |
|---|---|---|
| Potential difference | Volt | V |
| Electric resistance | Ohms | $\Omega = VA^{-1}$ |
| Electric conductance | $Ohms^{-1}$ | $\Omega^{-1} = AV^{-1}$ |

**Table 3. Some useful conversion factors**

| | |
|---|---|
| 1 Å = $10^{-10}$ m | 1 atm = 101325 pa or $Nm^{-2}$ |
| | = $1.013 \times 10^6$ dynes/$cm^2$ |
| 1 a.m.u. = $1.66053 \times 10^{-27}$ kg | 1 mm or torr = 133.32 Pa or $Nm^{-2}$ |
| $t° = t + 273.16$ K $\simeq t + 273$ K | 1 calorie = 4.184 J |
| 1 litre = $10^{-3}$ $m^3$ = 1 $dm^3$ | 1 erg = $10^{-7}$ J |
| 1 dyne = $10^{-5}$ N | 1 electron volt (eV) = $1.6022 \times 10^{-19}$ J |

**Table 4. Some common prefixes used for S.I. units**

| *Prefix* | *Symbol* | *Multiplying factor* | *Example* |
|---|---|---|---|
| deci- | d | $10^{-1}$ | 1 decimetre (1 dm) = $10^{-1}$ m = 0.1 m |
| centi- | c | $10^{-2}$ | 1 centimetre (1 cm) = $10^{-2}$ m = 0.01 m |
| milli- | m | $10^{-3}$ | 1 millimetre (1 mm) = $10^{-3}$ m = 0.001 m |
| micro- | μ | $10^{-6}$ | 1 micrometre (μm) = $10^{-6}$ m |
| nano- | n | $10^{-9}$ | 1 nanometre (nm) = $10^{-9}$ m |
| pico- | p | $10^{-12}$ | 1 picometre (pm) = $10^{-12}$ m |
| femto- | f | $10^{-15}$ | 1 femtometre (fm) = $10^{-15}$ m |
| deka- | da | $10^{1}$ | 1 dekametre (dam) = $10^{1}$ m |
| hecto- | h | $10^{2}$ | 1 hectometre (hm) = 100 m |
| kilo- | k | $10^{3}$ | 1 kilometre (km) = $10^{3}$ m = 1000 m |
| mega- | M | $10^{6}$ | 1 megametre (Mm) = $10^{6}$ m |
| giga- | G | $10^{9}$ | 1 gigametre (Gm) = $10^{9}$ m |
| tera- | T | $10^{12}$ | 1 terametre = $10^{12}$ m |
| peta- | P | $10^{15}$ | 1 petametre = $10^{15}$ m |

**Table 5. Values of some common physical and chemical constants**

| *Quantity* | *Symbol* | *Values in CGS units* | *Values in SI units* |
|---|---|---|---|
| Velocity of light | $c$ | $2.997925 \times 10^{10}$ cm/sec | $2.997925 \times 10^{8}$ m $sec^{-1}$ |
| (in vacuum) | | $\simeq 3.0 \times 10^{10}$ cm/sec | $\simeq 3.0 \times 10^{8}$ m sec |
| Planck's constant | $h$ | $6.6262 \times 10^{-27}$ erg sec | $6.6262 \times 10^{-34}$ J sec |
| | | $\simeq 6.62 \times 10^{-27}$ erg˙sec | $\simeq 6.62 \times 10^{-34}$ J sec |

(*Contd.*)

| *Quantity* | *Symbol* | *Values in CGS units* | *Values in SI units* |
|---|---|---|---|
| Avogadro number | $N_0$ or N | $6.022169 \times 10^{23}$ mole$^{-1}$ | $6.022169 \times 10^{23}$ mol$^{-1}$ |
| | | $\simeq 6.022 \times 10^{23}$ mole$^{-1}$ | $\simeq 6.022 \times 10^{23}$ mol$^{-1}$ |
| Gas constant | R | 0.082053 lit atm K$^{-1}$ mole$^{-1}$ | 8.3144 JK$^{-1}$ mol$^{-1}$ |
| | | $\simeq$ 0.0821 lit atm K$^{-1}$ mole$^{-1}$ | $\simeq$ 8.314 JK$^{-1}$ mol$^{-1}$ |
| | | or | |
| | | $8.3144 \times 10^{7}$ ergs K$^{-1}$ mole$^{-1}$ | |
| | | $\simeq$ 0.0821 lit atm K$^{-}$ mole$^{-1}$ | |
| Molar volume at | V | 22.414 litres | 0.022414 m$^{3}$ |
| NTP | | $\simeq$ 22.4 litres | $\simeq$ 0.0224 m$^{3}$ |
| Faraday's constant | F | 96487 coulombs/equivalent | 96487 C mol$^{-1}$ |
| | | $\simeq$ 96500 coulombs/equivalent | $\simeq$ 96500 C mol$^{-1}$ |
| Electronic charge | $e$ | $1.6022 \times 10^{-19}$ coulomb | $1.6022 \times 10^{-19}$ C |
| Electron rest mass | $m_e$ | $9.109558 \times 10^{-28}$ g | $9.109558 \times 10^{-31}$ kg |
| | | $\simeq 9.11 \times 10^{-28}$ g | $\simeq 9.11 \times 10^{-31}$ kg |
| Proton rest mass | $m_p$ | $1.672614 \times 10^{-24}$ g | $1.672614 \times 10^{-27}$ kg |
| | | $\simeq 1.67 \times 10^{-24}$ g | $\simeq 1.67 \times 10^{-27}$ kg |
| Neutron rest mass | $m_n$ | $1.67492 \times 10^{-24}$ g | $1.67492 \times 10^{-27}$ kg |
| | | $\simeq 1.67 \times 10^{-24}$ g | $\simeq 1.67 \times 10^{-27}$ kg |

# Appendix 2 : Greek Letters

| *Alphabets* | *Pronunciation* | *English equivalent* |
|---|---|---|
| α | alpha | a |
| β | beta | b |
| γ | gamma | g, n |
| δ | delta | d |
| ε | epsilon | e |
| ζ | zeta | z |
| η | eta | e |
| θ | theta | th |
| ι | iota | i |
| κ | kappa | k |
| λ | lambda | l |
| μ | mu | m |
| ν | nu | n |
| ξ | xi | x |
| ο | omicron | o |
| π | pi | p |
| ρ | rho | r, rh |
| σ | sigma | s |
| τ | tau | t |
| υ | upsilon | y, u |
| φ | phi | ph |
| χ | chi | ch |
| Ψ | psi | ps |
| ω | omega | o |

# Bibliography


| Sl. No. | Name of the Books | Author |
|---|---|---|
| 1. | Analysis of Vertebrate Structure (4th Ed.) | Milton Hilderbrand |
| 2. | Animal Physiology (2002) | P.S. Verma, B.S. Tyagi, V.K. Agarwal |
| 3. | Animal Physiology (3rd Ed.) | S.C. Rastogi |
| 4. | Atomic and Nuclear Physics | A.B. Gupta and Dipak Ghosh |
| 5. | A Text Book of Biophysics | Dr. R.N. Roy |
| 6. | A Text Book of Microbiology | R.C. Dubey and D.K. Maheshwari |
| 7. | A Text Book of Physical Chemistry | K.L. Kapoor |
| 8. | A Text Book of Quantitative Chemical Analysis (5th Ed.) | Jeffery GH Bassett J. Mendham J. Denny R.C. |
| 9. | Biophysics | Mohan P. Arora |
| 10. | Biophysics | G.R. Chatwal |
| 11. | Biophysics | Vasantha Pottabhi N. Gautam |
| 12. | Basic Thermodynamics | Evelyn Guha |
| 13. | Biophysics & Biophysical Chemistry | D. Das |
| 14. | Biophysical Chemistry | A. Upadhyay K. Upadhyay N. Nath |
| 15. | Biological Instrumentation & Methodology | P.K. Bajpai |
| 16. | Fundamentals of Molecular Spectroscopy (4th Ed.) | Colin N. Banwell Elaine M. McCash |
| 17. | Fundamentals of Ecology | M.C. Dash |
| 18. | Fundamentals of Solid Sate Physics (Pragati Prakashan) | B.S. Saxena R.C. Gupta, P.N. Saxena |

| | | |
|---|---|---|
| 19. | General and Inorganic Chemistry | R.P. Sarkar |
| 20. | Instrumental Methods of Analysis (6th Ed.) | Hobart H. Willard<br>Lynne L Merrit<br>J.R John Dean Frank A Settle |
| 21. | I.S.C. Physics (II) | P. Vivekanandan &<br>D.K. Banerjee |
| 22. | I.S.C. Chemistry (I & II) | R. Madan & B.S. Bisht |
| 23. | Modern Inorganic Chemistry | Dr. R.D. Madan. |
| 24. | Pharmaceutical Analysis (Instrumental Method) | A.V.Kasture, K.R. Mahadik,<br>S.G. Wadodkar, H.N. More |
| 25. | Physical Chemistry (7th Ed.) | P.C. Rakshit |
| 26. | Principles and Techniques Biochemistry and Molecular Biology (7th Ed.) | Edited by<br>Keith Wilson<br>John Walker |
| 27. | Physical Chemistry (Vol. II) | Hrishikesh Chatterjee |
| 28 | Problems on Physical Chemistry | S. Pahari<br>D. Pahari |
| 29. | Refresher Course in Physics (Vol. VIII) | C.L. Arora |
| 30. | Remington is Pharmaceutical Science (17th Ed.) | Alfonso R. Gennaro (Edited) |
| 31. | Review of Medical Physiology (14th Ed.) | William F. Ganong |
| 32. | Spectroscopic Methods in Organic Chemistry (5th Ed.) | Dudley H. Williams<br>Ian Fleming |
| 33. | Solid State Physics<br>(New Revised 6th & 8th New Age International) | S.O. Pillai |
| 34. | Technique in Microscopy & Cell Biology | V.K. Sharma |
| 35. | The Chordates | RMcNeill Alexander |
| 36. | Vertebrates : Comparative anatomy, Function & Evolution (4th Ed.) | Kenneth V. Kardong |